Sixth International Multi-Conference on Engineering and Technology (IMETI 2017)

MATEC Web of Conferences Volume 169 (2018)

Hualien, Taiwan
27 - 31 October 2017

Editor:

Wen-Hsiang Hsieh

ISBN: 978-1-5108-6388-0

Printed from e-media with permission by:

Curran Associates, Inc.
57 Morehouse Lane
Red Hook, NY 12571

proceedings.com

Some format issues inherent in the e-media version may also appear in this print version.

Printed by Curran Associates, Inc. (2018)

For additional information, please contact EDP Sciences – Web of Conferences at the address below.

EDP Sciences – Web of Conferences
17, Avenue du Hoggar
Parc d'Activité de Courtabœuf
BP 112
F-91944 Les Ulis Cedex A
France

Phone: +33 (0) 1 69 18 75 75
Fax: +33 (0) 1 69 28 84 91

contact-edps@webofconferences.org

Additional copies of this publication are available from:

Curran Associates, Inc.
57 Morehouse Lane
Red Hook, NY 12571 USA
Phone: 845-758-0400
Fax: 845-758-2633
Email: curran@proceedings.com
Web: www.proceedings.com

TABLE OF CONTENTS

Preface

The Proceedings of thc IMETI2017 includes 46 papers presented in the Sixth International Multi-Conference on Engineering and Technology Innovation 2017(IMETI2017) held in Hualien, Taiwan on Oct. 27 - Oct. 31, 2017. All papers were reviewed by at least two of the members of the technical committee and the accessory review team. The papers accepted for publication have been considered to be sufficiently original and to contain new developed data or technologies that deserve to be published in the Proceedings.

During the conference, cutting edge knowledge from the industry, academia and government to discourse on research and development, professional practice, business and management in the science and engineering fields was presented. Thank all of the attendees for sharing their research results.

We would like to thank the organization staff, the members of the program committees and reviewers. Thank them for their hard working to make the proceedings and conference come true.

Wen-Hsiang Hsieh

Yunlin, Taiwan, R.O.C.

May, 2018

Committees

Program Chairman

- Wen-Hsiang Hsieh, President, Taiwan Association of Engineering and Technology Innovation National Formosa University, Taiwan
- Zongyu Chang, Ocean University of China, P.R. China

Program Co-Chairman

- Chun-Pin Lin, Asia University, Taiwan
- Ing-Jr Ding, National Formosa University, Taiwan
- Jia-Shing Sheu, National Taipei University of Education, Taiwan
- Van-Tsai Liu, National Formosa University, Taiwan

Technical Committee Chairman

- Chih-Ta Yen, National Formosa University, Taiwan
- G.Y. Tzou, Chung Chou University of Science and Technology, Taiwan
- Tao-Hsing Chen, National Kaohsiung University of Applied Sciences, Taiwan
- Yi-Chang Wu, National Yunlin University of Science & Technology, Taiwan

Program Committee

- An Cheng, National Ilan University, Taiwan
- Chan Young Park, Hallym University, South Korea
- Chang-Chun Lee, National Tsing Hua University, Taiwan
- Chang-Huang Chen, Tungnan University, Taiwan
- Chen Lu, Beihang University, China
- Chun-Hsiung Tseng, Nanhua University, Taiwan
- Chung-Chi Huang, Far East University, Taiwan
- David Pei-Cheng Lin, Chung Shan Medical University, Taiwan
- Dae-Jin Kim, Kyung Hee University, South Korea
- Hsiang-Chen Hsu, I-Shou Univerity, Taiwan
- Hsiao-Han Liu, I-Shou University, Taiwan
- Hsin-Hung Lee, Center for Measurement Standards, Industrial Technology Research Institute, Taiwan
- Kuan-Chen Lin, Far East University, Taiwan
- Jeong-Dong Kim, Korea University, South Korea
- Jie-Tong Zou, National Formosa University, Taiwan
- Jih-Gau Juang, National Taiwan Ocean University, Taiwan
- Jium-Ming Lin, Chung Hwa University, Taiwan
- Minvydas Ragulskis, Kaunas University of Technology, Lithuania
- Shih-Sung Lin, National Defense University, Taiwan
- Tai-Ping Chang, National Kaohsiung first University of Science and Technology, Taiwan
- Te-Kwei Wang, Ming Chi University of Technology, Taiwan
- Wei-Ting Lin, National Ilan University, Taiwan
- Yi-Cheng Huang, National Changhua University of Education, Taiwan
- Yih-Chuan Lin, National Formosa University, Taiwan
- Yit-Jin Chen, Chung Yuan Christian University, Taiwan
- Yu Guo, Kunming University of Science and Technology, China
- Yung-Cheng Lee, WuFeng University, Taiwan
- Yung-Hui Chen, Lunghwa University, Taiwan

International Scientific Committee

- A. Gottscheber, University of Heidelberg, Germany
- Ali Awaludin, Gadjah Mada University, Indonesia
- Amir Abidov, Tashkent State Technical University, Uzbekistan
- Aylie Han, University of Diponegoro, Indonesia
- A. Walton, University of Edinburgh, UK
- Buntara S. Gan, Nihon University, Japan
- R. S. Beniwal, National Institute of Science Communication and Information Resources, India
- C. Warwick, University College London, UK
- E. Manske, TU Ilmenau, Germany
- G. Jäger,TU Ilmenau, Germany
- Igor Meglinski, University of Oulu, Finland
- J. Wood, Goldsmiths College, University of London,UK
- J. Carlos, Metrolho Instituto Politecnico de Castelo Branco Castelo Branco, Portugal
- Jong-Han Lee, Daegu University, South Korea
- J.S. Dai, Kings of college London, UK
- Kihong Shin, Andong National University, South Korea
- M. Woolley, University of Coventry, UK
- P.F. Savanna, College of Art and Design, USA
- R.X. Du, The Chinese University of Hong Kong, HK
- S.G. Hanson, Technical University of Denmark, Denmark
- S.D. Prior, University of Middlesex, UK
- T. Froelich, TU Ilmenau, Germany
- Uldis Strautins, University of Latvia, Latvia
- WooSeok Kim, Chungnam National University, South Korea
- Zhengchao Xie, University of Macau, Macau

Advisory Committee

- B.W. Huang, Cheng Shiu University, Taiwan
- C.C. Wen, Metal Industries Research & Development Centre, Taiwan
- C.H. Huang, Wufeng University, Taiwan
- Chun-Huy Wang, Nan-Jeon University of Technology, Taiwan
- C.M. Lin, National Taitung University, Taiwan
- C.T. Lin, Metal Industries Research & Development Centre, Taiwan
- F. Long, Xiamen University, China
- H.S. Lin, National Kaohsiung University of Applied Sciences, Taiwan
- I.T. Tang , National University of Tainan, Taiwan
- J.C. Zhong, Changchun University of Science and Technology, China
- J.F. Yao, Xiamen University, China
- J.X. Xie University of Science and Technology, Beijing, China
- K.H. Wen, Fuzhou University, China
- K.S. Shu, Chia Nan University of Pharmacy and Science, Taiwan
- L.W. Ji, National Formosa University, Taiwan
- M. Chang, Chung Yuan Christian University, Taiwan
- M.H. Wang, National Chin-Yi University of Technology, Taiwan
- Q.K. Xue, Tsinghua University, Beijing, China
- S.J. Chang, National Cheng Kung University, Taiwan
- S.J. Lin, National Kaohsiung University of Applied Sciences, Taiwan
- S.T. Shen, National Formosa University, Taiwan
- Sungjin Kim, Kumoh National Institute of Technology, South Korea
- T.H. Meen, National Formosa University, Taiwan
- W. Water, National Formosa University, Taiwan
- W.Z. Guo, Shanghai Chia-Tung University, China
- Wonho Suh, Hanyang University, South Korea
- X.L. Du, Beijing National Laboratory for Condensed Matter Physics, China
- Y.C. Wang, National Yunlin University of Science & Technology, Taiwan
- Y.L. Hwang, National Formosa University, Taiwan
- Y.L. Yeh, Far East University, Taiwan
- Y.Q. Zhou, China University of Petroleum, China
- Y.X. Sun, China University of Petroleum, China

Statement of Peer review

In submitting conference proceedings to MATEC Web of Conferences, I certify to the Publisher that I adhere to the **Policy on Publishing Integrity** of the journal in order to safeguard good scientific practice in publishing.

1. All articles have been subjected to peer review administered by the proceedings editors.
2. Reviews have been conducted by expert referees, who have been requested to provide unbiased and constructive comments aimed, whenever possible, at improving the work.
3. Proceedings editors have taken all reasonable steps to ensure the quality of the materials they publish and their decision to accept or reject a paper for publication has been based only on the merits of the work and the relevance to the journal.

Title, date and place of the conference :

The Sixth International Multi-Conference on Engineering and Technology Innovation 2017(IMETI2017), Oct. 27 - Oct. 31, 2017 in Hualien, Taiwan.

Proceedings editor(s): Wen-Hsiang Hsieh

Date and editor's signature *May 08, 2018*

MATEC Web of Conferences **169**, 01001 (2018) https://doi.org/10.1051/matecconf/201816901001
IMETI 2017

Experiment and CFD simulation of exhaust tube in high-voltage circuit breaker

Xiangyang Ye[a], Francesco Pisu, Stephan Grob, Mahesh Dhotre and Javier Mantilla

Interrupter Development, ABB Ltd., 5401 Baden, Switzerland

Abstract. In a high-voltage circuit breaker, the exhaust tube connects the arc zone with the exhaust volume. During the arc interruption process, the exhaust tube transports the hot gas from the arc interruption zone to the exhaust volume through its distributed holes. The design of a high performance exhaust tube in the circuit breaker development aims for well controlled hot gas evacuation mass flow and pressure waves. In this paper, the exhaust tube behaviour is investigated using Computational Fluid Dynamics (CFD). To verify the CFD simulation, a basic experimental study with pressure measurements at different positions of the exhaust tube is performed. Further, the design parameters influencing the exhaust tube behaviour and circuit breaker performance are investigated and discussed.

1 Introduction

In an electrical transmission network, high-voltage circuit breakers interrupt short circuit currents in order to protect the network and apparatus. Switching gas with good dielectric and thermodynamic properties such as SF6, or recently, a mixture of $C_5F_{10}O/CO_2/O_2$ [8] with low Global Warming Potential (GWP), are used to extinguish the electric arc, which occurs after separation of the electrical contacts.

In a high-voltage circuit breaker, the exhaust tube is a tube connecting the arc zone with the exhaust volume as shown in Figure 1. During the arc interruption process, an electric arc occurs between the arcing contacts (tulip finger and plug). Polytetrafluoroethylene (PTFE) vapour, resulting from the nozzle ablation, is driven by the high pressures between the arcing contacts as an energy carrier with high temperature into the heating volume, so that a pressure is built up in the heating volume. When the current approaches zero, the arc pressure between the contacts collapses, allowing the gas to flow back from the heating volume into the opening gap between arcing contacts at high velocity to extinguish the arc. The exhaust tube transports the hot gas from the arc interruption zone to the exhaust volume through its distributed holes. As, The hot gas mass flow, pressure distribution, and pressure waves traveling along the length of the exhaust tube at the downstream region of the arcing zone play a big role for the upstream arc zone region and on the cooling of the arc in the current interruption. Further, it influences the metal erosion behaviour of the upstream arcing contacts. Therefore, the design of a high performance exhaust tube aims for a well-controlled hot gas evacuation mass flow and pressure wave.

[a] Corresponding author : xiangyang.ye@ch.abb.com

In this paper, the effects of different geometry parameters, e.g., position and area of the exhaust tube holes, on the exhaust tube performance are investigated using Computational Fluid Dynamics (CFD). To verify the CFD simulation, basic experimental study with pressure measurement at different position of exhaust tube is performed, and the result compared with the simulation. Based on the CFD simulation and the experimental study, the behaviour of an exhaust tube and its effects on the circuit breaker performance is investigated. Further, the design parameters influencing the exhaust tube behaviour and circuit breaker performance are investigated and discussed.

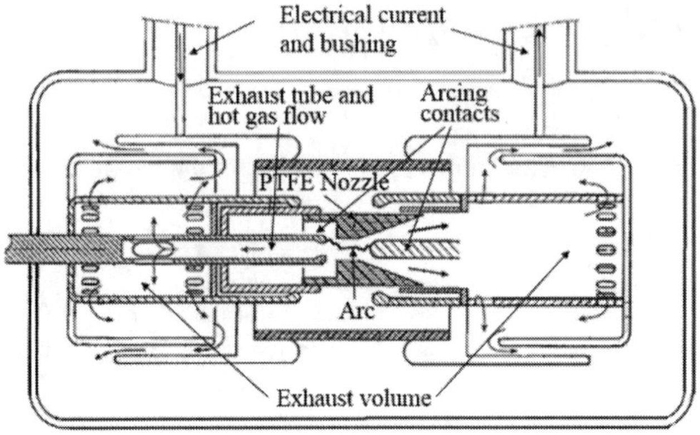

Figure 1. High-voltage circuit breaker and exhaust tube.

2 Experimental set-up

2.1 Experimental set-up for the exhaust tube study

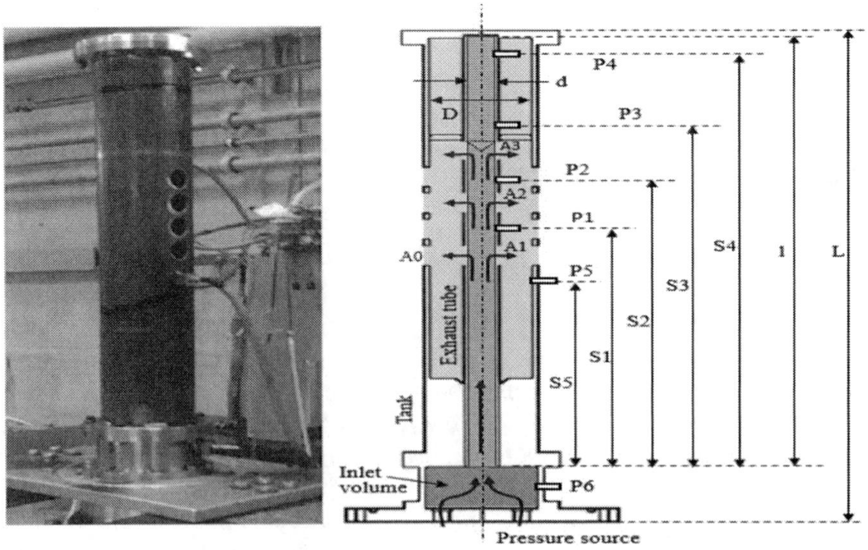

Figure 2. Experimental set-up of exhaust tube study.

Figure 2 shows the experimental set-up for the exhaust tube study. The tank has the length of L and inner diameter of D. There are 2 rows with 4 circular holes in each row in the middle of the tank and the two rows are placed in an angle of 180°. On the bottom side, the tank is connected through several holes to a CO_2 vessel as the pressure source. The exhaust tube is located in the tank with inner diameter d, and length l. It has 3 columns of holes and each row has two holes placed in an angle of 180°. The exhaust hole columns with area of A1, A2 and A3 are distributed along the exhaust tube;

pressure sensors P1 – P5 are put along the tube inner surface with different distance from bottom S1-S5. Pressure sensor P6 is put in the middle of inlet volume. In Test 1, the exhaust tube is blocked with a conical body at end of hole A1 as shown in Figure 2 with orange colour. In all other tests, the exhaust tube is not blocked with the conical body, hence, there is a dead volume above the hole column A3.

Table 1 shows the position and area normalized against the exhaust tube diameter and its area for each test. In Test 1, since the dead volume above the hole column A3 is blocked, no pressure measurement at position S3 and S4 is available, while Test 2 to 4 have the pressure measurement in the dead volume. Test 1 and Test 2 have the same uniform hole column area, while Test 3 has small (S), medium (M) and large (L) hole column distribution (SML) and Test 4 has LMS hole column distribution along the flow direction.

2.2 Testing procedure

In every test, the whole test device is placed under ambient air, i.e., is filled with air with atmosphere pressure initially. At the start of the test, the CO_2 pressure vessel valve is opened, so that the CO_2 gas flows through several connecting holes into the initial volume and further into the exhaust tube. Finally the gas flows into the tank and through the tank holes to the atmosphere. During the test, the pressure at different positions is measured and saved with a transient recorder.

Table 1. Test number and the corresponding geometry parameters normalized against exhaust tube area (A) or exhaust tube diameter.

Test number	A1/A	A2/A	A3/A	A0/A	S1/d	S2/d	S3/d	S4/d	S5/d
1	0.58	0.58	0.58	0.86	21.8	26.7	none	none	16.6
2	0.58	0.58	0.58	0.86	21.8	26.7	31.6	38.2	16.6
3	0.14	0.72	0.86	0.86	21.8	26.7	31.6	38.2	16.6
4	0.86	0.72	0.14	0.86	21.8	26.7	31.6	38.2	16.6

3 CFD simulation

3.1 Mathematical model

The main parameters of CFD simulation, namely pressure, density, temperature and gas velocity, are solved with the Navier-Stokes equations for the conservation of mass, momentum and energy. These equations are discretized and solved iteratively using a commercial finite volume solver. The application and implementation of arc model are described in detail [1-5] and are not discussed in this work. For the sake of completeness, a brief model description and numerical details are provided:

Energy source terms are added to represent the effect of the arc on the flow: ohmic heating is added based on the solution of the electric field distribution inside the arc and the radiative transfer is computed based on discrete ordinate method (DOM) approximating the radiation transfer equation on 12 ordinate directions [5]. Electric field is computed statically using a simple Laplace equation for the electric potential. The mass source term is added to represent the effect of PTFE nozzle material ablation and the erosion of the arcing contacts as in [6] and [7]. In order to take turbulence effects into account the realizable k-ε model [6] is used. A numerical scheme is used for the pressure – velocity coupling due to its suitability for transient calculations. Thermodynamic and transport properties of the mixture in pressure and temperature range are calculated assuming local thermal equilibrium conditions [8] and tabulated in the form of look-up tables, retrieved by the solver during calculation.

https://doi.org/10.1051/matecconf/201816901001

3.2 Numerical discretization

Thermodynamic and transport properties of the mixture in the pressure and temperature range are calculated assuming local thermal equilibrium conditions, tabulated in the form of look-up tables and retrieved by the solver during calculation. The model equations are discretized and solved on a hybrid block-structured and unstructured mesh. The numbers of cells in the discretization are in the range of 150-200 K for a 2D model. More details about the model and numerical methods are given in [5].

To limit the calculation time, two-dimensional (2D) simulation is selected, instead of three-dimensional (3D) simulation and the breaker model is simplified to a smaller part to focus on the main problem of the study.

4 Result and discussion

4.1 Validation of CFD method

Figure 3 shows the CFD set-up for the validation calculation. Because of the complex flow conditions in the apparatus connecting the test device and the CO_2 pressure vessel with valve and pipe, the measured pressure in the CO_2 vessel cannot be used as the inlet pressure boundary condition, hence, the measured pressure in the inlet volume is used as the total inlet pressure with CO_2 species. The atmosphere pressure 1 bar is taken as the outlet pressure boundary condition. The initial condition is set with 1 bar filled with air as in the experiment.

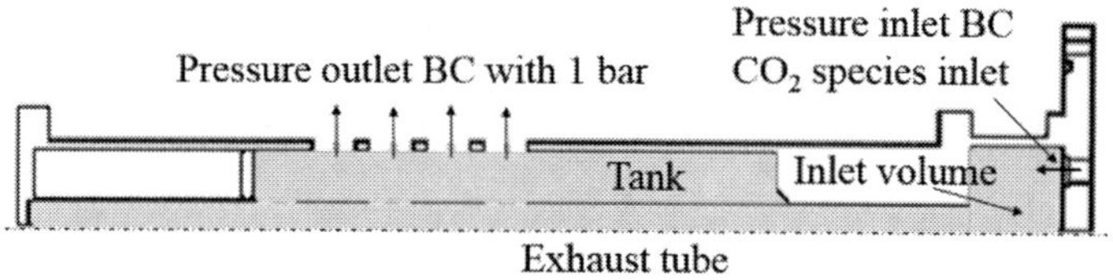

Figure 3. CFD simulation set-up and meshing with the yellow region for the validation.

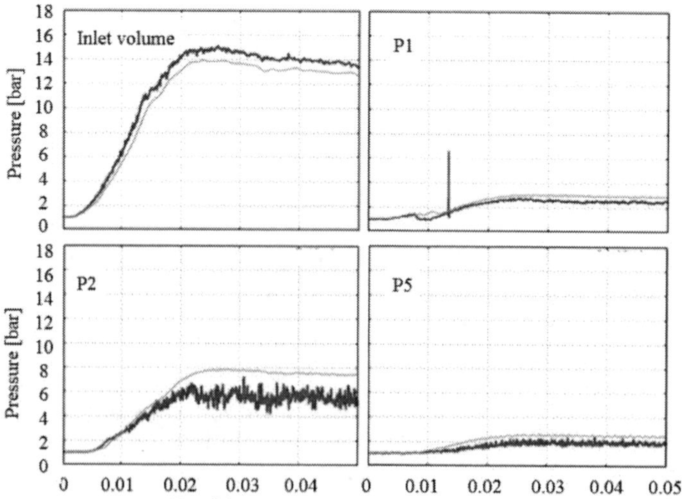

Figure 4. Comparison of simulated pressure distribution to the measurement: Test 1 with red lines, green lines of simulation.

As shown in Figure 4, for Test 1 with uniform hole column area and blocked dead volume, the simulated pressure is compared against the measured pressure in the inlet volume, at position P1, P2 and P5. The spike in the measured pressure signal of P1 is a disturbance or noise in the pressure measuring system and have normally no influence on the measurement in general. The agreement between simulation and experiment is reasonably good. Since the result differ by less than 5%, the use of the pressure in the inlet volume as the inlet pressure boundary condition is justified.

The agreement between measurement and simulation at P1 is better than at P2, suggesting that the 3D geometrical configuration at hole column A3 has much less influence on the pressure P1. The bigger deviation at P2 comes from the 2-dimensional modelling of the flow area between the hole column A3 and the conical body as shown in Figure 5. If the resulting minimum area is too small, the local flow velocity will be too low, and consequently the local pressure at P2, too high. Therefore, a more accurate result can be only achieved with 3D simulation.

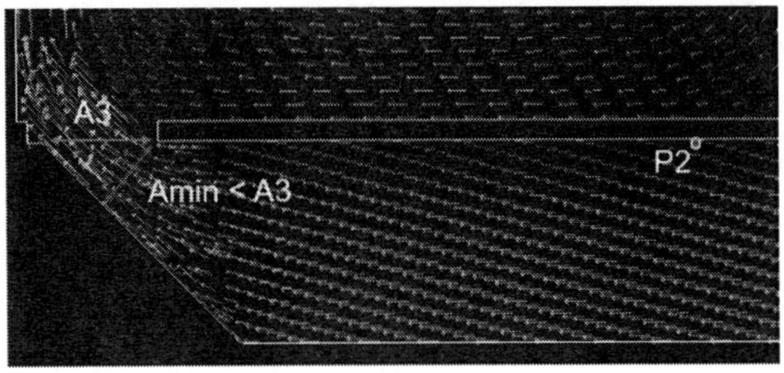

Figure 5. 2-dimensional modelling of 3D geometrical configuration of hole A3 and the conical body.

Figure 6 shows the comparison of measurement and simulation for Test 2 with uniform hole column area and dead volume. The agreement in general is satisfactory especially the measurement of P3 that confirms the pressure oscillation in the dead volume above the hole column A3. The simulated pressure at P2 shows a transient effect of 2-dimensional modelling in the first 30 ms, after which the pressure recovers to the measured value. Besides this, the 2-dimensional modelling also shows an overall pronounced pressure oscillation which is induced through the Helmholtz oscillation in the dead volume. This can be seen clearly by comparing the simulation results in Figure 5 of Test1 and Figure 6 of Test 2. In practice the oscillation is damped for P1 and P5 as is suggested by 3D simulation.

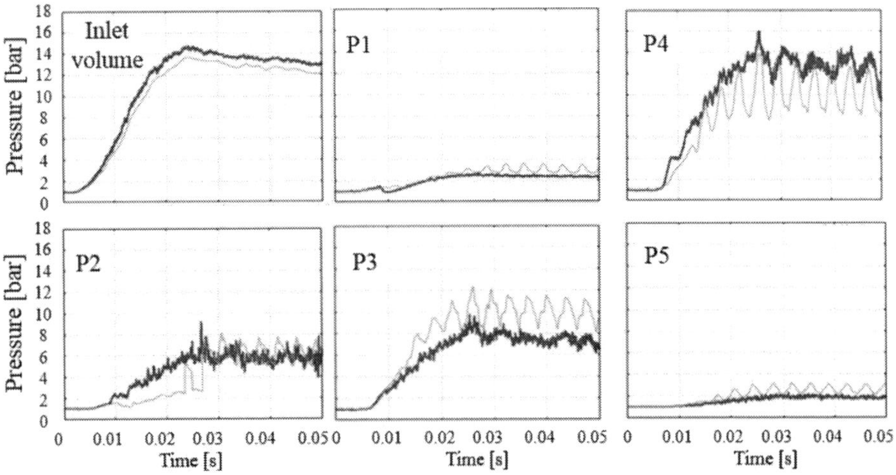

Figure 6. Comparison of simulated pressure distribution to the measurement: Test 2 results with red lines, green lines of simulation.

Figure 7 shows the results of Test 3 with SML hole column distribution. Due to a transient effect, the simulated pressure at P2 reaches the measured value after 40 ms. P1 indicates a good agreement without pronounced pressure oscillation, while the simulation for P3, P4, and P5 show a pronounced pressure oscillation.

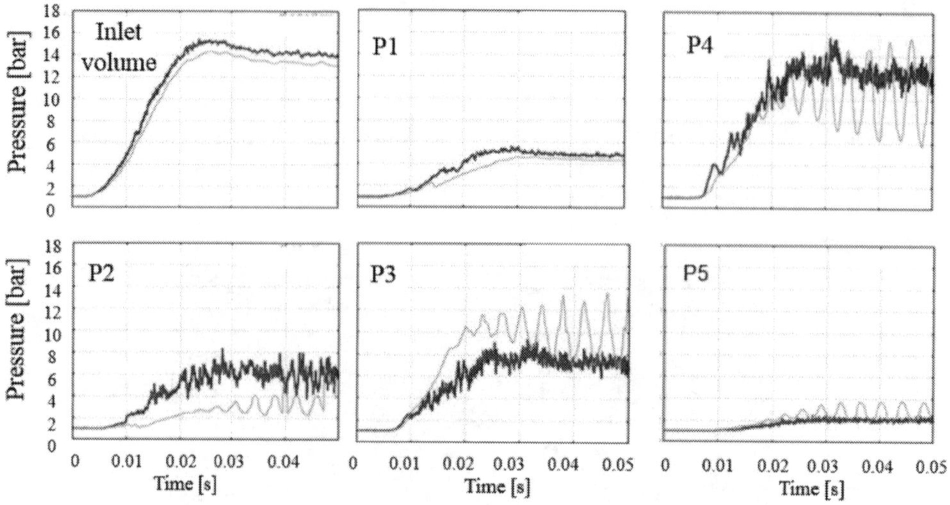

Figure 7. Comparison of simulated pressure distribution to the measurement: Test 3 results with red lines, green lines of simulation.

Figure 8 shows the results for Test 4 with LMS circular hole distribution. Here the measured pressure at P1 shows a pronounced pressure oscillation, which is absent in the simulation. The pressure measurement of P2, P3 and P4 all show a moderate pressure oscillation, but the simulation gives a more pronounced pressure oscillation. This behaviour can be explained again with 2-dimensional modelling for the 3D geometrical holes. The highly asymmetrical hole configuration exhibits different pressure oscillation behaviour. This should be checked further with 3D simulation.

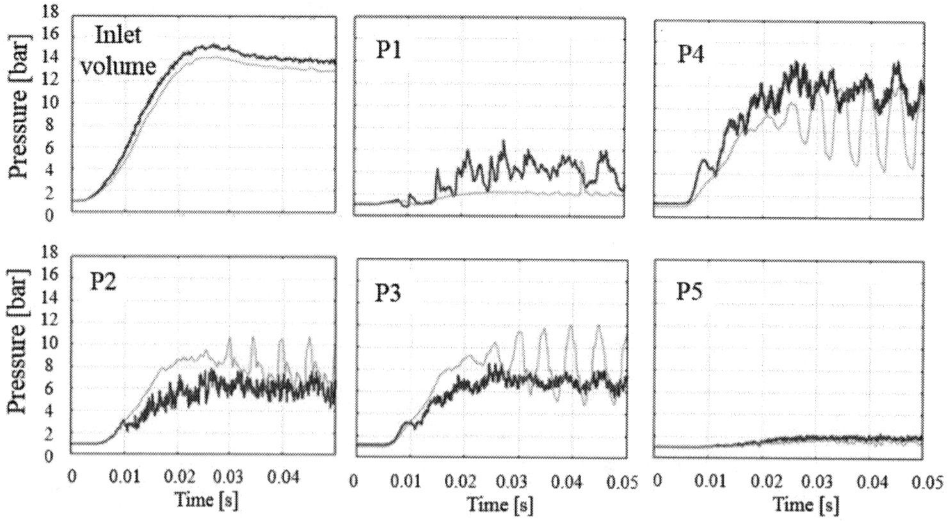

Figure 8. Comparison of simulated pressure distribution to the measurement: Test 4 results with red lines, green lines of simulation.

The validation confirms that the 2D simulation reflects the general behaviour of exhaust tube, although with some deviation. This confirmation significance for further investigations, since the exhaust tube simulation with arcing zone represents the exhaust tube in a real breaker. 3D arc simulation requires much longer calculation time and often suffers problems with numerical instability, therefore, 2D arc simulation is generally preferred.

4.2 Arc simulation of exhaust tube

Figure 9 shows a typical CFD set-up for the arc simulation of exhaust tube. The exhaust tube is connected to the arcing zone which consists of arcing contacts, heating volume, and nozzles. The temperature generated by the arc affects the flow field in the breaker, they are coupled and interact with each other and are considered in the governing equation system of the mathematical model of the CFD code. To simplify the simulation, arcing contacts have a constant opening gap with no movement. The model is further simplified by setting the outlet pressure boundary condition of the filling pressure, to between 6 and 10 bar depending on the switching gas.

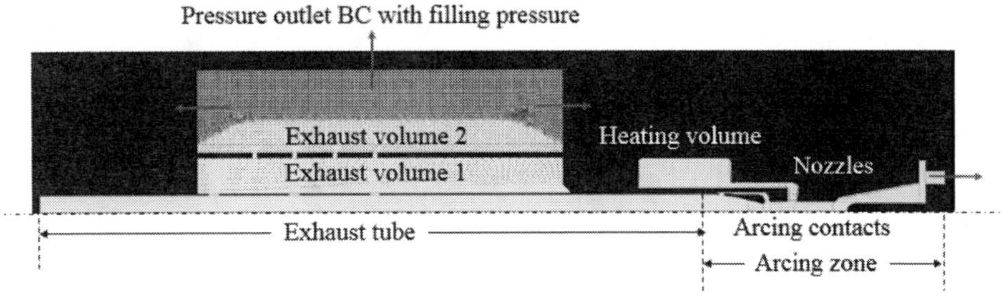

Figure 9. CFD set-up for the arc simulation of exhaust tub.

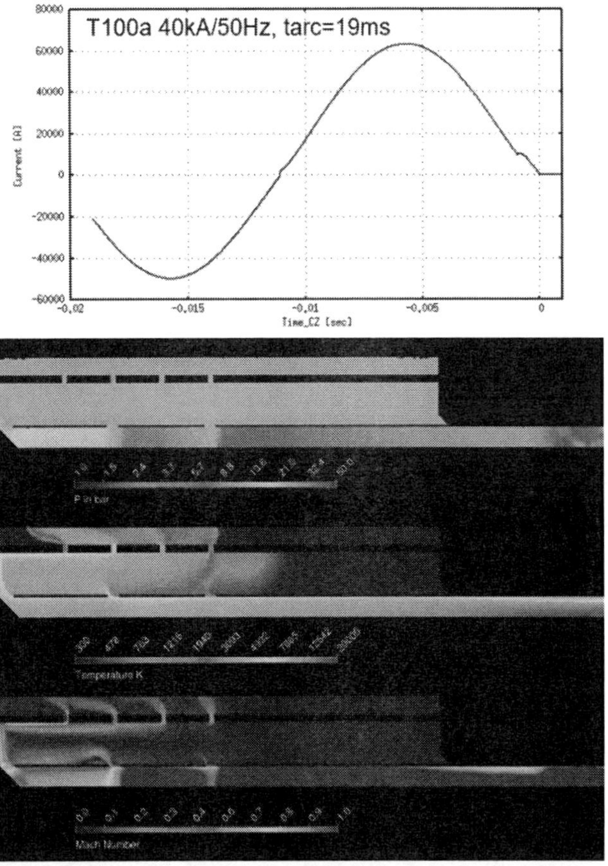

Figure 10. Typical current and flow pattern of arc simulation for an exhaust tube with CO_2.

Figure 10 shows a typical flow pattern from the arc simulation of exhaust tube at current peak of T100a long arcing time case. It can be seen that the Mach number and pressure can change locally with the hole position. This affects the mass flow through holes.

Two parameters can be used to describe the hot gas evacuation in the exhaust tube:

- Total mass through holes: It is the sum of the integral of the mass flow through all exhaust holes. Figure 11 shows the total mass through holes of different exhaust tube. It can be seen that the lowest mass through holes is the variant with LMS configuration. The highest mass through holes can be obtained with the variant consisting stepwise expansion tube.

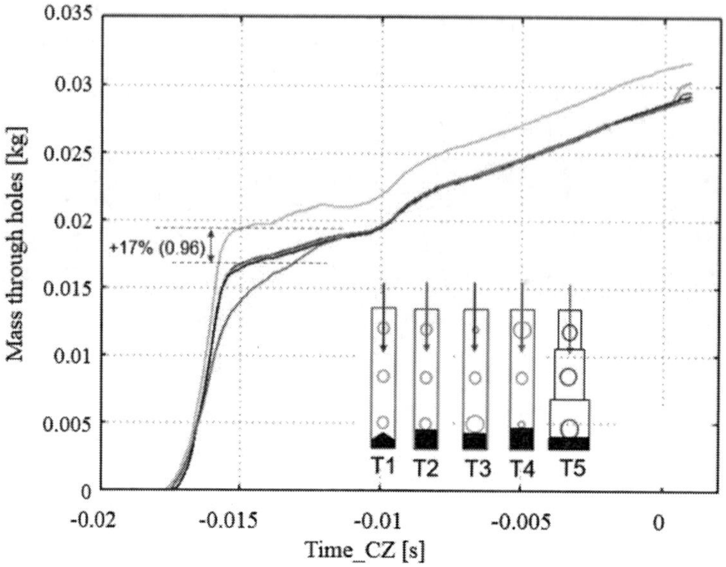

Figure 11. Total mass through holes of different exhaust tube, T100a, 19ms arcing time.

- Exhaust tube efficiency: It is defined as the ratio of the inlet mass flow and the outlet mass flow $\eta = \dot{m}_{out} / \dot{m}_{in}$. The exhaust tube efficiency of different variants is compared in Figure 12. It can be seen that again the lowest exhaust tube efficiency is the variant with LMS configuration. The highest efficiency can be obtained with the variant consisting stepwise expansion tube

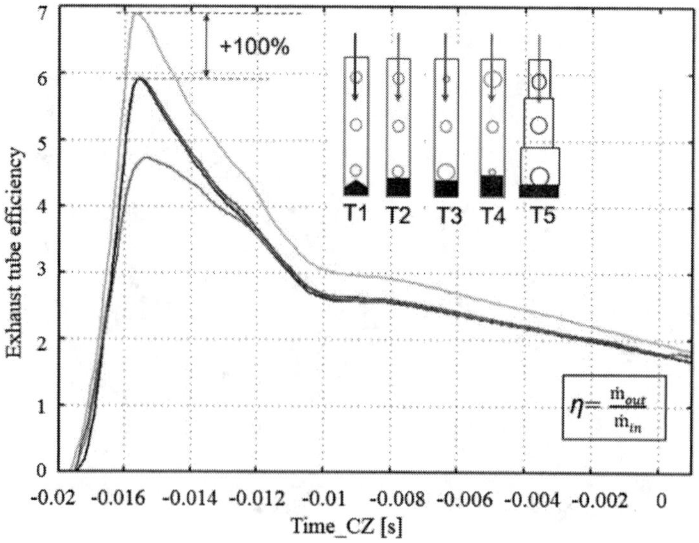

Figure 12. Exhaust tube efficiency of different exhaust tube, T100a, 19ms arcing time.

With these two parameters, exhaust tube design can be done essentially to fulfil the requirement for the current interruption.

5 Conclusion

Based on the comparison against the experimental study, the CFD simulation is validated and its accuracy and limitation in the 2D modelling of the 3-dimensional configuration is established.

The total mass through holes and the exhaust tube efficiency can be used to quantify the exhaust tube performance according to the requirement of current interruption. Expressed in terms of these two performance metric parameters, the exhaust tube with expanding sections seems to be the variant that delivers the best exhaust tube performance.

References

1. C.M. Claessens and G. Speckhofer, *CFD-Simulation of electrical arcs in circuit-breakers, Proceedings of ASME/JSME Conference*, San Diego (1998)
2. X. Ye, A. Dahlquist, J. Abrahamsson and C. Franck, *CFD simulation of heat transfer and gas mixing in exhaust parts of high-voltage circuit breaker, Proceedings of IEEE International Symposium on Electrical Insulation*, Vancouver, Canada, June 9-12 (2008)
3. H. Nordborg and A. Iordanidis, Journal of Physics D: Applied Physics, **41**(13), 135205 (2008)
4. M. Dhotre, X. Ye, S. Kotilainen, M. Schwinne and R. Bini, *CFD simulation for self-blast high voltage circuit breaker: Mixing and Heat transfer, Proceedings of the Electrical Insulation Conference*, Annapolis, Maryland, USA, June 5-8 (2011)
5. A. A. Iordanidis and C. M. Franck, Journal of Physics D: Applied Physics, **41**(13), 135206 (2008)
6. L. Niemeyer, IEEE Transactions on Power Apparatus and Systems, **PAS-97**(3), 950-958 (1978)
7. C.B. Ruchti and L. Niemeyer, IEEE Trans. Plasma Sci., **PS-14**(4), 423-434 (1986)
8. X. Ye, M.T. Dhotre, J.D. Mantilla and S. Kotilainen, *CFD Analysis of the Thermal Interruption Process of Gases with Low Environmental Impact in High Voltage Circuit Breakers, Proceedings of the Electrical Insulation Conference*, Seattle, WA, USA June 7-10 (2014)

A proficiency-based learning algorithm applied to e-learning system for knowledge-oriented course

Cheng-Hsiu Li[a]

Department of Information Management, National Taitung Junior College, Taiwan

Abstract. In the cognitive psychology domain, the memorization of knowledge is one kind of representation of memory capability. Humans often get the knowledge with knowledge-oriented courses. These courses are composed of the question banks with the fixed knowledge-based content. Learners can obtain the required cognitive knowledge through the repeated memory process. This study proposed a proficiency-based learning algorithm to simulate this kind of learning behavior and then implement an instruction-assisted system called proficiency-based adaptive and guided e-learning system (PAGE system), in which learners can study independently through a guided mechanism. The PAGE system is quite suitable for the knowledge-oriented courses, which are designed by several items with standard answers. Obviously the study indicated the PAGE system is validated in the learning process of memorization. The experimental result proved that the PAGE system can help learners to remember efficiently. All participants got positive progress and 80% of them achieved their criterion for cognitive knowledge growth. In addition, the experiment also showed that the learning attitude and active performance are crucial factors that affect learners' learning growth.

1 Introduction

The instruction objectives include the cognitive domain, affective domain, and psychomotor domain [1]. The cognitive domain can be separated into memory, thought, recognition, and application, and it is generally classified as essential knowledge and a higher level of intelligence and skill. The memory performance belongs to the cognitive domain [2-4]. The knowledge-oriented courses are the cognitive learning foundation in which they integrate the cognition with practice and skill. Hence learners can absorb the basic knowledge in the cognitive domain.

The knowledge-oriented courses belong to cognitive learning. The application of knowledge-oriented courses is popular with the certification fields, and learners gain the basic knowledge from the knowledge-oriented course of the specific field. The certification of International Project Management Association (IPMA) assesses the knowledge of project-management field by cognitive exam for the basic project-management ability in the level C and D [15]. The certification of General English Proficiency Test (GEPT) includes five levels to evaluate the English proficiency level, and the different fixed range of vocabulary is the basic cognition in levels [16]. The Council of Labor Affairs (LAC) in Taiwan currently holds around 202 occupational classifications for certification tests. Each occupational classification has three categories. Each class requires passing the practical and cognitive

[a] Corresponding author : lcs@ntc.edu.tw; chsiu.lee@gmail.com

tests to obtain the technician certificate [5]. The mentioned cognitive certification test covers the essential knowledge aspect, and it is constructed in the form of questions and answers. Learners need to demonstrate the master cognition based on the questions. The cognitive test belongs to criterion-based style, and learners need to satisfy the level to pass. The LAC in Taiwan has implemented a regulation on "Technician Certification and Licensing" to improve the quality of technicians in September 1972. The LAC is formed with professional teams to the council for promoting certification and licensing training, and also to ensure the occupational classification of certification tests is suitable for business requirements and international standards [5]. Currently, almost vocational schools in Taiwan have integrated the regulation on technician certification in the skills education. The practical tests are adopted as a part of the skill training courses, and the cognitive tests are also used to knowledge-oriented learning courses. In general, the knowledge-oriented courses consist of standard questions with answers in the assigned cognitive content. Learners can remember and obtain the cognitive knowledge by the continuous assessment and practice.

Hence, this study observed the proficient learning cases and proposed a proficiency-based learning (PBL) algorithm which could be used to design the knowledge-oriented course in the e-learning system, and then a web-based instruction-assisted system called proficiency-based adaptive and guided e-learning system (PAGE system) was implemented. In the PAGE system, learners can learn independently by the PBL algorithm and without teachers' involvement. Finally, learners can master their cognitive knowledge in the proficient learning process.

2 Design of the PAGE system

This paper presents the design of the PAGE system based on the aspects of system structure, knowledge-based learning course, and proficiency-based learning algorithm.

2.1 Design of the system structure

The study uses a web-based interface to implement the PAGE system. Learners can conveniently learn regardless the limitation of time and distance by utilizing the Internet and www-style operating interface.

The proposed PAGE system emphasizes that the system can guide learners to study independently and proficiently. In terms of system components, the PAGE system consists of a registration subsystem, guided learning subsystem, proficiency-based adaptive and guided learning subsystem, question and answer subsystem, and a learning portfolio subsystem. The system components are illustrated with Figure 1.

The registration subsystem mainly provides the interface needed for registration, including the columns of personal account and password, basic information, and target grade setting. The target grade setting will affect the analysis result of the personal learning path and the critical judgment in every stage-examination. After registration, the PAGE system guides students into the pre-test and generates a personal learning path based on the pre-test result.

The function of the guided learning subsystem is to remember the last learning progress and guide learners to study with their customized personal-learning paths until they obtain enough cognitive knowledge.

The proficiency-based adaptive and guided learning subsystem mainly guides learners to adaptively learn by the proficiency-based learning algorithm. When learners already finish learning one of subjects, the subsystem will hold the stage-examination to them. If the exam result does not correspond to the target setting, students must to study again the same stage until passing the criterion. When learners completed all subjects, they must to attend the final-examination to test their knowledge level.

The main function of the question and answer subsystem is to provide the standard questions and answers interface recorded and identified learners' proficiency.

The learning portfolio subsystem provides learners' study tracks, including the target grade, pre-test and post-test result, personal learning path, and learning effectiveness.

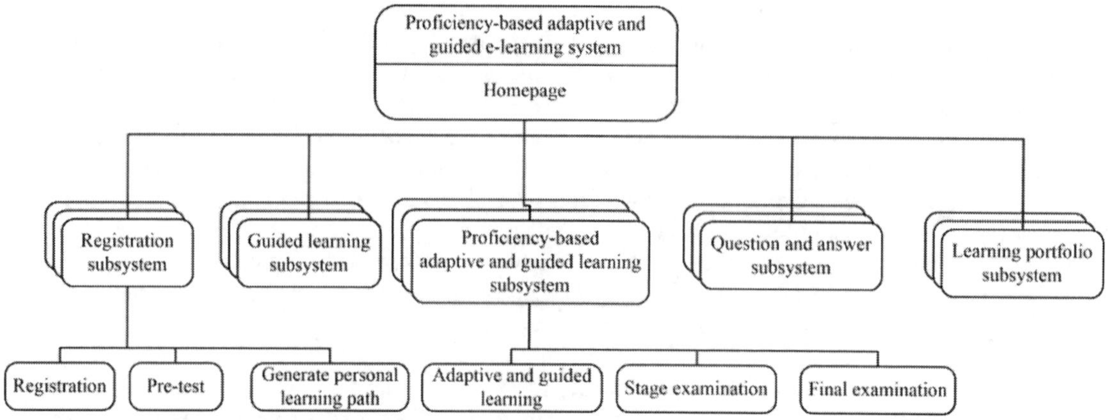

Figure 1. Description of system components.

When learners connect to the PAGE system server on the Internet, the server would respond with a web-based interface to learners. The operating process as followings:

First, learners must register their basic information and target grade in the registration subsystem, and the PAGE system will analyze their personal learning paths according to the pre-test result. The learning paths will be stored into the database as learning and guiding in the future.

When learners access the system, it would guide them to learn by the proficiency-based adaptive and guided learning subsystem, which will remain the last learning record and check the personal learning path. The question and answer subsystem immediately provides the learning content with Q&A style to students, and it also carries out the stage-examination, feedbacks, and final examination. Finally, learners' learning records are stored into the learning portfolio database, and they can check their records and learning status anytime and anywhere. The operating system structure is shown as Figure 2.

2.2 Design of the knowledge-oriented course

The study defines the cognitive course as the knowledge-oriented course. The study uses a web-based interface to implement the PAGE system. Learners can conveniently learn regardless the limitation of time and distance by utilizing the Internet and www-style operating interface.

Because the proficient learning character of the PAGE system, it is quite suitable to cognitive learning in the knowledge-oriented courses. Hence the construction of knowledge-oriented courses can be separated with sub-courses by different subjects, and the each sub-course context is designed on the Q&A format. For identifying learners' guided learning path, this study suggests containing at least a half of questions in the pre-test of every subject. The structural decomposition diagram of the knowledge-oriented course is shown in Figure 3.

The study suggests designing four entities including the subject entity, question entity, answer_sel entity, and question_pic entity to construct the knowledge-oriented course by data modeling process.

The subject entity describes all subjects of the knowledge-oriented course. The subject entity uses the primary key (PK) of subject_id attribute as the index for the question entity, and it also can link to the subject by foreign key 1 (FK1) in the question entity.

The question entity describes the character of standard questions in the knowledge-oriented course. The question entity is constructed of PK included by question_id attribute, FKs included by subject_id, quespic_id, and ans_id, and other attributes included by type attribute, ques_num attribute, and question attribute.

The quespic_id entity describes the question's picture source. The quespic_id entity uses the quespic_id attribute as the index for the question entity, and it associates with the question's pictures by FK2.

The answer_sel entity describes the answer and detail explanation of question. The answer_sel entity uses the ans_id attribute as the index for the question entity, and it gets the answer and detail explanation of question in the answer_sel entity by FK3.

This paper constructs the data modeling of knowledge-oriented course using the mentioned four entities, and it is illustrated with Figure 4.

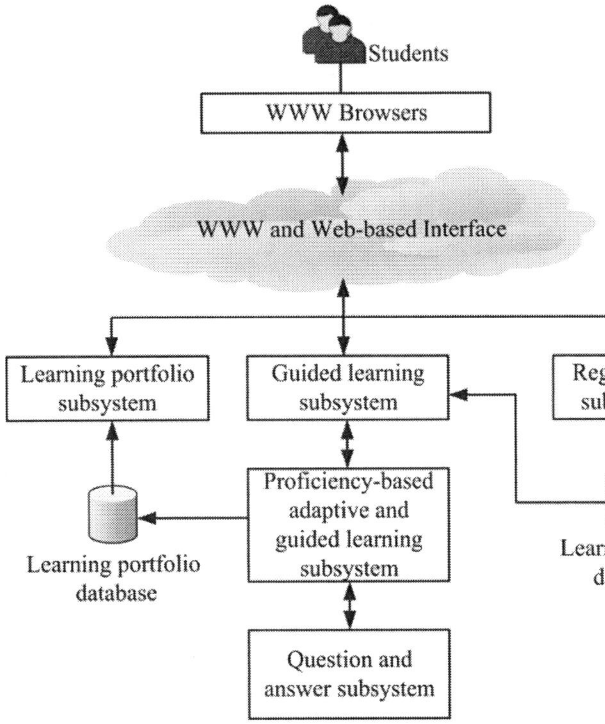

Figure 2. Description of the system structure.

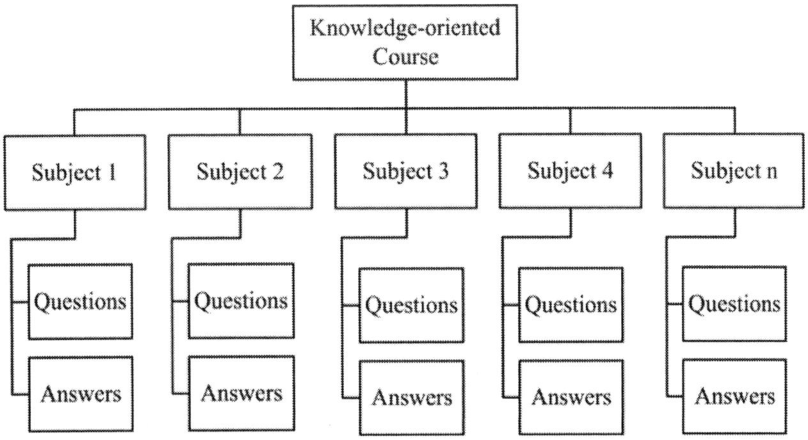

Figure 3. The structural decomposition diagram of knowledge-oriented course.

This study conducted the Taiwan LAC cognitive test of class C technician certification of computer software applications as the experimental course. It was designed by standard questions and answers. This course had 700 cognitive questions and was divided into five subjects based on different characters.

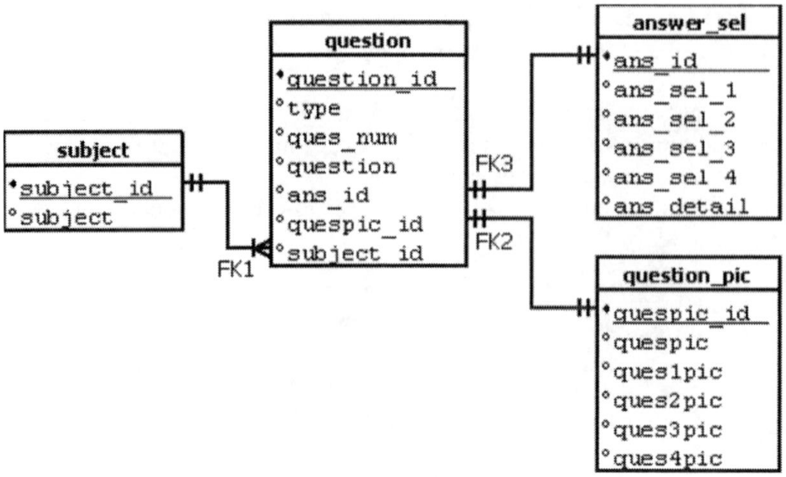

Figure 4. The data modeling of the knowledge-oriented course.

2.3 Design of PBL algorithm

The competency-based education (CBE) theory is an outcome-based teaching method which focuses on the learner and emphasizes the cultivation and mastery of specific and measurable skills throughout various stages of development [6-8]. The standard CBE instruction process is a proficient instruction process following the processes of continuous assessment, feedback, and correction [9-14]. This study observes the character of proficient instruction process and applies it to the proficiency-based learning (PBL) algorithm. Figure 5 shows the PBL algorithm proposed in the study.

The cognitive content of knowledge-oriented course is randomly classified to different subjects as the personal learning path according the learner's pre-test result, termed as subject k, k=1~n. Every learner owns the different customized path, and the PAGE system will guide to study.

When learners finish studying each subject, PAGE system will guide student to test the stage-examination in the current subject, and use (1) to randomly generate 50% questions from the question entity for testing.

$$Questions\_Num(k) ::= Count(Questions(k)) * 0.5$$

$$Questions(k)' = RAND(Questions(k))|num=Questions\_Num(k) \quad (1)$$

After testing, the PAGE system calculates counts of right answers and conducts (2) to calculate the grade of the stage-examination.

$$stage\_question\_right(k) ::= \text{the answer of } Questions(k) \text{ is right}$$

$$stage\_exam\_grade(k)=(COUNT(stage\_question\_right(k))/Questions\_Num(k))*100 \quad (2)$$

The PBL algorithm uses the method of symbol determination which judges the keyword 'test_result(k)' to determine the proficiency degree and guided consideration. First, calculating the arithmetic difference of test_check(k) using (3); it subtracts the grade of stage-examination from the predicted learning grade in the registration.

$$test\_check(k) = (predicted\_grade) - (stage\_exam\_grade(k)) \quad (3)$$

Next, assigning the result of test_check(k) to test_result(k) and deciding the modulus judgement of test_result(k) to generate the necessary decision symbol of PBL algorithm. This algorithm uses the threshold limit value by 'zero' to judge the relationship of test_check(k). If the test_check(k) is big or equal to zero, the decision symbol is appointed to '+'; otherwise the decision symbol is '-'.

$$test\_result(k) = test\_check(k)$$

$$\left|test\_result_{(k)}\right| = \begin{cases} +(test\_result_{(k)}),\ test\_check_{(k)} \geq 0 \\ -(test\_result_{(k)}),\ test\_check_{(k)} < 0 \end{cases} \tag{4}$$

After the decision symbol is judged, the PBL algorithm will enter the proficiency assessment procedure shown as Figure 5; this is the core process in the PBL algorithm. The proficiency assessment procedure conducts the variable 'SIGN(test_check(k))' to be the decision symbol. If the SIGN(test_check(k)) is '+', it means that the leaner is proficient in the subject and is permitted studying next subject; otherwise if the SIGN(test_check(k)) is '-', it means that the learner's proficiency is insufficient and needs to study the same subject again.

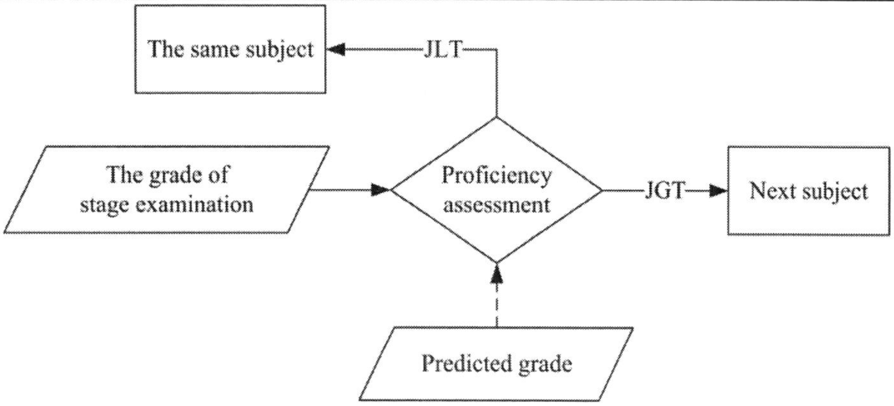

[Proficiency assessment procedure]
Assume predict-grade ::= the grade is the learner's predictive setting
Assume stage-grade ::= the grade is the result of stage-examination
If SIGN(test_check(k)) = = '+' then
  Pass and study next subject
Elseif SIGN(test_check(k)) = = '-' then
  Fail and study again
End IF

Figure 5. Proficiency-based learning (PBL) algorithm.

According to the PBL algorithm, when learners login the system, it will compare the last learning progress and guide to study by the guided learning subsystem.

The system proficiently guides learners to study the cognitive knowledge. When they obtain enough proficiency in the subject, the system will guide to the stage-examination for checking the proficient degree. If passing the stage-examination, the learners will move on next subject until all subjects are passed; otherwise, they need to study the failure subject again.

Finally, when learners complete all subjects and pass all examinations, the system will guide to the final examination and judge the final proficient degree.

2.4 System implementation

The PAGE system was implemented with the PHP version 5.1.6, MySQL version 5.0.45, and CentOS Linux 5.3 operating system.

The interface of the question and answer subsystem consists of the current subject title and standard questions and answers. In addition, the interface designs a check-button below every question for checking whether learner is proficient at the question. The implemented learning interface is shown in Figure 6.

When learners finish the learning procedure, the PAGE system stores all learning records into the learning portfolio database and permits learners to track their study status through the learning portfolio subsystem. The learner's portfolio interface shown in Figure 7 includes the adaptive guided path, predictive goal, all examination records, and study performance.

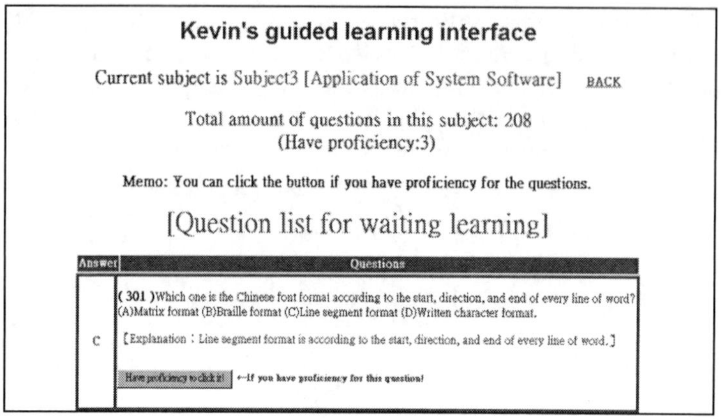

Figure 6. Courses learning interface with questions and answers (Q&A).

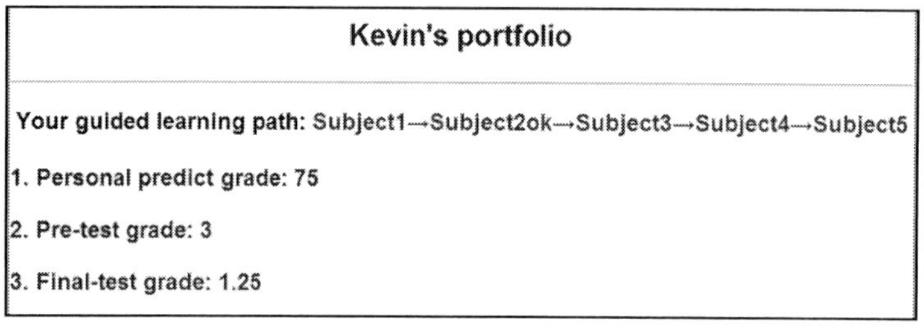

Figure 7. Students' portfolio interface.

3 Research methodology

3.1 Research limitation

This study considers the scope and duration of experiment and then set the research limitation on the following aspects: knowledge-oriented material, research participant, and experiment duration.

3.1.1 Scope of the knowledge-oriented material

The scope of the knowledge-oriented material in the study was the Taiwan LAC cognitive test for class C technician certification in computer software applications. It covered 700 knowledge-oriented questions and was classified to five subjects.

3.1.2 Limitation of the research participant

15 third-year participants in National Taitung Junior College were randomly selected from 261 members in the PAGE system. For the reason, all of them attended the 2010 class C technician certification in computer software applications, and their certification tests would be considered in the study. The study expected to analyze the system validity and participants' learning effectiveness by comparing the certification test result.

3.1.3 Limitation of the experiment duration

The experiment duration was held two months from September 1, 2010. During this period, participants learned the experimental knowledge-oriented course in the PAGE system independently without teachers' assistance. After the experiment, all participants attended the final examination for analysis.

3.2 Research design

Considering research limitations, the study designed the experiment process, PAGE system structure, teaching course, and adaptive and guided learning.

On the first step, the study chose research participant randomly from participants who attended the 2010 class C technician certification in computer software applications. The next step, the research participants studied independently in the PAGE system according the proficiency-based adaptive and guided learning function.

The experiment duration was two months, and the research participants had the final examination to check the proficient degree in the end of experiment. The experiment process is shown as Figure 8.

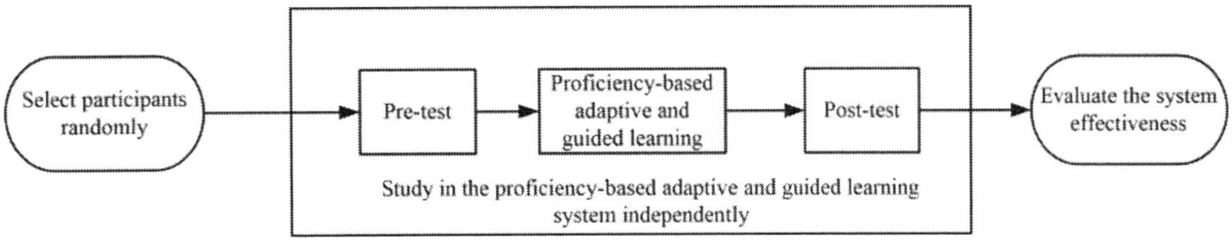

Figure 8. The experiment process.

4 Experiment and discussion

The study discusses and analyzes research results based on three aspects including the system validity, required learning time, and learning effectiveness. Finally the study expects to verify the PAGE system's proficient guiding ability.

4.1 Analysis of system validity

The analysis of system validity compares post-test with pre-test and determines the positive or negative growth of validity.

The grade growth comparison among participants is shown in Table 1. The Grade-growth is calculated by subtracting Pre-grade from Post-grade. If the Grade-growth is bigger than zero, the calculated result is marked by the symbol '+'; otherwise if the Grade-growth is smaller than zero, the calculated result is marked by the symbol '-'.

In Table 1, the analysis showed all participants had positive growth using the PAGE system, and 87% of them had positive growth above 10 points. This result indicated the system implemented by the proficiency-based learning algorithm could guide participants to study efficiently with proficient procedure. Participants got grade from the highest grade of 69.7 points to the lowest grade of 3.2 points; only two participants, who were Participant ID 1 and 14, were less than 10 points.

Table 1. Grade growth comparison among participants.

Participant ID	Pre-grade	Post-grade	Grade growth (Post-grade – Pre-grade)
1	88	91.2	+3.2
2	61	88.7	+27.7

Participant ID	Pre-grade	Post-grade	Grade growth (Post-grade – Pre-grade)
3	67	83.7	+16.7
4	41	82.5	+41.5
5	13	70	+57
6	31	78.7	+47.7
7	41	76.2	+35.2
8	27	82.5	+55.5
9	21	63.7	+42.7
10	33	95	+62
11	18	56.2	+38.2
12	30	82.5	+52.5
13	24	93.7	+69.7
14	35	40	+5
15	26	83.7	+57.7

In Table 2, the pre-test grade of Participant ID 1 was 88 points, and he was limited to improve his performance, but he only used 1.13 days to improve his grade to 91.2 points in the final post-test.

The pre-test grade of Participant ID 14 was too low against to the criterion. He spent all experiment duration 49 days to finish the study procedure, and his attitude was poor and lack motivation during experiment. The above reason would cause his low learning performance.

Table 2. Grade comparison among Participant ID 1 and 14.

Participant ID	Pre-grade	Post-grade	Learning days	Grade growth	Growth
1	88	91.2	1.13	3.2	+
14	35	40	48.96	5	+

Hence, the result showed the PAGE system had good performance on the proficient guided learning. The study also discovered the participant's active learning attitude would affect the learning growth when studying in the PAGE system.

4.2 Analysis of required learning time

Based on the comparison of results in Table 3, the system provided customized subjects to participants. If participants were proficient in some subjects, the system marked symbol "ok" in the learning path and did not guide to study to simply the learning time. In the experiment, Participant ID 1 to 3 were proficient in some subjects, and they do not need to study all subjects of the learning path. Participant ID 1 only needed to learn subject 2, and he finished learning and got the proficiency in 1.13 days. Participant ID 2 and 3 only needed to learn two subjects, and they finished learning and got the proficiency in 17.47 days and 11.11 days.

Table 3. Comparison of learning paths, learning period, learning subjects and goal achieved.

Participant ID	Procedure 1	Procedure 2	Procedure 3	Procedure 4	Procedure 5	Learning days	Goal achieved?	Subjects
1	Subject1	Subject2ok	Subject3ok	Subject4ok	Subject5ok	1.13	Yes	1
2	Subject1	Subject2	Subject3ok	Subject4ok	Subject5ok	17.47	Yes	2
3	Subject1ok	Subject2ok	Subject3	Subject4ok	Subject5	11.11	Yes	2
4	Subject1	Subject2	Subject3	Subject4	Subject5	38.24	Yes	5
5	Subject1	Subject2	Subject3	Subject4	Subject5	41.97	Yes	5

Participant ID	Procedure 1	Procedure 2	Procedure 3	Procedure 4	Procedure 5	Learning days	Goal achieved?	Subjects
6	Subject1	Subject2	Subject3	Subject4	Subject5	41.99	Yes	5
7	Subject1	Subject2	Subject3	Subject4	Subject5	42.01	Yes	5
8	Subject1	Subject2	Subject3	Subject4	Subject5	42.02	Yes	5
9	Subject1	Subject2	Subject3	Subject4	Subject5	47.13	*No*	5
10	Subject1	Subject2	Subject3	Subject4	Subject5	47.14	Yes	5
11	Subject1	Subject2	Subject3	Subject4	Subject5	47.28	*No*	5
12	Subject1	Subject2	Subject3	Subject4	Subject5	47.36	Yes	5
13	Subject1	Subject2	Subject3	Subject4	Subject5	48.45	Yes	5
14	Subject1	Subject2	Subject3	Subject4	Subject5	48.96	*No*	5
15	Subject1	Subject2	Subject3	Subject4	Subject5	48.96	Yes	5
Average						38.08	12	4.33

In order to clearly compare the difference with the learning subjects and learning period, the study figures out the learning time difference which is subtracted the average learning time from learning time, and the statistic is shown in Figure 9.

Hence, Figure 9 shows that the Participant ID 1 to 3 reached the proficiency in a shorter period and were guided by a simplified personal learning path. These participants' learning effectiveness was still positive on the simplified learning period. This result represented the PAGE system could guide participants to study adaptively and effectively.

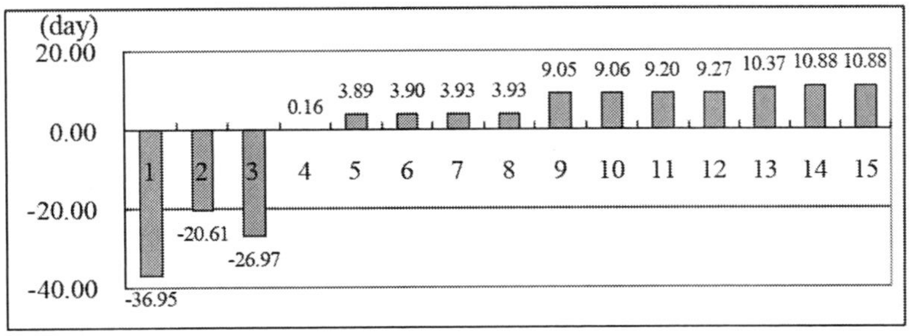

Figure 9. Learning days compared with average.

4.3 Analysis of learning effectiveness

The study analyzed participants' learning effectiveness by considering the participants' achieving on the target grade, shown in Table 4. Table 4 shows that there were 12 participants to reach the predict goal and get the proficiency, but there were also three participants not yet.

Table 4. Comparison of participants' learning effectiveness.

Participant ID	Predictive grade	Post-grade	Get goal?
1	80	91.2	Yes
2	70	88.7	Yes
3	70	83.7	Yes
4	70	82.5	Yes
5	70	70	Yes
6	70	78.7	Yes
7	70	76.2	Yes
8	70	82.5	Yes

Participant ID	Predictive grade	Post-grade	Get goal?
9	70	63.7	No
10	70	95	Yes
11	70	56.2	No
12	70	82.5	Yes
13	80	93.7	Yes
14	70	40	No
15	70	83.7	Yes
Goal achieved		Unable to goal achieved	
12		3	

Table 4 shows that 80% of the participants could achieve their target through the PAGE system. Even though there were 20% of participants could not, but they still had a positive learning performance from 5 points to 42.7 points as shown in Table 1. The result shows that all participants had positive learning effectiveness through the PAGE system.

5 Conclusions

The application of knowledge-oriented course is popular with the certification fields, and learners obtain the cognitive knowledge from the knowledge-oriented course in the specific field. The knowledge-oriented courses always offer a fixed learning format. Learners absorb knowledge through the memory process with continuous assessment and feedback proficiently.

Hence, this study observed the proficient learning cases and proposed the PBL algorithm which could be used to design the knowledge-oriented course in the e-learning system, and then a web-based instruction-assisted system called the PAGE system was implemented. In the PAGE system, learners can learn independently by the PBL algorithm and without teachers' involvement. Finally, learners can master their cognitive knowledge in the proficient learning process.

The study proved that the PAGE system had good performance on the proficient guided learning. All participants had positive growth with using the PAGE system and 80% of them achieved their proficient target. In addition, the simplified personal learning path did not affect participants' learning effectiveness and could help them to reach the proficiency in a shorter period. The result represented the PAGE system could guide participants to study adaptively and effectively. The study also discovered the participant's active learning attitude was the crucial factor that affected their learning growth.

The proposed PAGE system successfully corresponds to provide a proficiency-based adaptive and guided learning procedure to guide learners to study independently and proficiently, and all learners have the positive learning effectiveness through the PAGE system.

References

1. A. Gordon, Writing Instruction Objectives and Tests, Retrieved from http://instructionaldesign.gordoncomputer.com/Objectives.html (2006)
2. B. S. Bloom, *Taxonomy of Educational Objectives Book I: Cognitive Domain*, New York: David McKay Co, Inc (1956)
3. E. J. Simpson, The Classification of Educational Objectives in the Psychomotor Domain, Washington, DC: Gryphon House (1972)
4. D. R. Krathwohl, B. S. Bloom and B. M. Bertram, *Taxonomy of Educational Objectives: The Classification of Educational Goals, Handbook II: Affective Domain*, New York: David McKay Co, Inc (1973)
5. Council of Labor Affairs, Retrieved from http://www.labor.gov.tw/english/index.jsp (2012)
6. C. C. Chang and P. N. Chou, Chinese Journal of Science Education, **14**(2), 209-235 (2006)

7. G. E. Hall and H. L. Jones, Competency-based education: a process for the improvement of education, Englewood Cliffs, NJ: Prentice-Hall (1976)
8. M. Ritzen and J. Kosters, Combining different aims in a portfolio ways in which it can serve the student, 14th World Conference on Educational Multimedia, Hypermedia and Telecommunications, Denver, Colorado, USA (2002)
9. C. C. Chang, Interactive Learning Environments, **14**(3), 265-286 (2006)
10. B. Cheng, M. Wang, J. H. Yang, Kinshuk and J. Peng, Computers & Education, **57**(1), 1317-1333 (2011)
11. W. M. Su, B. Herron and P. J. Osisek, Nursing Clinics of North America, **46**(3), 291-298 (2011)
12. S. E. Weinberger, A. G. Pereira, W. F. Lobst, A. J. Mechaber and M. S. Bronze, Annals of Internal Medicine, **153**(11), 751-756 (2010)
13. R. M. H. Wong, Porta Linguarum, **9**, 179-198 (2008)
14. W. C. Hsu and C. H. Li, A competency-based guided learning algorithm applied on adaptively guiding e-learning, interactive learning environments, DOI:10.1080/10494820.2012.745432 (2012)
15. IPMA Certification. http://ipma.ch/certification/certify-individuals/ (2014)
16. GEPT. https://www.lttc.ntu.edu.tw/E_LTTC/E_GEPT/intermediate.htm (2014)

IPTV program recommendation based on combination strate gies

Hao Li, Huan Xia, Yan kang and Mohammad Nashir Uddin[a]

School of Software, Yunnan University, Kunming, China

Abstract. As a new interactive service technology, IPTV has been extensively studying in the field of TV pro-gram recommendation, but the sparse of the user-program rating matrix and the cold-start problem is a bottleneck that the program recommended accurately. In this paper, a flexible combination of two recommendation strategies proposed, which explored the sparse and cold-start problem as well as the issue of user interest change over time. This paper achieved content-based filtering section and collaborative filtering section according to the two combination strategies, which effectively solved the cold-start program and over the sparse problem and the problem of users interest change over time. The experimental results showed that this combinational recommendation system in optimal parameters compared by using any one of two combination strategies or not using any combination strategy at all, and the reducing range of MAE is [2.7%,3%].The increasing range of precision and recall is [13.8%95.5%] and [0,97.8%], respectively. The experiment showed better results when using combinational recommendation system in optimal parameters than using each combination strategies individually or not using any combination strategy.

1 Introduction

Currently, IPTV, DTV, and the Internet have been using widely, and the number of TV programs is rapidly growing. It usually takes a long time for us to find out what we prefer from hundreds of TV programs. Previously there were not too many programs due to various limitations and restrictions on TV channels. Then only few TV programs are selectable for the audience. Also, EPG (Electronic Program Guide) technology relieves the information overload of the TV program. With the number of TV programs increasing, EPG has become more and more inadequate. Therefore, some recommendation systems based on EPG have proposed 0, 0.

IPTV (Internet Protocol Television) is also known as network TV. It takes home TV as the primary terminal equipment through using the infrastructure of a broadband cable television network, to provide a variety of digital media services including television programs. The provider involved in deploying IPTV services that range from cable and satellite TV carriers to large telephone companies and private network operators.

With the rapid development of the Internet, programs and movies (from now on collectively referred to as TV programs) emerge in large numbers. In this background, if the user cannot quickly find their favorite programs, the ratings or demand rate will decline, and users will gradually reduce

[a] Corresponding author : mun2003dhk@gmail.com

their reliance on IPTV, which eventually leads to decline in service quality and some users. Therefore, how to quickly find the user's favorite TV programs has great significance for the healthy development of IPTV. Recommendation system as an information filtering technology can efficiently solve above problems. However, IPTV unlike conventional television and its characteristics presented as follows.

Interactivity: differently from traditional TV, where the communication is unidirectional, the two-way capabilities of IPTV systems allow the user to interact with the system. In particular by allowing IPTV systems to explicitly or implicitly collect user preferences.

Accuracy: IPTV to deliver multimedia content to the user IP address, then the service provided by IPTV service provider can be accurate to individuals.

Multi-Media: IPTV services include scheduled television programs and video on demand (VoD) contents.

Freedom: Users can choose which TV program they want to watch according to their wishes at any time.

As above characteristics of IPTV shown, first of all, the accuracy of IPTV provides the implementation condition for the recommendation system. Secondly, the IPTV provides the data support for the recommendation system, which guarantees the quality of the recommendation. Finally, IPTV's multimedia and the freedom of the IPTV recommendation system provide the necessary support. Therefore, the recommendation system is integrated into IPTV infrastructure, and the recommended results are calculated by using the explicit or implicit data collected by the IP network. The guidance will give users faster and more exciting way to select TV programs and will positively promote the healthy development of IPTV business.

IPTV recommendation system is still facing many challenges. In the field of IPTV, due to the increasing number of users and the difference of the users' choice, so that the difference of user's rating is enormous. At the same time, the problem of user-rating matrix sparsity caused by lacking user ratings because of the substantial increasing of challenges and the cold-start problem caused by newly added items or users have a significant influence on the quality of the recommendation result.

The recommendation algorithm currently used in IPTV is collaborative filtering algorithm and content-based recommendation algorithm. Collaborative filtering algorithm is divided into user-based collaborative filtering algorithm 0 and item-based collaborative filtering algorithm [8]. User-based collaborative filtering algorithm is based on the assumption that if the users' rating for the items is similar, their rating will be similar for other items as well. Item-based collaborative filtering algorithm holds that users' rating for the items has similarity 0.

Content-based recommendation tried to recommend items that are similar to the items user liked in the past, through matching user interest model and contents of an item. According to the characteristics of content-based filtering, we can know it does not have a sparse problem and item cold-start problem. But its recommendation results are too customized to simply find items which are different from the items the user liked in the past 0, 0.

Relevant researchers propose some other recommended algorithm in IPTV based on collaborative filtering algorithm and content-based filtering algorithm. Such as LSA (latent semantic analysis) content-based algorithm, SVD content-based algorithm, weighted combination algorithm for based filtering and collaborative filtering algorithm, switch combination algorithm for content-based filtering and collaborative filtering algorithm. LSA content-based algorithm increases additional latent semantic analysis for the plot of different TV programs. SVD content-based algorithm takes advantage of singular value decomposition to decompose user-program metadata matrix W, whose elements W_{ci} represent the relevance of characteristic (metadata) c for item i. These two kinds of improved algorithm based on the content-based recommendation can improve the quality of recommendation, but there is still no solution to the fundamental problem, user cold-start problem, over-customization problem, and the problem of user interest change over time. The following two combination algorithms, just only put together two kinds of algorithms, is not obvious to the quality of the recommendation results.

This paper combined the advantages and disadvantages of collaborative filtering, content-based filtering, SVD, combination algorithm to solve above problems. The original data is separated into two pieces. The data which is favorable to the collaborative filtering algorithm is handed to collaborative filtering algorithm, and then the content-based filtering algorithm processes the remaining data. The two pieces of recommendation results were integrated into the final recommendation result. The result can restrain the non-ideal problem that comes from sparsity problem of collaborative filtering algorithm for recommendation results. However, when data volumes and data sparsity are particularly significant, the data handed to collaborative filtering algorithm may be still sparse. SVD has been performing well for the sparse problem. Nevertheless, the paper [12] proves that SVD instead is not as good as collaborative filtering algorithms when the sparse degree is still gigantic. Its main reason is that the data integrity is required in the simple SVD algorithm, and the general approach is to use 0 to fill up, but this simple way of filling does not accurately represent the user's particular rating. That leads to the SVD calculation error is enormous. Therefore, can fill the predicted data outputting from the collaborative filtering to the missing data in SVD to reduce the error caused by the SVD in the missing data occupied by 0? If the hypothesis (hypothesis 1) is true, then separation of data improves collaborative filtering once again enhances the quality of collaborative filtering, the quality of SVD. Therefore, the process of collaborative filtering is transferred to the process of SVD. Secondly, [17] pointed out that the SVD may not be as collaborative filtering algorithm to behave well when the data is very dense. Then, can the compact data calculated by SVD take advantage of again collaborative filtering algorithm to (or "intending to") further enhance the quality of recommendation? If the hypothesis (hypothesis 2) is true, the SVD process shifts to the collaborative filtering process. Finally, the data handed to content-based filtering will face to the massive sparsity, but content-based filtering is not sensitive for the sparsity of data, and the over-customization problem also will be restrained considerably in collaborative filtering. Meanwhile, there are two issues. Issue 1: how to build user interest model in the section of content-based filtering? Issue 2: how to update user interest model?

This paper carried out research based on the above idea and some of the problems. This System was named ICR (IPTV Combined Recommender) according to two combination recommendation strategies proposed by this paper. ICR includes the section of collaborative filtering and the section of content-based filtering. The elaboration of two kinds of combination strategies was placed in Chapter III.

2 Related work

With a broad range of information overloads resulting in the need for information filtering. In the field of television, here comes the emergence of personal TV program recommendation system: Based on Metadata TV-anytime recommendation system [13], PDPR [14], etc.

With the broad use of cloud computing technology, PDPR has adopted the cloud computing technology in the field of TV recommendation. PDPR based on cloud computing technology to collect and analyze the user's view mode (such as the view mode of mobile phones or personal computers, etc.) to achieve the purpose of recommending digital TV programs. For example, PDPR can analysis the user's interest through analyzing the data generated from communications services, mobile phone, and personal computer. The recommendation method determining the weight value of the time characteristic by analyzing the proportion of user's viewing time in the total time of the program is still based on the content filtering recommendation.

There is another disadvantage to this method besides the problem proposed in the content-based recommendation. It is quite uncertain to determine the degree of user's interest only depending on the proportion of user's viewing. The users, for example, go to do other things with the TV is on. Except that, the user left due to some other things with seeing a little time for the program. The system will regard the user's action as a moderate preference for this program. However, the truth may be that the user fond of the program very much. PDPR does not provide adequate solutions for the above problems.

The paper 0 proposed a method of content recommendation based on context awareness to provide users a more advanced IPTV services. Its background comes from users, devices, services, networks, and each setting characteristic is assigned a corresponding weight value. The user is divided into N group, and the similarity between each group and the content is calculated by measuring the average weight of its context characteristic. At last, the highest N content is recommended to the user group.

The paper [4] proposed three algorithms for IPTV, which is based on the substance of the latent semantic analysis algorithm, the SVD algorithm, and the content based SVD algorithm. LSA content-based algorithm increases additional latent semantic analysis for the plot of different TV programs. SVD content-based algorithm takes advantage of singular value decomposition to decompose user-program metadata matrix W, whose elements w_{ci} represent the relevance of characteristic (metadata) c for item i. Firstly, these three algorithms are not useful in combination, but for a particular class of problems (such as cold start) in isolation. These algorithms cannot solve many problems simultaneously, and the recommendation quality is not ideal when the simple SVD algorithm is in the calculation of heavy sparse. To compare with the method proposed by this paper, it also solves the problem of cold-start, sparsity, and the problems of recommendation results are too customized. Secondly, LSA content-based algorithm of paper [4] does not consider the factor of user interest, which will lead to the deviation of the user interest model and the real situation. This paper proposed a new method to update the user profile to minimize the difference from the actual situation. Finally, differently from the SVD algorithm the paper 0 propose, in the first combination strategy this paper's propose to combine SVD and collaborative filtering algorithm. SVD is not a simple decomposition of the original matrix, but the characteristic increment combination of collaborative filtering and SVD is used to solve the influence of the matrix over sparse for the accuracy of the decomposition.

The paper 0 proposed a prediction method of item keywords in the content-based recommendation. This method uses the window algorithm to update the user profile, but this paper is different from the user profile update rule of 0 where the original weight is just multiplied by or divided by a fixed value, two. The value does not take into account the impact of the user's rating, the number of keywords in the target program, and the occurrence number of keywords on the weight change, which leads to the increase of the calculation error.

Some other methods are also proposed, for example, the paper [5, 6] regard time and external burst news event as a contextual factor to improve the quality of the recommendation algorithm.

3 Combination strategy and architecture

Collaborative filtering item based have been proved that it has an advantage over user-based collaborative filtering [18] in the area of interest is fixed (such as books, movies, e-commerce, TV shows, etc.), so this paper adopts a collaborative filtering algorithm for item-based collaborative filtering algorithm in this paper.

Based on the idea proposed in Chapter 1, the issue 1 is how to build a user interest model, and that needs to solve first. The use of program content data comes from the EPG content or other content providers in ICR systems proposed in this paper. ICR provides a user interface, where users can register. The registration requires the user to provide specific interest preference, and the preference will be stored in the user profile database in the form of user profile (user profile is a structured user interest model shown). The user interface also allows a user to provide explicit rating through interaction with IPTV for a particular TV program, and the score data will be stored in user rating database. After each time, the user rate a program, a user profile will be updated according to the content provided by the content provider.

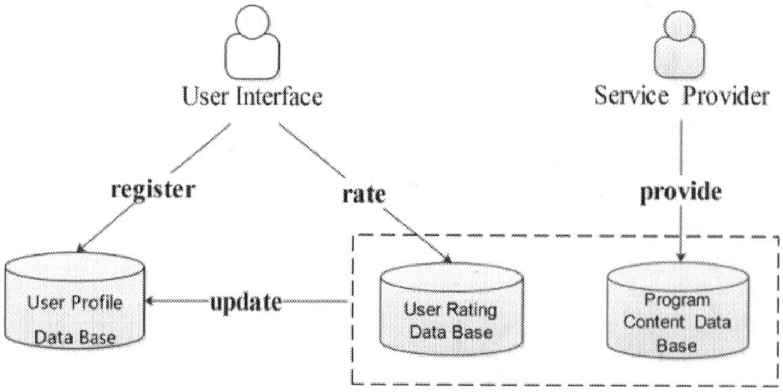

Figure 1. Business Process.

These data will be used as input data of recommendation algorithm. It's business processes as shown in Fig.1.

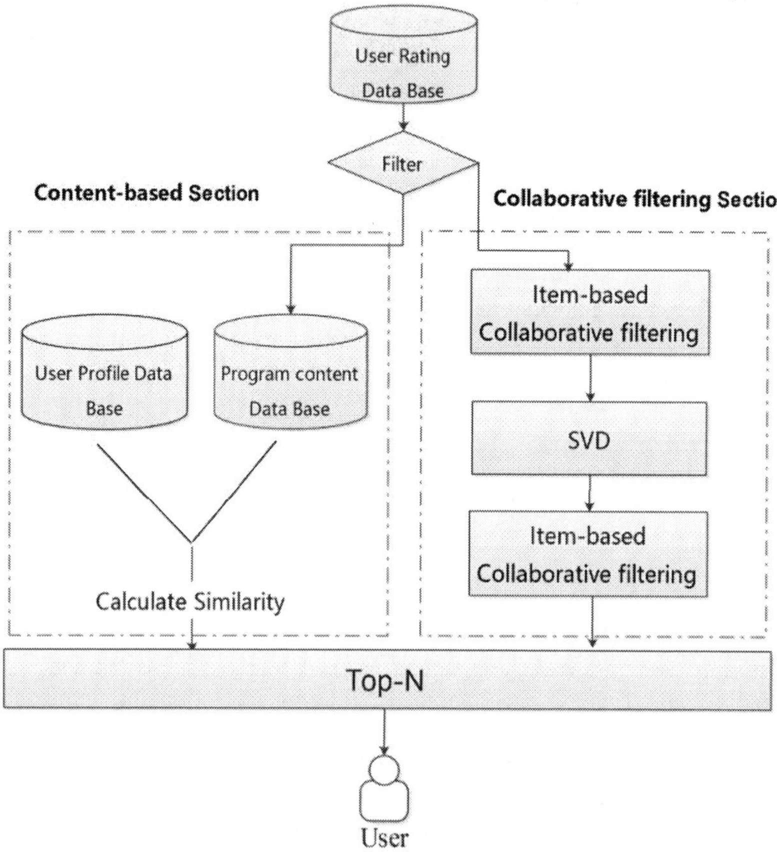

Figure 2. Architecture

In front of Chapter I idea, the data is separated, which is the second recommendation strategy that proposed in this paper. The data filter section as shown in Fig. 2 implement the second recommendation strategy. Its role is to determine which recommendation section the data belongs. When the number of user's rating of a program accounted for a proportion less, the parameter m in the total amount of users. This program will be recommended for content-based filtering section, otherwise through collaborative filtering section. The basis of the judgment rule is: In the related similarity measure method if the user set which rated the item I is represented by U_I. The user intersection U_{ab} of two users set (U_a *and* U_b) which respectively measured item a and item b need calculating.

$$U_{ab} = U_a \cap U_b \quad (1)$$

The similarity between item a and b is calculated by using the method of correlation similarity measure on the set U_{ab}. According to the idea of item-based collaborative filtering, only the rating is similar in more number of a user; then we have a high determinacy for similarity between the items. However, in the case of user - program rating data extremely sparse, the set of elements in U_{ab} is microscopic. Even if the similarity of the two items calculated in such a situation is very high, we cannot be sure that the similarity between them is very high. When the number of user's rating of a program accounted for a proportion less than parameter m in the total amount of users, the user value of the co-rating between this program and other programs will be a better chance of little. The following experiments will optimize the parameter m:

3.1 Content-based filtering section

Unlike Web-based IPTV field, such as the content of news sites which are text type can get desired content by keywords and information classification, etc., and also obtain information through simple mouse and keyboard actions.

IPTV content for video content and video content is not very efficient for all video content including every frame of the picture. In this context, at present, the content data (actor, type, Director, etc.) of TV program is used to describe TV shows.

The online registration of ICR system requires an explicit user preference, and all choices are divided into three types of characteristic values, respectively actor, and director.

Each type of features is allowed to have one or more keywords, and each keyword has the light to represent the relative importance of keywords, as well as has the keyword occurrence in the program rated by the current user. At the time of registration for each keyword of each type of feature, weight is given initial value.

3.1.1 User profile update method

Then solve the second problem (issue 2), that is, after building user interests model, to adapt to the users ' interests and to improve quality content-based recommendation, how to update a user profile? A user profile update method is proposed in this paper. If each type of domain of the three types of characteristics is regarded as related k-keywords, then each type can be described by the k-keywords. For example, type (action, Sci-Fi, comedy, horror), Actor (Jackie Chan, Bruce Lee, Jet Li, etc.). A user profile is defined as triples {key, w, times}, where the key represents a feature value keyword, w accounts for the equal weights, times represent the current occurrences of the keyword. Assume a user profile of the actor class shape, actor (Jackie Chan, 1,2), which accounts for a weight value of particular actor "Jackie" is 1, occurrences are 2. When a TV program which a user has rated, the show's actors for "Jackie" and "Andy Lau" is matching the program associated with the user profile type keywords.

$$w_{new} = w_{pre} * \frac{rating}{\sum w_i x_i} * f(x) \quad (2)$$

If the keyword exists in the user profile is in the program, then using the formula (2) updates the keyword weight, otherwise, add the keywords and assign initial values.

If a user likes a higher rating program or specific keywords to find a program, then the corresponding keyword weight would increase more or less. In this paper proposed a keywords weight update formula based on this idea.

w_{new} represents the target keyword weights updated. w_{pre} represents the weight of the target keywords before the update. $rating$ represents the rating the user rated on the current program. $\sum w_i x_i$ represents the sum of keywords weight extracted from this program content, which i represent i-th keyword in this type of feature. $f(x)$ is one of the independent variable x (x accounts for the target keyword occurrences) of the nonlinear function. Is mainly based on the fact that the more occurrences of targeted keyword? And the corresponding keyword weight to increase exponentially. The αβ values may be determined according to specific application needs, f (x) is defined as:

$$f(x) = \alpha e^{\beta x} \tag{3}$$

3.2 Content-based filtering recommendation algorithm

Content-based Algorithm is named CB Algorithm, CB Algorithm is shown as following:

Algorithm 1: Contend-based
Input: user ID, item ID
Output: predicted rating matrix

Process:

a) Building vector space model: The user profile gotten from user profile according to the user ID and program content information got from program content database according to the program ID is constructed as a vector space model.
b) Calculation of the similarity between the two vector space model: Assume p, u, on behalf of the user vector and program vector, respectively. i represents i-th element in the vector.

$$sim(u,p) = \frac{\sum_i \left(u(i) \times p(i)\right)}{\sqrt{\sum_i u(i)^2} \times \sqrt{\sum_i p(i)^2}} \tag{4}$$

Formula (4) shows how to calculate the cosine similarity between program vector and user model vector.

The recommendation system will calculate the similarity between each programming vector and target user vector. And the similarity of the highest N program as the recommendation results of a content-based section.

A. Collaborative filtering Section

Hypothesis 1 presented in Chapter I is that it be able to fill the predicted data from the collaborative filtering to the SVD part of missing data to reduce the error caused by the SVD in the missing data occupied by 0. Open GroupLens movie data set is used, the question was verified, the similar experiment was arranged in Chapter IV. Experiments show that this hypothesis is valid. Then the reasoning is in hypothesis 2, that is, the dense data calculated by SVD using collaborative filtering algorithm can interact to (or "intending to") further enhance the quality of recommendation. The paper [17] pointed out the SVD may not be as collaborative filtering algorithm to behave well when the data is very dense. Item-based collaborative filtering algorithm needs to calculate the similarity between the items. If a rating comes from two items never being the same user, which will lead to the similarity between the two items cannot be calculated. Therefore, the quality of recommendation results will be increased, if the density data derived from the SVD take advantage of collaborative filtering again. The second hypothesis was also established. So far, the feasibility of this idea was entirely justified.

The first combination strategy is proposed in collaborative filtering section of this paper. First of all, a user-program rating matrix will be created for those programs handed to collaborative filtering section. Secondly, the original rating matrix will be filled based on item-based collaborative filtering. Then, to put the user-program rating matrix filled as an input of singular value decomposition, and SVD will fill original user-program rating matrix again, and get a user-program rating matrix without missing value. Thirdly, the user-program rating matrix without losing value will be used item-based collaborative filter again, and the missing item of original will be predicted finally, regarding N programs with highest ratings as recommendation results of the collaborative filter section. Using algorithm among this process is named CF-SVD-CF.

Algorithm 2: CF-SVD-CF

Input: original rating matrix

Output: predicted rating matrix

Process:

a) Item-based Collaborative Filtering Predictions：original rating matrix R is defined to be a $m \times n$ matrix, which represents the rating of m users on the n items. Using the method of item-based collaborative filtering and measure method of the modified cosine similarity to calculate the similarity between items. The calculating formula is as follows:

$$sim(i,j) = \frac{\sum_{u \in U}(R_{u,i}-\overline{R_u})(R_{u,j}-\overline{R_u})}{\sqrt{\sum_{u \in U}(R_{u,i}-\overline{R_u})^2}\sqrt{\sum_{u \in U}(R_{u,j}-\overline{R_u})^2}} \tag{5}$$

In the formula (5), $\overline{R_u}$ represents an average of all ratings the target user u has rated, $R_{u,i}$ accounts for the rating the user u has rated on item i in the original matrix, U represent user set. To find the target, user u to the program j rating prediction, the formula is as follows:

$$Pre_{u,j} = \frac{\sum_{i \in ratedItem(u)} sim(i,j) \times R_{u,i}}{\sum_{i \in ratedItem(u)} sim(i,j)} \tag{6}$$

In formula (6) ratedItem (u) represent a set of the user u has rated. $sim(i,j)$ represent the similarity between program i and program j. Finally, the matrix R_{pre} is obtained by filing the prediction rating into original matrix R.

b) Standardized R_{pre}: If a rating comes from two programs never being the same user, the similarity between the items will not be calculated. So R_{pre} may not be filled entirely. If not, first calculating the average value $\overline{R_j}$ of each column in the R_{pre}, then fill the average column value $\overline{R_j}$ into the vacancy in the current column, and each column of the ratings of all items minus the target program where the average row value of $\overline{R_i}$ to get the R_{norm} matrix of standardized. The reason of standardized treatment is there is some different influence for calculating similarity in users who rated different number of items

c) Calculate for SVD:S, U, V, is obtained by using the singular value decomposition method for R_{norm}. Their sizes are $\times m\, m \times n, n \times n$. The relationship between them is $R_{norm} = U \times S \times V^T$.

d) Dimension Reduction: A $k \times k$ matrix S_k is obtained by preserving k-diagonal element for matrix S. Matrix U, and matrix V are also reduced to $U_k$and V_k accordingly. Their sizes are $m \times k$ and $n \times k$. The experiment will optimize the parameter of K.

e) Users and programs in k-dimensional space said: m a user in a k-dimensional feature space representation for K U * tick (s k) and n a program in a k-dimensional feature space representation for tick (s k) x (v k) ^ t.

f) SVD prediction rating: according to the [28], the target user u in the program I on the prediction rating is calculated:

$$Pre_{u,i} = \overline{R_u} + U_k\sqrt{S_k}(u) \times \sqrt{S_k}V_k^T(i) \tag{7}$$

Which $\overline{R_u}$ the said target user u all has hit rating project rating average, $\sqrt{S_k}V_k^T(i)$ on behalf of column i of the matrix $\sqrt{S_k}V_k^T$. $U_k\sqrt{S_k}(u)$ on behalf of the row u in the matrix $U_k\sqrt{S_k}$. A rating matrix R_{filled} without missing values is obtained by filling the predicting to the original rating matrix R.

Matrix R_{filled} is just a matrix obtained by using SVD method to predict the predicted matrix, which is not based on the method of collaborative filtering.

g) Similarity calculation: Using the modified cosine similarity measure method to calculate, the similarity between program I and program J in matrix R_{filled}, the formula is as follows:

$$sim(i,j) = \frac{\sum_{u\in U}(R_{u,i}-\overline{R_u})(R_{u,j}-\overline{R_u})}{\sqrt{\sum_{u\in U}(R_{u,i}-\overline{R_u})^2}\sqrt{\sum_{u\in U}(R_{u,j}-\overline{R_u})^2}} \quad (8)$$

$\overline{R_u}$ represents an average of all ratings the user u has rated $R_{u,i}$ represents the rating user u has rated on item i, U represents the set of users.

h) Final prediction rating of the original matrix: We conclude with prediction generation, achieved by the following weighted sum:

$$Pre_{u,j} = \frac{\sum_{i\in ratedItem(u)} sim(i,j) \times R_{u,i}}{\sum_{i\in ratedItem(u)} sim(i,j)} \quad (9)$$

i) The prediction calculates for user u on the item j. $ratedItem(u)$ represents the set of items user u has rated. $R_{u,i}$ represents the rating user u has rated on item i. $Sim(i,j)$ represents the similarity between item i and item j.

4 Evaluation

Evaluation is a core aspect of recommender systems design and deployment. The measurement of the quality of the assessment rating is divided into two categories: coverage rate and accuracy measure while the efficiency test includes two kinds of methods: statistical precision measurement method and decision support accuracy measurement method. MAE (Mean Absolute Error) is one of the commonly used recommendation quality measurement methods in statistical precision measurement method. MAE is predicted by calculating the deviation between the prediction and the actual user's rating. MAE is smaller, and the prediction for the user is more accurate [16], and the MAE formula is as follows:

$$MAE = \frac{1}{N}\sum_{i=1}^{N}|P_i - T_i| \quad (10)$$

In the formula (9), N is a set of all the missing rating items. P_i represents the prediction for the user on target item. T_i represents the user's real rating on the target item. MAE is smaller, and the prediction for the user is more accurate. But MAE in this system is not related to that data which is handed to content-based section because the content-based section does not predict the user's rating. So the MAE values are calculated only for that part of the data in the collaborative filtering section.

Also, the recommendation quality is usually evaluated by recall and precision in traditional information retrieval field. The two measurement methods are still suitable for TV program recommendation system. But it is needed to modify the formula to obtain a more accurate measurement, $Set_{preferred}$ represents the set of users' preference, which involves the programs where the rating is equal or greater than the sum of all ratings the target user has rated and the standard error of all ratings the target user has rated. //Assumes that $item$ is one of the programs in $Set_{preferred}$; P_{ave} is the average of all items rated. P_{stdDev} is the standard deviation of all items rated. Then, the rating, $item_{rating}$, of $item$ must meet following conditions:

$$\text{If } item \in Set_{preferred} \quad (11)$$

$$\text{Then } item_{rating} \geq P_{ave} + P_{stdDev} \quad (12)$$

$Set_{recommend}$ represents the recommendation result set which is generated by the recommendation system. Recall and precision can be defined as follows:

$$recall = \frac{Set_{preferred} \cap Set_{recommend}}{Set_{preferred}} \tag{13}$$

$$recision = \frac{Set_{preferred} \cap Set_{recommend}}{Set_{recommend}} \tag{14}$$

When using the first recommendation strategy, due to the addition of content- based recommendation results, so $Set_{preferred}$ must also contain content-based recommendation result set.

B. Data Set

In the following experiments, this paper uses MAE, recall, precision as a measure of the quality of the system to assess the quality of the system. Adopting the public GroupLens movie data set, and using web crawler program to collect all the video content data for the experiments. GroupLens data set contains 1000209 rating data, from 6040 users in 3900 films on the rating. The value of the rating range is from 1 (dislike) to 5 (like). In this paper, the original data is divided into training set and test set, the training set is 80%, and the test set is 20%.

C. The experiment and analysis

In this paper, the collaborative filtering and content-based recommendation are implemented, and the quality of each recommendation strategy is evaluated through the switch or combination of the two sections. The experimental process of the development of programming language for the Java, the database is Mysql. In the collaborative filtering section, the implementation of item-based collaborative filtering algorithm and the SVD algorithm make the use of rewriting and combination of the algorithm in the open source project. Regarding content-based implementation, firstly, the collection of the film content data is used to establish program content table in [movieId, Mysql, genre, actors, director] on behalf of each program Id, type, actor and director fields. Secondly, the establishment of the user profile table to [userId, genre, actor, director] on behalf of the field of each user's Id, types and their corresponding weights, actors and their corresponding weights, directors and their corresponding weights, and using the training set to update the table. Finally, the use of the table and the programs handed to content-based filtering section to match and calculate, so to get the user for a program of similarity.

As mentioned in Chapter IV, the parameter m represents the proportion of the number of rating rated by the target user in the total user rating for a program. The parameter K accounts for the reservation count of singular value of the singular matrix S. First of all; the first implementation is carried out to prove the correctness of the first hypothesis (hypothesis one).

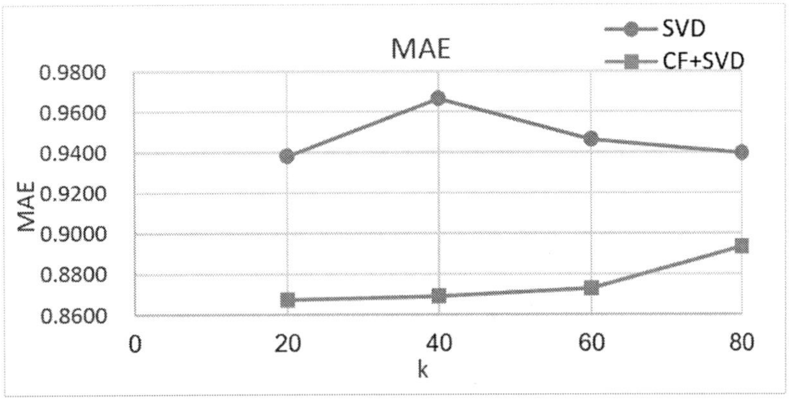

Figure 3. MAE of SVD and CF+SVD.

Fig.3 shows in the simple SVD algorithm with the parameter k={20, 40, 60,80} MAE are higher than the algorithm where collaborative filtering is used first to fill the original matrix. And then use SVD to predict the missing value of the original matrix. A conclusion, the first hypothesis is upright, can be drawn. That is, it can fill the predicted data outputting from the collaborative filtering to the missing data in SVD to reduce the error caused by the SVD in the missing data occupied by zero.

All of the following experiments were used to evaluate the recommendation quality of two kinds of combination recommendation strategies in different parameter m and K, respectively. To determine the optimal parameters m and k through experiments and observe the value of the corresponding parameters of MAE, Precision, and Recall.

The experimental results are shown in Fig. 4, Fig. 5, Fig. 6 with two combination recommendation strategies. The abscissa represents the value of k,50 as a unit. Parameter k is set to {5, 20, 40, 60, 80, 100, 150, 250}. Ordinate respectively represents MAE, Precision, Recall with the value of the corresponding k. Each polyline corresponds to an absolute value of m, m is set to {0.01, 0.005, 0.007}.

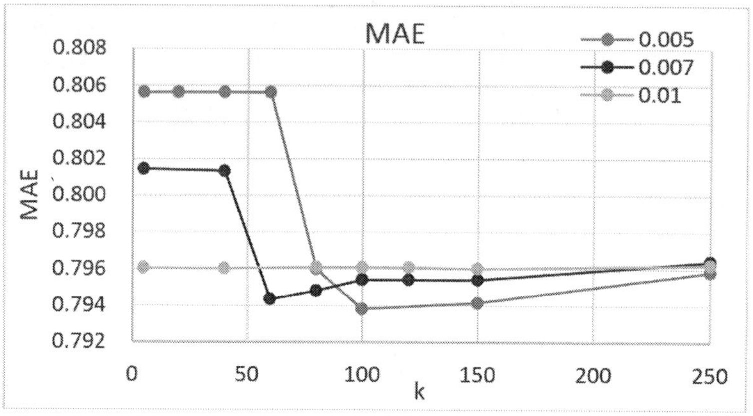

Figure 4. MAE with two recommendation strategies.

Fig.4 shows that, firstly, when m is 0.005, 0.007, and k is small, the MAE values decreased very slowly. But with increasing value of k, MAE dropped rapidly, and when k=60(m=0.007), k=100(m=0.005), the results reach the optimum. After that, with the increase of k, the results gradually become awful. Secondly, when m=0.01, MAE has maintained a gradual change, the minimum value is only 0.79602. Thirdly, when m=0.005, average recommendation quality is better than the other two.

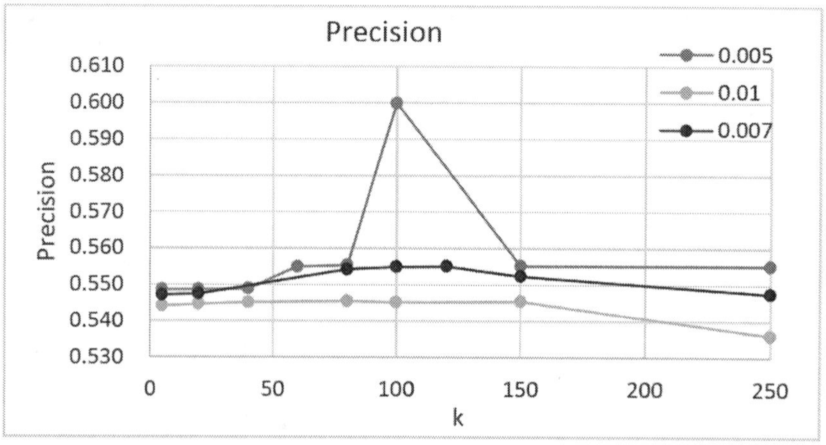

Figure 5. Precision with two recommendation strategies.

Fig 5. indicate that when m value of 0.005, 0.007, 0.01, with the increase of k, precision has small fluctuation. But when m=0.005, k∈ [80, 150], precision increases rapidly and k=100 reach highest value 0.599884.

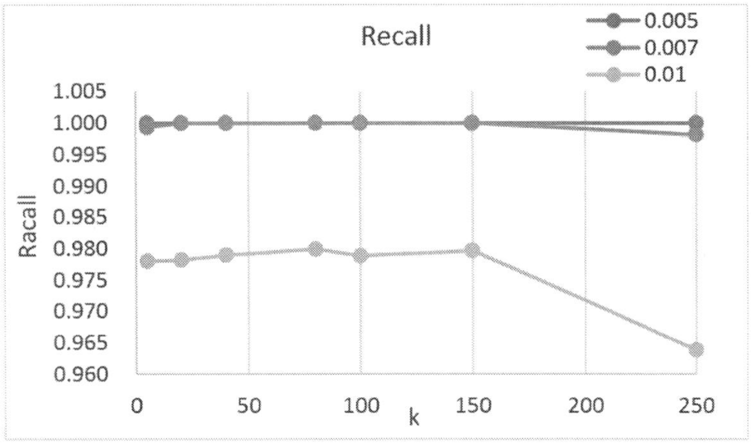

Figure 6. Recall with two recommendation strategies.

In Fig. 6, when m=0.005, Recall remained at 1.0, m=0.007, Recall has a gradual fluctuation about 0.999. But when m=0.01, and Recall rather dramatically reduce, the highest value is only 0.97988. When k ∈ [5, 150], the fluctuation is small, but when k increases to more than 150, Recall value dropped sharply, and k=250 reach the lowest value 0.96385.

This paper retained the second recommendation strategy to evaluate recommendation quality of the first recommendation strategy and canceled the first recommendation strategy. Parameter k takes value collection for {5, 20, 40, 60, 80, 100, 150, 50}, and an experiment is proceeded in same training set and test set. The first recommendation strategy involves the content-based recommendation to test MAE values, so MAE of two recommendation strategies is same as only the second recommendation strategy. Then the group experiment only tested the precision, recall. The experimental results as shown in Fig.7, Fig.8 Fig.9:

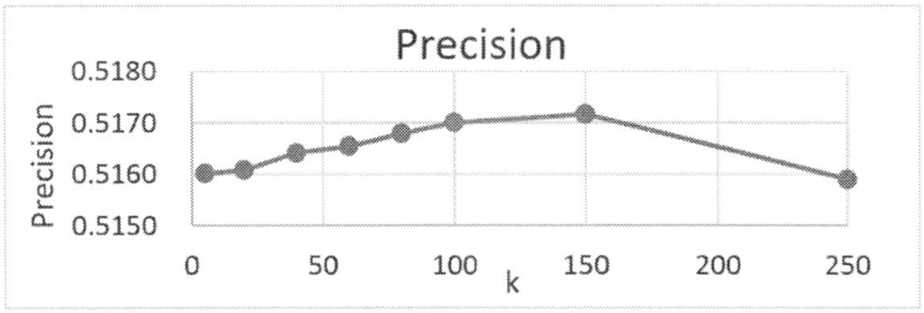

Figure 7. Precision with the first recommendation strategy.

In Fig. 7, with the increasing k, Precision continue to rise, and the highest value 0.517167 is reached in the k=150.

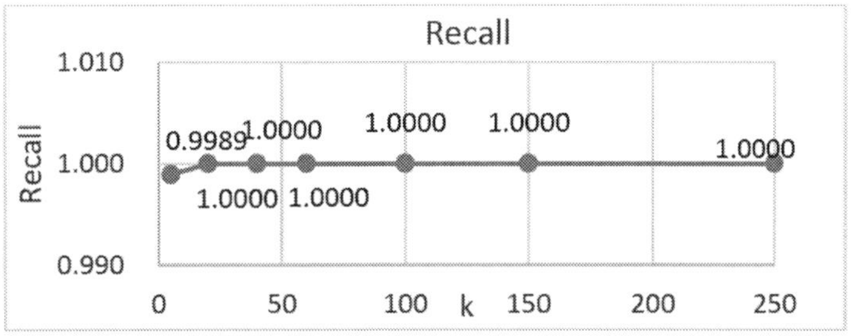

Figure 8. Recall with the first recommendation strategy.

In Fig. 8, the Recall has no obvious change, only in the k=5, Recall is 0.9989. In the other k, Recall is 1.0. To evaluate the quality of the second recommended strategy this paper retained the first recommendation strategy, and the simple collaborative filtering replaced second recommended strategy.

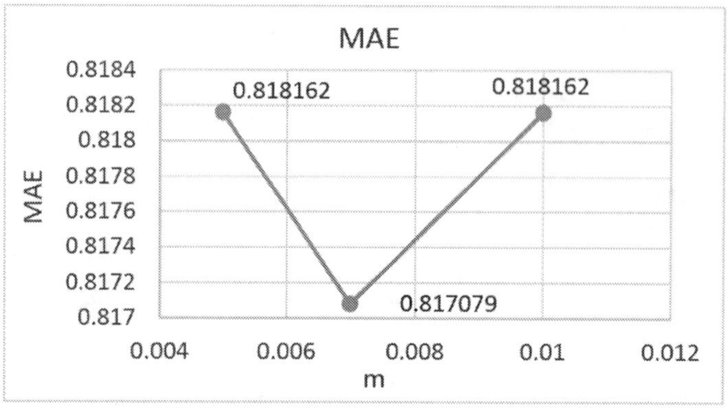

Figure 9. MAE with the second recommendation strategy.

The parameter m is set to {0.01, 0.005, 0.007}, and the experiment is processed in the same training set and test set. The MAE, Precision, Recall experimental results are as shown in Fig.9, Fig.10, Fig.11:

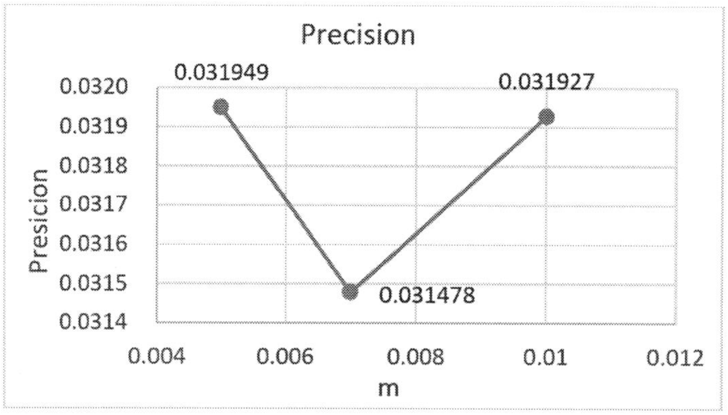

Figure 10. Precision with the second recommendation strategy.

Fig 9, Fig.10, Fig.11 indicate that the recommendation quality with the pure second recommended approach is awful, the highest Precision is only 0.031949, the largest recall is only 0.022173, the lowest value of MAE is only 0.817079.

When two recommendation strategies are not used, only using collaborative filtering algorithm, the experimental result in the same data set is MAE=0.817595, recall=0.029644, precision=0.027475.

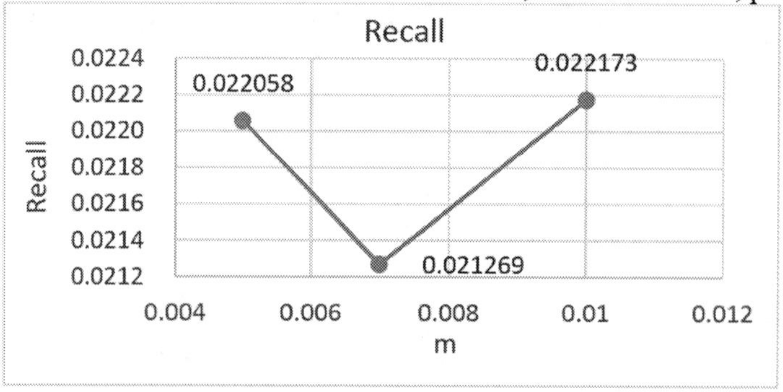

Figure 11. Recall with the second recommendation strategy.

5 Conclusions and future work

According to the validation of the two strategies mentioned above, it is proved that the two strategies are effective. A large number of experimental results shows that when using two kinds of recommendation strategies, m=100, k=0.005, MAE, Precision and Recall have reached the optimal value. And Precision, MAE, Recall compared by using any one of two combination strategies or not using any combination strategy at all have been significantly improved. Here comes to a conclusion which is that the utilization of the ICR system for IPTV recommendation system provides a very efficient solution to the problem of sparsity and cold-start.

The ICR system proposed in this paper is still insufficient. In the experimental process, due to the limitations of the SVD technology—the calculation is large, the running time is long so that the efficiency of the system is significantly affected. It can be anticipated that ICR system is facing a more serious problem of effectiveness with larger data. In this case, the author will put forward a distributed computing solution through using the Hadoop distributed computing framework to build a distributed architecture ICR system to improve the efficiency of recommendation. At the same time, it is also noted that another problem in the CF-SVD-CF algorithm, the original matrix for the first time to use collaborative filtering to predict. Although at this moment the data is denser than the data not processed by the data filter, the user-program matrix is still very sparse in a large number of programs. Therefore, the use of collaborative filtering for the original matrix prediction will make the prediction error is relatively large. The basis of this problem, the author will further use the combination method of feature value increment and the user profile. And filling a rating which is transformed into by the similarity between the program and the user profile. Then using collaborative filtering method to alleviate the sparsity of the rating matrix, and enhance the quality of recommendation ICR.$itemSet_{preferred}P_{ave}$.

References

1. Z. Yu and X. Zhou, IEEE transaction on Consumer Electronics, **50**(1), 393-399 (2004)
2. T. Isobe, M. Fujiwara, and H. Kaneta. et al., IEEE Trans Consumer Electronics, **51**(2), 665-674 (2005)
3. S. Song, H. Moustafa, and H. Afifi, IEEE Transactions on Multimedia, **14**(6), 1528-1537 (2012)
4. R. Bambini, P. Cremonesi, and R. Turrin, *Recommender Systems Handbook*, 299-331 (2011)
5. P.G. Campos, F. Díez, and I. Cantador, User Modeling and User-Adapted Interaction, **24**(1-2), 67-119 (2014)
6. N. Pervin, F. Fang, and A. Datta, et al., Acm Transactions on Management Information Systems, **3** (4), 19 (2013)
7. J.S. Breese, D. Heckerman, and C. Kadie, *Proceedings of the Conference on Uncertainty in Artificial Intelligence*, **7**(7), 43-52 (1998)
8. B. Sarwar, G. Karypis, and J. Konstan, et al., *Proceedings of the 10th international conference on World Wide Web. ACM*, 285-295 (2001)
9. B. Arwar, G. Karypis, and J. Konstan, et al., *Processing of 2nd ACM Conference on Electronic Commerce*, 158-167 (2000)
10. A. B. Barragáns-Martínez, E. Costa-Montenegro, and J.C. Burguillo, et al., Information Sciences, **180** (22), 4290-4311 (2010)
11. G. Wang and H. Liu, Computer Engineering and Applications, **48**(7), 66-76, DOI:10.3778/ j.issn.1002-8331.2012.07.018 (2012)
12. B. Sarwar, G. Karypis, and J. Konstan, et al., In Acm Webkdd Workshop (2000)
13. H. Zhang and S. Zheng, *Consumer Electronics (ISCE 2005) Proceedings of the Ninth International Symposium on. IEEE*, 242-246 (2005)
14. S. Lee, D. Lee, and S. Lee, IEEE Transactions on Consumer Electronics, **56**(2), 1034-1042 (2010)

15. Y. Guo, Recommender system collaborative filtering algorithm and application [D], Dalian University of Technology (2008)
16. A.L. Deng, Y.Y. Zhu, B.L. Shi, Journal of Software, **14**(9), 1621-1628 (2003)
17. X. Sun. Study on Sparseness and Cold Start of Collaborative Filtration System [D], Zhejiang University (2005)
18. X. Liang, Recommended system practice [M], Beijing, People's Posts and Telecommunications Press, 2, 51, 60 (2012)

A green building information modelling approach: building energy performance analysis and design optimization

Shang-yuan Chen[a]
School of Architecture, Feng Chia University, Taiwan.

Abstract. This study concerning green BIM focuses on the integrated application of Building Information Modelling (BIM) and building performance analysis (BPA) software as tools for the design and analysis of building projects, and employs a sequential decision-making cycle and continuously improving design to achieve an optimal proposal consistent with environmental effectiveness. Taking a new Taichung hotel construction project energy consumption design optimization as an example, this study relied on the steps of (1) Determination of the scope of discussion of the proposal within the building life cycle, (2) Setting of energy conservation targets, (3) Accessing to external climate data, (4) Entering internal settings, (5) Implementation of energy conservation calculation module, (6) Visualization analysis and hot spot tracking (7) Proposal Revision, and (8) Optimal proposal, to verify the green BIM concept. With regard to the setting of energy conservation targets, this study recommends that building energy use intensity (EUI) be used as an energy load measurement unit of integrated performance indicators, and employs performance optimization percentage as a rating criterion. In accordance with this method, green BIM combined with assessment of green building indicators is as a means of facilitating integrated design and analysis decision-making.

1 Motivation and goals

In the face of a rapidly changing climate and global energy crises, how to use building information modelling (BIM) tools to obtain architectural designs offering environmental effectiveness has become a leading issue in the contemporary architectural and construction industries. BIM has two implications, one of which is information modelling, and the other of which is information management. Along related lines, building performance analysis (BPA) software provides building performance visualization and data processing analysis capabilities. Green BIM consequently emphasizes the effective combination of BIM and BPA technologies in integrated design, and seeks to promote a rational building design and analysis decision-making cycle and continuously improving design to ensure the optimized development of designs offering better environmental effectiveness. However, the implementation of green BIM is not a trivial task, and further research and investigation is needed in the areas of software tool selection and integration, and establishment of integrated design procedures and optimal criteria.

[a] Corresponding author : shangyuanc@gmail.com

2 Literature review

The chief categories of green BIM research include BIM and BPA, as well as the integration and application of BIM and BPA in sustainable building design.

The book "Green BIM—Successful Sustainable Design with Building Information Modelling" proposed the "green BIM" concept for the first time, explored the influence of BIM on the improvement of design methods and various construction industry participants, and employed BIM in conjunction with BPA to promote the development of sustainable design [1]. McGraw-Hill Construction suggests that green BIM can employ energy modelling together with BIM tools to greatly enhance sustainable design results [2].

Building Product Models, published by Eastman in 1999 defined building product model concepts, technologies, and standards, and set the stage for BIM [3]. Eastman's BIM Handbook defined BIM and related technologies, and provided BIM applications and illustrative cases for various types of participants (project owners, project managers, designers, engineers, and contractors, etc.) [4].

BPA, also known as building performance simulation (BPS), relies on computer software to predict building performance and output visualized images, data, statistical analysis charts, and forms resulting from simulation. BPA can help users to understand the performance of their design proposals, which will facilitate design decisions and provide a basis for the continuing optimization of design proposals. The focus of building simulation work consisted of building heat dissipation performance during the 1960s and '70s, and simulation was employed to perform thermal load calculations and energy conservation analysis [5, 6]. The focus shifted from energy consumption to many other building performance characteristics during the early 1990s [7], including the use of integrated modelling to assess heat and mass transfer, air flow, and visual and acoustic performance. To date, the use of BPA in conjunction with BIM tools try to establish integrated design and analysis decision-making procedures [8].

Green BIM emphasizes the combination of BIM and BPA software technologies in order to perform integrated design, and seeks to promote a rational building design and analysis decision-making cycle in order to ensure the optimized development of designs offering better environmental effectiveness. In the article "Building Performance Analytics: Establishing a Methodology of Design Integration," Andrasik (2015)[9] further stresses that the mission of BPA lies at the intersection of BIM and sustainable design, and the implementation of BPA tasks begins during the early design stage in sustainable design. When employed in conjunction with BIM-based design methods, BPA enables the assessment and comparison of the building performance of different design proposals, can help guide decisions, and can effectively improve construction projects [10] (Figure 1).

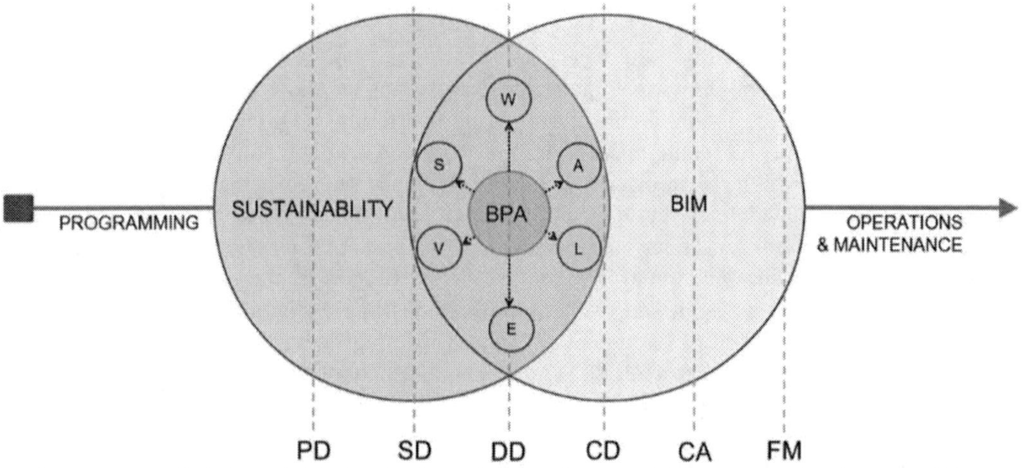

Note: Pre-Design (PD), Schematic Design (SD), Design Development (DD), Construction Detailing (CD), Construction Administration (CA), and Facility Management (FM)

Figure 1. Linkage between sustainable design, BPA, and BIM.

Apart from use in obtaining 3-D geometric information, BIM can also be used to formulate that part of non-geometric information that can be easily transmitted to BPA software. BIM system software companies are also incorporating BPA tools within the scope of their software development and integration efforts, and hope that they can help designers and other personnel to precisely control building performance within a 3-D operating environment in order to achieve sustainability. Examples include Autodesk® Ecotect Analysis [11] and even easier-to-use Energy Analysis for Revit [12]; Bently® Hevacomp [13] series and Tas® building modelling and simulation tool [14]; and Graphisoft® Eco Designer Star [15]. Table 1 summarizes and compares software programs, supported file formats, visualization items, and performance analysis items. The compared BPA items include natural and artificial lighting, indoor lighting, overshadowing and shading analysis, solar radiation and thermal performance analysis, wind ventilation analysis, acoustic analysis, visual access, energy load, carbon emission analysis, and whole building energy performance simulation, etc.

Table 1. BPA software comparison

Software tools	Engine	Import file formats	Visualization items	BPA items	export file formats
Autodesk® Energy Analysis for Revit	DOE-2	form Revit	Energy Analysis Result	Energy Analysis	gbXML
Autodesk® Ecotect Analysis	CIBSE, Split-Flux	dxf/ 3ds/ rvt-gbXML/ skp-obj	Solar radiation, shadows, natural lighting, thermal simulation and analysis, building energy load analysis	Exports to Green Building Studio, can perform reporting of carbon emissions, water usage, and costs	gbXML/ idf/ inp/ dxf
Bently® Hevacomp Dynamic Simulation	EnergyPlus	dxf/ dgn/ dwg/ ifc/ skp/ 3ds/ cis2/ gbXML	Energy analysis, three-dimensional visualization of external shading, indoor sunlit areas	Calculation of building heating and cooling loads, calculation of pipe and duct system dimensions, carbon emission analysis	Idf/ dwg/ dgn/ skp/ IFC
Tas® building modeling and simulation tool	TAG simulator	dwg/ gbXML/ inp/ idf	Daylighting, wind tunnel analysis, simulation of hot zones	Carbon emissions, energy consumption analysis, cost analysis, indoor comfort indices	gbXML
Graphisoft® EcoDesigner STAR	StruSoft's VIPcore calculation engine	from ArchiCAD	Meteorological data analysis, energy load analysis, thermal bridge analysis	Building energy load analysis, energy optimization settings, renewable energy programs	gbXML/ PHPP/ Isbem/ VIP-Energy

This study uses assessment of building energy consumption as an example to illustrate the application of green BIM in integrated design procedures and employees Autodesk®'s Energy Analysis for Revit as a validation tool. This software was chosen for the following three reasons:

(1) Energy Analysis for Revit consists of a BIM tool combined with a BPA tool, and possesses a user-friendly interface suitable for architects and designers

(2) The validation of its DOE-2 engine has earned considerable credibility

(3) The virtual weather station technology developed by Autodesk® can overcome the past problem that local climate data is not easy to acquire (this is described in detail in the third section)

Most of the BPA software and 3D modeling software are separated; files' interoperation and integration between them are inconvenient. With regard to energy consumption analysis software, because of the complexity of building energy forecasting, energy simulation tools usually consist of two parts. The first part is the engine, which includes formulas and procedures, and the second part is

 https://doi.org/10.1051/matecconf/201816901004

the user interface, which allows the input of parametric data and displays results. Architectural design simulation engines chiefly serve to support building design and analysis via the comparison of the energy consumption of different design proposals. While most simulation engines have been developed by academic or research organizations, most user interfaces have been realized by private software companies [16] and the combination of the two parts forms the main BPA software on the market (ex. RIUSKA, eQUEST, and Design Builder, etc.) (Figure 2). However, Autodesk®'s Energy Analysis for Revit integrates energy performance analysis with BIM software. In the Revit platform, energy analysis includes a set of program commands, and Revit relies on the Green Building Studio (GBS) cloud service to transmit information input or created on the Revit platform, including (1) building geometric information, (2) location and climate data, and (3) non-geometric attributes and parameters, to the GBS DOE-2 simulation engine in gbXML format, and also transmit results back to the Revit platform for display. In other words, the BIM-Revit platform integrates BIM modeling and the GBS performance simulation service, and thus provides a replacement for the BPA software used as a user interface in the past (Figure 3) [17].

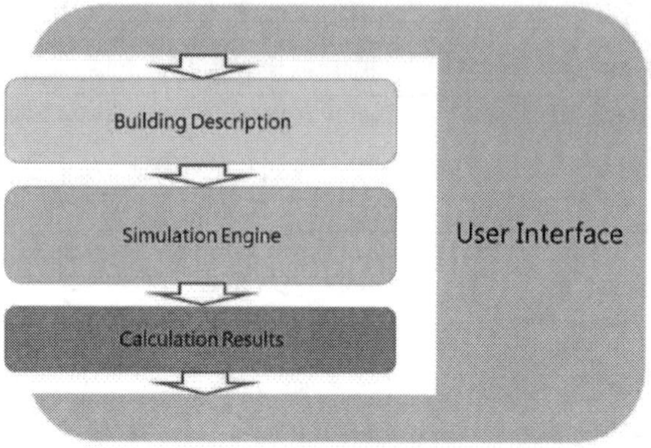

Figure 2. A BPA tool architecture [16].

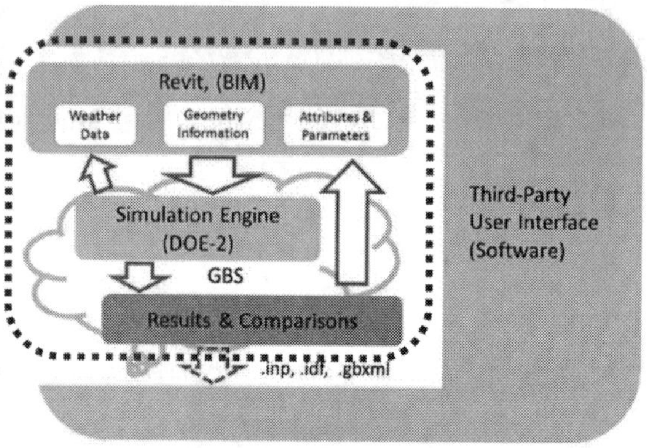

Figure 3. Relationship between Revit, Energy Analysis, GBS, and third party user interface.

3 Theory and method

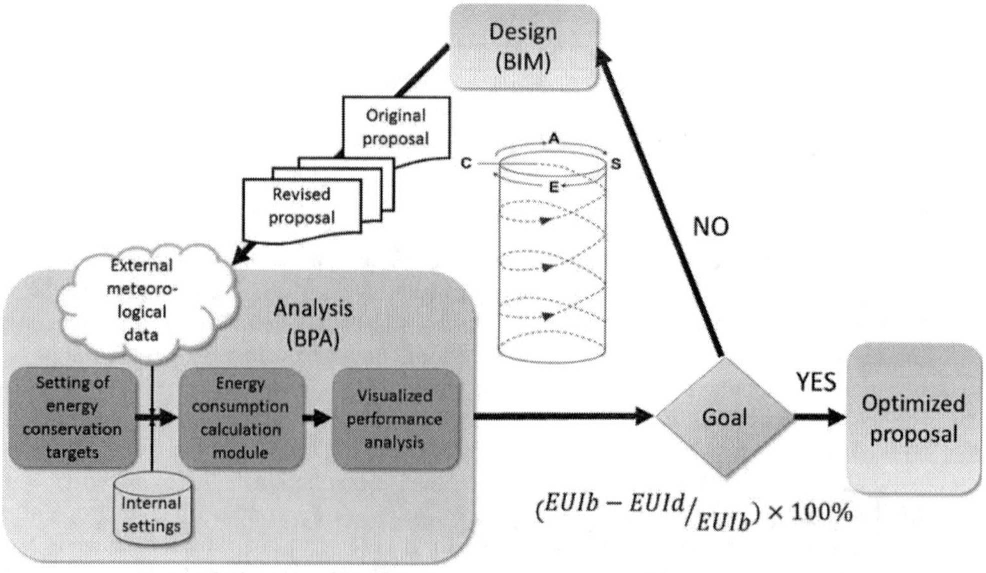

Figure 4. A green BIM-based decision-making cycle model.

As summarized above, green BIM emphasizes the use of BIM as a basic tool from the start of the design process. Responding to localized climate conditions, BPA was implemented via a decision-making cycle consisting of design and analysis steps, and continued design optimization performed to obtain an optimal proposal meeting the need for environmental effectiveness, which ultimately achieved the goal of environmental sustainability. A model of the green BIM-based decision-making cycle is shown in Figure 4.

Energy consumption analysis consisted of the following processes:

(1) Determination of the scope of discussion of the proposal within the building life cycle:
The simulation software employed by this study calculated the building's energy consumption during the operating stage. And the decision-making cycle for performance analysis and optimization design occurred in the early design stage (including schematic design (SD) and design development (DD) stages). The first step was to find the optimal proposal for different configurations of the conceptual massing during the SD stage in accordance with the target of performance optimization percentage. Detailed building elements were added during the DD stage, and element attributes and parameters were adjusted to ensure the continued optimization of performance (Figure 5).

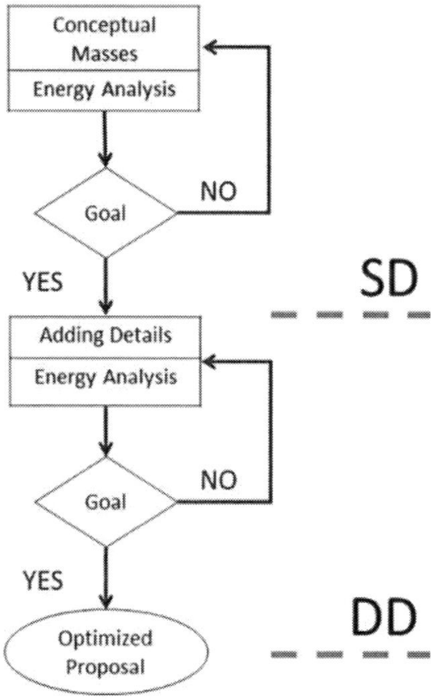

Figure 5. Design decision-making cycle (SD~DD).

(2) Setting of energy conservation targets:
After consulting reference information Leadership in Energy and Environmental Design (LEED) [18] and the Taiwan Power Cost and Price tips [19], this study suggest using energy use intensity (EUI), which is annual power consumption per unit area of the building, as an energy load measurement unit and integrated energy conservation performance indicator. In accordance with the corresponding building use classification and type in the "EUI values for various types of Buildings" (Table 2) issued by the Bureau of Energy, Ministry of Economic Affairs (MOEA), R.O.C., the study decided to take the relevant average value as a common benchmark value for project design [20]. The performance optimization percentage, which served as a target setting or rating criterion, was obtained from the ratio of the difference between the common benchmark EUIb value and optimized project EUId value of the common benchmark EUIb.

$$Performance\ optimization\ percentage = \left(EUIb - EUId /_{EUIb}\right) \times 100\% \qquad (1)$$

 https://doi.org/10.1051/matecconf/201816901004

Table 2. EUI Values for Various types of Buildings.

Building use classification		**EUI**		
		Average value	**Minimum value**	**Maximum value**
Main category	Subcategory	(kWh/m2·yr)	(kWh/m2·yr)	(kWh/m2·yr)
Government agencies	Central (general administration)	114.1	36.6	189.8
	Local (general administration)	87.1	37.7	134.4
Schools	General universities	84.4	45.7	123.4
	Technical universities	69.0	33.0	99.4
	High schools	65.4	30.4	107.0
	Vocational schools	59.3	30.4	98.4
Office high-rises		149.8	79.2	225.0
Hotels	International tourist hotels	207.6	117.6	299.2
	General tourist hotels	198.8	106.8	264.0
	General hotels	195.6	172.9	223.6

Notes: Energy consumption indicator calculations include indoor parking facilities. [21]

(3) Accessing to external meteorological data:
BPA software requires representative climate data as basic information for environmental simulation and energy calculations. Because of insufficient local climate data, the application of past BPA software was limited. Due to breakthroughs in cloud computing-based virtual weather stations technology, however, green BIM is not limited to application in certain regions. Autodesk®'s weather data database includes data from both physical weather stations and virtual weather stations. Autodesk®'s data is in the internationally-accepted TMY2 (typical meteorological year) format. Employing TMY2 data for each weather station, Autodesk® derives the corresponding data for virtual weather stations through simulations, which helps compensate for differences in data between real weather stations and shrinks the distance between meteorological grid points to less than 14 km, which boosts modeling accuracy [22].

(4) Entering Internal settings:
In accordance with Anderson (2014a)[23], factors influencing energy consumption during the building use stage include geometric shape information in the BIM model and non-geometric information (attributes and parameters) required by BIM or BPA, including building type, activity types, user density, shell attributes (structural materials, thermal conductivity, insulation factor), and active systems (air conditioning and lighting). Parameter settings for these items are entered in BIM or BPA software, and the parameter values input for the initial proposal must comply with the lowest standards of local building codes. The settings input for the later optimized proposal must also be within the scope permitted by law.

(5) Implementation of energy conservation calculation module:
As shown in the diagram (Fig.3), Revit and Energy Analysis for Revit software upload the model's geometric information and non-geometric internal settings (performance parameters), and climate data to GBS in the gbXML format, and GBS uses the DOE-2 engine to perform energy consumption analysis.

(6) Visualization analysis and hot spot tracking:
Following processing via the cloud DOE-2 simulation engine, the results of visualization analysis are transmitted back to the user, including such building performance values as performance simulation results and various numerical analyses and statistical tables. Hot spot tracking is performed by preliminarily determining variables that have a major influence on energy consumption from the results of visualization analysis, where these variables can provide a basis for subsequent proposal revision.

(7) Proposal Revision:
The controlling variables are adjusted on the basis of the results of visualization analysis and hot spot tracking, and a revised proposal submitted.

(8) Optimal proposal:
If a revised proposal fails to comply with the target settings, return to Step (4) and repeat steps (4)-(7), until an optimal proposal has been obtained target settings.

4 Empirical verification

This study used the BPA and design optimization process for a new Taichung hotel construction project to implement and verify a green BIM-based decision-making cycle model.

(1) Determination of the scope of project discussion within the building life cycle:
The green BIM-based decision-making cycle for the new Taichung hotel construction project took place during the initial design stage (from the SD to the DD stage). Optimization of conceptual massing was first performed during the SD stage, and the DD stage. Details were added to optimal proposals selected during the SD stage, and continued optimization performed during the DD stage.

(2) Setting of energy conservation targets:
In the study case, the energy conservation target set during the SD stage consisted of a performance optimization percentage of 21%, which implied a 21% or better improvement over the common benchmark. And during the DD stage, the performance optimization percentage was increased by a further 5%.

(3) Accessing to external meteorological data:
Taking a site in Taichung's Beitun District as an example after inputting the latitude and longitude of the site location, GBS. automatically transmits TMY2 climate data for the closest weather station (Figure 6).

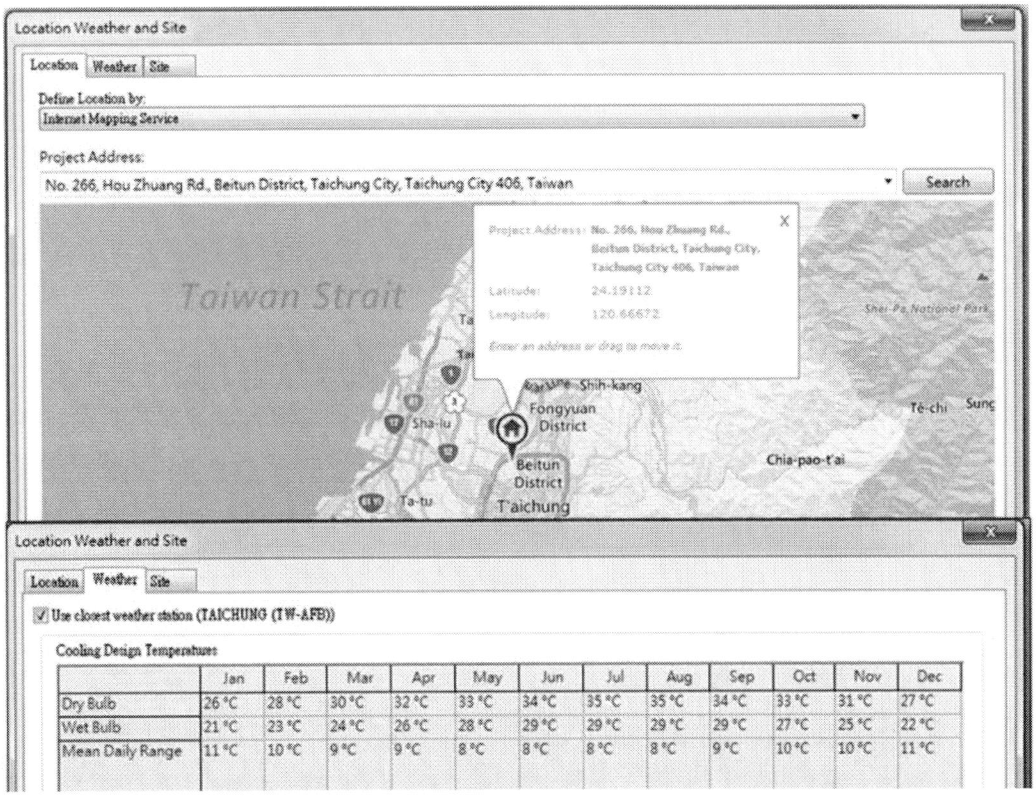

Figure 6. Site location settings and weather station TMY2 climate data.

(4) Entering Internal settings:
A model of the initial proposal was created (Create Mass/Place Mass) (Figure 7), and simple modeling was performed directly in Revit. After first establishing geometric information in the form of the

massing and floors, the initial proposal was modeled based on default values. The building type was "hotel building," the percentage glazing was set at 40%. Building operating details: refers to the number of hours of operation each week, where 12/7 indicates the building was used 12 hours per day, seven days a week. HVAC system: refers to the choice of HVAC system used in the model for energy analysis; the default choice was central VAV, hot water heating, 5.96 COP refrigerator, and boiler with an efficiency of 84.5. External ventilation information: refers to the amount of external air flow required by each person, air exchange rate per unit area, and ventilation rate per hour. The other construction material options employed the default values in the BIM model (Table 3). After performing settings, the model was converted to an energy analysis model (Figure 8).

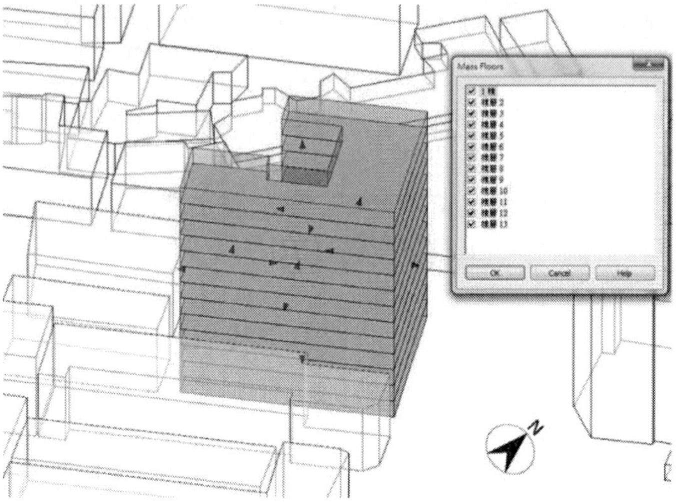

Figure 7. Mass model of project, including floors (geometric information).

Table 3. Internal settings of building.

Energy Settings

Parameter	Value
Report Type	Standard
Energy Model	
Analysis Mode	Use Conceptual Masses
Analytical Space Resolution	0.4572
Analytical Surface Resolution	0.3048
Core Offset	3.6000
Divide Perimeter Zones	☑
Conceptual Constructions	Edit...
Target Percentage Glazing	40%
Target Sill Height	0.7500
Glazing is Shaded	☐
Shade Depth	0.6000
Target Percentage Skylights	0%
Skylight Width & Depth	0.9144
Energy Model - Building Servic	
Building Operating Schedule	12/7 Facility
HVAC System	Central VAV, HW Heat, Chiller 5.9
Outdoor Air Information	Edit...

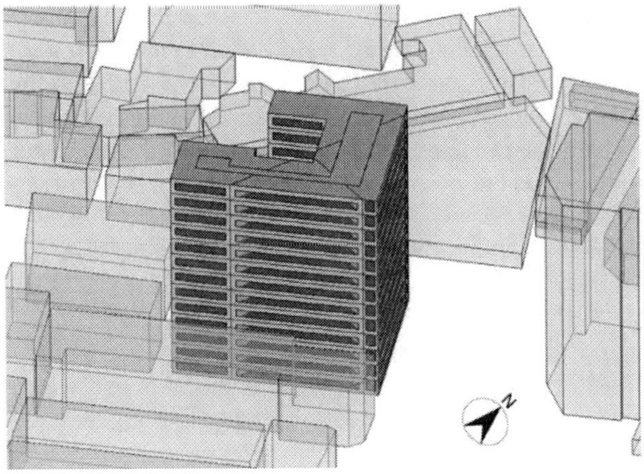

Figure 8. Conversion to an energy analysis model.

(5) Implementation of energy conservation calculation module:
The information model and internal settings and climate data are transmitted in gbXML format to the cloud GBS's DOE-2 engine for computation, and the visualized performance analysis results are transmitted back.

(6) Visualization analysis and hot spot tracking:

■ Visualization analysis :
Items includes analysis of site climate conditions, building energy use performance, and the building's physical environment. The results of analysis can serve as a reference for subsequent revisions of the proposal. Apart from the TMY2 climate data provided by weather stations in Step (3), site climate analysis also includes wind rose charts (Figure 9), which are provided within visualization analysis tables. Building energy use analysis and building physical environment analysis include such items as EUI, building life cycle (30 years) energy consumption and cost calculations, energy recovery/conservation potential, average carbon emissions, monthly air conditioning load, and peak power demand.

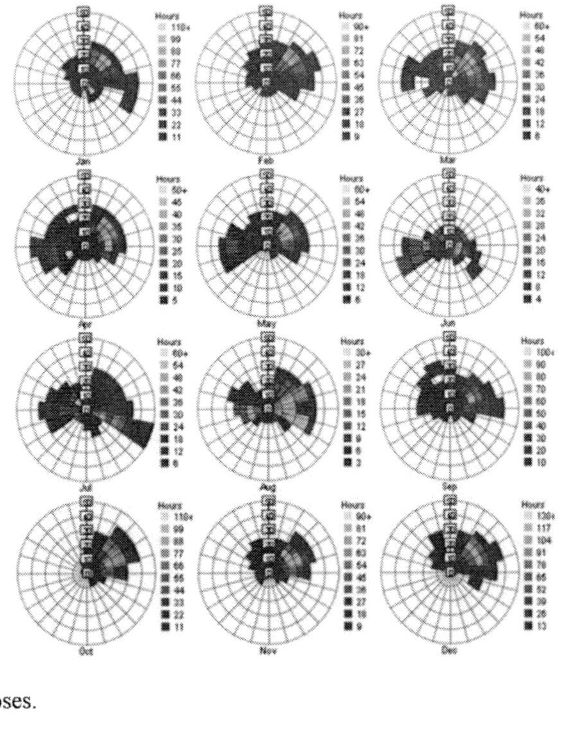

Figure 9. Mothy Wind Roses.

■ Hot spot tracking：

Building performance calculation results can be reviewed in connection with target settings, and key factors determined for use as a basis for revision. Different types of buildings have different energy consumption characteristics. The assessment of EUI must be based on a comparison with other buildings with the same operating mode to be fully meaningful. In accordance with simulation results, the initial proposal had an EUI value of 204 kWh/, m2·yr (Tables 4) ,which was less than common benchmark value of 198.8 kWh/m2*yr, general tourist hotels, after checking in the EUI values for various types of Buildings issued by the Bureau of Energy, MOEA (Tables 2). The building's optimized energy performance was calculated to be -2.9%, which was less than the preset target value of 21%. And according to the results of power usage ratio analysis, air conditioning equipment had the highest energy consumption (46%), followed by lighting equipment with 21% (Figure 10). Energy load composition analysis indicated that sunlight through windows and window heat conduction were the largest sources of air conditioning load, followed by lighting equipment and solar radiation through windows. Since the latter two sources were mutually correlated factors, experimentation during the revision process was used to refine the design proposal and achieve better energy conservation. In addition, monthly power usage distribution revealed that energy consumption was relatively high during the summer months of July and August (Figure 11), which indicated that improvement of power usage during summer could reduce operating load.

Table 4. The initial project (CD)

Building Performance Factors

Location:	T'aichung, Taiwan
Weather Station:	548644
Outdoor Temperature:	Max: 35°C/Min: 8°C
Floor Area:	8,298 m²
Exterior Wall Area:	5,840 m²
Average Lighting Power:	9.69 W / m²
People:	332 people
Exterior Window Ratio:	0.40
Electrical Cost:	$0.06 / kWh
Fuel Cost:	$1.20 / Therm

Energy Use Intensity

Electricity EUI:	204 kWh / sm / yr
Fuel EUI:	42 MJ / sm / yr
Total EUI:	778 MJ / sm / yr

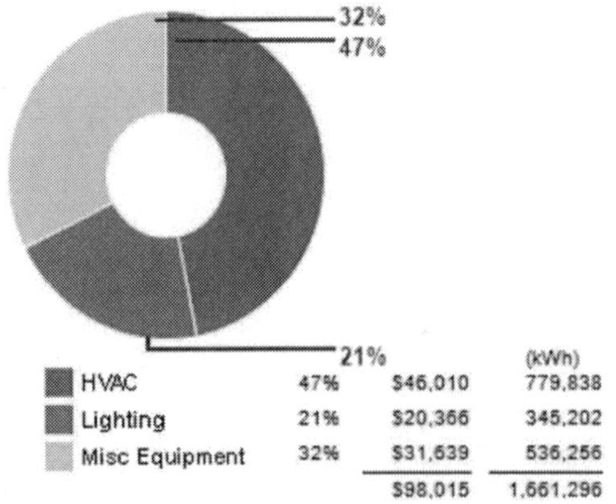

Figure 10. Analysis of Energy Use: Electricity.

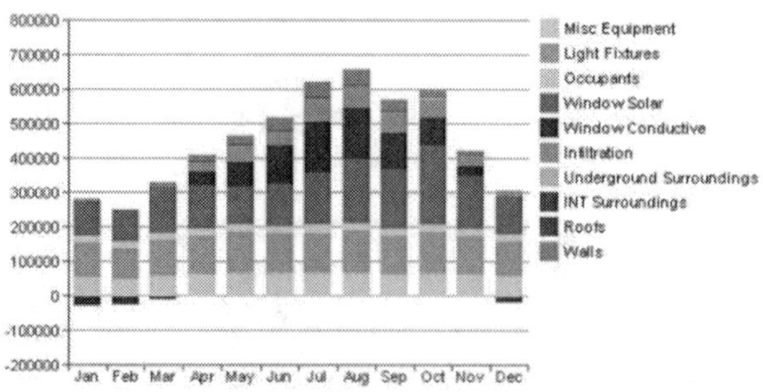

Figure 11. Review of monthly cooling load.

(7) Proposal revision:
According to passive building design principles, the monsoon seriously affects air convection and occupants' comfort of the building volume and its open space. According to the wind rose, the site would typically receive W and SW monsoon winds in the summer and NE monsoon winds in the winter (Figure 13). A discussion of the relationship between the building mass and outdoor space was conducted to ensure that the project's open spaces would face the wind in the summer, and be sheltered from the wind in the winter. Several rounds of revision and assessment were performed, which involved repeated implementation of steps (4)-(8), and revision continued until a proposal meeting the preset energy conservation targets was obtained (Figure 13).

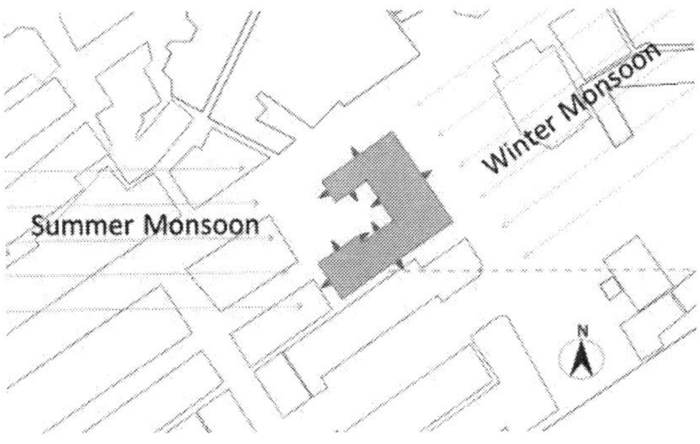

Figure 12. Schematic diagram of relationship between yearly monsoon winds and building configuration.

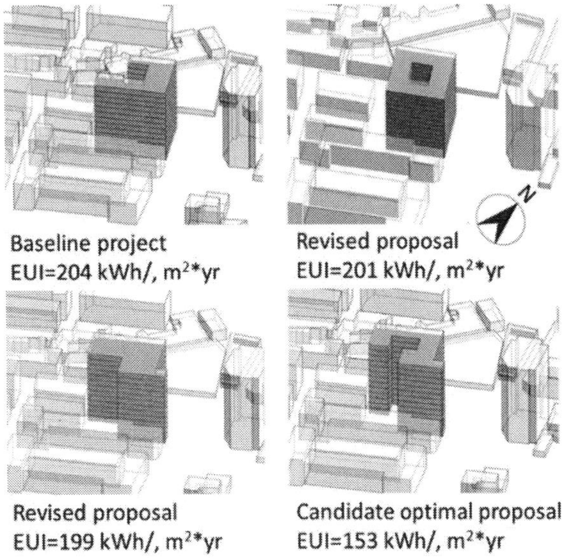

Figure 13. Continued optimization of proposal.

(8) Optimal proposal:
The candidate optimal proposal had EUI=153 kWh/, m2·yr, which represented a reduction of 51 kWh/, m2·yr compared with the common benchmark EUI of 198.8 kWh/m2·yr. This was equivalent to a performance optimization percentage of 22.9%, and met the performance optimization target of at least a 21% improvement over the ommon benchmark EUI value. A candidate optimal conceptual massing was found during the SD stage through the foregoing analysis. Taking the candidate optimal conceptual massing as a basis, the study then entered the DD stage, which involved a return to Revit modeling and repeated implementation of steps (4)-(8), with the difference that the analytical model in Step (4) employed the option of "use building elements." energy load review after revision in response to results of visualization analysis and hot spot tracking found that heat conduction through the windows and outer walls was the largest source of air conditioning load.

The following are the results of steps (7) and (8) during the DD stage:

(9) Proposal revision:
A monthly energy load review after revision in response to results of hot spot tracking found that heat conduction through the windows and outer walls was the largest source of air conditioning load. While the study originally employed a 40% percentage glazing, this was revised and reduced to 29% after the addition and revision of window elements (Figure 14). In addition, the outer walls were set as thermally insulated walls, windows were set as having two-layer LowE glass, horizontal shading eaves were placed on the southern and northern facades, and vertical shading eaves were placed on the eastern and western facades (Figure 15).

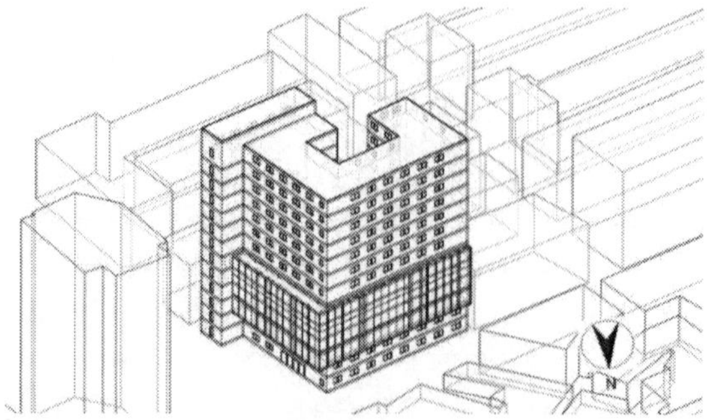

Figure 14. Addition of window elements, thermally insulated wall, and Low-E glass.

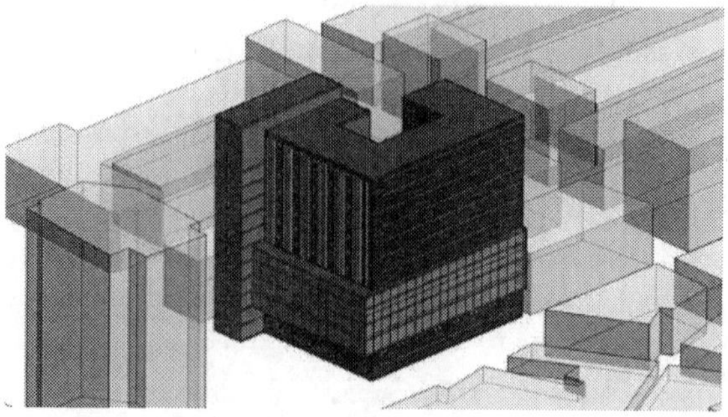

Figure 15. Addition of shading eaves.

(10) Optimal proposal:
After the optimal proposal from the conceptual massing stage entered the DD stage, and continuing improvement with regard to hot spot tracking, comparative analysis of the optimal proposal in the following tables revealed that the optimal proposal's EUI of 140 kWh/m2·yr represented a reduction of 13 kWh/m2·yr compared with the candidate optimal proposal in CD stage had EUI=153 kWh/m2·yr, which was equivalent to a performance optimization percentage 8.5% (Tables 4 & 5). The performance optimization percentage met the taget setting, equal or better than 5% during the DD stage.

Table 5. Optimal proposal (SD).

Energy Use Intensity

Electricity EUI:	140 kWh / sm / yr
Fuel EUI:	132 MJ / sm / yr
Total EUI:	636 MJ / sm / yr

5 Recommendations and conclusion

Taking energy consumption throughout the building life cycle as an example, the study verified the practical effectiveness of green BIM in the case of a new Taichung hotel construction project. The following are recommendations:

A. Setting of targets:
Setting standards for energy conservation targets are not necessarily irreplaceable. In view of the fact that it is difficult to change to Taiwan Power Co. as the central power supply and billing model, this study consequently proposed the use of EUI as a target output unit. Furthermore, the "EUI values for various types of Buildings" issued by the Bureau of Energy, MOEA provides one of small number of common bechmark reference values for power usage in the Taiwan area. In theory, depending on the project's ideas and goals, there are other possible energy target standards; the applicability of these standards will depend on the direction in which the government seeks to lead industry and the feasibility and maturity of relevant technologies. Other feasible energy conservation target standards include: (1) Energy Cost Budget, (2) Net Zero Energy standards, (3) Passivhaus, (4) Peak heating or cooling [24] .

B. Validation:
The DOE-2 engine employed by GBS was developed by the Lawrence Berkeley National Laboratory, and fully passed ANSI/ASHRAE Standard 140 testing and certification in 2008. It is currently one of the most widely used optimization engines. ANSI/ASHRAE Standard 140 refers to the standard testing method for the evaluation of building energy analysis computer programs issued by the American Society of Heating and Air-Conditioning Engineers [25] .

Green BIM emphasizes the use of building information modeling as a basic tool from the initial design stage. In addition, use of a design and analysis decision-making cycle when performing BPA in response to local climate conditions can yield optimized design proposals with good environmental effectiveness, and ultimately achieve the goal of environmental sustainability.

References

1. E. Krygiel, B. Nies, *Green BIM-Successful Sustainable Design with BIM* (Wiley, 2008)
2. McGraw-Hill Construction, *Green BIM: How BIM is Contributing to Green Design and Construction* (Smart Market Report, 2010)
3. C. M. Eastman, *Building Product Models* (CRC., 1999)
4. C. M. Eastman, *BIM handbook: a guide to BIM for owners, managers, designers, engineers, and contractors*, 2nd ed (Wiley 2008)
5. T. Kusuda,. Early history and future prospects of buildings system simulation, Proceedings of Building Simulation, **1**, 3 (1999)
6. JA. Clarke, *Energy Simulation in Building Design*, 2nd ed, (Clarke, JA. Butterworth-Heinemann, Oxford, 2001)
7. G. Augenbroe, Building and Environment, **27**(2), 149 (1992)
8. M. R. Asl, S. Zarrinmehr, M. Bergin, W. Yan, Energy and Buildings, **108**, 1 (2015)
9. P. Andrasik, JNIBS, **3**(4), 28 (2015)
10. J. Scade, Construction Management & Economics, **29**(4), 371 (2011)
11. Autodesk® Ecotect Analysis 2011, Available online: http://usa-stg.autodesk.com/adsk/servlet/download/item?siteID=123112&id=13140033
12. Autodesk® Energy Analysis for Revit 2016, Available online: http://www.autodesk.com.tw/products/energy-analysis-revit/overview
13. Bently® Hevacomp 2016, Available online: https://www.bentley.com/en/products/brands/hevacomp
14. Tas® building modeling and simulation tool, Available online: http://www.buildup.eu/tools/34288
15. Graphisoft® Eco Designer STAR 2016, Available online: http://www.graphisoft.com/archicad/ecodesigner_star/
16. T. Maile, M. Fischer, V. Bazjanac, *Building Energy Performance Simulation Tools - a Life-Cycle and Interoperable Perspective*, (Stanford University, Available online: http://cife.stanford.edu/node/169, 2007)
17. Autodesk® 2016, Energy Analysis Workflows, Autodesk® Building Performance Analysis Help, Available online: http://help.autodesk.com/view/BUILDING_PERFORMANCE_ANALYSIS/ENU/?guid=GUID-E85A114E-BA0D-4811-B1A5-4EE26462708A
18. LEED 2016. Available online: http://www.usgbc.org/leed
19. Taiwan Power Cost and Price tips, Available online: http://www.taipower.com.tw/content/new_info/new_info-d11.aspx?LinkID=14
20. Chung, W., Applied Energy, **8**(5), 1470 (2011)
21. R. L. Huang, *EUI Values for Various types of Buildings, Building Energy Conservation Applied Technical Handbook* (Bureau of Energy, MOEA, 2013) , Available online: http://ecct.tgpf.org.tw/print/files/1021231%E5%BB%BA%E7%AF%89%E7%AF%80%E8%83%BD%E6%8A%80%E8%A1%93%E6%89%8B%E5%86%8A.pdf
22. S. Malkin, Weather data for Building Energy Analysis, (Autodesk, Inc. 2008), Available online: https://sustainabilityworkshop.autodesk.com/sites/default/files/core-page-files/weather_data_greenbuildingstudio_adsk_white_paper.pdf
23. K. Anderson, Energy Model, *Design Energy Simulation for Architects Guild to 3D graphics,* 1st ed., (Fuller, W., Ed; Routledge: New York and Landon, 2014)

24. K. Anderson, Planning and Goal-Setting, *Design Energy Simulation for Architects Guild to 3D graphics*, 1st ed., (Fuller, W., Ed; Routledge: New York and Landon, 2014)
25. Autodesk® 2016, Green Building Studio Validation, Autodesk BPA Help, Available online: http://help.autodesk.com/view/BUILDING_PERFORMANCE_ANALYSIS/ENU/?guid=GUID-EF68E7D5-C0A5-4805-BFE5-7C74C57B712E

The influence of application processes on performance reliability

Yue Zhang, Ning Huang[a], Weiqiang Wu and Shuo Zhang

School of Reliability and Systems Engineering, Beihang University, Beijing, China

Abstract. Application process has significant influence on network performance reliability, however, previous studies mainly focus on the impact of network characteristics (such as routing and traffic) on network performance reliability and ignore the influence of application process. In this paper, a multi-tiers model is proposed to capture the process of application. Furthermore, an application paths matrix is proposed to describe all the transmission paths of the application process in the network. Based on the application paths matrix, the boundary of the critical point of the traffic generation rate is derived theoretically, below the lower limit of the boundary the network stays in free-flow state and is reliable, while above the upper limit of the boundary the congestion occurs and the network performance reliability decreases. Furthermore, the effect of the boundary is verified in numerical simulations. The simulation results also show that congestions occurred in middle-tier host nodes could bring more influence on the network performance reliability.

1 Introduction

Network performance reliability is determined by both the network infrastructures and applications [1]. The decrease of network performance reliability is usually caused by congestion, which is reasoned by the heavy use of network applications [2]. Numerous works related to the performance of part nodes in the network are usually based on the queueing theory or network calculus theory, which is difficult to evaluate the performance of the total network due to the computational complexity [3,4]. By the end of the 20th century, the researches in complex network have proved that the network collapses above a certain traffic threshold and a traffic dynamics analysis method based on the mean-field theory is proposed to capture the performance of the whole network [5,6]. Since then numerous works focused on the performance analysis of whole network.

In particular, Ohira and Sawatari proposed the initial model to infer the network congestion in 1998 [6], in which the traffic is generated according to the packets generation rate λ and congestion occurs when λ is above the certain packet generation critical point λ_c. Since then numerous works follow this basic idea and concentrate on the shifting of critical points that the network turns from a free-flow state to congestion state [7]. Researchers firstly find that the network structure is important to the congestion of networks [8, 9], such as Toroczkai et.al have found that the scale-free networks are less prone to congestion than the random networks [10]. Since improving the routing strategy is a more practical and effective way to improve network performance than changing the physical

[a] Corresponding author : hn@buaa.edu.cn

structure of the network, so some researchers have focused on the influence of routing on the congestion [11,12], such as Wang et al. have proposed an efficient local routing strategy to improve the network performance [13]. Furthermore, in a specific network structure with particular routing strategy, more researches about the mitigating and controlling traffic congestion have been related to the resource allocation methods, especially, the node delivery capacity allocation has attracted much interest [14-16]. Recently, some researchers realized that dependence relations between networks is important factor to the network performance and some works have analyzed the influence of dependence relations on the congestion [17-19].

The above works have analyzed the influence of traffic or the network characteristics (such as routing or structure) on performance. However, modern Internet applications actually shows process feature, which means that the application is consisted by multiple tiers and each application tier provides certain function to its preceding tier and uses the functionality provided by its successor to carry out its part of the overall application process [20,21], but the above works ignored the multi-tiers application process. Although some researchers analyzed the performance of multi-tiers application [22,23], they are basically based on queuing theory and difficult to evaluate the performance of whole network.

In this paper, we focus on the influence of application process on the network performance reliability. The multi-tiers application model and the application paths matrix are proposed to help to analyze the influence of application process on network performance reliability. The boundary of the critical point of packets generation rate is estimated. We studied the influence of application process on the network performance reliability with a Lattice network topology. Some interesting results are observed in this paper and could be utilized to provide advice to network administrators and application providers.

2 MULTI-TIERS APPLICATION AND TRAFFIC

The following concepts are defined to help understand the relationship between the applications [24]:

A Service: is a function that a network provides to users;

An Application: is a procedure or service used by customers with some demanded performance requirements.

In the following, the service particularly describe the function that host node could provide, while the application describe the procedure provided to the end user, which is consisted by multiple services.

2.1 Multi-Tiers application model

In current, most Web applications are designed as multi-tier systems [21], for example, the ecommerce Websites e-bay and the web-based information systems such as Wikipedia. The multi-tier applications consist of multiple tiers though each tier hosts server(s) with identical functionality. Each tier provides a predefined service to the successor tier and receives services from its preceding tier. An 3-tiers application used in [23] is illustrated as an example in Fig.1. The 3-tiers application consist of the web-tier, application-tier and data-tier: the web-tier receives requests from end-users and provide them to access the applications; the application-tier receive requests from the Web-tier and look up information in the database; The data-tier basically stores applications data.

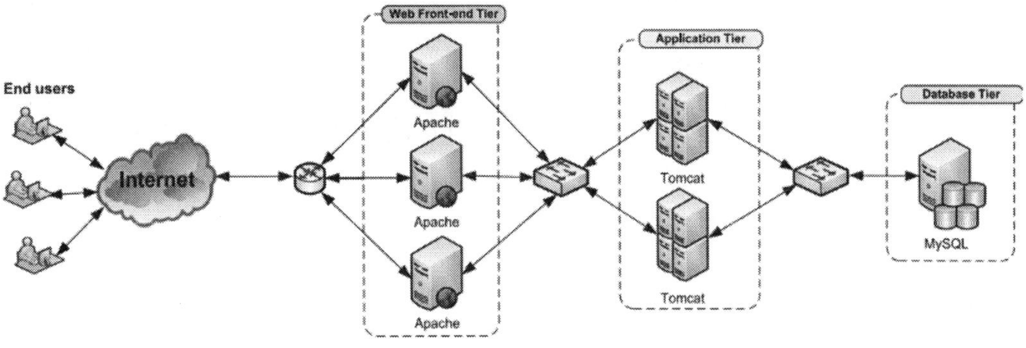

Figure 1. An example of 3-tiers application in internet [23].

For an application deployed in the network with M tiers denoted by T_1, ..., T_M, the packet is processed exactly once by tier T_i and then forwarded to tier T_i+1 for further processing until it reaches T_M. In particular, T_1 is the origin tier of application packet and T_M is the terminal tier of application packet. If let $\boldsymbol{APP_M}$ denote the process of application with M tiers, then it could be describe as $\boldsymbol{APP_M}=\{T_1, T_2,\ldots, T_M\}$. Moreover, each tier could corresponds to multiple host nodes, let S_i denote the set of host nodes for T_i and card(S_i) denote the cardinal number of set S_i.

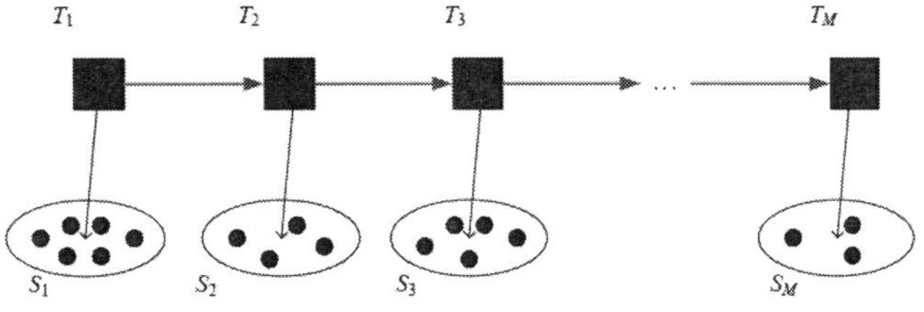

Figure 2. An example of application process of $\boldsymbol{APP_M}$ with $\boldsymbol{M}$-tiers.

Fig.2 present an illustrative example of the application process with M tiers. It shows that each tier T_i corresponds to a node set S_i which could provide identical functionality in the network, such as a set consist by 6 nodes S_1 could provide the service of T_1.

2.2 Application paths matrix

In order to better understand the influence of application process on the performance reliability, in this section, we introduce the application paths matrix **P**. Let us consider the application process $\boldsymbol{APP_M}$ with M different tiers. The corresponding host nodes set is $\{S_1, S_2,\ldots, S_M\}$. Let ***path*** denote the path of application process in the network, ***path*** could be described as:

$$path=\begin{bmatrix} a_{1,1} & a_{1,2} & \cdots & a_{1,card(S_1)} & a_{2,1} & a_{2,2} & \cdots & a_{2,card(S_2)} & \cdots & a_{M,1} & a_{M,2} & \cdots & a_{M,card(S_M)} \end{bmatrix} \quad (1)$$

where $a_{k,l}(1\le k\le M, 1\le l\le card(S_i))$ is the host node for T_k. If this path pass through host node $a_{k,l}$, then $a_{k,l}=1$, else $a_{k,l}=0$. The application process matrix **P** represents for all paths of application process in the network, so **P** could be described as follows:

$$\mathrm{P}=\begin{bmatrix} path_1 \\ path_2 \\ \cdots \\ path_{card(S_1)} \end{bmatrix}_{card(S_1)\times(\sum_{i=1}^{M}card(S_i))} = \begin{bmatrix} a(1)_{1,1} & a(1)_{1,2} & \cdots & a(1)_{M,card(S_M)} \\ a(2)_{1,1} & a(2)_{1,2} & \cdots & a(2)_{M,card(S_M)} \\ \cdots & \cdots & & \cdots \\ a(card(S_1))_{1,1} & a(card(S_1))_{1,2} & \cdots & a(card(S_1))_{M,card(S_M)} \end{bmatrix}_{card(S_1)\times(\sum_{i=1}^{M}card(S_i))} \quad (2)$$

If $path_i$ pass through host node $a_{k,l}$, then $a(i)_{k,l}=1$, else $a(i)_{k,l}=0$. Let λ_i denote the packets generation rate of $path_i$. In this paper, we assumed that $\lambda_i=\lambda_j=\lambda,(1\le i,j\le M)$. If we use **Z** denote the load of host node, then **Z** could be described as:

$$Z=(z_{1,1},z_{1,2},\dots,z_{1,card(S_1)},z_{2,1},z_{2,2},\dots,z_{2,card(S_2)},z_{M,1},z_{M,2},\dots,z_{M,card(S_M)}) \tag{3}$$

where $z_{i,j}=\sum_{l=1}^{card(S_i)}\lambda_l a(l)_{i,j}=\sum_{l=1}^{card(S_i)}\lambda a(l)_{i,j}$.

2.3 Network traffic model

In fact, the generation and transmission of traffic is influenced by the application process in the network. So in this paper, we have improved the traffic model used in [9,13] by considering the application process. The details of the traffic model in a network with $\boldsymbol{APP_M}$ could be:

1) Traffic generation process: in a network with application process $\boldsymbol{APP_M}=\{T_1, T_2,\dots, T_M\}$, the packet is originated in the host nodes set S_1 of T_1. We assume the packets are generated in the nodes set S_1 according to the packets generation rate λ, below which a packet is generated while above which no packet is generated each simulation time.

2) Traffic delivery process: the packets are delivered along with the application process. In particular, the packets generated in S_1 will be transported to S_2 and wait for the service from T_2. After the service of T_2, it will be transported to S_3 and wait for the service of T_3. Repeat the above process, until it reaches to S_M. Moreover, packets are delivered from S_i to S_{i+1} along the shortest path routing strategy. If there are more than one shortest path, then we choose one shortest path randomly. All nodes have the same delivering capacity C_d, that is, the maximal number of packets each node can deliver during one simulation time step. All services have the server capacity C_s, that is, the maximal number of packets each server can serve during one simulation time. The "first-in-first-out" rule is applied at each queue of the nodes and at each serve queue of server nodes in the network.

3) Traffic removing process: the packet will be removed from the network after it served by the terminal tier T_M.

3 Network performance reliability model

In this paper, the basic idea of the network performance reliability is that: Let T_k denote the travel time of the packet k and T^m denote the max acceptable delay that user could tolerance. If $T_k>T^m$, then this packet is past its acceptable time, which means that the packet transmission process is unreliable. The travel time T is consist by the delivery time between every two tiers and the service time at each tier. So in a network with M-tiered application process $\boldsymbol{APP_M}$, the travel time T could be measured as:

$$T=\sum_{i=1}^{M}T_s(i)+\sum_{i=1}^{M-1}T_d(i,i+1) \tag{4}$$

where $T_s(i)$ denote the service time at tier i and $T_d(i, i+1)$ denote the delivery time between tier i and $i+1$. Let $P(t)$ denote the set of the packets in the network at time t, $card(P, t)$ is cardinal number of the set $P(t)$; Let $\{S(t)\mid s\in P(t),\ T_s>T_s^m\}$ denote the set of the "unreliable" packets in the network at time t, $card(S, t)$ is the cardinal number of the set $S(t)$. Let $R(t)$ denote the network performance reliability at time t, $R(t)$ could be described as:

$$R(t)=1-\frac{card(S,t)}{card(P,t)} \tag{5}$$

Furthermore, if we assumed that the max acceptable delay T^m could be the max travel time when the network in free-flow state in this paper, then the network is reliable when it stays in free-flow state and obviously congestion have great influence on the network performance reliability.

Actually, the application process could be divided into the delivery process between every two consecutive tiers and the service process at each tier. So if congestion occurs in the delivery process or the service process, the network reliability will decrease. The order parameter η is widely used to measure the network states in complex network science. If η=0, it indicates that the network system is in the free-flow state; While if η>0, network turns into congested state 9:

$$\eta = \lim_{t\to\infty} \frac{1}{\lambda S} \frac{<\Delta Z_p>}{\Delta t} \tag{6}$$

where $Z_p(t)$ denote the number of packets in the system at time t, $\Delta Z_p = Z_p(t+\Delta t) - Z_p(t)$, <...> indicates average over time windows of width Δt , λ is the traffic generation rate and S is the number of nodes generating packets.

Related works in complex networks focus on the delivery process of packets in the network, and a critical point of packets generation rate λ_c^d is observed, below which $\eta \to 0$ and the packets are delivered freely in the network, while above which $\eta > 0$ and congestion occurs in the delivery process of packets. In the following, we focus on the relation between the packets generation rate λ and the service process in the network and discuss the critical point of packets generation rate λ_c^s from the service process view. The basic idea is that: if the average load on host node exceeds its server capacity, then congestion occurs. So the critical point of packets generation rate λ_c^s could be calculated as follows:

$$\max\{z_{i,j}\} = C_s \tag{7}$$

substituting (3) to (7), we could have:

$$\lambda_c^s \max\{\sum_{l=1}^{card(S_1)} a(l)_{i,j}\} = C_s \tag{8}$$

Considering two limiting cases: 1) if the load are uniformly distributed on host nodes, then we could calculate the maximum value of λ_c^s . The average load z_i of host node for tier T_i could be:

$$z_i = \lambda_c^s card(S_1) / card(S_i) \quad \{1 \le i \le M\} \tag{9}$$

Equation (9) could be simplified as:

$$\lambda_{c,\max}^s card(S_1) / \min\{card(S_i)\} = C_s$$
$$\lambda_{c,\max}^s = \frac{C_s \min\{card(S_i)\}}{card(S_1)} \tag{10}$$

2) for the tier T_i with minimum card(S_i), if the load are centralized in one particular host node j and other host nodes all support load generated by one host node for T_1 respectively, then we could calculate the minimum value of λ_c^s . The average load $z_{i,j}$ of host node could be:

$$z_i = \lambda_c^s (card(S_1) - \min\{card(S_i)\} + 1) \tag{11}$$

Equation (9) could be simplified as:

$$\lambda_c^s(card(S_1)-\min\{card(S_i)\}+1)=C_s$$
$$\lambda_{c,\min}^s=\frac{C_s}{(card(S_1)-\min\{card(S_i)\}+1)} \quad (12)$$

So in a network with application process $\boldsymbol{APP_M}$, with host nodes set$\{S_1, S_2,\ldots, S_m\}$. Specially, if $card(S_1)=\min\{card(S_i)\}$ so we could find that $\lambda_{c,\max}^s=\lambda_{c,\min}^s$. In summary, if $\lambda>\lambda_{c,\max}^s$, the congestion occurs at the server hosts nodes. However, if $\lambda<\lambda_{c,\min}^s$, the server host nodes stay in free-flow state and the network state is determined by the delivery process of packets. Finally, the boundary of the critical point λ_c could be given by $(\lambda_{c,\min},\lambda_{c,\max})$, where $\lambda_{c,\max}$ could be $\min\{\lambda_c^d,\lambda_{c,\max}^s\}$, above which the network turns into congestion state; and $\lambda_{c,\min}$ could be min{ λ_c^d, $\lambda_{c,\max}^s$ }, below which the network stays in free-flow state. In the following, we focus on the influence of application process on network reliability when packets generation rate λ approximate to λ_c.

4 Numerical results and analysis

In this section, we analyze the network reliability considering the 3-tiers application on $L\times L$ lattice network (L=32) with Periodic boundary conditions. The host nodes set for each tier could be S_1, S_2 and S_3. Moreover, we assumed that $card(S_1)\geq card(S_2)\geq card(S_3)$. The hosts nodes are distributed in the network randomly and all hosts nodes have same server ability C_s=2. According to 9, the critical point λ_c^d of the delivery process $\lambda_c^d=2/L=0.125$. In the following, we assume the max acceptable delay $T^m=(M-1)\times D$. Finally, all nodes have same delivering capacity C_d=2 and the buffer size is assumed as infinity.

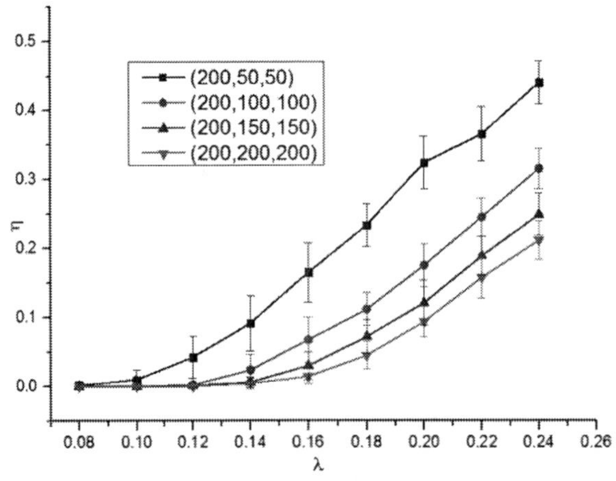

Figure 3. The relation between η and λ (each point is averaged from 30 independent simulations).

According to (9) and (12) we could find out that the critical point λ_c is related to the minimum value of $card(S_i)$. In Fig.3, we analyzed the critical point λ_c with different $card(S_1)$, $card(S_2)$ and $card(S_3)$. The curves with ($card(S_1)$, $card(S_2)$, $card(S_3)$) represent for the simulation results when the cardinal number of hosts set for each tier is $card(S_1)$, $card(S_2)$ and $card(S_3)$. So according to (9) and (12), we could calculate the maximum and minimum value of the critical point λ_c for each ($card(S_1)$, $card(S_2)$, $card(S_3)$) simulation scenario. The results are (0.013, 0.5), (0.02, 1), (0.04, 1.3) and (0.12, 2). Fig.3 shows the critical point λ_c stays in the boundary set illustrated in section 3, which confirmed the validity of boundary set. Furthermore, the results in section 2 also show that the critical point increases with the increase of $card(S_2)$ and $card(S_3)$.

In the following, we discussed the network performance reliability and the packets generation rate. Fig.4 shows network performance reliability against simulation time with different packets generation rate λ. The cardinal number of hosts set for each tier is (200,200,200) and the critical point λ_c is 0.14 according to Fig.3. Fig.4 shows that if $\lambda < \lambda_c$, the network stays in free-flow state and the performance reliability approximate to 1, while if $\lambda > \lambda_c$, the congestion occurs in the network and the network performance reliability falls quickly. Moreover, the larger λ is, the more unreliable the network is.

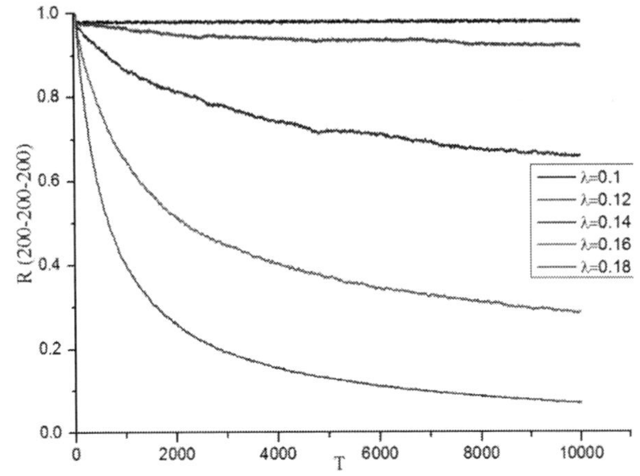

Figure 4. The relation between the R and λ (each point is averaged from 30 independent simulations).

Usually deploying more servers in the network could improve the network reliability. Since the network becomes unreliable when $\lambda > \lambda_c$,so in Fig.5 we discussed the relation between the reliability and the server hosts numbers with fixed server hosts numbers and terminal hosts numbers. The packets generation rate λ is 0.14. It shows that the network performance reliability increase with the increase of the server hosts numbers for T_2.

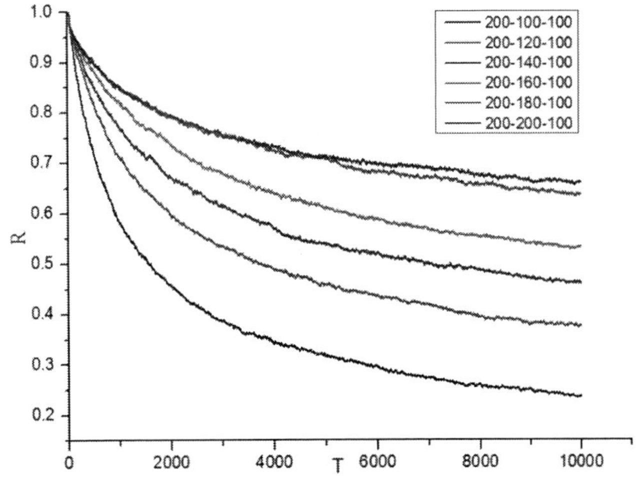

Figure 5. The relation between R and middle server host numbers(each point is averaged from 30 independent simulations).

A general problem that network administrators and service providers care about is that how to allocate the server hosts to each tiers when the total server hosts and the server hosts for T_1 is fixed. Since the network becomes unreliable when $\lambda > \lambda_c$, we discussed the influence of the server hosts number on the network performance reliability when $\lambda > \lambda_c$ in Fig.6. The total server hosts number for T_2 and T_3 is 250. Fig.6 shows that the network performance reliability decreases as the decrease of the server hosts numbers for T_2. Since the less the server hosts number for T_i is, the heavier congestion in T_i is. Fig.6 shows that the congestion occurred in middle Tier could bring more influence on the reliability than the terminal tier.

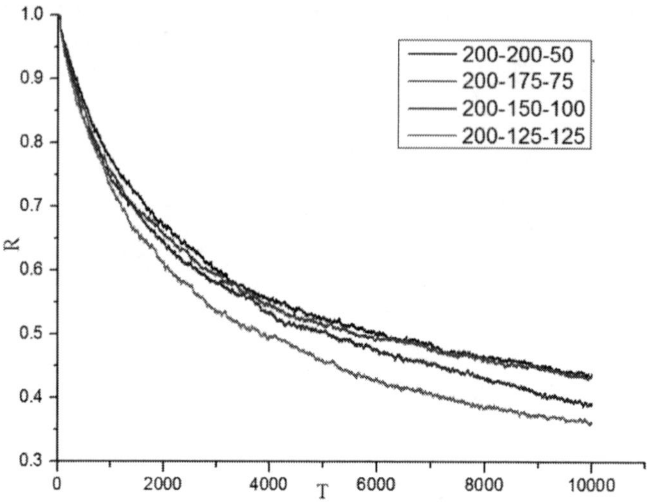

Figure 6. The relation between R and server hosts numbers (each point is averaged from 30 independent simulations).

5 Conclusion

In this paper, a general multi-tiers application model is proposed to capture the multi-tier process of application and an application paths matrix is proposed to capture all paths of the application process. The boundary of the critical point of the packets generation rate is estimated, below the lower limit of the boundary the network stays in free-flow state and the network is reliable, while above the upper limit of the boundary the network stays in congestion state and the network performance reliability decreases. Furthermore, the effective of the boundary is verified in numerical simulations on 32×32 lattice network with 3-tiers applications. The simulation results also shows that the congestion occurs in the middle tier could bring more influence on the network performance reliability. We believe that the results and methods in this paper are helpful to optimize the application processes in network. An open question to be solved in future work is that in this paper we considered the lattice network topology, more network structures could be studied in future, for example, the BA network or the ER network

References

1. N. Huang and Z.T. Wu, J. Syst. Eng. Electron, **35**(12), 2651-2660 (2013)
2. R. Li, N. Huang and R. Kang, *IEEE Proceedings Reliability and Maintainability Symposium* (*RAMS*) (San Jose, USA, 2010)
3. J.Y.L. Boudec and P. Thiran, *Network calculus: a theory of deterministic queuing systems for the internet* (Berlin, German, 2001)
4. T.G. Robertazzi and G. Thomas, *Computer networks and systems: queueing theory and performance evaluation* (New York, USA, 2000)
5. A. Arenas, A. Díaz-Guilera and R. Guimera, Phys. Rev. Lett., **86**(14), 3196 (2001)
6. T. Ohira and R. Sawatari, Phys. Rev. E, **58**(1), 193 (1998)
7. S. Chen, et al., Math. Probl. Eng., 256-267 (2012)
8. P. Holme, Adv. Complex Syst., **6**(02), 163-176 (2003)
9. L. Zhao, et al., Phys. Rev. E, **71**(2), 026125 (2005)

10. Z. Toroczkai and K.E. Bassler, Nature, **428**(6984), 716 (2004)
11. P. Echenique, J. Gómez-Gardeñes and Y. Moreno, Phys. Rev. E, **70**(5), 056105 (2004)
12. G. Yan, T. Zhou, B. Hu, Z.Q. Fu and B.H. Wang, Phys. Rev. E, **73**(4), 046108 (2006)
13. W.X. Wang, C.Y. Yin and G. Yan, Phys. Rev. E, **74**(1), 016101 (2006)
14. J. Ma, W. Han, Q. Guo and Z. Wang, Physica A, **456**, 281-287 (2016)
15. Y. Xia and D. Hill, EPL, **89**(5), 58004 (2010)
16. G. Q. Zhang, et al., Physica A, **390**(2), 387-391 (2011)
17. C.G. Gu, et al., Phys. Rev. E, **84**(2), 026101 (2011)
18. R. G. Morris and M. Barthelemy, Phys. Rev. Lett., **109**(12), 128703 (2012)
19. F. Tan, J. Wu, Y. X. Xia and C.K. Tse, Phys. Rev. E, **89**(6), 062813 (2014)
20. D. Huang, B. He and C. Miao, IEEE Commun. Surv. Tutor., **16**(3), 1574-1590 (2014)
21. D.K. Barry and P.J.Gannon, *Web services and service-oriented architectures: the savvy manager's guide* (San Francisco, USA ,2003)
22. B. Urgaonkar, G. Pacifici, P. Shenoy, M. Spreitzer and A. Tantawi, *ACM SIGMETRICS Performance Evaluation Review* (Banff, canada, 2005)
23. K. RahimiZadeh, M. Analoui, P. kabiri and B. Javadi, J. Netw. Comput. Appl., **56**, 166-187 (2015)
24. N. Huang, Y. Chen, D. Hou and L.D. Xing, J. Syst. Eng. Electron., **22**(6), 1030-1036 (2011)

A research on the calculation method of car proportion based on image processing

Hong-Wei Xiao[a], Chang-Jiang Jiang, Jia-Xing Zhai and Jing-Wei Chen

Department of Vehicle Engineering, Jilin University, China

Abstract. The proportion of the car is one of the important parts of car styling design. The study of the car proportion is of great significance. The proportion of the car is determined by the technical conditions, functions and aesthetics. At present, there're not many studies on the area proportion of the car .There're studies on the line relationship and length, width, height proportion of the car. This paper uses computer image processing method to calculate the projected area proportion of the hatchback car. And it provides a new thinking and method for the car proportion study. The result shows that the projected area proportion values of the hatchback cars of European, American, Japanese, Korean and Chinese series on the market have a difference, but are close. And the total average projected area proportion values of car window and car body of the hatchbacks have been calculated.

1 Introduction

The proportion of car is an important part of the car styling, whether a car is beautiful, depends on whether the proportion of the car is coordinate. The proportion of car is determined by the development of automobile industry technology and aesthetic laws. The car styling should meet the function of use, also should meet the aesthetic needs of people. To meet the use function, ergonomics and aerodynamics requires, the body length, height, Width, wheelbase etc. will have some limits. And after meeting the basic use requires, it needs to meet the aesthetic requirements.

At present, the study of the proportion of car is mainly concentrated on the relationship between the lines on the body and the proportions on the length, width and height directions.Computer image processing technology has been widely used in scientific researches, industrial and agricultural productions, biomedical engineering, military and other fields.

Ravi Ranjan et al. proposed a classification and identification method for surface defects in friction stir welding based on image processing [1]. Paula Cristina Oliveira et al. proposed a non-destructive method based on digital image processing for measuring the vitality and nutritional status of vines [2]. Hua-wei Cui et al. proposed a rapid measurement method of relative lesion area based on Photoshop software [3]. Lu Fu et al. studied the technical factors such as ergonomics, aerodynamics and body structure in car styling [4]. Zhen Wang and others studied the car styling features from the view of holidtic cognition, proposed that the ratio, space and stance together constitute the overall car styling feature frame [5]. Zhuan Wei and others studied the side proportions on the sedan car body, also the

[a] Corresponding author : xiaohw@jlu.edu.cn

characteristics, contents and application of digital image processing are analyzed [6]. Yu-mi Huang analyzes the characteristics, contents and applications of digital image processing [7].

2 Calculation principle

A pixel is the minimum unit of a digital picture: a point. Theoretically, the face is composed of numerous lines, and the line is composed of numerous points, that is, the surface is composed of numerous points. But In actual use, the number of pixels of the image is limited, and the single pixel has a certain area. So calculating the ratio of the number of pixels in different parts of a picture can get the area proportion between them.

This paper calculates the area proportions in a single car picture, then making classification to the statistics, dividing them into three categories: projected area proportion between the front window and the front view of the car body, the side windows and side view of the car body, the rear window and the rear view of the car body. This method is a rapid measurement method of projected area proportional. The requirements of the samples in this experiment are the pictures are clear, the shooting distance is moderate, and the shooting directions are the front, the side and the rear. And the sample car pictures come from professional car website: www.pcauto.com.cn.

The projected area proportions between the window and the body of the car are calculated as follows:

$$R=P_w/P_b \tag{1}$$

Where, R is the projected area proportion value between the window and the body, P_w is the number of pixels of the window in the picture, and P_b is the number of pixels of the body in the picture.

3 Method validation

According to the ratio of 1: 1 ,in mm Drawing a 50 × 50 square, 40 × 70 rectangular, circle of radius 25, bottom side length 30, high 40 isosceles triangle, and bottom side length 40, high 50 isosceles triangles. And then print it in a 1: 1 ratio on an A4 paper .After that use a digital camera to take picture of it with the camera lens vertical to the paper, then put the picture into Photoshop software as shown in Figure.1, the graphics number are①,②,③,④,⑤.

Then use the pen tool to create five paths along the profile of the five graphs, then use these paths to create the select area, as shown in the dotted area in Figure.2, and then view the number of pixels within the five selection area respectively through the histogram expansion window in Photoshop as shown in Figure.2. The results are as shown in Table 1 and Table 2:

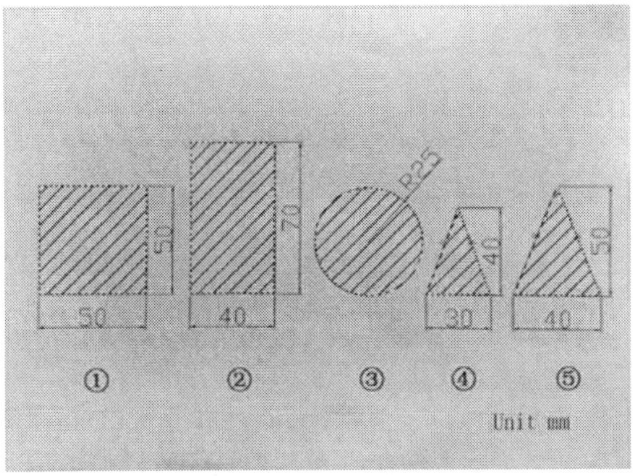

Figure 1. Method test drawing

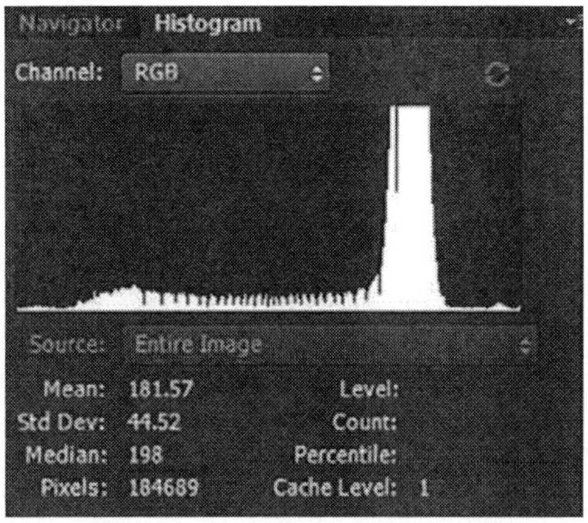

Figure 2. Method test drawing

Table 1. The measured values of the pixels and the true area of each graph

Graphics No.	Pixel value	True area (cm2)
①	184689	25.00
②	201154	28.00
③	143535	19.63
④	43510	6.00
⑤	76195	10.00

Table 2. The ratio between each graph

Computing Objects No.	Pixel ratio	True area ratio	Absolute error	Relative error/%
①/②	0.9181	0.8929	0.0252	2.82
①/③	1.2867	1.2736	0.0131	1.03
①/④	4.2447	4.1667	0.0780	1.87
①/⑤	2.4239	2.5000	0.0761	3.04
②/③	1.4014	1.4264	0.0250	1.75
②/④	4.6232	4.6667	0.0435	0.93
②/⑤	2.6400	2.8000	0.1600	5.71
③/④	3.2989	3.2717	0.0272	0.83
③/⑤	1.8838	1.9630	0.0792	4.03
④/⑤	0.5710	0.6000	0.0290	4.38

The relative error is the ratio between the absolute error caused by the measurement and the true value (conventional) of the measured, multiplied by 100%, the average relative error of the measurement method is 2.64%, indicating that the method is effective when the measurement accuracy requirement is not high. And this method can be used to calculate the area proportions between the graphics in a single photo.

4 Calculation and statistics of projected area proportion between the car windows and car body

4.1 Sample collection

Widely collecting the front, side and rear view pictures of the mainstream hatchback on the market as test samples, of which there are 18 European cars, 5 American cars, 12 Japanese cars, 6 Korean cars, 17 Chinese cars, a total of 58 models (the amount of the American and Korean hatchback cars is small on the market).The hatchback models are as shown in Table3.

Table 3. Model Table

Car Series	Car Models
Euro series	Audi A1,BMW 1,Mercedes Benz A, Renault Megan, DS 3, DS 4S, Peugeot 308s, Peugeot 208,Volkswagen Golf, Skoda Rapid, Skoda Fabia, Volvo V40,Seat Ibiza, Seat Leon, Fiat 500, Fiat Ottimo, Mini cooper, Smart for 4.
American series	Buick Verano, Ford Fiesta, Ford Focus, Chevrolet Aveo, Chevrolet Lova RV.
Japanese series	Honda Fit, Lexus CT, Toyota YARIS L, Toyota Prius, Nissan Tiida, Nissan Micra, Suzuki Alto, Suzuki SX4, Suzuki Swift, Mazda 3 Axela, Mazda 2, Mazda 3 Xincheng.
Korean series	Kia K2, Hyundai i30, Kia Seoul, Kia K3S, Hyundai Veloster, Hyundai Accent.
Chinese series	Bao Jun 310, Baic Senova D20, Geely Emgrand GS, Chery QQ, Chery Fulwin2,Chery A3, SAIC MG3, BYD F0, Qoros 3, Haima Aishang, Geely Panda, Venucia R50, FAW Oley, Changan CX20, Changan Benben, Changan Eado XT, Zhonghua H320

Figure 3. Selected hatchback samples

4.2 Calculation examples

Take the projected area proportion between the window and the body calculation of BMW 1 Series as example:

The three virtual projection planes are the front view plane, back view plane and the side view plane. Take the areas on the car in the three view photos as projected areas simply. So the projected area proportion is the size proportion between the areas on the car in the three view photos.

First, collect the front, the side and the rear view pictures of the target model, as shown in Figure.4. Next, put them into Photoshop separately, using the pen tool to separately hook out the paths along the profiles of the car body and the window (Considering the integral design of the side car body, the area of the wheel opening is taken into account). For the sake of clarity, the path is indicated by red dotted lines, as shown in Figure.4. And then use the body path and window path to establish the select area. After that, view the number of pixels of the selected area in the histogram extension window, as shown in Figure.5. The pixels of BMW 1 series car pictures are as shown in Table 4:

Figure 4. BMW 1 Series car three views

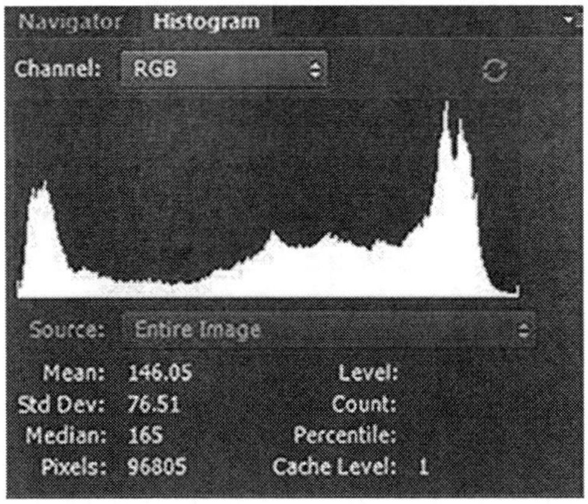

Figure 5. Histogram expansion window2

Table 4. The pixels of BMW 1 series car pictures

view direction of window	Pixel value	view direction of car body	Pixel value
Front view window	19114	Front view car body	96805
Side view window	13723	Side view car body	90995
Rear view window	17680	Rear view car body	95991

Put the value of pixels into equation (1), so the proportion of projected area between the car window and the car body can be obtained. The proportions of the projected area between the car window and car body of BMW 1 Series are here: Front view proportion value is 0.20, side view pro portion value is 0.15 and rear view proportion value is 0.18.

4.3 Statistic of the calculation results

In the following chart, the proportion 1 is the projected area proportion value between the front window and the front view car body. The proportion 2 is the projected area proportion value between the side window and the side view car body. The proportion 3 is the projected area proportion value between the rear window and the rear view car body.

Table 5. Comparison of the average proportion values of the five hatchback series

Items	Proportion 1	Proportion 2	Proportion3
European series	0.22	0.16	0.19
American series	0.22	0.17	0.19
Japanese series	0.22	0.16	0.20
Korean series	0.23	0.16	0.18
Chinese series	0.23	0.17	0.19
Total average	0.22	0.16	0.19
Standard deviation	0.00	0.00	0.01
Max	0.23	0.17	0.20
Min	0.22	0.16	0.18

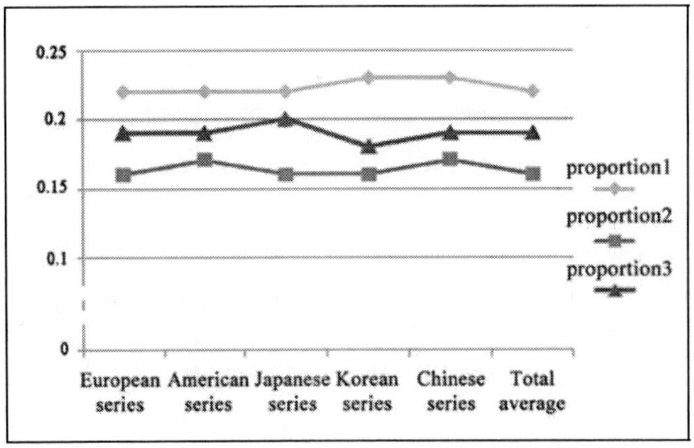

Figure 6. Line chart of the average proportion values of the five hatchback series

5 Conclusions

This paper presented a fast method based on image processing to measure the projected area proportion between the car window and the car body, and calculated the projected area proportions between the window and the body of current mainstream hatchbacks through this method. Here come the conclusions:

a) The total average projected area proportion values of the mainstream hatchbacks on the market : The total average projected area proportion value between the front window and the front view car body is about 0.22; the total average projected area proportion value between the side window and the side view car body is about 0.16; the total average projected area proportion values between the rear window and the rear view car body is about 0.19.

b) The changes of the proportion values between the five hatchback series:

As can be seen from Table 5, the standard deviations of proportion 1, proportion 2 and proportion 3 between the five hatchback series are about 0.00,0.00 and 0.01 respectively (keep two decimal places) .This means that the projected area proportions between the car window and car body between the five hatchback series do not change too much.

c) The order of the projected area proportion values of the five hatchback series :

Relatively speaking, the projected area proportions between the front window and the front view body of the mainstream hatchbacks from large to small order is: Chinese series> Korean series = American series=Japanese series=European series;

The projected area proportions between the side window and the side view body of the mainstream hatchbacks from large to small order is: American series = Chinese series> Japanese series = Korean series = European series;

The projected area proportions between the rear window and the rear view body of the mainstream hatchbacks from large to small order is: Japanese series> American series = Chinese series = European series> Korean series.

References

1. R. Ranjan, A.R. Khan, C. Parikh, R. Jain, R.P. Mahto, S. Pal and D. Chakravarty, Journal of Manufacturing Processes, **22**, 237-253 (2016)
2. P.C. Oliveira, J.P. Moura and L.F. Fernandes, Computers and Electronics in Agriculture, **124**, 289-294 (2016)
3. H.W. Cui, Y.L. Yang and J.T.Li, Journal of Anhui Agricultural Sciences, **37**, 10761-10762 (2009)
4. Lu Fu and Li-ming Fu, Packaging Engineering, *Research of the technical factors of automobile modeling*, 167-172 (2008)
5. Zhen Wang and Zheng-yu Tan, Packaging Engineering, *Research on vehicle modeling features based on holidtic cognition*, 51-54 (2013)
6. Zhan Wei and Jiang-hong Zhao, Art and Design, *The proportion research of automobile form design*, 155-157 (2008)
7. Yu-mi Huang, China Computer & Communication, *Digital image processing technology and its application*, 144-145 (2016)

The novel cloud application technology with virtual platform

Ying-Chuan Chen[a] and Yi-Lang Tsai

National Center for High-Performance Computing, Taiwan

Abstract. Virtualization technology of cloud application service is the major research issue in recent years. Our virtual platform can obtain a lot of advantages base on virtualization technology with suitable virtual machine solution that all system resources can deployed quickly and managed competently, and it has some features including the high capacity for reliability, elasticity and customization. In our virtual platform, there are many different application services are provided for cloud service style. For system computing resources, we need to share all of system and software resources using the virtualization technology is presented in related sections, and we proposed virtual and remote application technology to implement innovative, convenient and powerful cloud application service base on virtual platform. For users of our platform, they can easy to use any software and tools on demand with pay per use mode anytime, and they do not have specialized knowledge and ability of install and maintenance.

1 Introduction

Cloud computing [1-2] is a well-known Information Technology (IT) concept, user can convenient access to complete computing resources, which include servers, networks, storage, software, applications and services. All of computing resources are provisioned and released with minimal management effort or service provider interaction. Cloud service is the access to computers and their function through three common deployment styles, such as public cloud, private cloud, and hybrid cloud. Users of the cloud request this access from a set of web service interfaces that operate a pool of computing resources base on the virtualization technology with virtual machines, network devices, storage, etc. Cloud computing solution has many benefits like easy to management, on-demand, customization high scalability, high reliability, pay-per-use, and cost down.

The related features are showed like: First feature is rapid elasticity, services can be elastically deployed and released. In some cases, to scale quickly inward and outward appropriate with demand automatically. For consumer, the capabilities available for provisioning often appear to be unlimited and can be appropriated in any quantity at any time. The second feature is resource pooling, all of computing resources are pooled to serve multiple consumers using a multi-tenant model, with different physical and virtual computing resources which dynamically allocated and reallocated according to each consumer demand. There is a sense of location independence in that the customer generally has no control and knowledge over the exact location of the provided resources but may be able to specify location at a higher level of abstraction (such as country, state, or datacenter). The third

[a] Corresponding author : ycc0301@narlabs.org.tw

feature is broad network access, capabilities are available over the network and accessed by standard mechanisms that promote use by heterogeneous thin client devices (including smart phones, tablets and laptops) from anywhere. The fourth feature is measured service, cloud platforms control and optimize resource automatically use by leveraging a metering capability at some level of abstraction appropriate to the type of service like processing, bandwidth, and storage. Resource usage can be controlled and monitored, providing transparency for both the provider and consumer of the utilized service. The fifth feature is on-demand self-service, anytime all of consumers can use physical or virtual computing resources, such as server time and network storage, as needed automatically without requiring human interaction through cloud service providers.

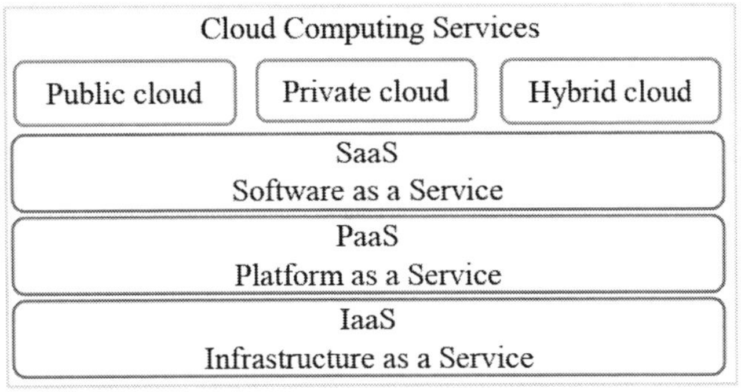

Figure 1. Cloud computing service.

The cloud computing service concept is shown in Figure 1, there are three types of the cloud service models [3-4] showed below:

- Software as a Service (SaaS): Related applications are available remotely. The SaaS services represent a software delivery method that provides access to software and its functions remotely as a web-based service. Providers offer users access to specific application programs controlled and executed on the provider's infrastructure. Among the providers we have Salesforce, Windows Live Mesh, GoogleApps, and facebook. For example, Salesforce is an industry leader in providing online CRM (Customer Relationship Management) cloud services. And Microsoft deliver a SaaS service that called Windows Live Mesh allows files and folders to be shared and synchronized across multiple devices from anywhere.
- Platform as a Service (PaaS): We can develop applications using the services provided. The PaaS services are used for applications, and other development, while providing cloud components to software. Developers gain by PaaS is a framework they can build upon to develop or customize applications. PaaS makes the development, testing, and deployment of applications quick, simple, and cost-effective. All applications and infrastructure are run and managed by services vendor. Google's App Engine is a well-known PaaS service, and it enables users to build Web applications on the same scalable systems that power Google applications.
- Infrastructure as a Service (IaaS): IaaS model are most common services in market, and it provides self-service models for remote accessing, monitoring, and managing datacenter infrastructures. Related equipment of IaaS is provided in the form of virtual machines. The client maintains the applications, runtimes, databases, server software while the supplier maintains the Cloud virtualization, hardware servers, networks and storage components. This service model allows the cloud to operate during high traffic and demanding situations as resources are dynamically increased as they are needed. Typical examples are Microsoft Azure, Amazon Web Services (AWS), Google Compute Engine (GCE) and S3 (Simple Storage Service) where compute and storage infrastructures are open to public access with a utility pricing model.

Virtualization Technology (VT) [5-6] is a very important role, it has the capability to deploy various virtual machines by a hypervisor. The concept for Virtualization Technology showed as Figure 2. A virtual machine (VM) is a software implementation of a machine that executes related programs like a physical machine. Related physical computing resources are divided as virtual computing resources in virtualization platforms via hypervisor, and those virtual computing resources are assigned to each VM for user's requirements. Hypervisor of VM is the most important component to control between physical resources and virtual resources. It also providing each Guest OS a set of virtual interfaces. All commands were delivered to hardware layer from virtual layer, which receives results from hypervisor of VMs. There have two different styles of virtualization for Guest OS, such as Full-Virtualization (FV) and Para-Virtualization (PV), which can be combined with Hardware-Assisted Virtualization technology for CPU of Intel or AMD. XEN

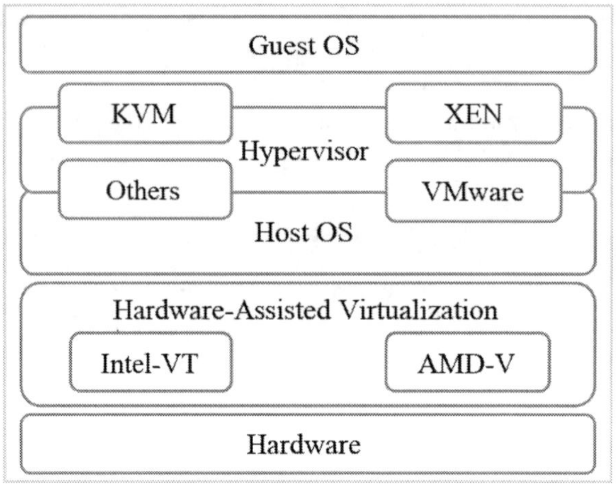

Figure 2. The architecture for virtualization technology.

The main goal of our paper is to develop an efficient, scalable and friendly Cloud Application Service on Virtual Platform that is scalable, efficient, and independent of the operating system. The Cloud Application Service adopts open architecture to provide virtualization applications with the heterogeneous operating systems to any client devices. Users can easily obtain virtualization applications in the cloud platform through any web browsers over Internet.

In cloud computing platform, users can use service of SaaS (software as a service) subscriptions instead of traditional software licenses model. In traditional system platform, users need to install operating systems and applications under a granted license. Users may be burdened with many complexes works in terms of software installation, configuration, updating and even troubleshooting issues. In cloud platform, users may access software on demand through the Internet without any installation and maintenance issues. For the system providers, there are two alternative models of making the SaaS service available. The first model is to develop software based on web technologies. This not only requires significant work, but may also encounter compatibility problems with the different browsers. The second model is based on the remote virtual applications that provides a transparent method to deliver software resources based on the virtualization technology. By virtualization technology, all cluster servers and the software which are maintained by the cloud providers and the users just only pay per use. Cloud service providers provide the resources sharing platform for accessing virtual applications, so users develop their programs and execute in the cloud platform.

The related technologies are showed in section II, and framework and implementation of the system is discussed in section III. Conclusion of our paper will be showed in section IV.

2 Technology explanation

The cloud application service is based on cluster architecture with the virtualizing application that is the most important technology. The virtual machine solution is installed with the cluster server. Each application shares the virtual system resources of the cluster platform. The cluster server is the important infrastructures with high speed network interfaces for the cloud computing technology. The main purpose of the cluster server is to formulate the responses which are sending by the clients, it also executes the high availability and load balancing while the jobs access to make the system reliable.

There have many virtual machine solutions exist for the deployment of open source clouds. Among these, it includes Eucalyptus, OpenNebula and OpenStack. We propose three famous solutions in order to select the most suitable solution to some conditions, such as open source, modularity, popularity, community use and friendly.

2.1 Eucalyptus

Eucalyptus [7] is an open source software platform for implementing Infrastructure as a Service (IaaS) in a cloud computing environment. It allows the installation of a private and hybrid cloud infrastructure. And it is developed in Java and C program language, with a main storage controller walrus and controllers on each node. The network is managed by the component cloud controller and each controller is authenticated by SSH key files and permission to authenticate transactions. There are some features like: (1) Eucalyptus works with various hypervisors including KVM, Xen and VMware. (2) It supports common Linux and Windows OS for virtual machines (VMs). (3) Compatible with Amazon Web Services (AWS) and Simple Storage Service (S3). (4) It is a cloud platform used by Amazon EC2, and the Application Program Interface (API) compatible with Amazon EC2 platform. (5) It can be installed and deployed from source code or DEB and RPM packages. (6) Internal processes communications are secured through SOAP and WS-Security. (7) Multiple clusters can be virtualized as a single cloud. (8) Administrative features such as user and group management and reports. However, Eucalyptus scalability is limited, compared to the massive scalability, and some source code of modules are closed.

2.2 OpenNebula

OpenNebula [8] is another well-known open source virtual machine solution, subject to the requirements of the Apache License version 2. It is developed using C++, Ruby and Shell program language, and it supports authentication via LDAP. The network component is managed manually by the administrator. Storage, the default copy of the virtual machines are accessed through SSH policy. It provides the simple flexible solution for the comprehensive management of virtualized data centers to enable private, public and hybrid IaaS clouds. For data center virtualization management, many users use OpenNebula to manage virtualization data center, consolidate servers, and integrate existing Information Technology (IT) resources including computing, networking and storage. OpenNebula directly integrates with hypervisors, such as KVM, Xen and VMware ESX. It also can complete control all of virtual and physical resources, and providing advanced capability for capacity management, high availability and resource optimization.

There are some features of OpenNebula showed below:

- Openness of the architecture, interfaces, and code.
- Flexibility to fit into any datacenter.
- Simplicity, easy to deploy, operate and use.
- Stability for use in production enterprise-class environments.
- Scalability for large scale infrastructures.
- Lightness for high efficiency.

2.3 OpenStack

The novel OpenStack project is recent and under active development. OpenStack project is licensed under Apache 2.0 license.

Instance Name	Image Name	IP Address	Size	Key Pair	Status	Availability Zone	Task	Power State	Time since created	Actions
vm3	CentOS7	• 10.0.0.13	m1.medium	demo	Active	nova	None	Running	10 months, 1 week	Create Snapshot ▾
win7_3	win7_ultimate	• 10.0.0.8 Floating IPs: • 172.24.4.229	win7	-	Active	nova	None	Running	10 months, 4 weeks	Create Snapshot ▾
win7_2	win7_ultimate	• 10.0.0.7 Floating IPs: • 172.24.4.238	win7	-	Shutoff	nova	None	Shut Down	10 months, 4 weeks	Start Instance ▾
win7	win7_ultimate	• 10.0.0.5 Floating IPs: • 172.24.4.228	win7	-	Active	nova	None	Running	10 months, 4 weeks	Create Snapshot ▾
vm1	-	• 10.0.0.9	m1.tiny	-	Shutoff	nova	None	Shut Down	11 months	Start Instance ▾

Figure 3. Web user interface of OpenStack's instance.

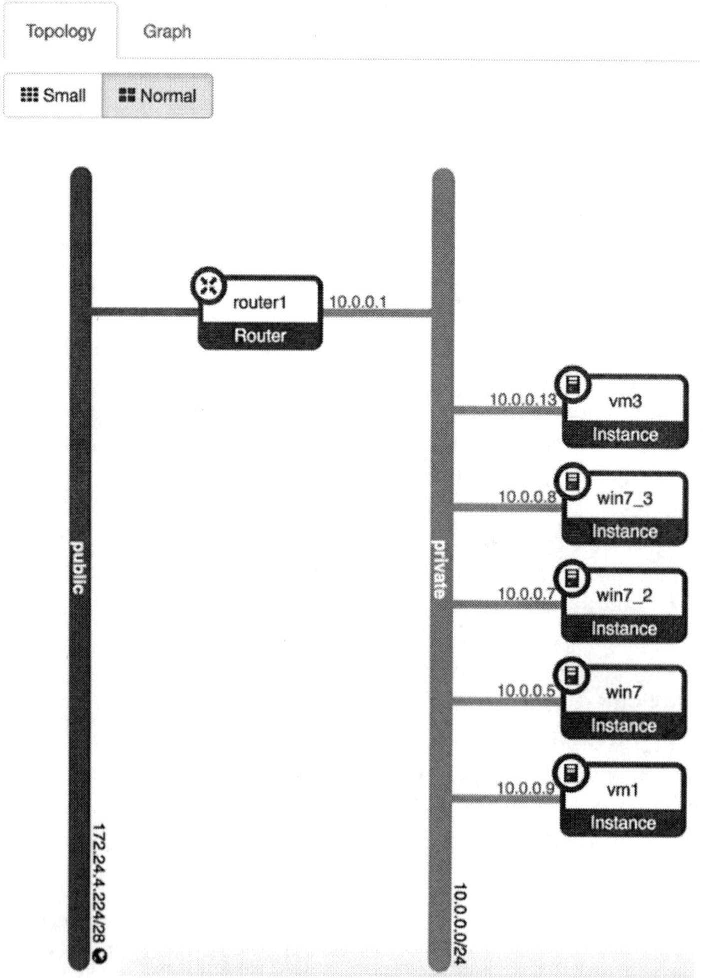

Figure 4. Web user interface of OpenStack's network.

In Figure 3 and Figure 4 show the web user interface of OpenStack, it is a virtualization platform developed by National Aeronautics and Space Administration (NASA) dedicated to massive infrastructures.

There are three main features showed below:

- Open: OpenStack is based on open source concept, the all of code can be modified and adapted if necessary. And there is a validation process for the adoption and development of new standards in the OpenStack project.
- High scalability: This solution is already deployed in a lot of companies whose associated data volumes is evaluated in petabytes of distributed architectures and massively scalable over 1 million physical machines, over about 60 million virtual machines and billions of stored objects.
- High flexibility and compatibility: This solution supports common virtualization hypervisors of the market: KVM, Xen, QEMU, Hyper-V, ESX, etc.

OpenStack solution is written in python program language and currently implements two control APIs, the EC2 API and Rackspace, and it is completely open source. OpenStack execute suitable drivers to interface with many hypervisors (KVM, Xen, QEMU, Hyper-V, and ESX). This project is dedicated to providing the computer industry with the opportunity to deploy a hosting architecture and massive scalability. It has great potential due to its architecture and community and the support of its partners. According to many-sided survey appears that the best virtualization solution is OpenStack.

Application virtualization [9-12] is layered on top of virtualization technologies. Application virtualization is the separation of an installation of an application from the client computer that is accessing it. Application virtualization can be divided into two types: remote and streaming. Remote applications run on a server. End users view and interact with their applications over a network via a remote display protocol. The remote applications can be completely integrated with the user's desktop so that they appear and behave the same as local applications. The remote applications can be shared with other users, or the application can be running on its own operating system instance on the server. With streaming applications, the virtualized application is executed on the end user's local computer. When an application is requested, components are downloaded to the local computer on demand. Once completely downloaded, a streamed application can work without a network connection. Both forms of application virtualization have benefits from centralized management. Applications can be installed, patched, and upgraded once for an entire environment. Application virtualization typically is combined with application streaming, so the software is not installed in the usual sense. Application virtualization allows software to be run without being installed on the machine and without interacting directly with the original operating system for which it was designed. Application streaming is an on-demand software distribution system. With application streaming, users receive only applications that they request from a master server. In addition, clients are given only the portion of the application that they need to launch the software and use its basic functions [13-15].

The virtual application is different from the virtual desktop sharing. In virtual desktop sharing, a server distributes any screen update. In virtual application sharing, the server distributes screen updates if they belong to the shared application's windows. Client users receive screen updates from the server and send keyboard and mouse events to the server. Application virtualization is also a software technology that encapsulates applications from the underlying operating system on which it is executed. With application virtualization, users can store and manage applications centrally, integrate with third party computer lifecycle management and software distribution systems. Application virtualization also enhances organizations' ability to control access to applications, track usage of virtual applications, and quickly test, deploy and update applications.

Full application virtualization requires a virtualization layer. The application virtualization layers are used to substitute part of the runtime environment which is provided by the operating system. The layer acquires registry operations of virtualized applications and transparently redirects them to a virtualized location. It makes the real application remaining to access the virtual resources. The pieces of the application's code, data, and settings are delivered when first used, and the application streaming are usually delivered over a protocol such as HTTP (Hypertext Transfer Protocol), CIFS (Common Internet File System) or RTSP (Real Time Streaming Protocol) [16-18]. Deploying and maintaining desktop and application level virtualization poses many challenges, but there are opportunities for improvements at the application and server layers to achieve the high-level of computing performance and scalability [19-20].

3 System design and implementation

In order to implement cloud application service, first we need to build virtual platform base on virtualization technology of OpenStack. Our virtual platform architecture is built using several core components: Identity, Compute, Image, Network and Object. In this section, we will explain in detail these components and we will demonstration the system architecture for our platform in Figure 5.

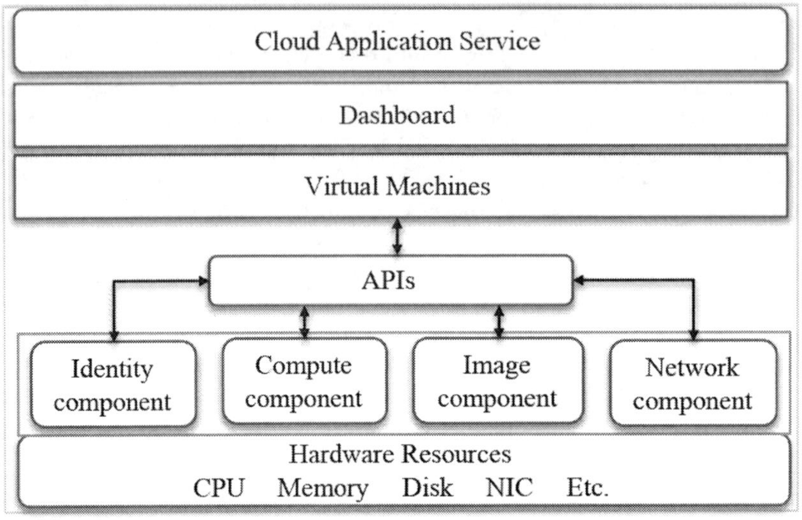

Figure 5. The system architecture for our platform.

3.1 Identity component

It acts as the key authentication mechanism of our virtual platform across the cloud operating system and can integrate with existing backend directory services like LDAP. It supports multiple forms of authentication including standard username and password credentials, token-based systems and AWS-style (i.e. Amazon Web Services) logins. Additionally, the catalog provides a queryable list of all of the services deployed in a cloud in a single registry. Users and third-party tools can programmatically determine which resources they can access.

3.2 Compute component

It is a management platform that controls the infrastructure to control IaaS clouds. It is designed to manage and automate pools of computer resources and can work with widely available virtualization technologies, as well as bare metal and high-performance computing (HPC) configurations. KVM, Xen, and VMware are available choices for hypervisor technology, as well as Virtual Machine Monitor (VMM), together with Hyper-V and Linux container technology such as LXC. It provides an administrative interface and an API needed for the orchestration of the Cloud. It includes instances management for servers and access control. Compute requires no prerequisite hardware and is completely independent of the hypervisor consists of seven main components. At first, we have the API Server, which acts as the Web front end service control of the hypervisor. Then, we find the Message Queue that implements the mechanism for dispatching the exchanged instructions to facilitate communication. After that, Compute Controller handles the lifecycle of instances, it is responsible for creating and manipulating virtual servers, while the component Object Store provides storage services.

3.3 Image component

It provides storage services, recording and distributing the images to virtual machine disks. The API of image component provides a standard REST interface for querying information about disk images and lets clients stream the images to new servers. An advance function of this component is the live migration of a running VM, from one physical server to another, without service interruption. Thus, it enables a dynamic and automated self-optimizing virtual platform, allowing hardware maintenance for the underperforming servers without downtimes. Also, it can present detail information about the images, and configure a variation on an image to produce an instance. We can add, delete, share, or duplicate images of VM through image component.

3.4 Network component

It is a system for managing networks and IP addresses. This component ensures the network is not a bottleneck or limiting factor in a cloud deployment, and gives users self-service ability, even over network configurations. It provides network models for different applications or user groups. Standard models include flat networks or VLANs that separate servers and traffic. It also manages IP addresses, allowing for dedicated static IP addresses or DHCP. Floating IP addresses let traffic be dynamically rerouted to any resources in the IT infrastructure, so users can redirect traffic during maintenance or in case of a failure. Besides, users can create their own networks, control traffic, and connect servers and devices to one or more networks. Administrators of platform can use software-defined networking (SDN) technologies like OpenFlow to support high levels of multi-tenancy and massive scale. It provides an extension framework that can deploy and manage additional network services—such as intrusion detection systems (IDS), load balancing, firewalls, and virtual private networks (VPN).

3.5 Dashboard component

It provides administrators and users with a graphical interface to access, provision, and automate deployment of virtual computing resources. The design accommodates third party products and services, such as billing, monitoring, and additional management tools. The dashboard is one of several ways users can interact with virtual machine resources.

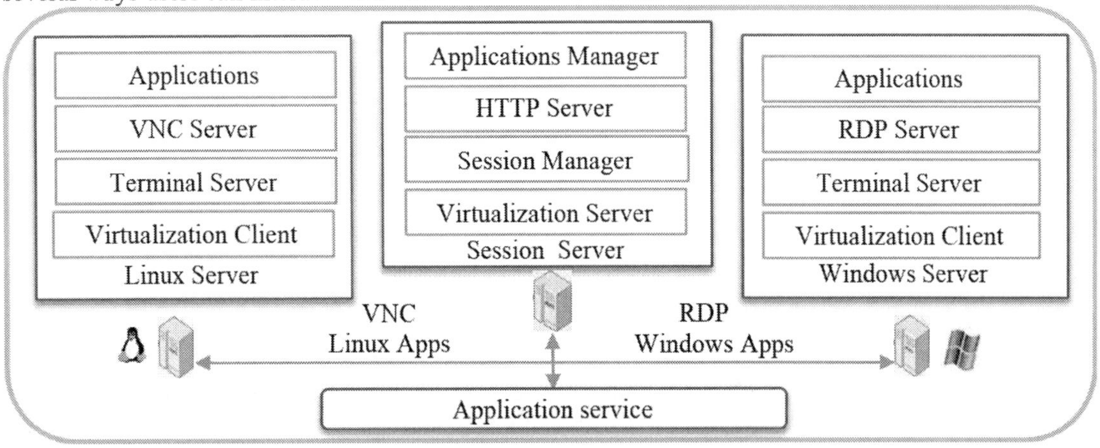

Figure 6. The cloud application service architecture.

The cloud application service architecture is showed as Figure 6. The cloud application service system is developed under cluster servers to provide more application services and reduce the software maintenance costs. Applications and shared data can be installed on the cluster server and its hypervisors, and users can request for some specific applications or data on demand from the server. The requested virtual applications will be delivered with dynamic streaming technologies to the client devices.

There are two ways to execute the virtual applications on the client devices: (1) Native mode: The remote control and streaming display software will be installed on the client device to execute the virtual desktops and applications. It is suitable for the homogeneous operating system between the server and clients. (2) Web-based mode: Only need a web browser with Java applet to install on the client device. It is suitable for the heterogeneous operating system between the server and clients.

The virtual application system has a session server to perform in the role of virtualization controller, and to manage user's connections and accountings information. The session server can be implemented with Linux or Windows operating system, and it maintains the virtualization, session and application manager functions. It is responsible for virtualizing the applications and sends them to the proper sessions for the application servers. The session server also communicates with application servers to deliver the virtual application streaming to client devices by RDP technology. Besides, the session server uses the HTTP server to provide the single web-based portal for user login. It also enhances the system security and elasticity by applying the centralized control of applications and files on the server cluster.

When a user requests the application, which installed on the Linux application server, the session server will manage the user account and its session and communicate with Linux application servers. When the Linux application server receives the request from session server, it will start VNC service and make the application virtualize to streaming data, and then delivers the data for the client user.

The virtual application system is composed of several modules: Application Virtualization Manager, User Session Manager and Data Synchronization Manager. The functions of each module are described as follows:

- Application Virtualization Manager: This module provides virtualization of applications and software resources. The development of this module is based on the open virtual application software, and designed according to the user requirements. The Full application virtualization also requires a virtualization layer with the operating system. Application virtualization layers replace part of the runtime environment normally provided by the operating system. The layer intercepts all file and registry operations of virtualized applications and transparently redirects them to a virtualized location.
- User Session Manager: This module is responsible for managing user connection sessions and authenticating accounts information. The user session begins when the user accesses the virtual desktop or application and ends when the user quits the virtual desktop or application from the web browser. It also plays a role of bridge between application servers and client devices. It applied the SSH (Secure Shell) and HTTP (Hyper Text Transfer Protocol) protocol to provide the single-entry website.
- Data Synchronization Manager: This module is responsible for managing the process of establishing consistency among data from cluster servers to client devices and the continuous harmonization of the data over time. Users and project developers can collaborate or develop on a single file without installing any relative application on their own client device. It also backups and shares the user's data to make the system reliable and elastic.

To enable collaborations among multiple virtual machines, an application sharing and migration mechanism will be applied. Through presentation streaming redirection and virtual machine cloning technology, an application can be easily shared or migrated. To realize remote application access in a cloud, RFB protocol is used to transfer the virtual application of a remote virtual machine. The RFB protocol works at the buffer frame layer and supports the remote access to graphical user interfaces, and the mouse or keyboard inputs can be transferred to the remote application, thus achieving a transparent access to the applications. In such a presentation streaming based software delivery mode, when a client wants to migrate or share an application to another client, the presentation streaming of this application should be redirected and the corresponding virtual machine will be cloned in case of application sharing.

The virtual application system supports both multicast and unicast transmissions. For unicast connections, either UDP or TCP can be used. Since TCP provides reliable communication and flow control, it is more suitable for unicast sessions. Multiple TCP clients sharing a single application may have different bandwidths, so an algorithm which sends the updates at the link speed of each client will be developed. For UDP clients, the system controls the transmission rate because UDP does not provide flow and congestion control. Several simultaneous multicast sessions with different transmission rates can be created at the system. The system can share an application to TCP clients, UDP clients, and several multicasts addresses in the same sharing session.

The application virtualization technology enables applications to be streamed from a centralized location into an isolation environment on the target device where they will execute. The application files, configuration, and settings are transmitted to the target device and the application execution at run time is controlled by the user session manager and virtualization layer. The virtual application system enables collaborative works, and provides an efficient, reliable and elastic service platform for cloud-based computing environment.

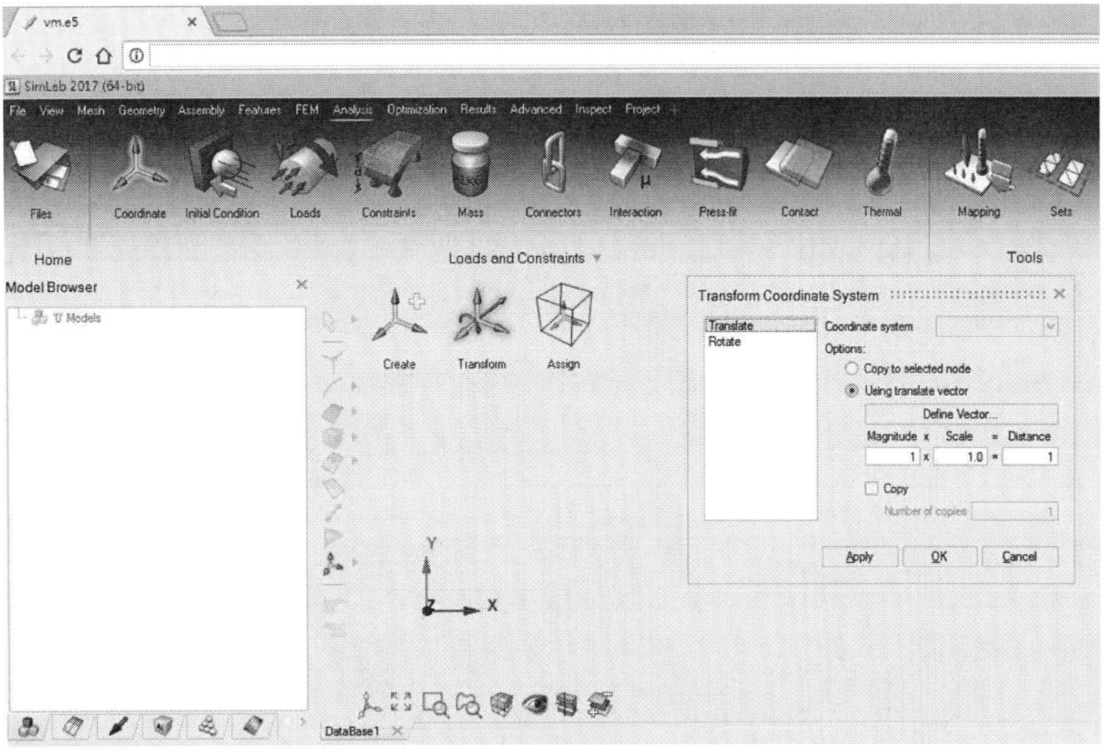

Figure 7. The HyperWorks is a cloud application services in our platform.

The cloud application services show as Figure 7 and Figure 8, user can use cloud application like HyperWorks software on our platform by any common web browsers including Internet Explorer, Google Chrome, Mozilla Firefox and Microsoft Edge. In Fig.7, the HyperWorks is a famous Computer-Aided Engineering(CAE) simulation software in the industry. The user interface of ProCAST software is shown in Fig.8, an advanced and complete casting processes simulation tool. In our platform, we can provide many software and tools to achieve user's requirement, even achieve customization function if necessary. For users of platform, they can easy to use any software and tools on demand with pay per use, and don't have related knowledge of install and maintenance anytime.

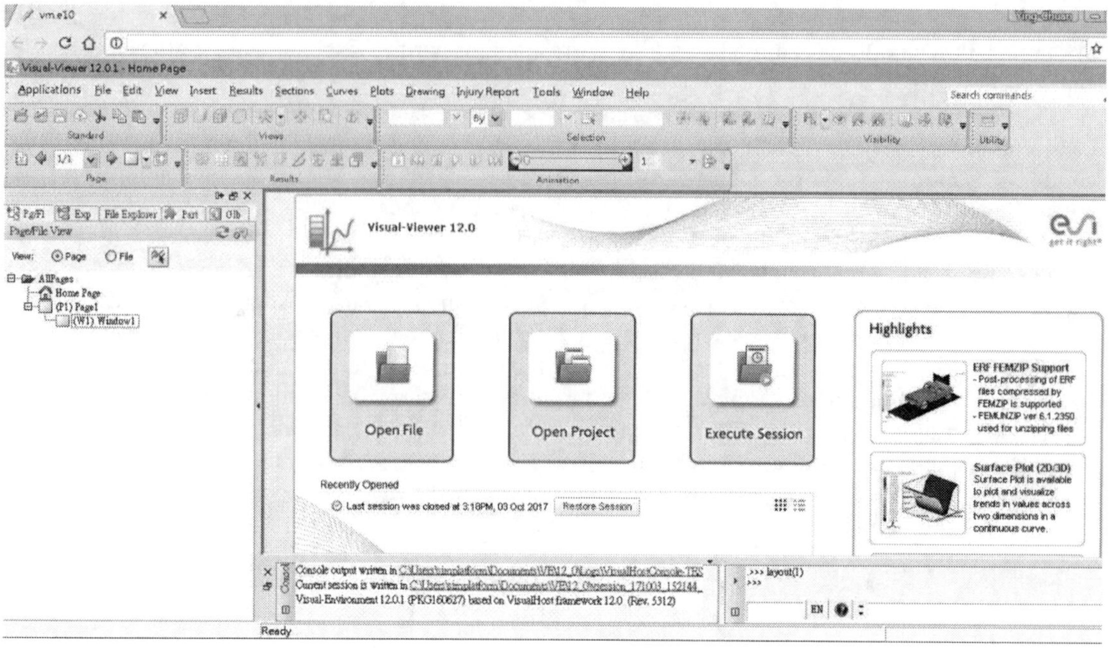

Figure 8. The ProCAST is a cloud application services in our platform.

4 Conclusion

In recent years, the cloud computing with virtualization technology is the best solution for the increasing efficiency of hardware and software. Virtualization technology acts a very important component to share and control all of virtual system resources in our cloud platform. In our paper, the development of cloud application technology with virtual platform will integrate related cloud computing technologies including IaaS, PaaS and SaaS to provide the innovative application services. Users have not need to worry about the costs of software license and maintenance of servers in our platform. They just only need some easy operations to obtain the various application and software resources on demand everywhere and anytime. In addition, all cloud application services are charge by the pay per use mode. By developing the cloud application technology, more innovative application services of cloud computing can provide to users, and that make the cluster server as a high economic benefits platform.

References

1. D.C. Marinescu, Cloud computing: theory and practice (2017)
2. J.W. Rittinghouse and J.F. Ransome, Cloud computing: implementation, management, and security (2016)
3. P. Mell and T. Grance, The NIST definition of cloud computing (2011)
4. I.A.T. Hashem, et al., Information Systems, **47**, 98-115 (2015)
5. R. Mijumbi, J. Serrat, J.L. Gorricho, N. Bouten, F. De Turck and R. Boutaba, IEEE Communications Surveys & Tutorials, **18**(1), 236-262 (2016)
6. A.S. Ibrahim, J. Hamlyn-Harris and J. Grundy, "Emerging Security Challenges of Cloud Virtual Infrastructure", *Proc. of The Asia Pacific Cloud Workshop (co-located with APSEC 2010)*
7. N. Singh and V. Dhir, International Journal of Advanced Research in Computer Science, **8**(4), 10-13 (2017)
8. A. Vogel, et al., *In: Parallel, Distributed, and Network-Based Processing (PDP), 2016 24th Euromicro International Conference on. IEEE*, 672-679 (2016)

9. Z. Shen, et al., International Journal of Modeling, Simulation, and Scientific Computing, **8**(2) (2017)
10. D. Liu, et al., *In: Audio, Language and Image Processing (ICALIP), 2016 International Conference on. IEEE*, 766-770 (2016)
11. L. Nema, A. Sharma and S. Jain, International Journal of Engineering Science and Computing, **6**(6), 7463-7468 (2016)
12. L. Yan, IEEE Communication Software and Networks ICCSN (2011)
13. K. Muthu Pandi and K. Somasundaram, Indian Journal of Science and Technology, **9**(11), 1-8 (2016)
14. J. Li, J. Huai, C. Hu and Y. Zhu, Elsevier, Information Sciences, **180**(17), 3086-3107 (2010)
15. F. Huang and D. J. Beveridge, Content or File Based Application Virtualization (U.S. Patent Application No. 15/178, 324., 2017)
16. T. Ilmonen, J. Giulio and M. Petri, Application sharing between devices in proximity to each other (U.S. Patent No. 9, 654, 611., 2017)
17. M. Chen, J. F. Liu and J. X. Ying, Application sharing by dynamic partitioning (U.S. Patent No. 9, 769, 516., 2017)
18. I. PSARAS, et al., *In: Proceedings of the Workshop on Mobility in the Evolving Internet Architecture, ACM*, 1-6 (2016)
19. Aihua Liang, Zhuo Liu, Limin Xiao, Li Ruan, *In: IEEE International Conference on Future Computer and Communication,* 591-594 (2010)
20. Alexander Gunka, Stepan Seycek and Harald Kühn, *In: Proceedings of the international workshop on Multi-cloud applications and federated clouds, ACM,* 35-42 (2013)

Big data analytics - towards the enrichment of content tourism for revitalization of Japanese rural area

Ali Bakdur[a], Fumito Masui and Michal Ptaszynski

Department of Computer Science, Kitami Institute of Technology, 090-8507 Kitami, Japan

Abstract. Japan's domestic travel and tourism industry expenditure has been declining gradually since 1998 (from 33.5 in 1998 to 21.6 trillion JPY in 2016). Our research purpose is to construct a data analysis model to transform the collected data to a meaningful graphical format by using big data analytics techniques to discover anomalies and sustainable development possibilities for economy and tourism of Japan's rural areas, with a particular focus on the prefecture of Hokkaido, subprefecture of Okhotsk. To strengthen the reliability of this model we apply popular Monte Carlo simulation combined with Bayesian statistic and implement it on an Apache Spark platform to acquire results within the span of the study. Through this research, we focus on observing and analyzing interests, expectations and tendencies of Japanese people living in rural areas. From such collected information, we can obtain reasons for the decline of this sector's impact on Japan's economy. Measuring public awareness has become more efficient since the content generator role has been passed on to ordinary people. Therefore, the analysis of Big Data with the use of data science techniques has become important to comprehend human behavior from multiple points of view, including the scientific, economic, political, historical and sociological.

1 Introduction

Following the commercial success of world's first generation jet airliners in the second half of the 1950's, passenger transport has begun to play a major role in the development of economy and globalization. Over the past six decades, tourism has experienced continued expansion and diversification to become one of the largest and fastest-growing economic sectors in the world. International tourist arrivals had increased from 25 million globally in 1950 to 278 million in 1980, 674 million in 2000, and 1186 million in 2015. Moreover, the number of global domestic tourists is estimated to attain 5 to 6 billion people as of 2015 [1].

There are several technical definitions of tourism but at this point, we will apply the definition of tourism considering the economy as a parameter, namely:

"Tourism can be defined as the science, art, and business of attracting and transporting visitors, accommodating them and graciously catering to their needs and wants." [2]

This economic approach to the definition of tourism is also appropriate in the case of Japan on which we focused.

[a] Corresponding author : alibakdur@ialab.cs.kitami-it.ac.jp

Japan is a unitary island country with a population of 126.8 million people. Although it's comparatively large population, Japan has a rapid population aging problem and a negative population growth rate. It is one of the most urbanized countries in the OECD, as 57% of its population live in predominantly urban areas [3]. Although it has almost 25 years of slow economic growth [4] (The notorious 'Japanese asset price bubble', an economic bubble in Japan which burst by the end of 1991, can be given as the reason for the country's economic decline), with 520 trillion JPY (4.8 trillion US Dollars) GDP nominal value in 2017 [5], Japan is still the 3rd largest economy in the world. Japan is also the first very highly developed country in Asia.

In addition to the above information regarding Japan, travel and tourism industry generated a total impact of 38 trillion JPY (343 billion US Dollars) of Japan's GDP in 2016. It makes Japan the 3rd country in the world regarding the total contribution of travel and tourism to the GDP after the United States and China. In view of Japan's island nature, 87.9% of the total contribution of the tourism industry to the GDP consists of domestic expenditure. Based on its direct, indirect, and induced GDP impact, travel and tourism generated 7.4% of Japan's GDP in 2016. This is nearly half the impact of the automotive manufacturing sector at 16.5%. Travel and tourism sustained a total of 4.5 million direct, indirect and induced jobs in Japan in 2016 [6].

It is obvious that Japan has a well-developed domestic travel and tourism industry. Its total contribution to Japanese GDP is making it one of the most influential sectors. However, both the country's GDP and the participation of the tourism industry in it had been declining gradually since 1998. Although Japanese domestic tourism industry has been in decline comparing to other countries, it is still on a very high level in 2016, but depending on the historical data it has lost great economical value since 1998 (from 33.5 in 1998 to 21.7 trillion JPY in 2016) [7].

A correct comprehension of the situation requires detailed knowledge and data analysis, which depends on scientific, economic, political, historical and sociological aspects. By the start of the digital era around 2002, the quality and quantity of available data have started to increase dramatically. Data have never before been as valuable as today. Not only data but also efficient processing of it has gained great importance. Nowadays many organizations [1, 3-9] have started to share their data with the public, which allowed us to collect required data samples for our research.

2 Domestic tourism overview of Japan

2.1 General overview of Japan

Simultaneously with the decline of Japanese domestic travel and tourism expenditure, the participation rate in domestic tourism activities in Japan is also in decline between 1986 and 2006. Participation in sightseeing has dropped from 65.9 % to 49.3 % of the overall population [8]. Our understanding from the statistical data is that today in Japan fewer people than in the past are traveling for leisure. Perhaps the reason for this is that people feel more comfortable in their usual habitat. Nevertheless comparing only historical statistical data values from past and present in a frequentist way may not give the correct comprehension of this phenomenon. The consideration of living habits changes throughout the time, the development of popular technologies such as the Internet, communication, home entertainment systems etc. is essential for a better interpretation and inference of the real state of Japanese daily routine and understanding of their insight on leisure.

As explained in the previous paragraph the necessity of interpreting different terms of time with several different kinds of parameters brought us to the Bayesian inference approach. From our point of view, the usage of Bayesian statistic, which we implemented in the technical part of this study to prove our concept, has given a better conceptual quality of probability.

2.2 Okhotsk subprefecture of Japan

To give an adequate understanding of human experience in the current situation in Japan, we are using Okhotsk subprefecture's statistical data as samples throughout this study. The Okhotsk subprefecture of Hokkaido, second largest island of Japan, consists of 3 cities: Abashiri, Kitami, and Monbetsu. It gets its name from the Sea of Okhotsk. The climate is much colder than in other parts of Hokkaido during the winter. The area is also famous for drift ice sightseeing. Agriculture is an important sector. Compared with other regions in Japan, the density of the population in Okhotsk subprefecture is low.

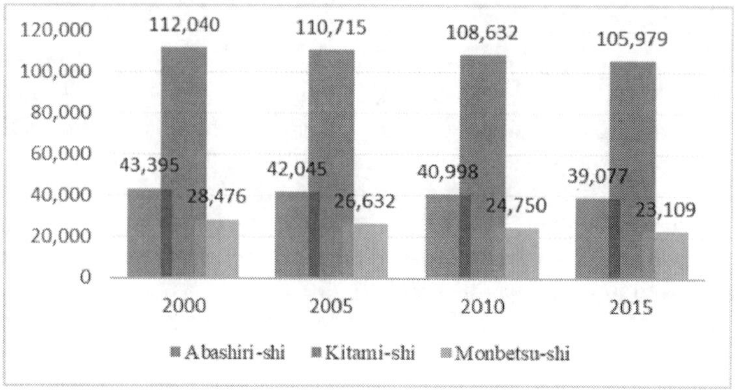

Figure 1. Population Census in Okhotsk (2000 -2005 – 2010- 2015) [9]

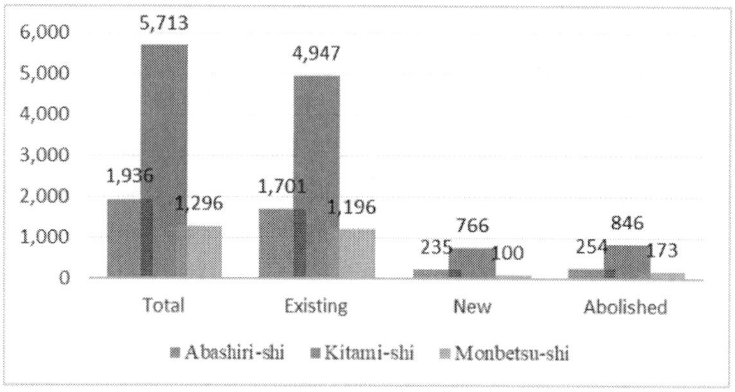

Figure 2. Establishment's Existence in Okhotsk (2014) [9]

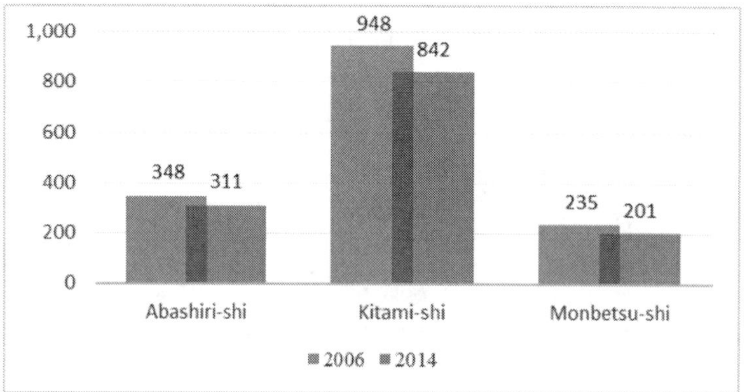

Figure 3. Accommodation, Food Service Businesses in Okhotsk (2006-2014) [9]

In this analysis, the character of Okhotsk statistical data, which is shown respectively in Figure 1, 2 and 3, indicates similarities with Japan's general characteristics. Okhotsk also has a negative population growth rate, but in contrary to the rest of the country all business establishments including tourism-related businesses are disappearing.

3 The purpose of applying the Bayesian approach to this study

Tourism information science combines several kinds of technology and knowledge, such as data science, or Big Data. In addition to a better understanding of the mechanisms behind various kinds of sightseeing spots, the data, from historical statistics and/or current streaming, will be processed in this research using data science approach. A statistical model will be produced by Monte Carlo simulation combined with Bayesian statistics and the result will be displayed with probability distribution [10].

Distributed computing with Apache Spark is essential for the performance issue. The principal working steps of this model will be revealed after conducting a preliminary research focused on tourism and demographic indicators in chronological statistic and prebuilt survey data in Okhotsk subprefecture of Japan. This observation is required to build a reliable model of local community characteristics and expectations.

The next step will be formulating a mathematical structure model with the detected factors and parameters relying on the previous studies. The analysis of data related to continuous events (social media networks) will be a part of the next phase for detecting present-day tendencies of the community.

We aim at collaborating with the local authorities in order to provide a better tomorrow for the community and to improve their quality of life-based on scientific proof earned through the completion of this study.

3.1 Bayesian statistics

The Bayesian statistics is a mathematical procedure that applies probabilities to statistical problems. It provides users with the tools to update their beliefs in the evidence of new data. If the occurrence of any incident depends on a variety of conditions then Bayesian interference combined with Monte Carlo method would be appropriate as an observation technique.

An important part of Bayesian inference is the establishment of parameters and models. The main mechanism is feeding the analysis model with repeated random sampling inputs to get numerical output. We assume that a repeating sampling input, depending on modeling complexity, may require high processing power and time, which could be achievable with Apache Spark components.

3.2 Monte Carlo method

Monte Carlo methods can be used to solve any problem having a probabilistic interpretation. A very powerful class of Monte Carlo techniques is the so-called Markov chain Monte Carlo (MCMC) algorithms. Monte Carlo methods are often necessary for the implementation of optimal Bayesian estimators. Monte Carlo methods are mainly used in three distinct problem classes [11]:

- Optimization,
- Numerical integration,
- Generating draws from a probability distribution.

3.3 Apache Spark

Apache Spark is an in-memory framework that is an alternative to MapReduce. Spark has a higher data processing engine for large datasets and in-memory computing. It can run programs up to 100 times faster than Hadoop MapReduce. The Spark core is designed to scale up from one to thousands of nodes. Spark powers a stack of libraries including SQL and Data Frames, MLlib for machine learning, GraphX, and Spark Streaming. It is possible to combine these libraries within the same application [12]. We used IBM Big Data University Virtual Lab Environment for faster processing [13].

3.4 PyMC

PyMC is a Python package for Bayesian statistical modeling and Probabilistic Machine Learning, which focuses on advanced Markov chain Monte Carlo and various fitting algorithms. Its flexibility and extensibility make it applicable to a large array of problems [14].

4 Case study and solution steps

Before proceeding to the case study assignment we should mention that an important element of statistical knowledge and eventually probabilistic program development are random variables. Random variables are basically divided into two sub-categories:

- Observed Random Variables,
- Unobserved Random Variables

A data scientist who disposes of a dataset sample can use observed random variables that can come in handy in likelihood distributions. On the other hand, unobserved random variables have no sample dataset values and are defined by prior distributions.

If we define random variables more specifically in terms of probability programming and probability distribution aspects, we need to mention at least two types:

- Discrete Random Variables

Discrete random variables find usage alongside the probability mass function when the output of the function data value is finite numbers. Examples include discrete uniform distribution, Bernoulli distribution, binomial distribution, Poisson distribution, and geometric distribution, etc.

- Continuous Random Variables

Continuous random variables are defined by a probability density function and the output of the data value can be defined as infinite numbers. Examples include normal distribution, uniform distribution, beta distribution and gamma distribution, etc. In the following steps, we will explain the usage of this information in more detail.

4.1 Problem statement

Japanese domestic travel and tourism industry has been declining gradually since 1998. What are the expectations for its future?

4.2 Data

We obtained Japanese domestic travel and tourism expenditure data for the years in between 1995 and 2015 (local currency in trillion JPY), which we then used for this analysis [7].

Table 1. Travel and Tourism expenditure/year data in trillion JPY by year (1995- 2015)

Year	Value	Year	Value	Year	Value
1995	26	2002	31	2009	24
1996	29	2003	29	2010	22
1997	32	2004	28	2011	21
1998	34	2005	27	2012	21
1999	32	2006	28	2013	20
2000	32	2007	26	2014	22
2001	32	2008	26	2015	22

As seen in Table 1, we have some prior states dataset for intuitive perspective. Since we have a sample dataset it makes our case suitable for defining observed random variables.

Using the bivariate numerical data in Table 1, we create the following regression line graphic with a negative linear relationship in Figure 4.

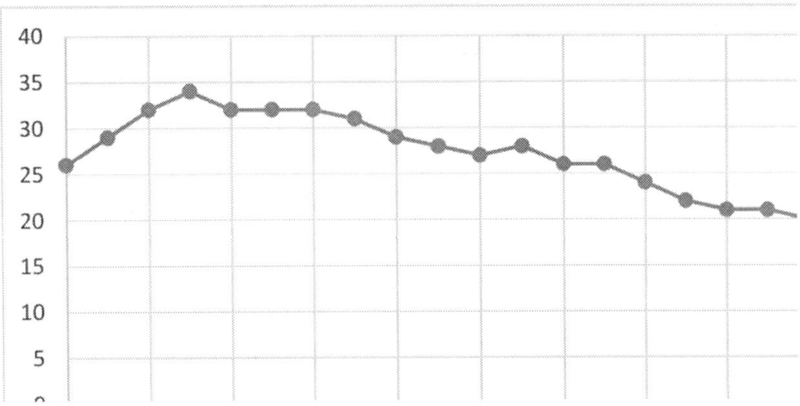

Figure 4. Travel and Tourism Expenditure (trillion JPY) /Yearly regression line graph.

4.3 Model

One big advantage of the Bayesian interpretation is that it can be used to model our uncertainty about events that do not have long-term frequencies [15]. In our case, we have a dataset with prior beliefs which gives us observed random variable opportunity. Our dataset values are also discrete and their value is an integer. This makes it possible for us to use the probability mass function for modeling. For this reason, we used Poisson distribution for probability mass function which is the best fit for our purpose.

For Poisson distribution (1), we can say that:

- x events that occur at times.
- λ is event occurrence rate.
- e is the constant number equal to 2.71828… (Euler's Number)

$$Poi(x|\lambda) = e^{-\lambda}\frac{\lambda^x}{x!}, x = 0,1,2\ldots \quad (1)$$

From the equation, the probability is dependent on λ. The value of λ determines the shape of the probability distribution. Two very different λ s will produce two very different distributions. We can plot a Poisson distribution with two different values of λ [13]. Observation of Figure 4, can help us to determine a turning point in the time period of event occurrence if there is any. In other words, if there is any unusual behavior on the observed random variables regression line graph the mean of which is our distribution governed by two different values of λ one before the turning point and one after the turning point.

```
import pymc as pm

def tourism_model(data):

    alpha = 1.0 / data.mean()
    lambda_1 = pm.Exponential("lambda_1", alpha)
    lambda_2 = pm.Exponential("lambda_2", alpha)
tau = pm.DiscreteUniform("tau", lower=0,
      upper=len(data))
observation = pm.Poisson("obs", lambda_,
              value=data.values, observed=True)
 model = pm.Model([observation, lambda_1, lambda_2, tau])
```

Figure 5. Sample Python code block responsible for the building of a probabilistic model with our data.

While the concept is complicated, to make it simple we initially attempted to solve this problem using Python programing language and its probability programming specified package- PyMC. PyMC allows building a probabilistic model in a short time. With the given sample Python code block as seen in Figure 5, we can define the explained model in a short time. τ (tau) represents the turning point which we are looking for.

4.4 Inference

After creating a model definition we can start to use the MCMC sampling algorithm.

- In the first step, we are introducing our observed data set to the model.
- In the second step, we are assigning the MCMC (Markov chain Monte Carlo) fitting algorithm for the sampling of our model. With MCMC algorithms, we are sampling our model 50000 times with random values.
- At the end of the sampling process, we got the required set of $\lambda1$, $\lambda2$, and τ (tau) data.

Next, a code inside the model creates a graph set for the interpretation of this problem. The key purposes of our research to provide a more user-friendly output for people in our community that do not have a technical background.

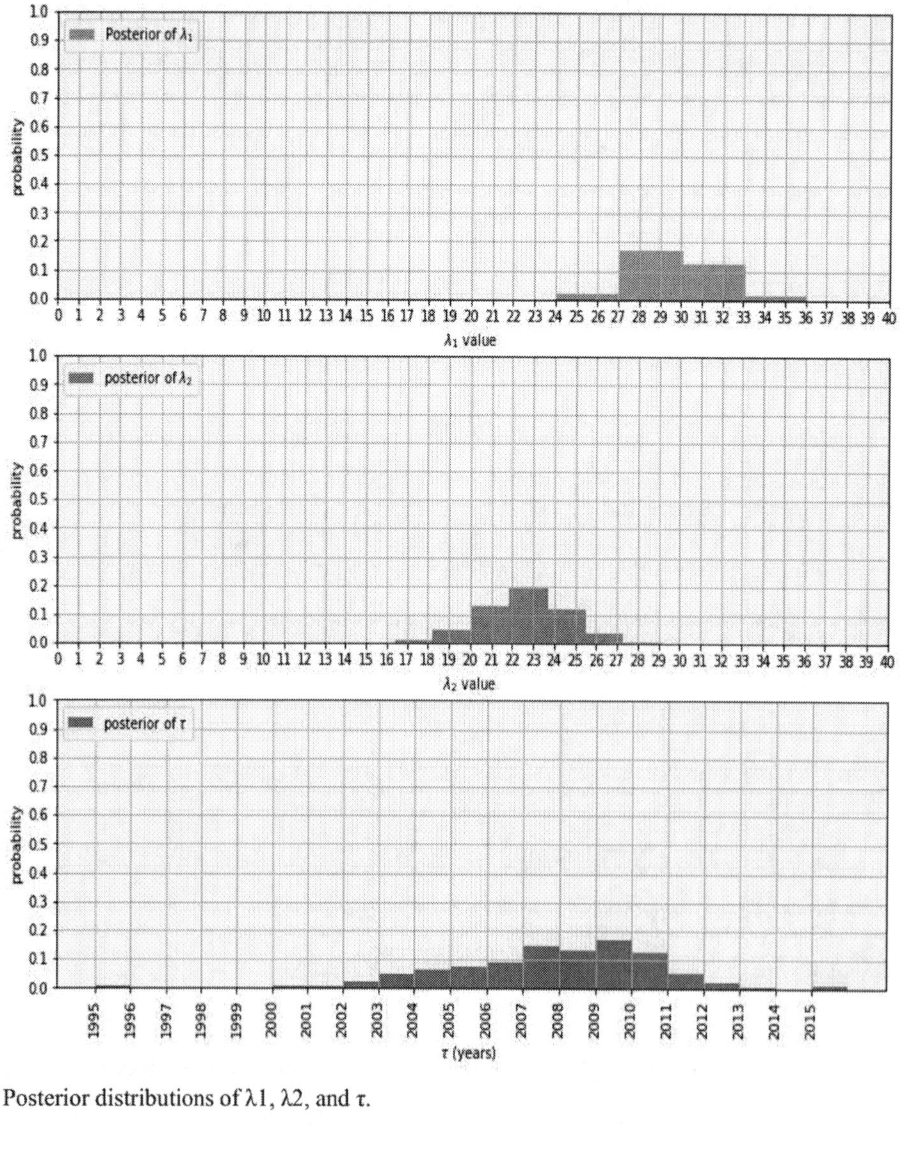

Figure 6. Posterior distributions of $\lambda1$, $\lambda2$, and τ.

4.5 Result

In Figure 6, the first graph λ1 shows the distribution before τ, while the second graph shows λ2 after τ. It should be noted that τ graphs have not had any certain turning point because of our prior data showing a continuous decline of the domestic travel and tourism expenditure. Since there was no change, λ1 and λ2 showed similar possibilities. We can conclude that posterior distribution of domestic travel and tourism spending will change in between two groups of data first realistic 22 up to 23 trillion JPY or optimistic 28 up to 30 trillion JPY for the following years.

Indeed, the expenditure for domestic travel and tourism in 2016 was 21.7 trillion yen. Another important point is that from the beginning of 2015, which was relatively a very good year for tourism in Japan, this trend has started to change in a positive way.

5 Conclusion

The result of this study will create an impact so as to increase the efficiency of the everyday life of the community living in the Okhotsk area of Japan and with the same approach based on Bayesian statistics the study could help understand better both prefectural (Hokkaido) and country level (Japan) human experience.

The value distribution of our surroundings is changing at every moment, and the impact it creates requires it to be measured as quickly as possible for fluent comprehension in modern society. Never before has data been as valuable as today, but processing the data in time is of crucial importance everywhere in the world. Big Data enforced various processing solutions, including data cleaning, and analysis to derive value from it. Nevertheless, the vast majority of the people needs for Big Data to be presented in a user-friendly manner so that they could understand it clearly and make their decisions.

Our main interest will be to support the Okhotsk sub-prefecture of Japan and the people who live there through the data collected and processed in this research.

References

1. UNWTO (United Nations World Tourism Organization), Tourism Highlights 2016 Edition, http://www.e-unwto.org/doi/pdf/10.18111/9789284418145
2. McIntosh, R. and Goeldner, C. (1977) "Tourism, 2nd edition." Wiley, New York.
3. OECD (Organization for Economic Co-operation and Development), Regional Outlook 2016: Productive Regions for Inclusive Societies, Japan, http://www.oecd-ilibrary.org/urban-rural-and-regional-development/oecd-regional-outlook-2016/japan_9789264260245-32-en
4. OECD (Organization for Economic Co-operation and Development), Japan: Boosting Growth and Well-being in an Ageing Society, http://www.oecd-ilibrary.org/economics/japan-boosting-growth-and-well-being-in-an-ageing-society_9789264256507-en
5. IMF (International Monetary Fund), World Economic Outlook (April 2017), http://www.imf.org/external/pubs/ft/weo/2017/01/weodata/index.aspx
6. WTTC (World Travel & Tourism Council), Japan Benchmarking Report 2017, https://www.wttc.org/-/media/files/reports/benchmark-reports/country-reports-2017/japan.pdf
7. WTTC (World Travel & Tourism Council), WTTC Data Gateway, Japan Domestic Tourism Spending, https://www.wttc.org/datagateway/
8. Historical Statistics of Japan, Japan Statistic Bureau, http://www.stat.go.jp/english/data/chouki/index.htm
9. MIC (Ministry of Internal affairs and Communications), Japan Statistic Bureau, http://www.stat.go.jp/english/data/index.htm
10. S.L. Scott, A.W. Blocker, F.V. Bonassi, H.A. Chipman, E.I. George and R.E. McCulloch, International Journal of Management Science and Engineering Management, **11**(2), 78-88 (2016)

11. D.P. Kroese, T. Brereton, T. Taimre and Z.I. Botev, Wiley Interdisciplinary Reviews: Computational Statistics, **6**(6), 386-392 (2014)
12. Apache Foundation, Apache Spark, http://spark.apache.org/
13. IBM Big Data University Virtual Lab Environment, https://my.datascientistworkbench.com/
14. PyMC3, https://pymc-devs.github.io/pymc3/index.html
15. Kevin P. Murphy, *Machine Learning A Probabilistic Perspective* (Chapter 2, 2012)

Thermal analysis on potential risk of cosmetic materials by DSC

Yi-Feng Chen, An-Chi Huang, Chung-Fu Huang and Chi-Min Shu[a]

[1]*Graduate School of Engineer Science and Technology, National Yunlin University of Science and Technology, Taiwan*
[2]*Department of Safety, Health, and Environmental Engineering, National Yunlin University of Science and Technology, Taiwan*

Abstract. To explore the potential risk of usage on daily chemicals, the product which most contact for human directly, cosmetics, were selected as study object. In this study, common cosmetic materials, such as propylene glycol, ethanolamine, silicon dioxide, iron oxide, and copper oxide were discussed for potential hazard. According to results of differential scanning calorimetry experiments, the apparent activation energy and SADT were calculated out as 779.22 kJ mol^{-1} and 45°C, respectively.

1 Introduction

The desire to look attractive is universally accepted. The origin of cosmetics can be tracked back to ancient Egyptians nearly 4,000 years ago. It was a badge of an honourable position. Today it has evolved into a way of decoration. In the mid-1970s, mineral make-up began revolutionizing cosmetics, especially when nanomaterial technology matured. Using nano-grade materials as additions is the latest approach and is an effective way to cover flaws on face and other parts of the body. However, this modern technology has had doubts about its safety for human health within last decade [1]. Even though cosmetic products are supervised strictly by the World Health Organization, it is still worrisome since they are interrelated closely in life. There are numerous cases in which people have been burned or developed an allergy by cosmetic mask, gel, toner, creams, lotion, and powder. That is not only because of their allergies, but also from unsafe applications and storage or unknown potential risk. Generally, approximately 50 kinds of composition are added to a cosmetic product. Furthermore, several products are usually used at the same time. A tiny interaction may inevitably cause the above hurt.

In this study, common cosmetic materials, such as propylene glycol, ethanolamine (ETA), SiO_2, Fe_2O_3, and CuO were discussed for potential hazard. Differential scanning calorimetry (DSC) was used to ascertain the matrix relation and DSC curve of above substances. It is believed that the results of this study can contribute to recommendations or references for cosmetic composition for future applications.

2 Experiments and methods

[a] Corresponding author: shucm@yuntech.edu.tw

Nano graded TiO2, propylene glycol, ETA, and benzophenone (Ph2CO) are the most common materials used in sun block, moisturizer, color make-up, and so on. ETA is a common material which is a wetting agent in lotions and creams.

Thermogravimetry (TG) and TA8000 system DSC821e were used to for non-isothermal experiments out. TG has the ability to acquire the parameters of pyrolysis substances. It is a way to select the scope of temperature in DSC tests, which are from 30.0 to 300.0°C. Through DSC experiments, various thermokinetic parameters can be obtained, such as onset temperature, maximum temperature, heat of decomposition, and self-heating rate. Moreover, considering assessment theory, reproducibility in each experiment of DSC was performed and validated.

Reactions in some cosmetic materials generally belong to an autocatalytic reaction, such as benzoyl peroxide [2]. To determine a non-isothermal kinetic model on the above cosmetic materials, the activation energy is an indispensable parameter. Therefore, the Arrhenius equation was used to calculate for nth order reaction. The Arrhenius equation describes the temperature versus reaction rate. An indefinite integral type Arrhenius equation is shown in Eq. (1).

$$k = Ae^{-E_a (RT)^{-1}} \tag{1}$$

where k is reaction rate constant, A is Arrhenius constant, E_a is activation energy which is usually displayed by kilojoule per mol (kJ mol^{-1}), R is air constant, and T is absolute temperature, which is displayed by degree Kelvin (K).

According to the Arrhenius equation, k is proportional to T and E_a. That is, when E_a receives a higher value, k rises faster as T is augmented; k is more sensitive with T. For different T, the plot of ln k versus T^{-1} presents a straight line, with the slope and intercept $-E_a R^{-1}$ and ln A, respectively. Therefore, E_a can also be expressed as in Eq. (2).

$$E_a = RT^2 \, d \ln k \, (dT)^{-1} \tag{2}$$

3 Results and discussion

Common cosmetic materials were tested by DSC at a heating rate of 4°C min–1 with matrix relation randomly. Overall, the results on thermal stability of materials have been positive. To summarize the salient features of the tests, several findings are of interest. Different substances seem to be closely connected to onset temperature and their reaction intensities. The pure materials and materials mixed with mineral make-up powder for DSC tests are shown in Fig. 1.

As expected, some reactions happened when the pure materials were mixed with others. The findings suggest that these materials are not necessarily mutually exclusive. Not surprisingly, the results of study did not show extreme danger since these substances are used on body and skin closely. Nevertheless, a phase change may cause the quality variation to become lower, even inferior. Interestingly, when the Ph_2CO mixed with propylene glycol and ETA, respectively, onset temperatures decreased. In regards to mining-make up materials with Ph_2CO, onset temperatures also decreased. The difference is that the onset temperature of Ph_2CO mixed with CuO compared higher with pure Ph_2CO. The initial reaction of Ph_2CO mixed with propylene glycol was low to 46.08°C. It means that the quality deterioration will occur because this temperature is as low as ambient surroundings. To explore more deeply concerning different situations, Ph_2CO mixed with propylene glycol was tested at heating rates of 0.5, 1.0, 2.0, 4.0, and 8.0°C min^{-1}.

$$d\alpha \, (dt)^{-1} = Ae^{-E_a (RT)^{-1}} f(\alpha) \tag{3}$$

Figure 2 illustrates that when heating rate became higher, the endothermic onset temperatures on DSC were oppositely getting lower. It is noteworthy that the endothermic peak heat flow of 0.5 and 8.0°C min^{-1} is –0.54 and –1.43 W g^{-1}, respectively. Apparently, the maximum endothermic peak heat flow is approximately three-fold higher than the minimum one. To obtain E_a and other parameters, the method from the American Society for Testing and Materials E698-5 was adopted, as shown in Eq. (3) [3].

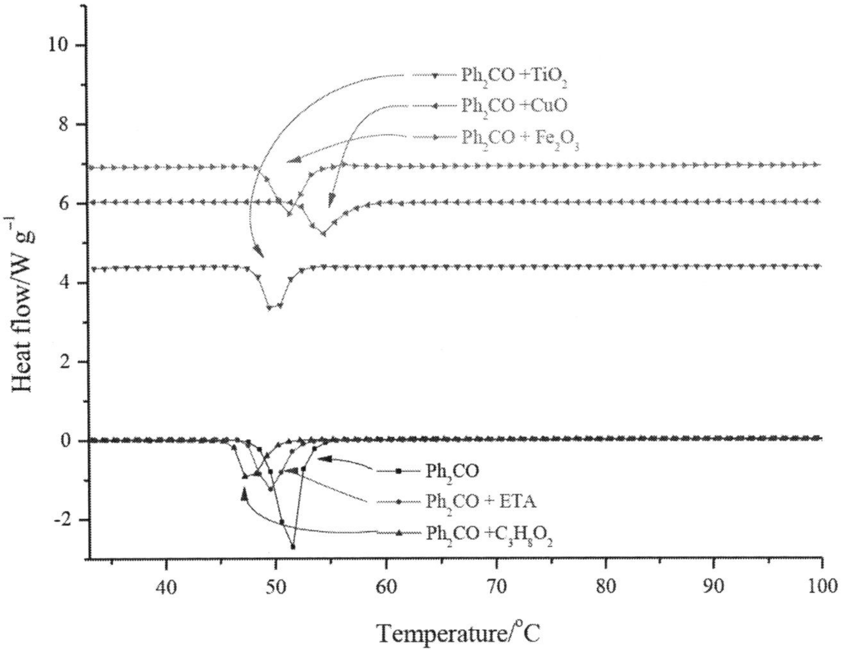

Figure 1. DSC curve during heating process for pure Ph_2CO and Ph_2CO mixed with common ingredients of cosmetic, respectively, at heating rate of 4.0°C min^{-1}.

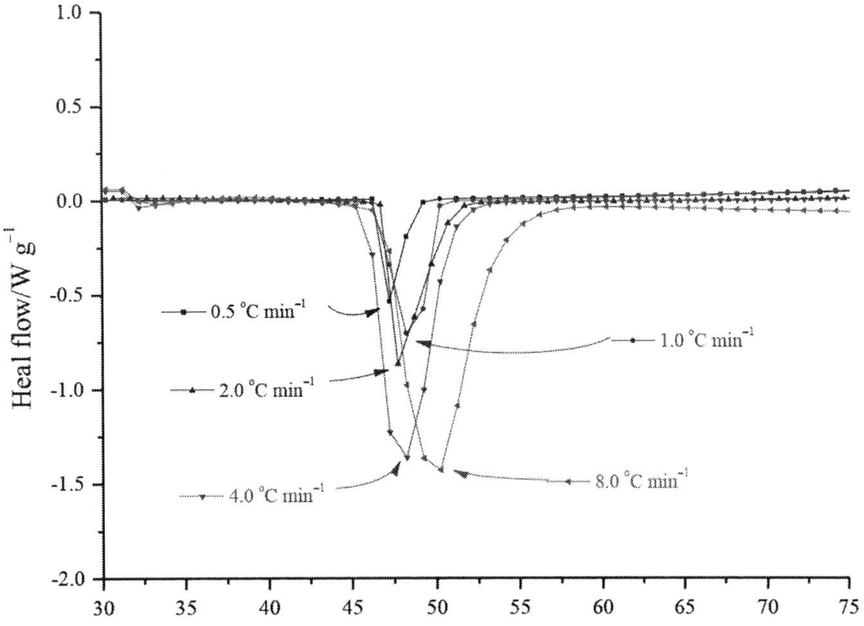

Figure 2. DSC curve during heating process for Ph_2CO mixed with $C_3H_8O_2$ at heating rates of 0.5, 1.0, 2.0, 4.0, and 8.0°C min^{-1}.

To refer to Eq. (3), the E_a between different heating rates was calculated. The linear regression was adopted to figure out the R-square of relationship in different heating rates, as shown in Fig. 3.

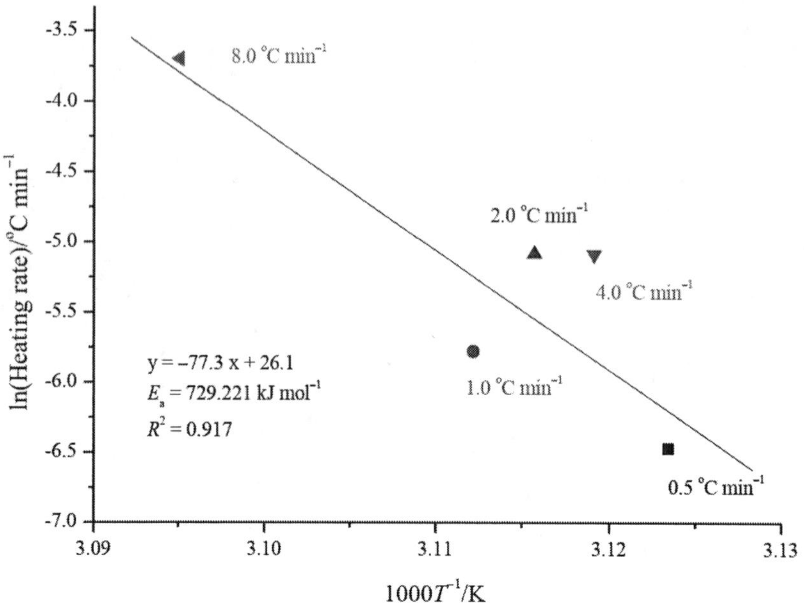

Figure 3. Apparent activation energy calculation with Ph_2CO mixed with $C_3H_8O_2$.

In the end, SADT was assessed in this study to consider the thermal stability of Ph_2CO mixed with propylene glycol. The SADT is steady at 45.0°C for 50.0, 100.0, 200.0, and 500.0 kg.

4 Conclusion

The findings of this study pointed out that Ph_2CO has an advance reaction and phase alternation when it is mixed with propylene glycol. It should be noted that the heat flow between different heating rates is a several-fold gap. It is conceivable that when this kind of combined cosmetic product deteriorates in quality it becomes indisposed for people to reuse. The parameters of thermokinetic technology used in this study can be applied in usage and storage mechanism for cosmetics of interest.

Acknowledgments

Financial support for this study was graciously provided by the Minister of Science and Technology (MOST), Taiwan, ROC (MOST-104-2922-I-224-015).

References

1. A. Gustafsson, et al., Genetic variation influences immune responses in sensitive rats following exposure to TiO_2 nanoparticles, Toxicol, **326**, 74-85 (2014)
2. S.H. Liu, H.Y. Hou, and C.M. Shu, Thermal hazard evaluation of the autocatalytic reaction of benzoyl peroxide using DSC and TAM III, Thermochim Acta, **605**, 68-76 (2015)
3. Y.T. Tsai, M.L. You, X.M. Qian, and C.M. Shu, Calorimetric techniques combined with various thermokinetic models to evaluate incompatible hazard oftert-butyl peroxy-2-ethyl hexanoate mixed with metal ions, Ind Eng Chem Res, **52**, 8206-8215 (2013)

Strength characteristics of artificial organic soils stabilized with copolymer stabilizer

Chia-Wen Law[1], Felix Ngee-Leh Ling[1,a] and Boon-Khiang Ng[2]

[1]*Faculty of Civil & Environmental Engineering, University Tun Hussein Onn Malaysia, 86400 Parit Raja, Batu Pahat, Johor, Malaysia*
[2]*Dairen Chemical (M) Sdn. Bhd., Malaysia*

Abstract. Organic soil is known as low strength material, and chemical stabilization is widely used to increase its bearing capacity. However, the use of traditional stabilizer has some limitations. Therefore, stabilization was carried out by using non-traditional stabilizer - Vinyl acetate-ethylene (VAE) copolymer emulsion in this study with the aim to determine its suitability to stabilize soil mixed with organic matter. Two types of artificial organic soil with kaolin: organic acid ratio of 5:5 (K5HA5) and 7:3 (K7HA3) were utilized. Control specimens were tested using pure kaolin. Different percentages of VAE (5%, 7.5%, 10%) were added in order to determine the minimum amount of stabilizer required to achieve a minimum strength increment of a 345 kPa. The strength of samples was determined with automated unconfined compressive test device. Specimens were air cured for 7 days prior to testing. Both K7HA3 with 7.5% VAE and K5HA5 with 10% VAE had achieved the minimum strength increment to be considered as effective stabilization. The strength of the artificial organic soil was found to be increasing with the increment of percentages of VAE used. Hence, it can be concluded that stabilizing mechanism of the artificial organic soils with VAE is not affected by organic matter.

1 Introduction

Modification of the engineering characteristics of organic soil is required to meet the design requirements before the development of the affected site is carried out. One of the usual practices of soil improvement method is soil stabilization. Soil stabilization improves the strength of soil through binding the soil particles together, proofing the particles or combination of both. There are several methods of soil stabilization, and these methods can be categorized into two broad categories, which are mechanical stabilization and chemical stabilization.

Mechanical stabilization achieved through mechanical process by induced vibration or compaction to alter the nature of native soil particles [1]. On the other hand, chemical stabilization depends mainly on the chemical reaction between stabilizer (cementitious material) and soil particles (pozzolanic material) to achieve desired strength [1]. Currently, the most traditional stabilizer that commonly used is lime. Lime stabilized soil effectively increases the strength, durability, workability, and compressibility of the soil [2]. However, the organic material decomposition blocked the polymerization of silicate and inhibited the pozzolanic reaction required in lime stabilized soil process

[a] Corresponding author : felix@uthm.edu.my

to gain strength [3]. Based on the findings of the research on cement stabilizer, which is another example of traditional stabilizer, it also not works well when there is an existence of organic matter and more admixtures needed for better soil strength [4]. Besides that, the organic content can affect the effectiveness of soil stabilization [5]. Due to the limitations of traditional stabilizers, technologies have increased the number of stabilizers. The polymer-based stabilizer is one of the non-traditional stabilizers which show a significant increase of unconfined compressive strength of the treated soil [6].

Currently, five types of polymers commonly used in concrete to improve flexural strength and tensile strength are acrylics, styrene-acrylics, vinyl acetate ethylene (VAE), polyvinyl acetate (PVA), and styrene-butadiene resin (SBR) [7]. Each of them has their unique characteristics such as adhesion, waterproof ability, and UV stability. Both vinyl acetate and acrylic copolymers are hydrophobic and showed moderate bonding efficiency in a range of soils in different case studies [8]. Besides that, previous study [9] also found that the vinyl acetate-acrylic copolymer can effectively increase the unconfined compressive strength of the peat soil. The peat soil is classified as the soil with high content of organic matter. The findings proved that the copolymer stabilizers have a positive impact in organic soil stabilization.

The polymer stabilizer coats soil particles and the physical bonds are formed when the emulsion water evaporates, leaving a soil polymer matrix. The improvement in strength is depends on the ability of the stabilizers to coat the soil particles adequately and physical properties of stabilizers due to the primary stabilization mechanism is physical bonding. Therefore, the polymer stabilizers are more suitable for the granular soils as compared to the fine grained soils due to the higher mixing efficiency resulting from their high specific surface area. Other than that, it is typically used in soil with excellent tensile and flexural strength and producing physical bonds with excellent strength [10]. Even though polymer stabilization demonstrates a great increases in unconfined compression strength in dry curing but is found to decrease after it is immersed in water [11]. This is due to the weakness of the hydrogen bonding to neutralize electrically clay mineral and also the ineffectiveness of the polymer to decrease the permeability of the mixture. Polar components present in polymer absorb strongly on the soil surfaces which promoting adhesion and resulting in the increases of strength [10]. In spite of growing interest on copolymer soil stabilization, there is limited number of researches has been published. Most of the investigation focused on the effectiveness of the soil stabilization rather than the mechanism of the process occurred during the soil stabilization.

Moreover, there are limited studies on the other types of the copolymer, namely VAE copolymer emulsion which also have the potential to increase the strength of stabilized soil. Hence, this study is carried out with the objective to stabilize organic soils with VAE stabilizer.

2 Material and methods

2.1 Material properties

The artificial organic soil was prepared by mixing inorganic soil with predetermined humified organic matter. In this study, two different ratios of inorganic soil vs. organic matter which was 5:5 and 7:3 were used to compare their effects on soil stabilization. The type of organic matter utilized in this study is organic acid which is categorized as humified organic matter while the inorganic soil used in this study is Kaolin grade S300. Some of the physical and chemical properties of Kaolin grade S300 and organic acid were determined by other researchers and are shown in Table 1 [12, 13], Table 2 [14] and Table 3 [15], respectively. Based on the USCS classification, kaolin S300 is categorized as low plasticity Silt (SL) which consists of 95% alumina and silicate. Whereas the organic acid utilized in this study was found to be consisted of humic acid and fulvic acid which is fall in the category of humified organic matter.

Table 1. Physical and chemical properties of kaolin grade S300 [12, 13].

Kaolin grade S300	
Properties	**Value**
% Sand	0.14 %
% Silt	94.94 %
% Clay	4.92 %
D_{60}	22.83μm
D_{30}	12.22μm
D_{10}	4.44μm
C_u	5.142
C_c	1.473
LL	33.75%
PL	23.01%
PI	10.74%
pH (at 25°c)	6.82

Table 2. The bulk chemical composition of the kaolin grade S300 [14].

Kaolin grade S300	
Bulk chemical composition	**Composition (%)**
SiO_2	54.2
Al_2O_3	40.8
SO_3	-
CaO	-
Fe2O3	1.51
TiO2	0.59
MgO	0.36
K2O	2.21
Na2O	-
Cl	-
BaO	-
SrO	-

Table 3. Composition of organic acid [15].

Humic acid	Fulvic Acid
61.68%	24.93%

Table 4. The physical properties of DA-524 [16].

Emulsifying system	PVA
Solid content (%)	55 min
Viscosity (cP)	1500-2500
pH	4.5-6.5
Free monomer (%)	0.3 max
Wet Tack	Good
Thickening response	Good
Borax stability	No
Tg (°c)	20
MFFT (°c)	3

Copolymer stabilizer VAE DA-524 was chosen to stabilize the artificial organic soil in this study. The VAE DA-524 emulsion was supplied by DAIREN Chemical Sdn. Bhd. VAE emulsion can be characterized as copolymer as it derived from two species monomers which are Vinyl Acetate monomer and Ethylene monomer. Vinyl acetate ethylene (VAE) emulsions are referred to emulsion derived from the copolymerization of vinyl acetate and ethylene. The total formulation of vinyl acetate is in the range between 60 and 95 percent, and the ethylene content ranges between 5 and 40 percent. VAEs appear as milky white water-based emulsions, and high-pressure reactors are required to synthesize VAE emulsions from vinyl acetate monomer and ethylene. The physical properties of DA-524 [16] are listed in Table 4.

2.2 Methods

The artificial organic soils utilized in this study were categorized as fine materials with majority particles smaller than 63 μm [12]. Therefore, small pycnometer test [17] was used to determine the specific gravity of all the soil samples. Kerosene was used as displacement liquid because distilled water is solvent for organic acid. Standard Proctor compaction test [18] was performed on all types of soil in order to identify its optimum moisture content to achieve the maximum dry density.

Cylindrical shaped test specimens of dimension 50mm in diameter and 100mm in height were prepared based on the predetermined maximum dry density. The contents of water utilized to remold specimens were based on its optimum moisture content as obtained from Standard Proctor compaction test. Extra water of 0.02% was added after considering the water losses due to hydration of VAE copolymer emulsion with soil during mixing. The losses of water through hydration were determined using Eqn. 1. A total of 36 samples were remolded in this study and its details are shown in Table 5. The specimens were air cured for seven days at room temperature. The water content of specimens was determined before and after curing. After air curing, the specimens were tested for its unconfined compression strength using automated Unconfined Compression Test device. A constant axial force was imposed on the test specimen with the constant strain rate of 1%/min until failure occurs (does not exceed 2%/min as specified in standard specification [19]). Failure was considered taken place when the axial stress showed decreasing with increasing strain or a strain of 20% was exceeded (whichever comes first) in accordance to standard specification[19]. The data of the test was utilized to calculate the unconfined compressive strength of the soil.

Table 5. Details of soil specimen.

Types of Soil	**Kaolin: Organic Acid ratio (Coding name of soil)**	**Percentage of VAE emulsion copolymer, %**		
		Number of replicates, n		
		5%	7.5%	10%
Artificial Organic Soil	5:5 (K5HA5)	3	3	3
	7:3 (K7HA3)	3	3	3
Inorganic Soil	10:0 (Pure kaolin)	3	3	3
Total specimens		9	9	9

$$\text{water loses due to hydration} = \left[\frac{\text{water content before mixing} - \text{water content after mixing}}{\text{water content before mixing}}\right] \quad (1)$$

3 Results and discussion

3.1 Physical properties of soil specimens

The optimum moisture content and maximum dry density of tested soil was determined using Standard Proctor Compaction test [18]. The compaction curves of the test specimens are shown in Figure 1 and its maximum dry density and optimum moisture content are listed in Table 6. The test results showed that the optimum moisture content was increased with contents of organic acid, whereas maximum dry density of tested soils was decreased with contents of organic acid. Besides that, the results of specific gravity (as listed in Table 6) also showed that the organic acid has significantly influenced the physical properties of soil. The specific gravity of soil was found to be decrease with increased contents of organic acid.

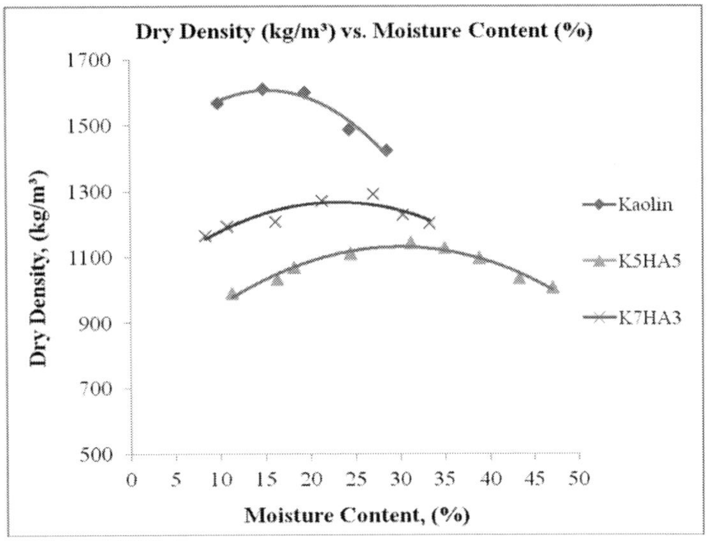

Figure 1. Compaction curves of kaolin S300 and artificial organic soils

Table 6. Specific gravity, maximum dry density and optimum moisture content of kaolin S300 and artificial organic soils.

Types of Soil	Kaolin: Organic Acid ratio (Coding name of soil)	Properties		
		Specific gravity	Maximum dry density (kg/m3)	Optimum moisture content (%)
Artificial Organic Soil	5:5 (K5HA5)	2.03	1130	31
	7:3 (K7HA3)	2.24	1260	23
Inorganic Soil	10:0 (Pure kaolin)	2.59	1605	15

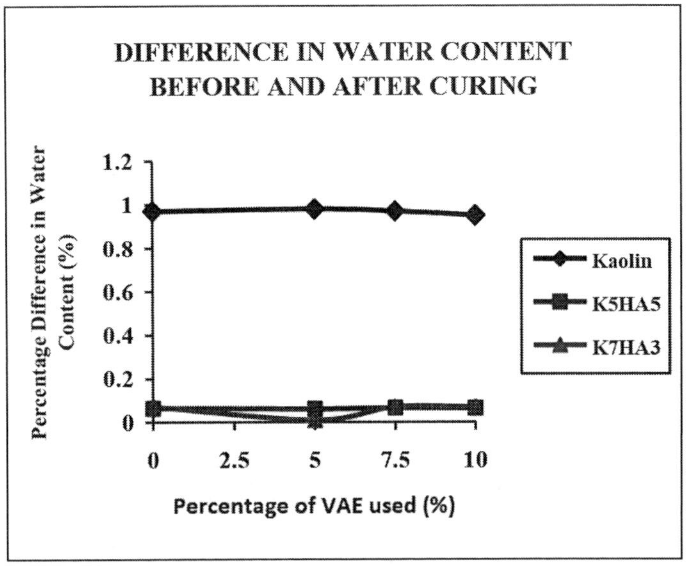

Figure 2. Loses of water throughout curing periods

The water content of specimens were determined before and after curing in order to determine other potential factors that may influence the strength of specimens. The losses of water throughout curing periods were determined and illustrated in Figure 2. The findings showed that the losses of water were relatively higher (almost 1%) for plain kaolin soil comparing with artificial organic soils which were around 0.1%. The findings showed that the contents of organic acid had increased the water holding capacity of soil. Organic acid is found to be capable in attracting water through adsorption process [20].

3.2 Effect of organic acid on unconfined compressive strength of soils

It is shown in Figure 3 that the UCS of both types of artificial organic soil K7HA3 (with 70% kaolin mixed with 30% organic acid) and K5HA5 (with 50% kaolin mixed with 50% organic acid) were higher than kaolin grade S300 with the use of the same amount of VAE copolymer emulsion. In comparison to the specimens of pure kaolin S300, the strength increment is due to the composition of the organic acid in the soil. In fact, the negatively charged organic acid itself is not a flocculating agent, and they are even an effective dispersing agents for Na-clays and soils [21]. However, with the presence of polyvalent cation, it can effectively facilitate the attachment of the polyanions from humic acid to clay surfaces and to bridge clay particles together, forming stable aggregates and increasing the strength of the soil [21]. It is thus proven that, with the presence of polyvalent cations in kaolin, humic substances in soil are likely to play a vital role in soil aggregate stabilization. In this study, the polycations which are Al from the kaolin possibly had facilitated the polyanions from the humic acid to increase the ability of soil aggregation which resulted in higher UCS of the soil specimens.

Besides that, as the humic substances become intimately associated with the mineral fraction of soil, colloidal complexes of humus-clay and humus silt aggregates are formed. These aggregated humus aggregates are formed by electrical processes which increase the cohesive forces that cause very fine soil particles and clay components to attract to each other and thus increasing its strength [22]. Hence, the unconfined compressive strength of K7HA3 and K5HA5 are higher than the pure kaolin grade S300 which is without the composition of organic content.

On the other hand, it is found that the UCS of K7HA3 is greater than K5HA5. This statement can be supported by the result of the previous research which showed that at low organic contents the soil particles tend to aggregate, whereas at higher organic contents the soil particles tend to disperse [23]. Also, the organic particles are stiff when compressed and act as rigid particles when dry, but when they absorb water, they become soft and sponge-like. Therefore, at higher water contents, specimens with higher levels of organic matter had lower soil strength because on the wetter side of the peak strength values, soil lumps had been slaked, and the excess water caused greater pore water pressure. Therefore, compaction effort was merely sheared or remolded the specimens rather than increased soil strength further [22]. The findings can explain the result of the K5HA5 with higher organic content obtained a lower strength than K7HA3 in this study.

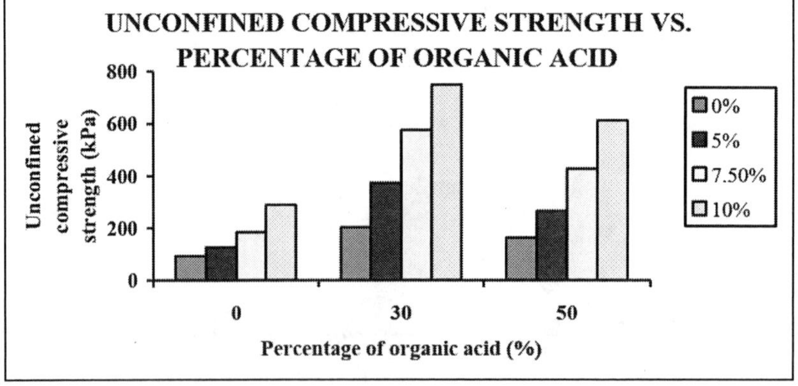

Figure 3. Effect of organic acid on unconfined compressive strength of soil

It is also proven that the physical properties of soil are not directly contributed to the unconfined compressive strength. As shown in Figure 3, the kaolin had achieved the highest dry density as compared to K7HA3 and K5HA5. However, the unconfined compressive strength of the K7HA3 is the highest and followed by K5HA5 and pure kaolin grade S300. The findings agreed well with the previous study [23, 24] in which the moisture content and density of specimens were found not to be statistically significant with the strength of stabilized soil.

3.3 Effects of VAE copolymer emulsion on unconfined compressive strength of soil

Figure 4 shows the experimental results on the effect of percentages of VAE copolymer emulsion on the unconfined compressive strength (UCS) of soils. It is evident from the findings that the contents of VAE copolymer emulsion influenced the UCS of the test specimens. The UCS of all three types of soil achieved the highest values when the percentages of the VAE copolymer emulsion were 10%. As stated in standard specification [25], an increase in UCS of 345kPa or more must be achieved for stabilization to be considered effective. Therefore, it is clearly found that for kaolin grade S300 with even up to 10% VAE, the strength of soil still below the target strength of 345kPa. Besides that, the minimum target of UCS for the K5HA5 and K7HA3 to achieve were 507.35kPa and 547.05kPa respectively. As shown in the graph, the usage of 10% of VAE copolymer emulsion in K5HA5 had met the minimum target of strength increment whereas the 7.5% usage of VAE copolymer emulsion in K7HA3 were also considered as the effective stabilizer.

The main reason that contributed to this result can be explained by the mechanism of reaction between polymer composed of polyvinyl acetate and soil. The clayey soil stabilization mechanism of organic polymer soil stabilizer includes filling, chemical reaction and enwrapping. A polymer comprised of polyvinyl acetate contains long-chain macromolecule and polarity carboxyl (–OOCCH3). When it is applied to clayey soil, a part of polymer fill up the voids of soil, and other parts stay on the soil aggregates surface. The hydrophilic groups ($-OOCCH_3$) in its molecular structure have a chemical reaction with positive ions of clay grain and create physicochemical bonds between molecules and soil aggregates with ionic, hydrogen, or Van der Waals bonds. Through these bonds, long-chain macromolecules of polymers enwrap the aggregate's surface and interlink to form an elastic and viscous membrane structure [9]. Therefore, the strength of the soil is improved. Besides that, the formation of the voids filling and physico-chemical bond in reactions of stabilized soil leads to the increases in bonding and interlocking forces between soil particles [10]. As a result, the unconfined compressive strength and cohesion of the clayey soil are improved.

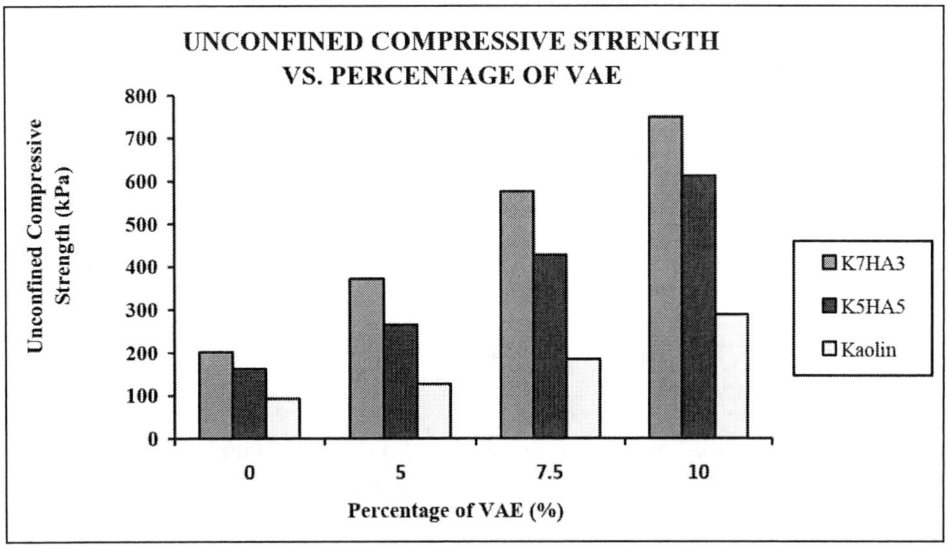

Figure 4. Effects of VAE copolymer emulsion on the unconfined compressive strength of the soil

Given the concentration of polymer on strength, more addition amount can fill up more voids and produce more bonds and hence result in the greater strength of the soil. However, the addition amount of polymer is related to the void ratio and clay content. At a certain concentration, the molecules of polymer might have already been enough to fill up most of the soil voids and the hydrophilic groups (-OOCCH3) in molecule might have fully reacted with the clay particle. Therefore, that predetermined concentration can be referred as optimum concentration. In this study, the optimum percentages of VAE copolymer that can use to achieve the maximum value of UCS are still undefined as the pattern of the graph is still increasing. Therefore, there must be an optimum usage of VAE copolymer emulsion to achieve the maximum UCS.

4 Conclusions

Artificial organic soil stabilized with VAE copolymer emulsion had been studied. It can be summarized from the experimental outcomes that VAE copolymer has made a substantial effect on the strength properties of the stabilized organic soil. Based on the results of laboratory investigation, the following concluding remarks were made:

1. The UCS of the artificial organic soil is increasing with the increment of copolymer stabilizer used.
2. According to ASTM D-4609, the stabilization considered as useful if there is an increment of 345kPa in unconfined compressive strength. In this study, stabilized kaolin grade S300 have not achieved the target with up to 10% VAE. In contrast, the K5HA5 with 10% VAE copolymer emulsion and K7HA3 with 7.5% had successfully achieved target strength increment.
3. The K7HA3 had reached the highest unconfined compressive strength for all the percentages of VAE copolymer emulsion used, and it is then followed by K5HA5 and kaolin grade S300. Therefore, it can be concluded that VAE copolymer emulsion is able to stabilize soils with organic matter (higher strength gained in artificial organic soils compared to pure kaolin). However, high amount of organic matters when comparing soil K5HA5 with K7HA3, the strength is slightly affected.

Acknowledgements

The authors wish to express gratitude and grateful appreciation to the financial support from University Tun Hussein Onn Malaysia and Dairen Chemical (M) Sdn. Bhd. The authors also thank all laboratory technicians in Research Centre for Soft Soil (RECESS), Department of Infrastructure and Geomatic Engineering, Department of Water and Environmental Engineering of Faculty of Civil and Environmental Engineering, Universiti Tun Hussein Onn Malaysia. Their advice, knowledge and technical support had made this research possible.

References

1. G.P. Makusa, *Soil stabilization methods and materials in engineering practice: State of the art review* (Lulea°, Sweden, 2013)
2. T.I. Jawad, M.R. Taha, Z.H. Majeed and T.A. Khan, Res. J. Appl. Sci. Eng. Technol., **8**(4), 510-520 (2014)
3. M.B. Hampton and T. B. Edil, *GSP No. 81* (New York, USA, 1998)
4. H. Chen and Q. Wang, Bull. Eng. Geol. Environ., **65**(4), 445-448 (2006)
5. E.O. Tastan, T.B. Edil, C.H. Benson, and A.H. Aydilek, J. Geo. Geoenv. Eng., **137**(9), 819-833 (2011)
6. A.R. Zandieh and S.S. Yasrobi, Geo. Geol. Eng., **28**(2), 139-145 (2010)
7. J. Nasvik, in *Concrete construction* (Addison, USA, Sept 2001)

8. GRT, Polymer Soil Stabilization (2015), available at https://globalroadtechnology.com/polymer-soil-stabilization.
9. P.F.I. Norazam, MATEC Web Conf., **103**, 07014 (2017)
10. J. Tingle, J. Newman, S. Larson, C. Weiss and J. Rushing, Transp. Res. Rec., **1**, 59-67 (1989)
11. A. Bayat and O. Farzaneh, in *Advances in transportation geotechnics 2*, edited by S. Miura, T. Ishikawa, N. Yoshida, Y. Hisari, and N. Abe (CRC, Boca Raton, FL, 2012), p.59
12. F.N.L. Ling, K.A. Kassim and A.T. Abdul Karim, Adv. Mater. Res., **341-342**, 108-112 (2012)
13. F.N.L. Ling, K.A. Kassim and A.T. Abdul Karim, in *Proceedings of peat technology-opportunities and challenges seminar* (UCTS, Sibu, 2014)
14. F.N.L. Ling, K.A. Kassim, A.T. Abdul Karim and T.W. Chan, Adv. Mater. Res, **723**, 985-992 (2013)
15. M.I. Jaimi, Report No. BK 35/14, Forest Research Institute Malaysia (FRIM) (2014)
16. Dairen Chemical Corporation (DCC), Technical Data Sheet VAE DA-524 (2016)
17. BSI, BS 1377: Part 2 (1990)
18. BSI, BS 1377: Part 4 (1990)
19. BSI, BS 1377: Part 7 (1990)
20. K.H. Tan, *Principles of Soil Chemistry (Third edition)* (CRC, Boca Raton, FL, 1998)
21. A.I. Husein Malkawi, A.S. Alawneh and O.T. Abu-Safaqah, Appl. Clay Sci., **14**(5-6), 257-278 (1999)
22. R.E. Pettit, Organic matter, humus, humate, humic acid, fulvic acid and humin: their importance in soil fertility and plant health (2004), available at http:// www.humates.com/pdf/ORGANICMATTERPettit.pdf
23. F.N.L. Ling, K.A. Kassim, A.T. Abdul Karim and S.C. Ho, Adv. Mater. Res., **905**, 305-308 (2014)
24. A. Bauer, Farm Res., **31**, 44-52 (1974)
25. ASTM Int., D4609-01, in *ASTM Standard 04.08* (2001) p.1-5

Affectivity of biological cement's application to sandy soil for geotechnical engineering

Hung Wen Yi[1], Emma Yuliani[2], Maytri Handayani[1,a], Hsu Chia Sseng[1] and Chen Ssu Ching[3]

[1]*Dept. of Civil Engineering, National Central University, Taiwan*
[2]*Dept. of Water Resources Engineering, Brawijaya University, Indonesia*
[3]*Dept. of Life Science, National Central University, Taiwan*

Abstract. In this research, the authors analyze the addition of bacterial producer of biological cement on sandy soil to increase its stability. *Pseudomonas sp* was injected into the soil to produce urease enzyme, which converts urea that reacted with water become ammonium and carbonate, causing calcite precipitation. Soil stability can be improved by the production of pore-filling materials and particle-binding materials through the calcite precipitation process. The aim of these applications is to improve the mechanical properties of soil that it will be more suitable for construction or environmental purposes. After bacterial inoculation, the experiment with variation of bacterial content concluded the best way to increase soil stability is inject 15% bacteria relative to total weight of soil. Highest permeability reduction is 73,73% and highest shear strength increation for cohesion is 6,84 with friction angle 44,46°. California Bearing Ratio test with 10 of hits generated that the treatment soil can proved a significant penetration stress changed. From Environmental Scanning Electron Microscopy test results of the inoculation sand test images and surface scan, the formed exopolysaccharide attached to the wall of the sand soil particles which expected will fill, patch, and close the pores between the sandy soil particles.

1 Introduction

In choosing the type of soil for materials of retaining construction such as river embankments or earth dams, the permeability is an important characteristic to consider in addition to the compaction characteristics. A soil shear strength value is required to calculate the soil bearing capacity or to calculate the earth pressure force on the retaining walls.

To increase soil stability, efforts to reduce the permeability and increase the shear strength is very important way because it can run on all types of foundations that consist of one or more types of rock. The last few years have been done a lot of research and studies about how to improve the stability of the soil. One of the most exciting research is soil stability improvement by adding bacteria that producing "biological cement" (exopolysaccharide) into the soil. This research has been tested in several countries, but is limited to a few types of soil material just in those countries.

[a] Corresponding author : hmaytri@gmail.com

If the general material used for material reduction in permeability and increased shear strength is chemicals, but in this study the authors will try to analyze the addition of bacterial producing exopolysaccharide in sandy soil as an alternative to the material. Exopolysaccharides have been defined as organic polymers of microbial origin, which in biofilm systems, are frequently responsible for binding cells and other particulate materials to each others and to the substrate [1].

In the process of bacteria inoculation in the soil, there will be three processes. They are bioclogging process where bacteria will produce a material that is capable of filling the pores of the soil to reduce tha value of permeability, biocementation process where bacteria produces a material that is able to increase the shear strength of the soil, and the process of gas production through bacterial activity which can lower the risk of liquefaction in sandy soil [2].

A frequent problem encountered in the repair of the foundation is to overcome seepage discharge and the value of cohesion that has not been controlled. In this case that can be cultivated is to reduce seepage discharge and the cohesion values until it reaches a certain threshold then conduct maintenance, inspection regularly, and improvements as necessary.

A decrease in the permeability of the chemicals used to hold seepage in rocks that are not too impermeable and temporary (approximately one year) although there is also an age more than 30 years (semi-permanent). A decrease in the permeability of these chemicals only used when the usual way cannot be used, because it is relatively more expensive and the process must be carried out carefully because it can damage human skin.

Given these problems, the author tries to give a solution in the form of an initial step the latest innovations in geotechnical engineering. This innovation is expected to be environmentally friendly, do not damage human skin, is also quite effective to reduce permeability and increase soil shear strength material, especially of sandy material.

To test the effectiveness of this study, the sand samples will be tested permeability with constant head test, shear strength with direct shear test, compaction with California Bearing Ratio test, and ESEM (Environmental Scanning Electron Microscopy) test.

2 Literature review

Bacterial that suitable for making biogrouting is non-pathogenic and environmentally friendly. Moreover, unlike the use of cement, soil or contaminated soil can be treated or improved even without disturbing the soil itself or the surrounding environment, it is because bacterial can penetrate and reproduce themselves in the soil. Many types of bacteria can be used for biogrouting, including *Pseudomonas sp*.

A genus of motile, polar-flagellate, non-spore forming, strictly aerobic bacteria (family Pseudomonadaceae) containing straight or curved, but not helical, gram-negative rods that occur singly. Most species are saprophytic. Saprophytes are characterized by the use of a certain type of digestive mechanism, called an extra-cellular digestion. This process involves the secretion of digestive substances into the surrounding environment, where they break down organic matter into simple substances. The resulting nutrients are then absorbed directly through the cell membrane of the organism, and metabolized. The metabolism is respiratory, never fermentative. They occur commonly in soil and fresh water and marine environments.

For this research, the authors use constant-head test to determination of hydraulic conductivity. In this laboratory setup, the water supply at the inlet is adjusted in such a way that the difference of head between the inlet and the outlet remains constant during the test period. After a constant flow rate is established, water is collected in a graduated flask for a known duration.

The total volume of water collected (where Q = volume of water collected, A = area of cross section of the soil specimen, t = duration of water collection) may be expressed as:

$$Q = Avt = A(ki)t \tag{1}$$

and because

$$i = \frac{h}{L} \tag{2}$$

Eq. (2) can be substituted into Eq. (1) to yield

$$Q = A\left(k\frac{h}{L}\right) t \text{ or } k = \frac{QL}{Aht} \tag{3}$$

The direct shear is the oldest and simplest form of shear test arrangement. The test equipment consist of a metal shear box in which the soil specimen is placed. Normal force on the specimen is applied from the top of the shear box. For most soil mechanics problems, it is sufficient to approximate the shear stress on the failure plane as a linear function of the normal stress (Coulomb, 1776). This linear function (where c = cohesion, σ = normal stress on the failure plane, $\emptyset$ = angle of internal friction, τ_f = shear strength) can be written as:

$$\tau_f = c + \sigma \tan \emptyset \tag{4}$$

The CBR value is then defined as the ratio of the measured force to that required for similar penetration into a standard sample of crushed California limestone rock. The only calculation necessary is to express the measured force for a certain penetration as a percentage of the 'standard' force for the same penetration. In equation below, F is the measured force and Fs is the force required (standard force).

$$\text{CBR} = (F / Fs) \times 100\% \tag{5}$$

SEM analysis is a test used to display the scanned electrons of a solid object. SEM Test using a high-energy electron beam to generate a signal different from the surface of an object. In the SEM test the output is produced is the magnification image of the electron signal generation, so there is a difference between objects of different material because of the different electron arrangements. Terms and provisions for SEM test is specified in ASTM E-986-97.

3 Research methodology

In probability theory and statistics, variance is the expectation of the squared deviation of a random variable from its mean. Informally, it measures how far a set of (random) numbers are spread out from their average value.

Researchers did the experiment with the variations of curing days, relative density, and nutrients type. The curing days in this experiments divided by two different types that 7 days and 28 days. The environment condition is outside building (with real conditions according to the site, temperature follows the existing conditions, exposed by sunshine and humidity). For the variation of relative density, researchers divided by two types, i.e. relative density 50% and relative density 70%. Researchers locked the relative density to uniform the sample also know the comparison and effectiveness from both relative density. For the type of nutrient, researchers also divided by two types, nutrient type 1 and nutrient type 2. Nutrient Type 1 consist of $CaCl_2$, $MgSO_4$, K_2HPO_4, and Saccharose [3]. Meanwhile, Nutrient type 2 consist of LB Broth Powder, Urea, $CaCl_2$, NH_4Cl, $NaHCO_3$, Saccharose.

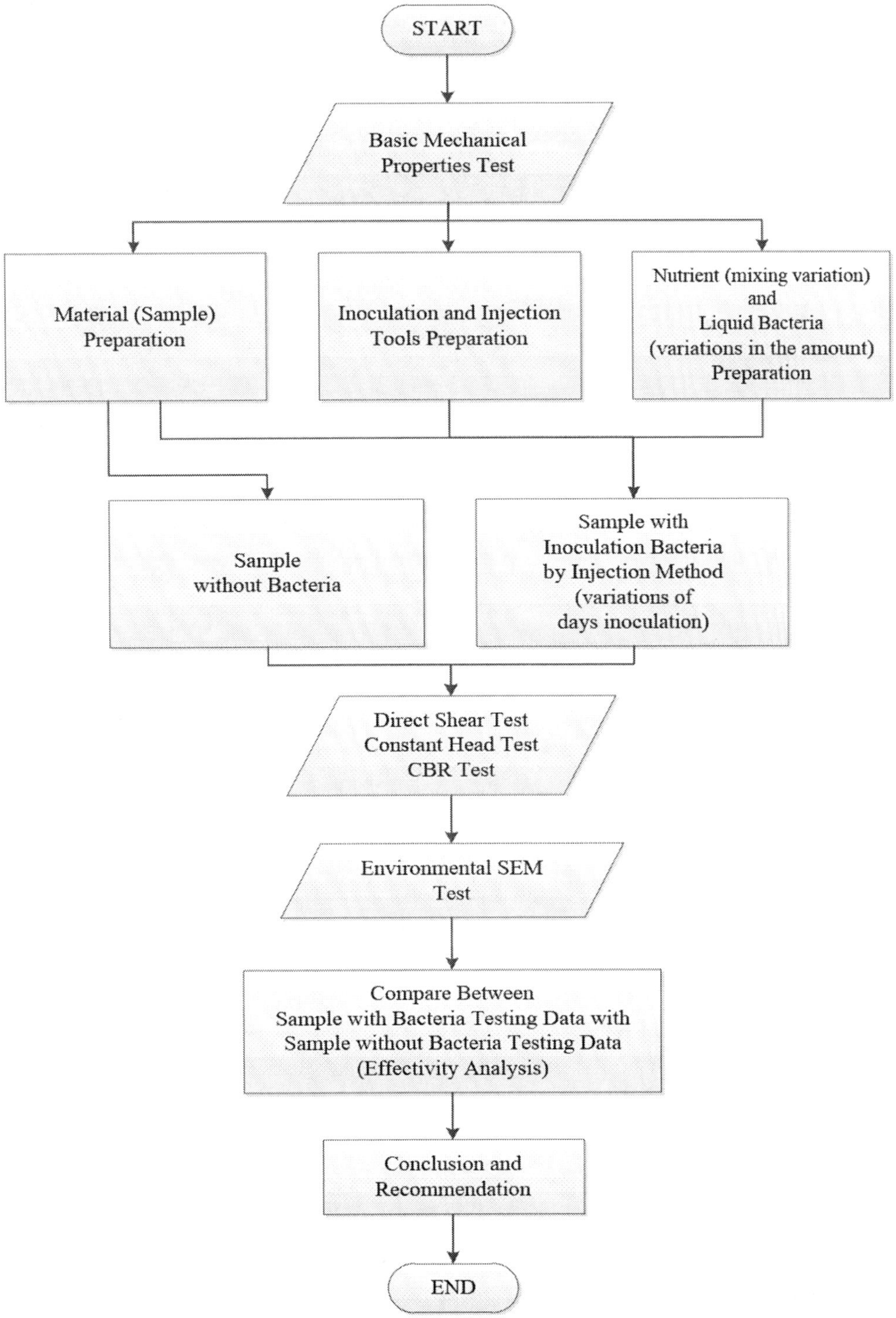

Figure 1. Research Stage

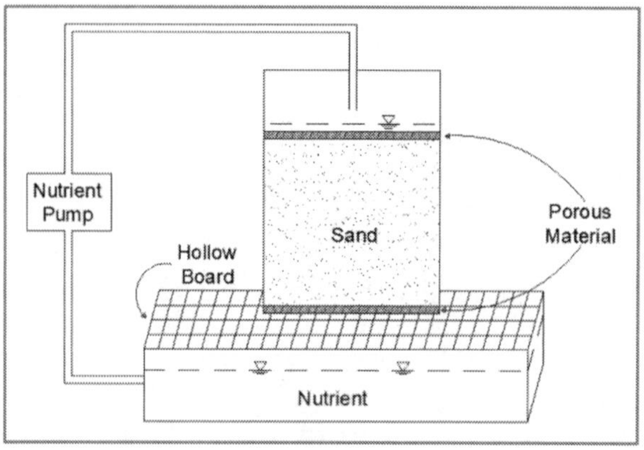

Figure 2. Research Methodology

Table 1. Fundamental Characteristic of Sandy Soil used in the tests

Sandy Soil Type	Silica Sand
Specific gravity, Gs	2.65
D_{60}(mm)	0.208
D_{30}(mm)	0.176
D_{10}(mm)	0.155
Coefficient of uniformity, C_u	1.344
Coefficient of curvature, C_c	0.963
γ_{dmax} (kN/m^3)	16.285
γ_{dmin} (kN/m^3)	13.830
Unified Soil Classification System	SP

4 Results and discussions

During inoculation, the microbacterials produced the synthesis exopolysaccharide different composition of chemical and physical properties. This exopolysaccharide production is affected by the growth phase and the development of microbacterial, nutrients available in its development media, also pH and temperature. This synthesis process is the process by which it is formed polymers and peptidoglycan on microbacterial cell walls. Unity the polymers and the peptidoglycan are then referred to as exopolysaccharide. Exopolysaccharides, which are secreted by communal microbacterials in the sample, can clog and glue the structure samples through the pores of the sand grains.

In the process of inoculation of microbacterials in the sand soil, it will there are 3 processes (Ivanov V and Chu J. 2008 [2]):

1. There is a bioclogging process that is the process by which microbacteria will producing both the material and the mass of the microbacterial cell itself which is capable of filling the interstellar cavities of the sand-soil can reduce porosity and sand permeability rate.
2. There is a biocementation process that is the process by which microbacteria produce a slime material extracellular polysaccharide able to increase the shear strength and stiffness of the sand soil.
3. There is a process that produces biogas which is the process where generated bubbles (air) through which microbacterial activity can be reduce the risk of liquefaction potential of sand soil.

The accumulation of cell mass that occurs in microbacteria, slime and gas produced microbacteria located on the sand soil will make the sand soil more impermeable to water [2]. This happens because extracellular polysaccharides produced by microbacteria will alter arrangement of sand structures. Production of these extracellular polysaccharides will bind the sand soil particles and fill the inter-cavities sand grains.

Based on Figure 3. located on the next page, it can be seen that in the process of biocementation material produced by microbacteria will bind the sand soil particles so that the sand soil become more dense and the ability to withstand the shear force will increased. Besides, the bioclogging process occurs clogging or blockage by material produced by microbacteria or by microbacterial activity that clogs or fills cavities between particles soil so that it can inhibit the flow of water as it passes through the particles sand soil and decrease the permeability of the sand soil.

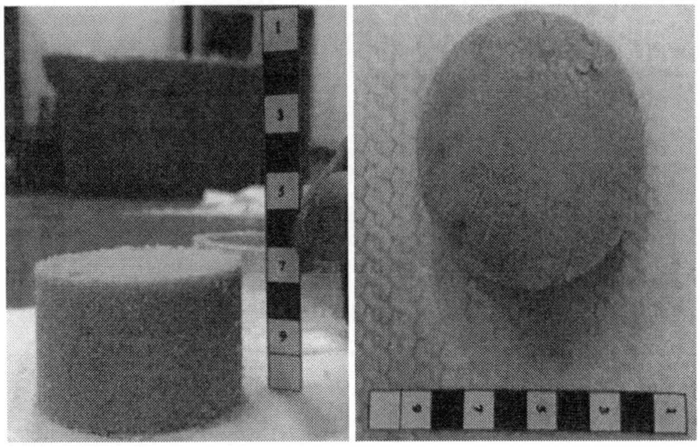

Figure 3. Sandy soil direct shear sample inoculated with *Pseudomonas sp*

In general, factors affecting extracellular production polysaccharide produced by microbacteria is growth medium, condition environment, and the formation of side reactions. Medium used to produce polysaccharide compounds usually contain carbon elements which is used to limit the comparison of nutrients with elements nitrogen. The change of 60-80% from the complex carbon source used into a simple polymer compound (hydrolysis) obtained high results from the fermentation process of the polysaccharide compound. An important step in the growth of microorganisms and the formation of products is the formulation of the medium the good one. Good growth and production medium should contain carbon, nitrogen, phosphorus, sulfur, potassium, and magnesium salts (Chawla et al. 2009 [4]). Besides, according to DeJong et al. 2006 [5], microorganisms are often lies in the cementation of sediments containing calcium, magnesium, iron, manganese and aluminum in the form of crystalline carbonates, silica, phosphates, sulfides and hydroxides, especially iron hydroxide.

In this research media used in the form of sugar (saccharose) and mineral salts, which are $CaCl_2$, $MgSO_4$, K_2HPO_4, and Saccharose. The carbon element which is required in the extracellular production process of the polysaccharide is obtained of saccharose ($C_{11}H_{22}O_{11}$). The required salts are potassium obtained of K_2HPO_4, magnesium is obtained from $MgSO_4$, Calcium is obtained from $CaCl_2$. Calcium (Ca) serves as a precipitation in calcite formation which will be influential in the bioclogging process or impairment permeability. Magnesium (Mg) functions in anaerobic fermentation play a role in biocementation. In addition, there are also Nitrogen (N2) substances obtained from the sample soil element.

Urease positive bacteria are commonly used for the biogrouting method and they are able to hydrolyse urea into ammonium and carbonate. The ammonium will increases the pH value, initiating the precipitation of calcium carbonate ($CaCO_3$) [6], which acts as a binding bridge between the sand particles. $CaCO_3$ precipitation occurs at the bacterial cell surface if there are sufficient concentration of Ca^{2+} and CO_3^{2-} in solution [7].

$$CO(NH_2)_2 + 2H_2O \rightarrow 2NH_4^+ + CO_3^{2-} \quad (6)$$

$$Ca^{2+} + CO_3^{2-} \rightarrow CaCO_3(s) \quad (7)$$

Compared with chemical or cement injection methods that are potentially damaging to the environment, MICP has been proposed as an environmentally friendly technique ([2, 8]), whereby the soil body can be treated in situ. Biogrouting technology can be used for applications such as increasing the stiffness and strength of sand ([9, 10]) and rocks to enhance slope stability and foundation bearing capacity, reducing foundation settlements, mitigating the liquefaction potential of saturated sand [5], reducing the hydraulic conductivity of dykes and dams by bioclogging, preventing soil erosion, reducing infiltration from ponds or to construction [2], strengthening of concrete and remediation of cracks [11].

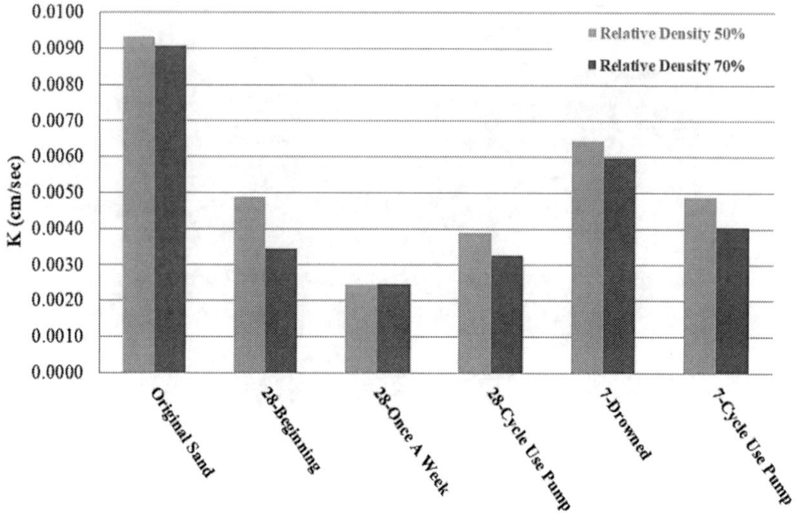

Figure 4. Chart of Constant Head Permeability Test Result

Table 2. Result Recapitulation of Direct Shear Strength Test

Treatments of Sample	Dr = 70%			Dr = 50%		
	Cohesion	Friction Angle	Shear Strength	Cohesion	Friction Angle	Shear Strength
	(kPa)	(Degrees)	(kPa)	(kPa)	(Degrees)	(kPa)
Original Sand	3.82	38.22	36.61	4.28	35.94	34.47
28-Beginning	5.03	34.02	33.13	5.07	34.27	33.44
28-Once A Week	7.28	35.95	37.47	6.05	33.53	33.64
7-Drowned	0.23	39.52	34.58	2.54	36.87	33.77
7-Cycle Use Pump	7.11	41.67	44.17	6.84	44.46	47.70

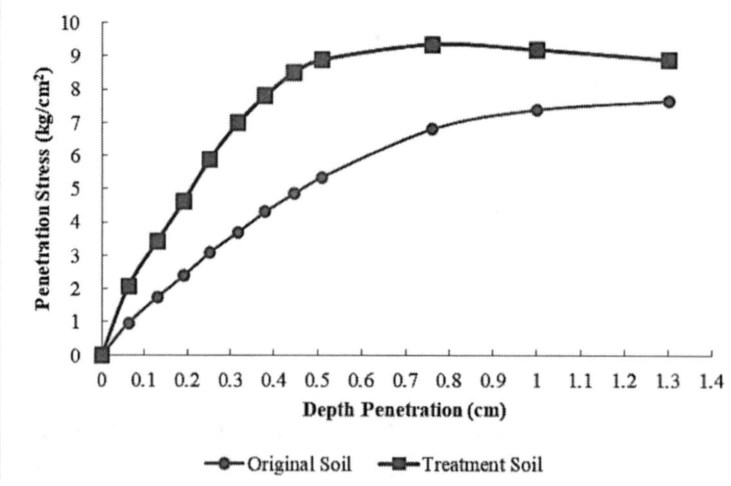

Figure 5. CBR Result with 10 Number of Hits

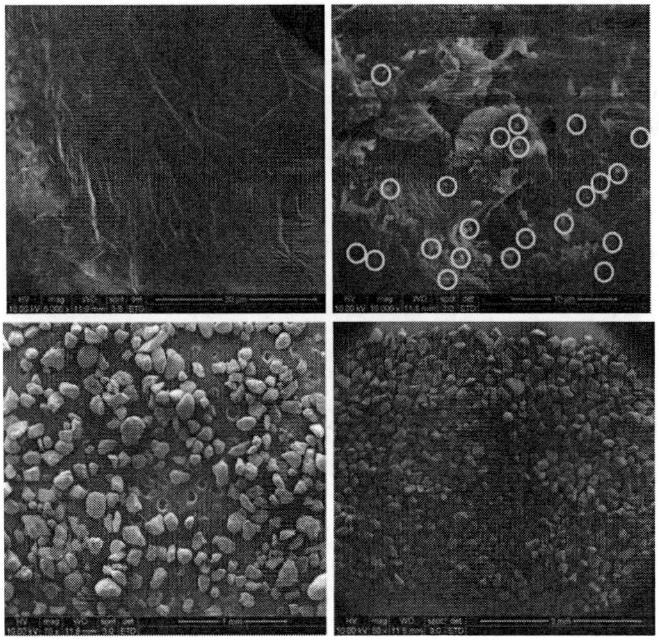

Figure 6. Comparison ESEM Test Result of Original Sand (left) with SEM Test Result of Sand inoculated with Pseudomonas sp. (right)

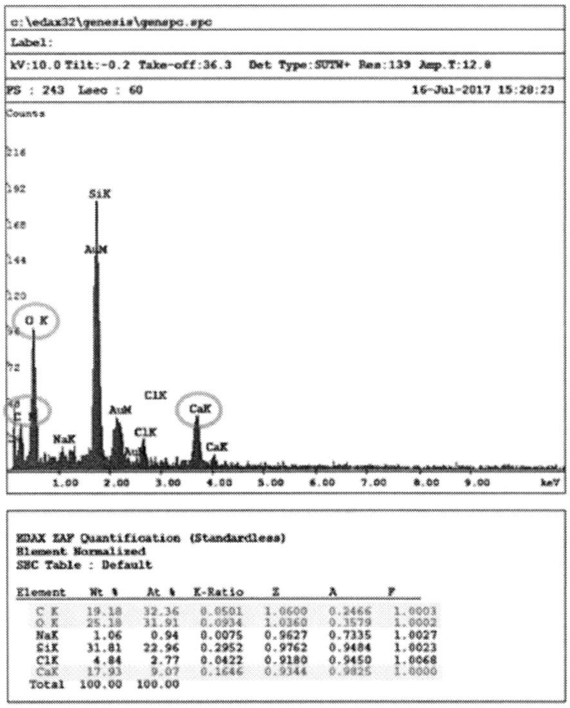

Element	Wt %	At %	K-Ratio	Z	A	F
C K	19.18	32.36	0.0501	1.0600	0.2466	1.0003
O K	25.18	31.91	0.0934	1.0360	0.3579	1.0002
NaK	1.06	0.94	0.0075	0.9627	0.7335	1.0027
SiK	31.81	22.96	0.2952	0.9762	0.9484	1.0023
ClK	4.84	2.77	0.0422	0.9180	0.9450	1.0068
CaK	17.93	9.07	0.1646	0.9344	0.9825	1.0000
Total	100.00	100.00				

Figure 7. Result of Surface's Scanning EDAX test for Sample Inoculated *Pseudomonas sp*

5 Conclusions and recommendations

The experiment with the variation of bacterial content, concluded that in constant head permeability test and direct shear strength test, the best way to make a highest permeability reduction in inoculation method is inject 15% bacteria relative to total weight of soil. Highest permeability reduction in inoculation method is the bacteria inoculation method with 28 curing days with relative density 70% and giving the nutrient once a week, which the value of reduction 73,73% relative to the value of original sandy soil. Highest shear value increation in inoculation method is the bacteria inoculation method with 7 curing days with relative density 50% and the inoculation process done with cycle treatment, which the value of cohesion is 6,84 and the value of friction angle is 44,46°.

Based on California bearing ratio test result, at CBR result with 10 number of hits, the treatment soil can proved a significant penetration stress changed. From Scanning Electron Microscope test result of the inoculation sand test images, it can be seen extracellular polysaccharide formed attached to the sand wall which will fill the pore space between the sand soil particles and embed the particles. This proves that micro bacteria that are cultured or inoculated into sand samples can live, experience growth and development. In addition, it also seen on the wall of samples picture of began to form exopolysaccharide. The formed exopolysaccharide attached to the wall of the sand soil particles which will fill, patch, and close the pores between the sand soil particles.

There are some notes need to be more attention, as well as: the tools used to support the growth and development of micro bacteria inside the soil should be in a sterile state in order to minimize the growth of other organisms; need to review for the bubbling effect on soil pores and the swelling effect against the accuracy of the results; and due to the complexity, the applications of biological microorganism to geotechnical engineering would require an integration of microbiology, ecology, geochemistry, and geotechnical engineering knowledge.

References

1. W.G. Characklis and P.A. Wilderer, Glossary. In: Characklis, W. G., Wilderer, P. A. (eds) "Structure and Function of Biofilms," Wiley, Chichester, United Kingdom of Great Britain, pp. 369-371 (1989).
2. V. Ivanov and J. Chu, Environ Sci. Biotechnol, **7**, 139-153 (2008)
3. A. Asyhari, M.R. Yunus, A.D. Haryani and M. Handayani, J. Appl. Technol. Environ. Sanitation, **4**, 47-52 (2014)
4. PR. Chawla, IB. Bajaj, AS. Shrikant and RS. Singhal, Food Technol. Biotechnol, **47**, 107-124 (2009)
5. J.T. DeJong, M.B. Fritzges and K. Nusslein, Journal of Geotechnical and Geoenviromental Engineering, **132**(11), 1381-1392 (2006)
6. DJ. Tobler, MO. Cuthbert, RB. Greswell, MS. Riley, JC. Renshaw, SH. Sidhu and VR. Phoenix, Geochim Cosmochim Acta, **75**, 3290-3301 (2011)
7. C. Qian, R. Wang, L. Cheng and J. Wang, Chin J Chem., **28**, 847-857 (2010)
8. RH. Karol, *Chemical Grouting and Soil Stabilization* (New York (NY): Marcel Dekker, 558, 2003)
9. L.A. Van Paassen, C.M. Daza, M. Staal, D.Y. Sorokin, W. van der Zon and M.C. van Loosdrecht, Ecol Eng., **36**, 168-175 (2010)
10. V.S. Whiffin, Microbial CaCO3 precipitation for the production of biocement (Doctoral dissertation, Murdoch University, 2004)
11. W. De Muynck, N. De Belie and W. Verstraete, Ecol Eng., **36**, 118-136 (2010)

The study for installing noise barrier on highway route no.9

Chaddanai Jiradecha[1] and Pruethipong Singhatiraj[2,a]

[1]*Department of Environmental Engineering, Kasetsart University, Thailand*
[2]*Department of Highways, Ministry of Transports, Thailand*

Abstract. Highway Route No. 9 is a Bangkok bypass motorway that aims to alleviate traffic congestion in greater Bangkok areas. Presently, it is found a rapid expansion of new communities residing along both side of Highway No. 9. To lessen the noise level, Department of Highways installed the noise barrier on the sensitive locations. However, people from communities living on both sides of highway still demand some additional noise pollution improvement. To improve the efficiency of using noise barriers and to promote environmental and social quality for the people on the sensitive communities along the highway route, the Department of Highways, therefore, designs a new technique for installing noise barrier on the sensitive areas, i.e. education institutions, temple, residential area. In addition, the mathematic model is applied to 12 selected locations of sensitive areas for monitoring the effectiveness of the new installation technique. The mathematic model is used for predict the future noise level after installation of noise barrier in year 2037. The results confirm that the appropriate height and width of noise barrier can improve the comfortable of the people on the sensitive area.

1 Introduction

Highway No.9 (Bangna - Bang Pa In) is a motorway network system starting at Wang Noi District, Ayutthaya Province and ending at Bang Phli District, Samut Prakarn Province. It is a Bangkok bypass road that aims to alleviate traffic problem in greater Bangkok areas. The project distance has 65.325 km. length. The alignment of the network is shown in Figure 1. From the air photo map shown in Figure 1, Most of the land currently used on the area around motorway route no.9 is for the purpose of agriculture products, mainly on rice planting. However, based on the analysis of land use data during 14 years (1999-2013), it shows that there is a trend of land use changing from agricultural purpose to be residential purpose. The percentage of rice field dropped from 43% in 1999 to 30% in 2013 while the percentage of residential area rose from 12% in 1999 to 30 % in 2013

As discussed above, it is found a rapid expansion of new residential community residing along the both side of Highway route no. 9. The previous study on the noise impact on the communities surrounding Highway 9 discloses that the noise level did not exceed the noise standard. Even though the noise level is within the standard, the Department of Highways installs the noise barrier to lessen traffic noise to the community. Nevertheless, People from communities living on both sides of the highway need some improvement on the noise pollution. So, a new installation technique of noise barrier is carry out in this study to reduce the traffic noise.

[a] Corresponding author : pruethi@gmail.com

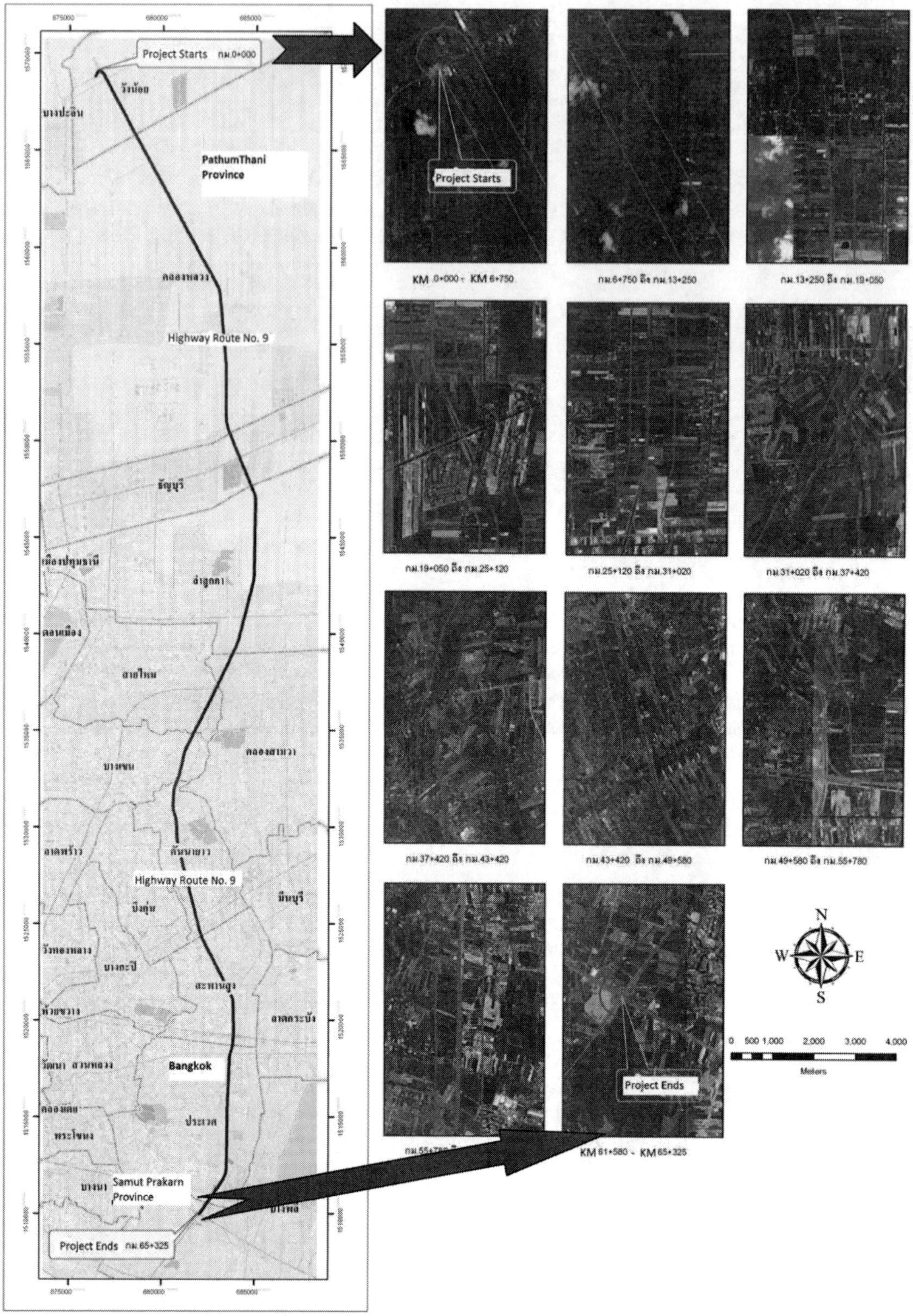

Figure 1. Project Alignment

2 Site investigation

2.1 Land uses

Most of the land currently used on the area around motorway route no.9 is for the purpose of agriculture products, mainly on rice. However, based on the analysis of land use data during 14 years (1999-2013), it shows that there is a trend of land use changing from agricultural purpose to be residential purpose. The percentage of rice field dropped from 43% in 1999 to 30% in 2013 while the percentage of residential area rose from 12% in 1999 to 30 % in 2013

2.2 Sensitive areas

Based on an investigation on the area within 500 meters from each side of the road throughout the project, it is found that there are 30 sensitive areas that may be affected by traffic noise. The 30 sensitive areas comprise of 8 educational institutions, 10 temples, 11 communities and 1 hospital.

2.3 Present condition of the existing noise barriers

Previously, there is some noise pollution complaints from the people living closed to the highway route no.9. To lessen the noise issue, the department of highways installed noise barriers on such area. So far, the Department of Highways installed noise barriers in 11 areas from 30 sensitive areas stated earlier. However, to fix systematically traffic noise problem, the detailed study of action plan for noise pollution was recognized by the Department of Highways and then carried out in this study. Presently, Glassfiber Reinforced Concrete (GRC) type and Fiber Reinforced Plastic (FRP) type are typically used by the Department of Highways.

2.4 Geological data

To design the foundation of noise barrier, the geological data on the project area is investigated. The soil characteristics along Highway No. 9 (Bangna - Bang Pa In) are mostly clay and sand and can be classified as very soft to soft clay. The thickness of the clay is about 4-5 meters depth in the first 7 kilometers and in the range of 5-15 meters from KM. 7 until the end of the project.

2.5 Acceptable noise level

Table 1. Acceptable Noise Level

Type of Land Use	Noise Level (dBA)	
	Day Time (06:00-22.00)	**Night Time (22:00-06:00)**
Sensitive Areas (School Hospital Temple)	55	50
Residential Area	60	55
Resident in Commercial Area	65	55
Entertainment Area and Agriculture	65	65
Commercial Area	70	60
Industrial Area	70	65

From various national standards, the acceptable noise level is normally set by the characteristic of land using. The noise level for each type of land using is mostly classified as daytime level (06:00-22:00 or 07:00-22:00) and night time level (22:00-06:00 or 22:00-07:00). So, in this study, the optimum values of noise level for each type of land use are classified as daytime and night time. Then, the public participation on the optimum values of acceptable noise level is conducted. This is done by interviewing people that are experimented on the noise comfortable test in 10-stations along the motorway route at different periods of time to determine the acceptable noise level. Then, the public hearing from various state agency and university's noise experts on the acceptable noise level are performed.

3 Traffic noise prediction model

The present condition of sound level and traffic volume are the key information necessary for developing traffic noise prediction model.

3.1 Present sound level

Based on the results of the sound level monitoring at 10 stations along the project route, the Leq 24 hr (average 24 hours A weighted Equivalent Continuous Sound Level) is higher than the standard sound level set by the National Environmental Board's Notification No. 15 (1997), which is set at 70 dBA. Moreover, the noise level in weekday is higher than the weekend. As a result, the average noise level in weekday is investigated for developing noise prediction model.

3.2 Traffic analysis

As stated earlier, the measured sound levels and their traffic volume are essential to develop traffic noise model. The two main purposes of the traffic data collection and traffic analysis are:
(a) to input to the traffic noise mathematical model (SoundPlan) for predicting sound level and then using sound level measured in section 3.1 to calibrate the predicted noise level result.
(b) to forecast future traffic volume in year 2037 and then using these future traffic volumes to predict the future noise level by using the developed mathematic model from (a).

From the investigation, it is noted that the average traffic volume in weekday is more than that in weekend and more than the annual average traffic volume (AADT). As a result, the average traffic volume in weekday is used in the mathematical model for noise level analysis. Based on an investigation on the area within 500 meters from each side of the road throughout the project, it is found that there are 30 sensitive areas that may be affected by traffic noise. The 30 sensitive areas comprise of 8 educational institutions, 10 temples, 11 communities and 1 hospital.

3.3 Future noise level

From the model, it is shown that the traffic volume increases at the average rate of 3.222 percent per annum. This leads the rate of noise level growth 0.153 percent per year. It can be summarized that increasing in traffic volume results slightly on noise level. Even though, the predicted future noise level on the project area increased slightly, the present noise volume exceeds the appropriate noise level. So, the application of noise barrier is essential on the project area to lessen noise pollution.

4 Noise barrier installation study

Presently, the noise barrier are made from 8 types of materials: 1) Concrete Products 2) Metals 3) Plastics 4) Recycled Materials 5) Fiberglass Reinforced Plastic (FRP) 6) Glass 7) Acrylic and 8) Wood. This study is aimed to investigate the most suitable type of noise barrier for current road conditions. The criteria for study are listed as follows:

(a) Noise barriers must be a sound-proof and suited with the highway works.
(b) Noise barriers must be certified internationally.
(c) Noise barriers should be appropriate for Thai Environment.
(d) Noise barriers should be available in Thailand

(e) Noise barriers must be consistent with local restrictions and the need of people in the community.
(f) Physical Property
(g) Easy to Erection
(h) Ease of Maintenance
(i) Price
(j) Safety

Based on the criteria set above, the most appropriate of noise barrier is Acrylic type and then follow by Fiberglass Reinforced Plastic (FRP), Glass Fiber Reinforced Cement (GRC), and Metal type, subsequently. However, the additional concepts were applied for the selection of noise barrier in this study.

(a) In area of pleasing roadside view, the transparent acrylic noise barrier should be selected.
(b) In area of displeasing roadside view, the non-transparent acrylic noise barrier should be selected.
(c) In area of high-rise buildings, the FRP noise barrier should be selected to absorb the reflection noise.

Due to the fact that most of the land use characteristic along the project area is agricultural and residential purpose. So, the acrylic type noise barrier is preferred for this study.

5 Pilot areas

To prove the effective of the above concept, some sensitive areas within a 100-meter from the roadway along the motorways route no.9 are selected as pilot areas. The SoundPlan program is used to predict the noise level. The pilot areas comprise of 12 locations, i.e. 6 education institutes, 5 residential areas, and 1 temple.

The following assumptions are used for the study.

(a) Acrylic type noise barrier is used for erection.
(b) SoundPlan Program is used for the design.
(c) The maximum height of noise barrier is 10 m.
(d) The target is to reduce noise level in year 2037 at least 10 dBs in each pilot locations, guided by The Federal Highway Authority (FHWA).

Prior to the application of new noise barrier, the noise level on all the pilot areas exceed the acceptable level. However, after the application of new noise barriers, 7 areas are within the acceptable level but 5 areas still beyond the satisfactory level. This is because these 5 areas are quite adjacent to the road, frontage road, or the zone closed to the major traffic intersections which are the source of traffic noise. However, the traffic volume decreases with increasing distance from the road.

Moreover, the results from SoundPlan program indicate that the new noise barrier can absorb noise level by average about of 10.7 and 10.9 decibels, in daytime and night time, respectively. The details are list on Table2.

Table 2. Noise Level Analysis Using SoundPlan (After Installation of Noise Barrier)

No.	Sensitive Area	Length (meter)	Height (meter)	Acceptable Noise Level (dBA)		Before Installed (dBA)		After Installed (dBA)		Noise level Reduction (dBA)	
				Day	Night	Day	Night	Day	Night	Day	Night
1	Education institute	230	5	55	50	59.9	58.4	49.1	48.2	-10.8	-10.2
2	Residential area	300	5	55	50	68.6	63.6	58.6	52.9	-10.0	-10.7
3	Temple	870	4.5	55	50	70.4	67.7	57.9	54.8	-12.5	-12.9
4	Education institute	235	5.5	55	50	75.5	69.9	64.6	59.5	-10.9	-10.4
5	Residential area	880	5.5/5	60	55	66.2	63.3	56	52.2	-10.2	-11.1
6	Education institute	1,130	5	55	50	65.6	61.7	55.6	51.1	-10.0	-10.6

7	Residential Area	560	4	60	55	69.5	64.1	59.5	54.8	-10.0	-9.3
8	Residential Area	300	4	65	55	71.7	68.2	60.4	56.7	-11.3	-11.5
9	Education institute	300	4	55	50	71.1	67.9	60.6	57	-10.5	-10.9
10	Education institute	900	5	55	50	65.8	61.2	55.8	51.9	-10.0	-9.3
11	Residential Area	300	4	65	55	68.4	62.9	57.8	52.4	-10.6	-10.5
12	Education institute	300	5.5	55	50	77.3	71.8	67	61.7	-10.3	-10.1

6 Detailed design of noise barrier

After the new noise barrier concept was verified by the mathematic model, the next step is to develop the detailed design of the noise barrier for erection.

(a) Acrylic Barrier. Acrylic type 12 mm. thickness is used. There are 3 levels. The lower level is 1m.x2m., blue color (Opacity 60%). The middle level is 2m.x2 m., blue color (Opacity 85%). The top level is 1m.x2 m., Opacity 92%. The vertical frame is WF 100 x 100 (17.2 Kg/m) @ 2 meters. The horizontal frame is aluminum made.

(b) Main Structure. The wide flange 250 x 125 (29.6 Kg/m) is used for the main structure to support 5 meter height noise barrier. These wide flanges are installed every 4 meters.

The detailed of this design is shown on Figure 2, 3 and 4.

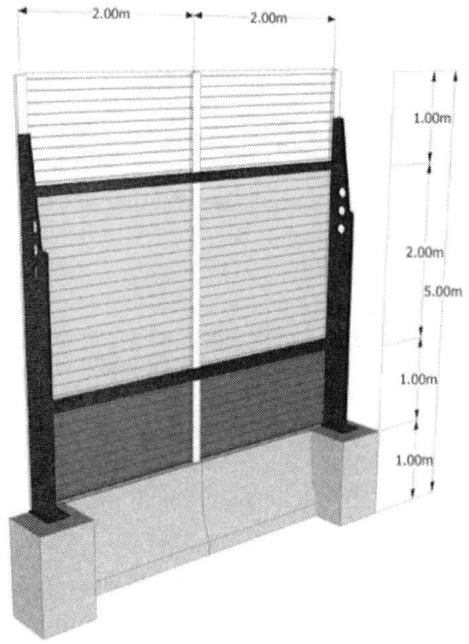

Figure 2. Noise Barrier on the Pilot Area

Figure 3. Noise Barrier on the Pilot Area

Figure 4. Noise Barrier on the Pilot Area

7 Conclusion

The traffic noise level is quite high at the locations closed to the roadway. This is because it closes to the source of traffic noise. The volume decreases as the distance from the roadway. The noise level increase at an average rate of 0.153 percent per year, while the traffic volume increases at an average rate of 3.222 percent per year.

For the study of noise barrier products, acrylic type is the most appropriate to implement in the project when considering the following factors, e.g. physical properties, easy to erection, ease of maintenance, price, safety, etc.

There are 12 locations that are selected as the pilot areas for noise barrier installation. These 12 locations are select by three main factors: current noise pollution situation, traffic noise impact, and the value of investment. It can be concluded from the pilot areas that the appropriate height and width of noise barriers can improve the comfortable of the people on the sensitive area. However, the installation of merely noise barrier is not possible to reduce the traffic noise to the appropriate noise level. The following measures shall be applied to support.

(a) Changing road pavement surface by using more porous material. This will reduce the noise caused by friction between the tire and the road surface.

(b) Limiting speed.

(c) Intelligence Traffic Management System to reduce the conglomerate of the vehicle. Then, it will alleviate the accumulation of noise pollution caused from the engine of the vehicle on that location.

(d) Dense leaf types trees planting, such as Ashoka Tree. This type of tree will help to absorb sound level from the roadway.

References

1. J. Chaddanai, The study on noise pollution and noise barrier on highway route no. 9 (Bangkok: Kasetsart Universtiy, 2016)
2. National Environmental Board's Notification No. 15 (1997)

Numerical study of the burner parameters on the thermal field in a sulfur recovery unit thermal reactor

Chun-Lang Yeh[a]

Department of Aeronautical Engineering, National Formosa University, Taiwan

Abstract. A sulfur recovery unit (SRU) thermal reactor is the most important equipment in a sulfur plant and is negatively affected by high temperature operations. In this paper, the effect of burner parameters, including the clearance of the acid gas tip and the inlet air swirler angle, on the thermal field in a SRU thermal reactor are investigated numerically, with the aim to reduce the high temperature inside the thermal reactor and to ensure an acceptable sulfur recovery. The simulation results show that the burner with a smaller clearance of the acid gas tip produces a lower temperature, a lower exit SO_2 mole fraction and higher exit S_2 and H_2S mole fractions. Among the clearancs of the acid gas tip investigated, the horizontal clearance of 152.4mm and vertical clearance of 240mm yield the lowest temperature, exit SO_2 mole fraction and highest exit S_2, H_2S mole fractions. The burner with a smaller inlet air swirler angle produces a higher temperature, a higher exit SO_2 mole fraction and lower exit S_2 and H_2S mole fractions. Among the swirler angles investigated, 60° yields the lowest temperature, exit SO_2 mole fraction and highest exit S_2 , H_2S mole fractions.

1 Introduction

Desulfurization is very important in the petroleum refining process because oxysulfides from the petroleum refining process are one of the major sources of air pollution. The most frequently used desulfurization process is the Claus process, which converts the H_2S in natural gas or crude oil into sulfur elements and reduces the formation of oxysulfides. A SRU thermal reactor is the most important equipment in a sulfur plant. It converts the NH_3, H_2S and hydrocarbons in the reactants into sulfur. Most of the sulfur elements are recovered from the SRU thermal reactor. The first section of a SRU that uses the Claus process is composed of a burner, a thermal reactor and a waste heat exchanger. The configuration and dimensions of the first section of a SRU for a typical petroleum refinery are shown in Fig.1. The inner surface of a SRU thermal reactor is fabricated using a refractory to protect its walls because a SRU thermal reactor operates at very high temperatures. SRU thermal reactors are negatively affected by high temperature operations because high temperature can damage the refractory and the heat exchanger tubes.

[a] Corresponding author : clyeh@nfu.edu.tw

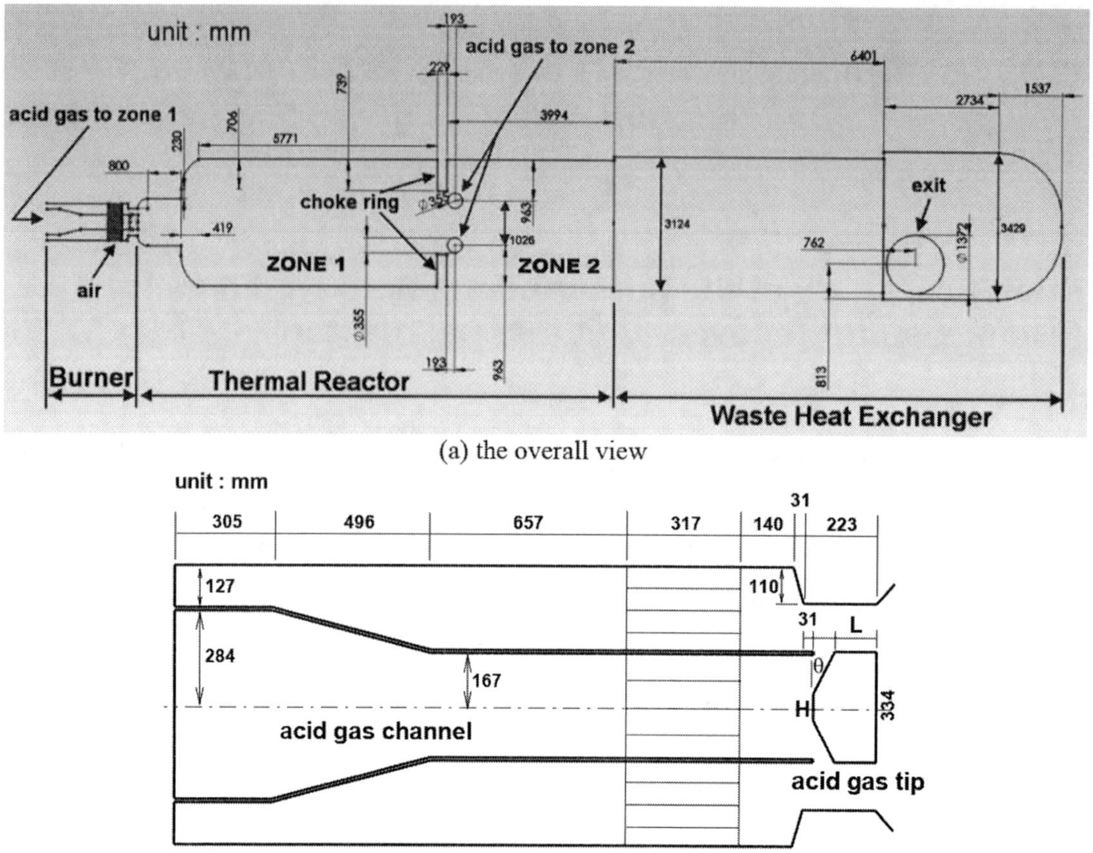

(a) the overall view

(b) enlarged view for the burner section

Figure 1. The configuration and dimensions of the first section of the SRU for a typical petroleum refinery

There have been theoretical and experimental studies of SRU thermal reactors. Adewale et al. [1] studied the thermal decomposition of H_2S into hydrogen and sulfur using a process simulator. Using the net fraction of the acid gas feed to the cracking coils as the controlling parameter, its effect on hydrogen production, the thermal reactor's energy requirement, the stability of the burner flame, steam production, the temperature of a Claus reactor and sulfur recovery of the primary SRU was studied. Chardonneaua et al. [2] presented experimental and simulation results for the addition of various amounts of toluene or carbon dioxide/toluene mixtures into the H_2S gas stream. The results show that there is a decrease in conversion efficiency when the amount of toluene or carbon dioxide/toluene added to the H_2S gas stream increases. The role of the reactor's operating temperature was also studied. The addition of toluene increases the optimum reactor temperature for enhanced sulfur recovery, but the presence of CO_2 reduces the optimum operating temperature. Manenti et al. [3] proposed a kinetic model with 2400 reactions and 140 species and implemented it in a proper reactor network to characterize the thermal reactor and the waste heat exchanger of sulfur recovery units. By doing so, reliable estimation of acid gas conversion, elemental sulfur recovery, and steam generation are achieved with the possibility to carry out an integrated process-energy optimization at the total plant scale. Selim et al. [4] examined the quality of sulfur deposits that were collected from H_2S combustion. Sulfur deposits from H_2S combustion under various conditions were captured and analyzed using X-ray powder diffraction and laser induced breakdown spectroscopy diagnostics. ZareNezhad and Hosseinpour [5] investigated different alternatives for increasing the reactor temperature of Claus SRUs by chemical equilibrium calculations. They found that the acid gas enrichment is a reliable technique for providing the required reactor temperature when a high flow of too lean acid gas is to be processed in a Claus unit. Monnery et al. [6] experimentally studied the reaction between H_2S and SO_2 using practical Claus thermal reactor temperatures between 850 and

1150°C and residence times between 0.05 and 1.2 seconds. The kinetic data obtained were used to develop a new reaction rate expression.

Our experience of operating a practical SRU thermal reactor in Taiwan shows that the refractories at the zone 1 corner and the choke ring are the parts of the thermal reactor that experience the greatest deterioration. The zone 1 corner has a suddenly expanded geometry and a recirculation zone forms behind it. The temperature at the zone 1 corner can exceed the maximum service temperature of the refractory and cause collapse or deformation. The choke ring is subjected to a bending moment from the rapid combustion gas stream and can collapse or deform. Several methods to alleviate this problem have been assessed, including (1) changing the choke ring dimensions [7], (2) changing the choke ring position [8], (3) modifying the geometry of the zone 1 corner to a streamlined geometry [9,10] and (4) replacing the choke ring by a vector wall [10,11]. However, these methods are either expensive or require an overhaul of the SRU thermal reactor. In the author's previous studies, two easier and more economical methods, including (1) changing the zone 1 fuel mass fraction [12] and (2) changing the inlet air quantity or the inlet O_2 mole fraction [13], were investigated numerically.

In a SRU, most of the acid gas is combusted in the burner which is also supplied with the necessary combustion air. A portion of the acid gas is bypassed around the burner to enter directly into the thermal reactor through nozzles. These nozzles are positioned at the start of the second zone of the thermal reactor. The amount of acid gas bypass is regulated to maintain an acceptable temperature in the first zone. The combustion and fluid flow in the burner has a significant influence on the thermal and fluid flow development in the downstream thermal reactor and heat exchganger. In this paper, the effects of burner parameters, including the clearance of the acid gas tip and the inlet air swirler angle, on the thermal field in a SRU thermal reactor are investigated numerically. The purpose of this study is to reduce the high temperature and to ensure an acceptable sulfur recovery in a thermal reactor. Practical operating conditions from a petrochemical corporation in Taiwan were used as the design conditions for the discussion.

2 Numerical methods and physical models

In this study, the FLUENT commercial code is used to simulate the reacting and fluid flow in a SRU thermal reactor. The SIMPLE algorithm by Patankar [14] is used to solve the governing equations. The discretizations of convection terms and diffusion terms are respectively performed using the power-law scheme and the central difference scheme. In terms of physical models, considering the accuracy and stability of the models and the evaluations of other researchers, the standard k-ε Model [15], the P-1 radiation model [16] and the non-premixed combustion model with β–type probability density function [17] are respectively used for the turbulence, radiation and combustion simulations. The standard wall functions [18] are used to resolve the flow quantities (velocity, the temperature, and the turbulence quantities) at the near-wall regions. The detailed governing equations and convergence criterion were described in a previous study by the author [19].

Figure 2. The numerical model for the SRU thermal reactor investigated

In this study, the numerical model of a SRU thermal reactor is constructed using an unstructured grid. Fig.2 shows the numerical model of the SRU thermal reactor investigated. In Fig.2, the heat exchanger section consists of 19 cooling tubes, which have a diameter of 0.5m, as shown schematically in Fig.3. The heat absorption rate for each heat exchanger tube is 40,000 W/m^2 and the other walls are adiabatic. No slip condition is applied on any of the solid walls. The exit of the heat exchanger section is connected to other equipment at 300 K and 1 atm by a pipe that is 1.372m in diameter and 11.5m in length.

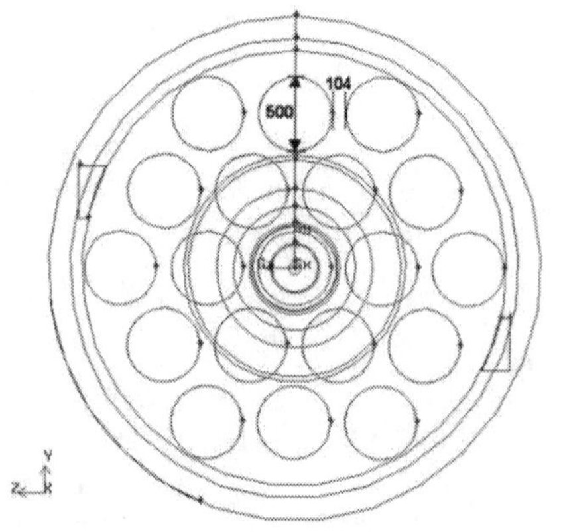

Figure 3. An illustration of the arrangement of heat exchanger tubes

Table 1. The design conditions at the acid gas inlet holes and the air inlet hole

	Oxygen-normal supply				Oxygen-rich supply			
x (%) location / species	Acid gas to zone 1	Acid gas to zone 2	Air inlet		Acid gas to zone 1	Acid gas to zone 2	Air inlet	
O_2	0	0	19.87		0	0	23.85	
N_2	0	0	74.98		0	0	71.26	
H_2O	7.83	4.12	5.15		4.12	27.97	4.89	
CO_2	1.27	1.5	0		1.48	0	0	
H_2S	82.06	89.88	0		89.9	39.61	0	
CH_4	2.28	2.7	0		2.7	0	0	
C_2H_6	1.52	1.8	0		1.8	0	0	
NH_3	5.04	0	0		0	32.42	0	
T (K)	319.92	316.15	403.15		313.15	316.15	397.15	
P (N/m^2)	76920	75068	74382		75068	75068	89572	
V (m/sec)	11.62	2.08	radial	tangential	11.46	1.88	radial	tangential
			12.4	34.1			10.8	29.8

In this study, two types of oxygen supplies are investigated: an oxygen-normal supply and an oxygen-rich supply. An oxygen-rich supply increases sulfur recovery. The design conditions (including the species compositions, the temperature, the pressure and the velocity) at the acid gas inlet holes of zone 1 and zone 2 and at the air inlet hole are listed in Table 1. These conditions are practical operating conditions that are used by a petrochemical corporation in Taiwan. The turbulence kinetic energy is 10% of the inlet mean flow kinetic energy and the turbulence dissipation rate is computed using Eq.(1).

$$\varepsilon = C_{\mu}^{3/4} \frac{k^{3/2}}{l} \qquad (1)$$

where l=0.07L and L is the hydraulic diameter.

The grid independence test and validation of the numerical methods have been performed in the author's previous studies mentioned above.

3 Results and discussion

To investigate the effects of burner parameters, parametric studies including the clearance of the acid gas tip and the inlet air swirler angle are performed. Their effects on the thermal field in a SRU thermal reactor are investigated numerically.

3.1 The effect of clearance of the acid gas tip

Figure 1(b) shows the configuration and dimension of the burner section, from which it is seen that length of the acid gas tip (L) controls the horizontal clearance between the acid gas tip and the acid gas channel. The horizontal clearance decreases with L.

Figure 4 shows a comparison of the cross-sectional average temperatures for SRU thermal reactors with different horizontal clearances of the acid gas tip. The vertical clearance is kept constant by keeping the height of the acid gas tip (H) at 80mm. Six lengths of the acid gas tip (L) are used and compared, including 25.4mm, 50.8mm, 76.2mm, 101.6mm, 127mm and 152.4mm, among which 127mm is the default L of a petrochemical corporation in Taiwan. It can be observed that a larger horizontal clearance (smaller L) tends to yield a higher temperature. Slower acid gas ejected from the acid gas tip produces a better mixing with the swirling air. In contrast, faster ejected acid gas may retard the mixing with the swirling air. The smallest horizontal clearance (L=152.4mm) leads to the lowest temperature.

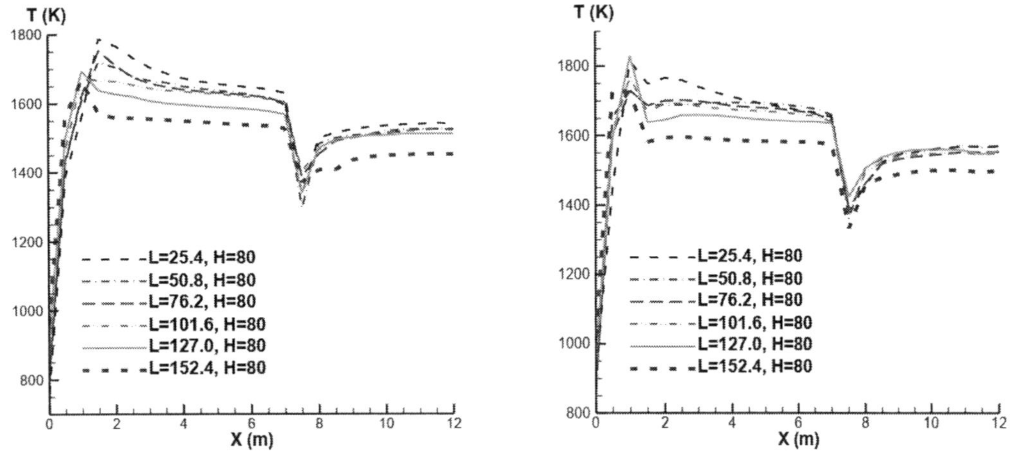

(a) oxygen-normal supply (b) oxygen-rich supply

Figure 4. A comparison of the cross-sectional average temperatures for SRU thermal reactors with different horizontal clearances of the acid gas tip

Figure 5 shows the temperature profiles on the symmetric plane for SRU thermal reactors with different horizontal clearances of the acid gas tip. The zone I average temperature and the exit S_2, SO_2 and H_2S mole fractions are labelled in the figures, too. It can be observed that a larger horizontal clearance tends to yield a higher zone I average temperature. This is similar to the result of Fig.4. In addition, the exit SO_2 mole fraction increases while the exit S_2 and H_2S mole fractions decrease with the horizontal clearance of the acid gas tip. This is because more H_2S is converted to S_2 due to the higher temperature when the horizontal clearance of the acid gas tip increases. However, S_2 can also react with the air to form SO_2 at higher temperatures. The burner with a smaller horizontal clearance of the acid gas tip produces a lower temperature, a lower exit SO_2 mole fraction and higher exit S_2 and H_2S mole fractions. Among the horizontal clearancs of the acid gas tip investigated, L=152.4mm yields the lowest temperature, exit SO_2 mole fraction and highest exit S_2 , H_2S mole fractions.

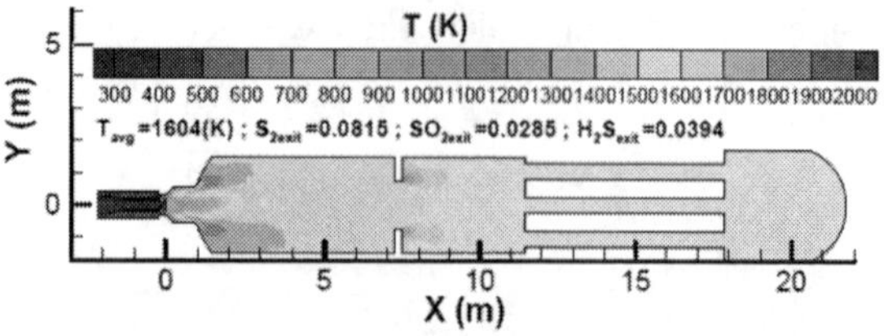

(a) L=25.4mm, H=80mm, oxygen-normal supply

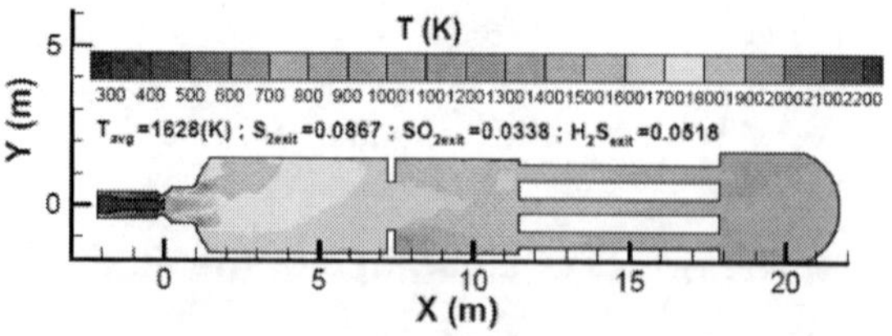

(b) L=25.4mm, H=80mm, oxygen-rich supply

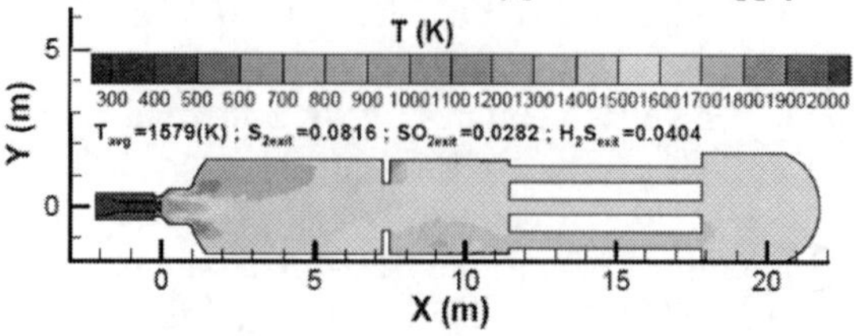

(c) L=50.8mm, H=80mm, oxygen-normal supply

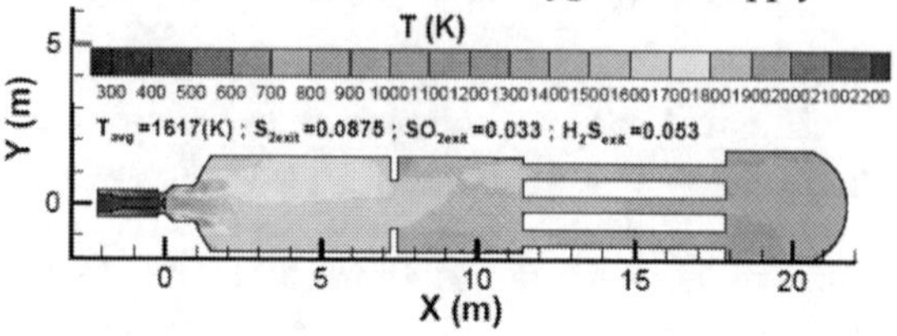

(d) L=50.8mm, H=80mm, oxygen-rich supply

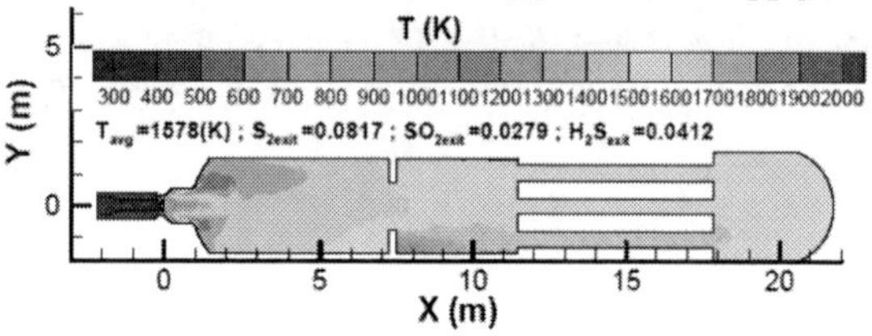

(e) L=76.2mm, H=80mm, oxygen-normal supply

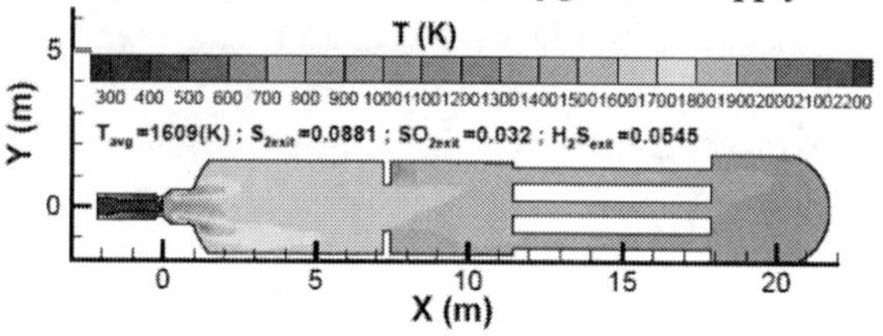

(f) L=76.2mm, H=80mm, oxygen-rich supply

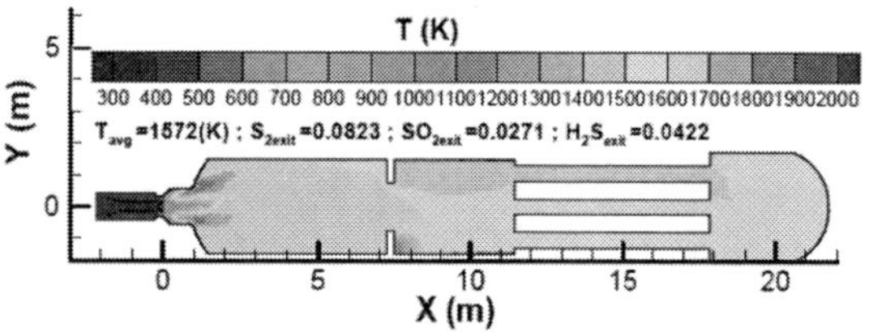

(g) L=101.6mm, H=80mm, oxygen-normal supply

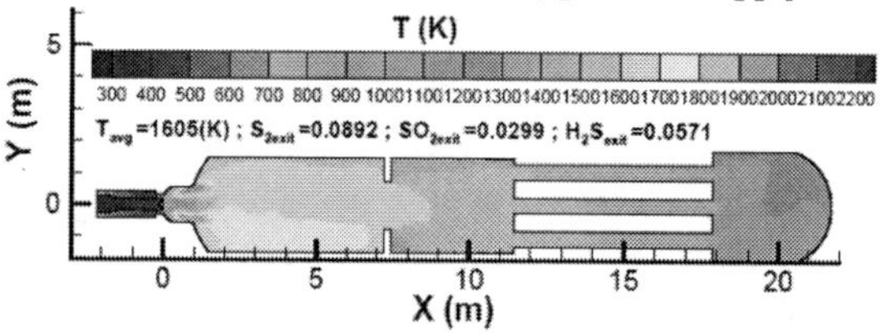

(h) L=101.6mm, H=80mm, oxygen-rich supply

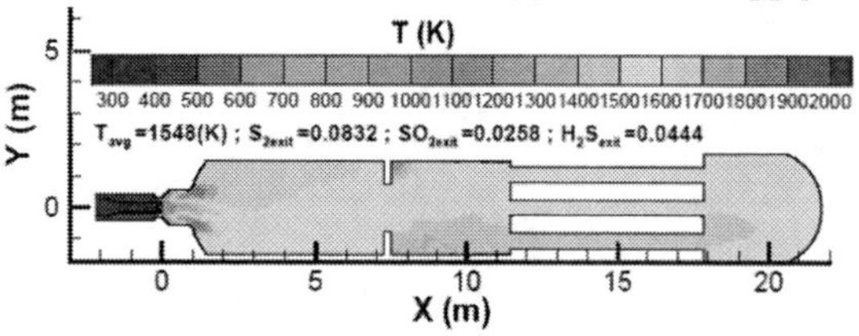

(i) L=127mm, H=80mm, oxygen-normal supply

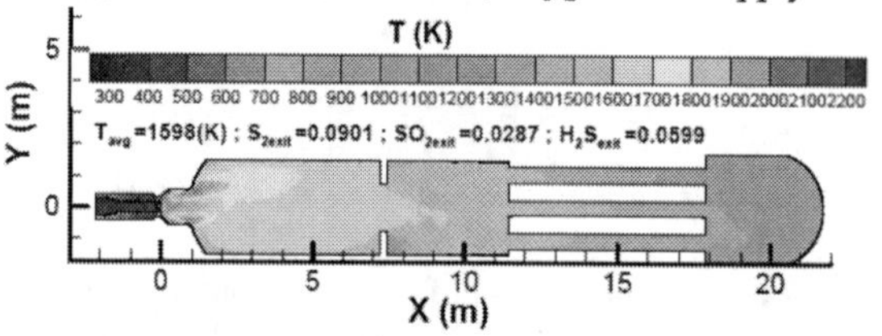

(j) L=127mm, H=80mm, oxygen-rich supply

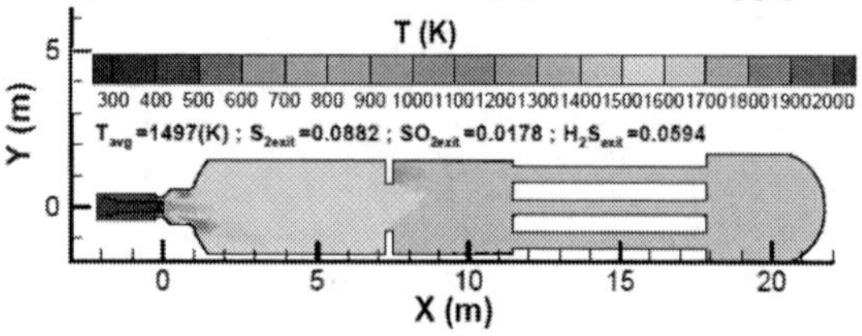

(k) L=152.4mm, H=80mm, oxygen-normal supply

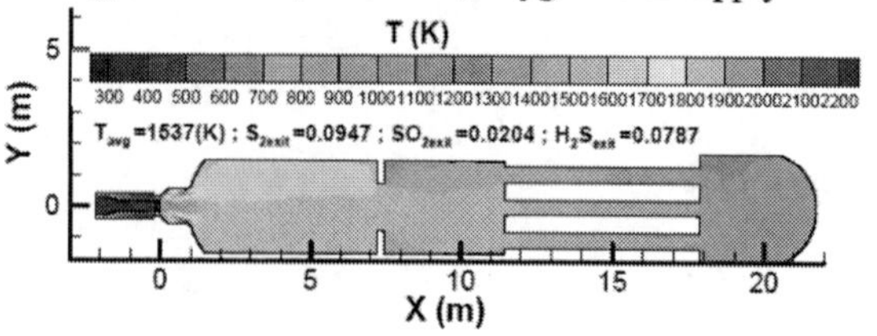

(l) L=152.4mm, H=80mm, oxygen-rich supply

Figure 5. The temperature profiles on the symmetric plane for SRU thermal reactors with different horizontal clearances of the acid gas tip

In Fig.1(b), the height of the acid gas tip (H) controls the vertical clearance of the acid gas tip. The vertical clearance decreases with H. Figure 6 shows a comparison of the cross-sectional average temperatures for SRU thermal reactors with different vertical clearancs of the acid gas tip. The horizontal clearance is kept constant by keeping the length of the acid gas tip (L) at 127mm. Four hights of the acid gas tip (H) are used and compared, including 40mm, 80mm, 160mm and 240mm, among which 80mm is the default H of a petrochemical corporation in Taiwan. It can be observed that a larger vertical clearance (smaller H) tends to yield a higher temperature. This result is similar to that of a larger horizontal clearance. The smallest vertical clearance (H=240mm) leads to the lowest temperature.

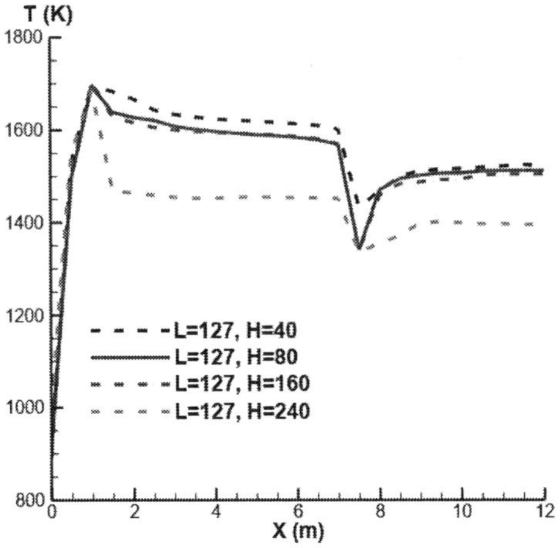

(a) oxygen-normal supply

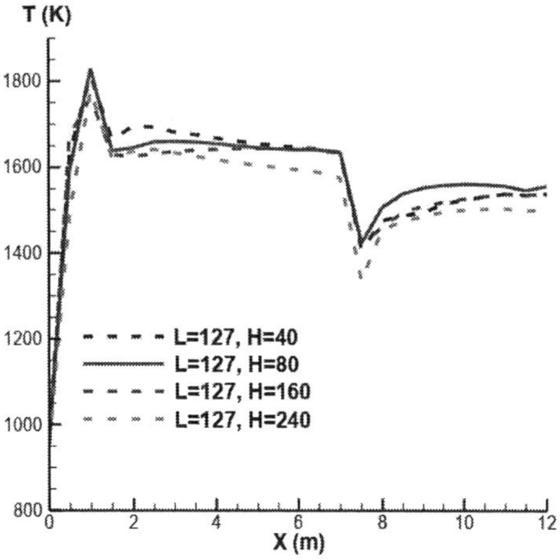

(b) oxygen-rich supply

Figure 6. A comparison of the cross-sectional average temperatures for a SRU thermal reactor with different vertical clearances of the acid gas tip

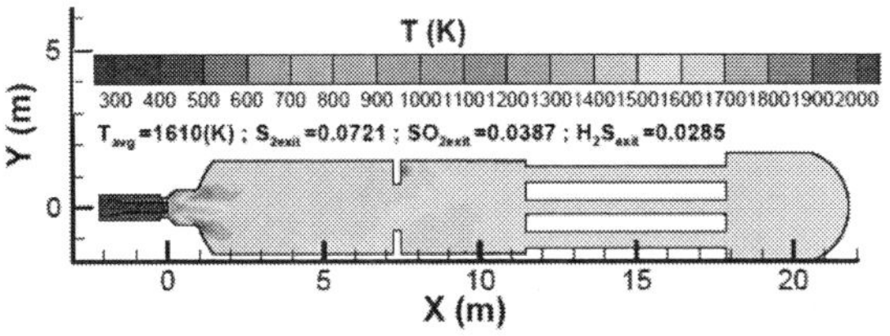

(a) L=127mm, H=40mm, oxygen-normal supply

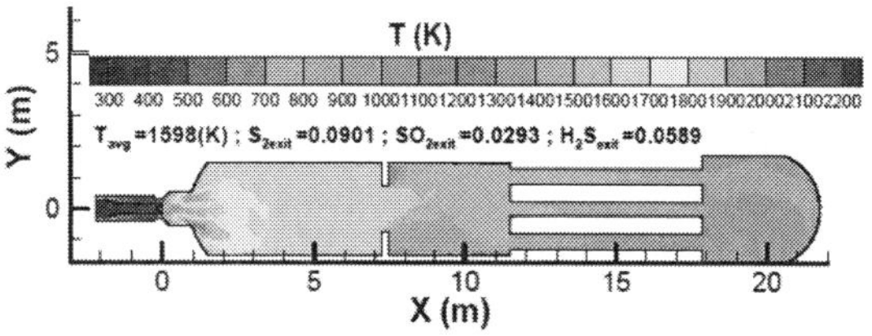

(b) L=127mm, H=40mm, oxygen-rich supply

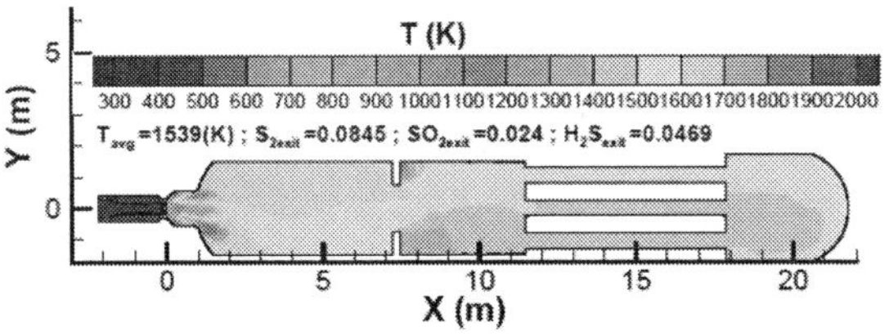

(c) L=127mm, H=160mm, oxygen-normal supply

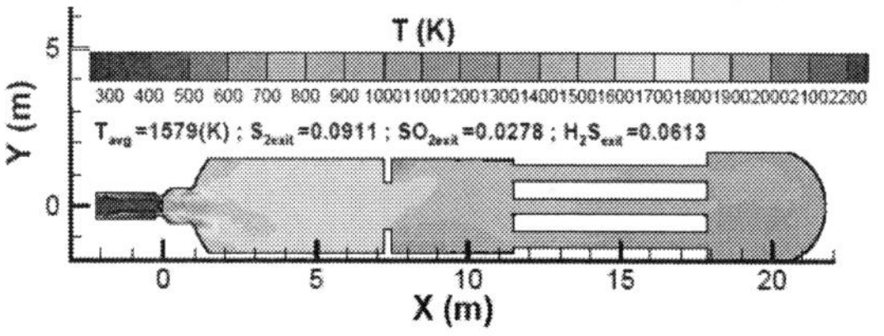

(d) L=127mm, H=160mm, oxygen-rich supply

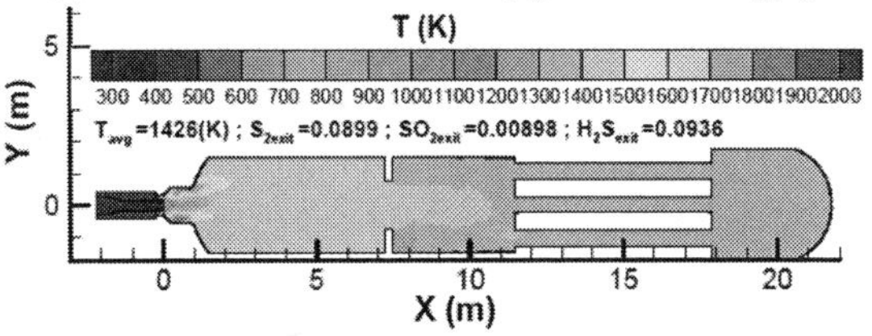

(e) L=127mm, H=240mm, oxygen-normal supply

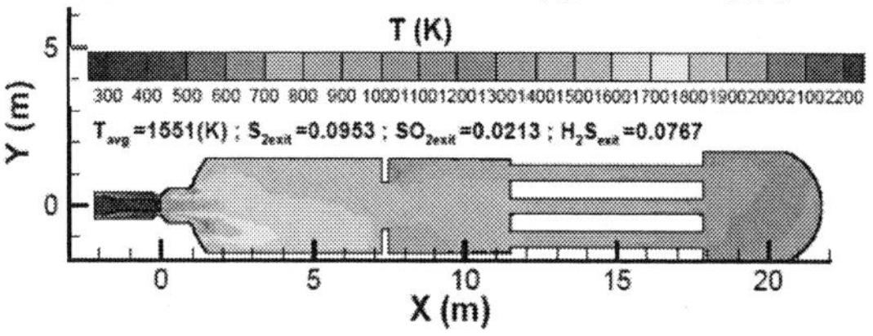

(f) L=127mm, H=240mm, oxygen-rich supply

Figure 7. The temperature profiles on the symmetric plane for SRU thermal reactors with different vertical clearancs of the acid gas tip

Figure 7 shows the temperature profiles for SRU thermal reactors with different vertical clearancs of the acid gas tip. Note that the case of L=127mm and H=80mm has been shown in Figs.5(i) and (j). It can be observed that a larger vertical clearance of the acid gas tip tends to yield a higher zone I average temperature. This is similar to the result of Fig.6. In addition, the exit SO_2 mole fraction increases while the exit S_2 and H_2S mole fractions decrease with the vertical clearance of the acid gas tip. This is similar to the result of Fig.5. From the results of Figs. 5 and 7, the burner with a smaller clearance of the acid gas tip produces a lower temperature, a lower exit SO_2 mole fraction and higher exit S_2 and H_2S mole fractions. Among the vertical clearancs of the acid gas tip investigated, H=240mm yields the lowest temperature, exit SO_2 mole fraction and highest exit S_2 , H_2S mole fractions.

3.2 The effect of air swirler angle

The inlet air swirler angle, θ_{sw}, controls the strength of the swirling motion. The swirling strength can be expressed in terms of the the radial (v_r) and tangential (v_θ) components of the inlet air velocity by the following equation.

$$v_r / v_\theta = \tan \theta_{sw} \quad (2)$$

A smaller θ_{sw} corresponds to a stronger swirling motion.

Figure 8 shows a comparison of the cross-sectional average temperatures for SRU thermal reactors with different inlet air swirler angles. The length (L) and height (H) of the acid gas tip are kept at 127mm and 80mm, respectively. Four inlet air swirler angles (θ_{sw}) are used and compared, including 10°, 20°, 30° and 60°. It can be observed that a smaller θ_{sw} tends to yield a higher zone I temperature. This is because a stronger swirling motion leads to a better mixing, a more complete chemical reaction and hence a higher temperature. The smallest θ_{sw} (10°) produces the highest zone I temperature.

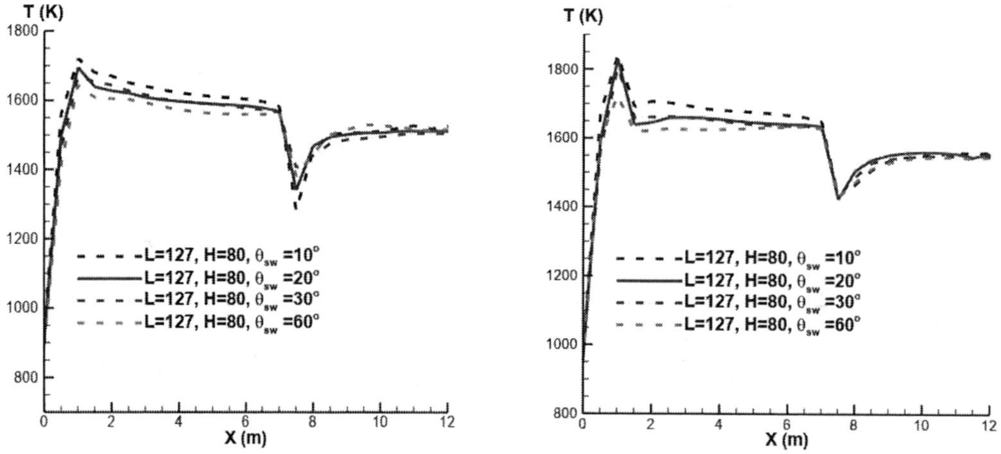

(a) oxygen-normal supply (b) oxygen-rich supply

Figure 8. A comparison of the cross-sectional average temperatures for SRU thermal reactors with different inlet air swirler angles

Figure 9 shows the temperature profiles for SRU thermal reactors with different inlet air swirler angles. Note that the case of θ_{sw} =20° has been shown in Figs.5(i) and (j). It can be observed that a smaller θ_{sw} tends to yield a higher zone I average temperature. This is similar to the result of Fig.8. In addition, for a smaller θ_{sw}, the exit S_2 and H_2S mole fractions are lower while the exit SO_2 mole fraction is higher. This is because more H_2S is converted to S_2 due to the higher temperature when θ_{sw} is smaller. However, S_2 can also react with air to form SO_2 at a higher temperature. The burner with a smaller θ_{sw} produces a higher temperature, a higher exit SO_2 mole fraction and lower exit S_2 and H_2S mole fractions. Among the swirler angles investigated, θ_{sw}=60° yields the lowest temperature, exit SO_2 mole fraction and highest exit S_2 , H_2S mole fractions.

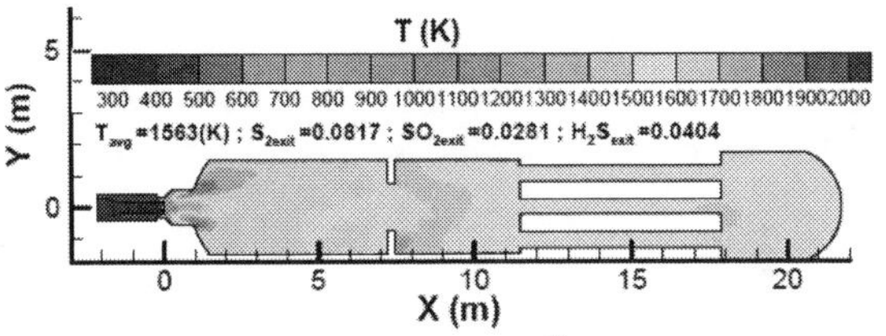

(a)L=127mm,H=80mm,θ_{sw}=10°,oxygen-normal

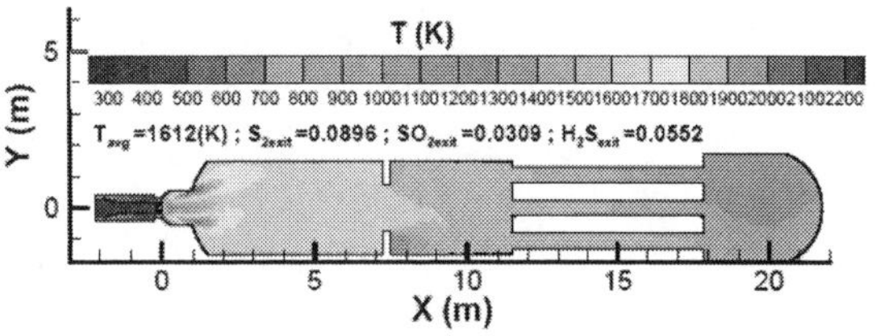

(b)L=127mm,H=80mm,θ_{sw}=10°,oxygen-rich

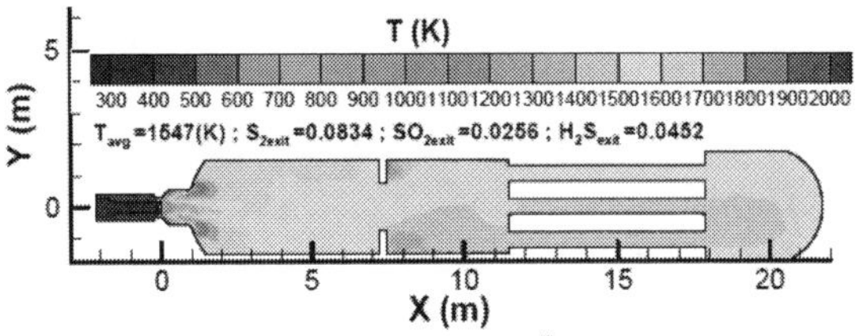

(c)L=127mm,H=80mm,θ_{sw}=30°,oxygen-normal

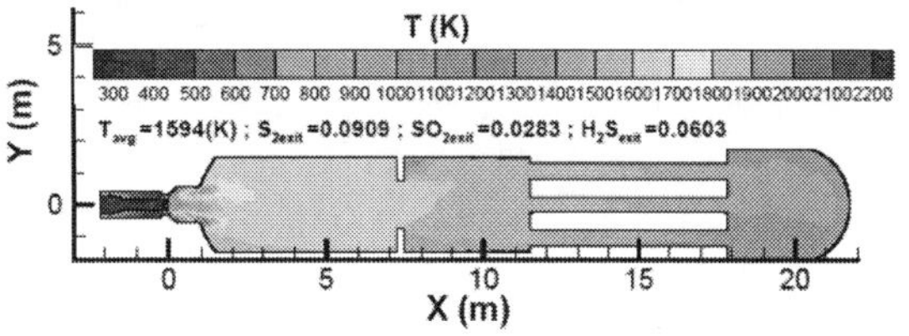

(d)L=127mm,H=80mm,θ_{sw}=30°,oxygen-rich

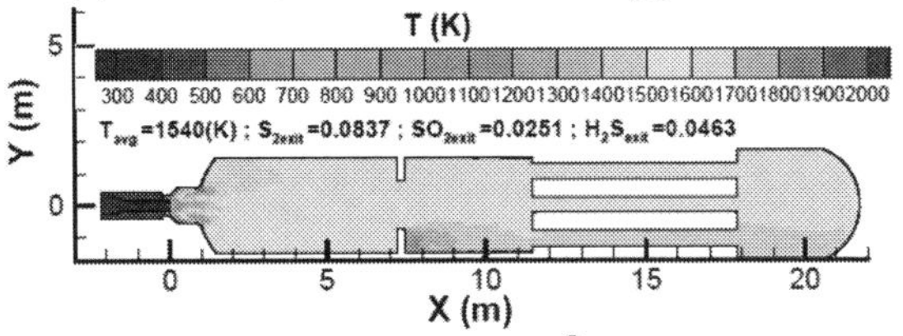

(e)L=127mm,H=80mm,θ_{sw}=60°,oxygen-normal

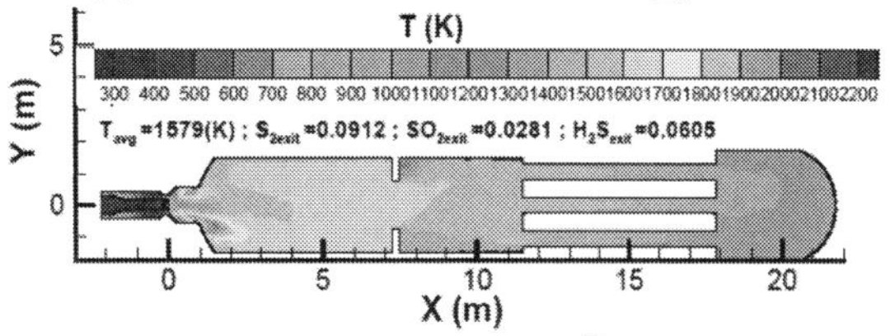

(f)L=127mm,H=80mm,θ_{sw}=60°,oxygen-rich

Figure 9. The temperature profiles on the symmetric plane for SRU thermal reactors with different inlet air swirler angles

4 Conclusion

This paper numerically investigates the the effect of burner parameters, including the clearance of the acid gas tip and the inlet air swirler angle, on the thermal field in a SRU thermal reactor, in order to reduce the high temperature inside the thermal reactor and to ensure an acceptable sulfur recovery. Practical operating conditions for a petrochemical corporation in Taiwan are used as the design conditions. The simulation results show that the burner with a smaller clearance of the acid gas tip produces a lower temperature, a lower exit SO_2 mole fraction and higher exit S_2 and H_2S mole fractions. Among the clearancs of the acid gas tip investigated, the horizontal clearance of 152.4mm and vertical clearance of 240mm yield the lowest temperature, exit SO_2 mole fraction and highest exit S_2, H_2S mole fractions. The burner with a smaller inlet air swirler angle produces a higher temperature, a higher exit SO_2 mole fraction and lower exit S_2 and H_2S mole fractions. Among the swirler angles investigated, 60° yields the lowest temperature, exit SO_2 mole fraction and highest exit S_2 , H_2S mole fractions.

Acknowledgements

The author gratefully acknowledges the grant support from the Ministry of Science and Technology, Taiwan, R.O.C., under contract MOST 105-2221-E-150-021.

References

1. R. Adewale, D.J. Salem, A.S. Berrouk and S. Dara, J. Clean Prod., **112**, 4815 (2016)
2. M. Chardonneaua, S. Ibrahim, A.K. Gupta and A. AlShoaibi, Energy Procedia, **75**, 3071 (2015)
3. F. Manenti, D. Papasidero, G. Bozzano and E. Ranzi, Comput. Chem. Eng., **66**, 244 (2014)

4. H. Selim, A.K. Gupta and A. AlShoaibi, Appl. Energy, **104**, 772 (2013)
5. B. ZareNezhad and N. Hosseinpour, Appl. Therm. Eng., **28**, 738 (2008)
6. W.D. Monnery, K.A. Hawboldt, A. Pollock and W.Y. Svrcek, Chem. Eng. Sci., **55**, 5141 (2000)
7. C.L.Yeh, T. Can. Soc. Mech. Eng., **40**(4), 511 (2016)
8. C.L. Yeh, Int. J. Mech. Eng. Rob. Res., **4**(3), 273 (2015)
9. C.L.Yeh, *The International Multi-Conference on Engineering and Technology Innovation 2015*, J5043 (2015)
10. C.L. Yeh, T. Can. Soc. Mech. Eng., **40**(5), 811 (2016)
11. C.L.Yeh, *The International Multi-Conference on Engineering and Technology Innovation 2015*, J5042 (2015)
12. C.L. Yeh, Appl. Sci., **6**(11), 331 (2016)
13. C.L. Yeh, T. Can. Soc. Mech. Eng., **41**(2), 293 (2017)
14. S.V. Patankar, *Numerical heat transfer and fluid flows* (New York: McGraw-Hill,1980)
15. B.E. Launder and D.B. Spalding, *Lectures in mathematical models of turbulence* (London England: Academic press, 1972)
16. R. Siegel and J.R. Howell, *Thermal radiation heat transfe* (Washington DC: Hemisphere publishing corporation; 1992)
17. Y.R. Sivathanu and G.M. Faeth, Combust. Flame, **82**, 211 (1990)
18. B.E. Launder and D.B. Spalding, Comput. Method Appl. Mech. Eng., **3**(2), 269 (1974)
19. C.L. Yeh, Int. J. Heat Mass Tran., **59**, 172 (2013)

New design of grab mechanism in plastic glove machine

Wei-Chun Hsu, Ming-Jin Jiang[a] and Gui-Liang Liu

Graduate School of OptoMechatronics and Materials, WuFeng University, Taiwan

Abstract. The aim of this study is to design a new type of grab mechanism for plastic bags machine in order to tear apart the plastic gloves from the plastic bags and stack them into a pile automatically. A grab mechanism with four bar linkage is developed and the main part of the grab mechanism is the clip which will pick up the plastic gloves and drag them to the other side. After a distance of drag, the clip will drop the plastic gloves to the collecting plate. The clips are fixed on the connecting rod of the four-bar linkage and the path of the clip can be traced by the position analysis of the four bar linkage. The path of the clip can be tracked by determination of the four links' lengths. This path tracking will be one of the advantages of this new mechanism because it provides the different drag distance to grab difference size of plastic gloves by simply adjusting the lengths of four links.

1 Preface

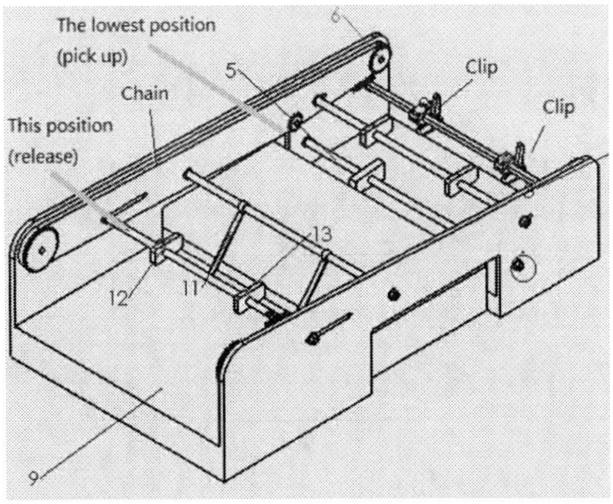

Figure 1. Chain conveyor belt mechanism.

In order to achieve the needs of customers and the globalization of the market considerations, the traditional machinery must develop more competitive products. In view of the medical use, the plastic gloves are usually made of PE and EVA materials. During the conventional production process, the plastic gloves are torn off the plastic bags by worker's hand and this is not good for medical use. In order to solve this problem, an automatic release mechanism should be invented. The general release

[a] Corresponding author : ingmar.hsu@wfu.edu.tw

mechanism [1] is composed of chain conveyor belt which connects the clips to tear off the plastic gloves. During the required process, the clip goes downward to pick up the plastic gloves from the plastic bags which have been cut with shapes of gloves. The clip drags the glove forward and upward for a distance and then the clip will drop the plastic gloves to the collecting plate as shown in figure 1. According to the design requirements of this path, we design a new grab mechanism. A new grab mechanism is designed by a four-bar linkage [2, 3] which provides high rigidity characteristics and is better than the low rigidity of chain.

2 New design of grab mechanism

A new grab mechanism design [4] is composed of a four-bar linkage and clips. The two clips are fixed in the connecting rod of the four-bar linkage as shown in figure 2. In this way, the path of the clip looks like the path of the coupling rod in the four-bar linkage. In order to determine the path of the clip, it can be obtained by means of the four-bar linkage position analysis. According to the design requirements of the clip path, the relationship between the length ratio of each link and the path of the clip is analyzed. The path length is adjusted by the link length to achieve the requirement of the production process.

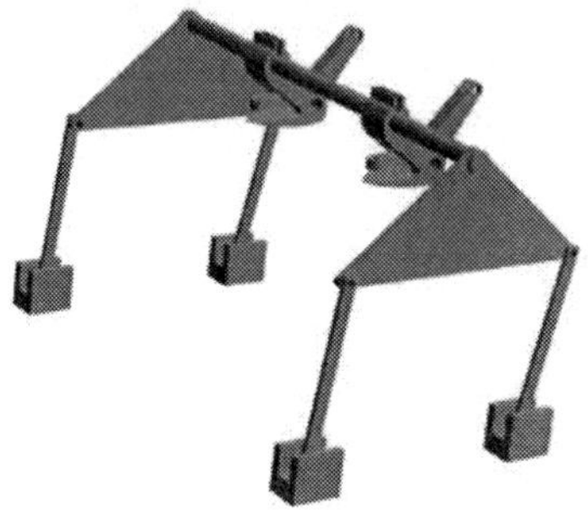

Figure 2. 3D diagram of grab mechanism.

3 Path analysis of clips

Before the size of the total mechanism is determined, the shape of the clip path should be determined in accordance with the requirement. Figure 3 shows a typical vector drawing of four-bar linkage. The vectors $\boldsymbol{r}_1$, $\boldsymbol{r}_2$, $\boldsymbol{r}_3$ and $\boldsymbol{r}_4$ represent the fixed link, drive link, connecting link and output link, respectively. The clip locates on point A of the connecting link. The sum of these vectors can be expressed as follow [5, 6]:

$$r_2 + r_3 - r_1 - r_4 = 0 \tag{1}$$

Thus

$$r_2 \sin\theta_2 + r_3 \sin\theta_3 - r_1 \sin\theta_1 - r_4 \sin\theta_4 = 0 \tag{2}$$

$$r_2 \cos\theta_2 + r_3 \cos\theta_3 - r_1 \cos\theta_1 - r_4 \cos\theta_4 = 0 \tag{3}$$

The lengths, r_1, r_2, r_3 and r_4, of the mechanism, position angle θ_1 and input angle θ_2 are given. There are only two unknowns, θ_3 and θ_4, in equation 2 and equation 3. Because there are trigonometric functions in two equations, these need application of iterative method to solve the problem [4, 6] to obtain the answers of θ_3 and θ_4.

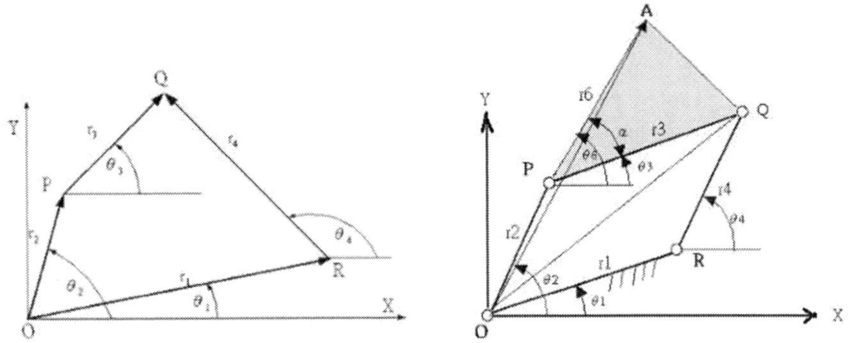

(a) Vector Loop for four-bar linkage (b) Vector for the couplar link

Figure 3. Vectors for the four-bar linkage.

After the angles of θ_3 and θ_4 are obtained, the horizontal and vertical components of the position vector r_A of point A are determined.

$$r_{Ax} = r_2\cos\theta_2 + r_6\cos\theta_6 \tag{4}$$

$$r_{Ay} = r_2\sin\theta_2 + r_6\sin\theta_6 \tag{5}$$

Where $\theta_6 = \theta_3 + \theta$ .Thus the position of the clip can be obtained. The loop position of the clip can be calculated at each angle of θ_2 when angleθ_2 varies from 0 to 360°. In this way the loop position of the clip can be displayed by connecting the calculated positions when angleθ_2 varies from 0° to 360°. Therefore different path will be obtained according to the different size of the grab mechanism and this will be a good advantage to achive the production requirement of different glove sizes.

4 Results and discussion

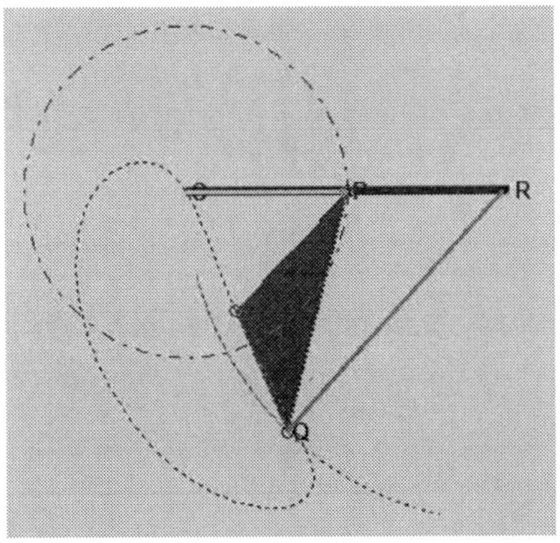

Figure 4. Trajectories for the four-bar linkage with length ratio 4: 2: 3: 4: 2 (θ_1=0°).

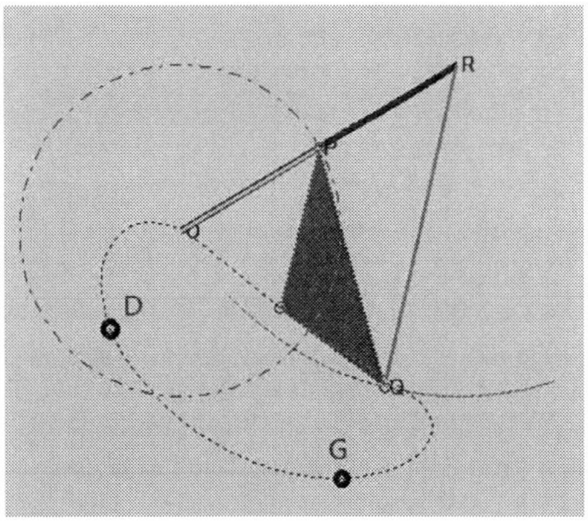

Figure 5. Trajectories for the four-bar linkage with length ratio 4: 2: 3: 4: 2 (θ_1=45°).

According to the mechanism design theory and automatic technique, a grab mechanism with four bar linkage is developed. The clip fixed on the mechanism goes downward to pick up the plastic gloves from the plastic bags which have been cut with shapes of gloves. The clip drags the glove forward and upward for a distance and then the clip will drop the plastic gloves to the collecting plate and then return to its original position to pick up another one. The shape of this clip path is determined with the size of the four-bar linkage. According to the kinematic trajectory analysis of the plastic glove release mechanism [4], it can be known that the path shape of the clip can be adjusted by changing the lengths ratio of the four-bar linkage. Therefore, the four rods 1, 2, 3, 4 and the length of the coupling rod is set to be 4: 2: 3: 4: 2. The drive link 2 is driven directly by the motor rotation, so link 2 needs to turn 360°. The angle is 30° and the angleθ_1 of the fixed link 1 is 0 degrees in this case. Figure 4 shows the blue, red and black trajectoriers represents the pathes of point P on link 2, point Q on link 4 and the clip, respectively. The black dotted trajectory for the clip path does not achieve the requirement of the movement of picking up the plastic gloves. This time the angleθ_1 of the fixed link 1 is switched to 45 degrees in this case. And the diagram is shown in figure 5.

This diagram shows a better trajectory for the clip. The clip can pick up the plastic glove on point G and drop it on point D. For a better choice of the clip trajectory, the clip pick up and drag the plastic glove left and upward from point G to point D and this route is the working process. The clip returns to point G to grab another glove after drop the last one and this route is the non-working process. In order to reduce the cycling time of the clip, the length of the trajectory should be reduced. After many times of trials, a better choice of length ratio for the four rods 1, 2, 3, 4 and the length of the coupling rod is set to be 4.5: 2: 3: 4: 3. And the angleθ_1 of the fixed link 1 is also 45 degrees in this case. The result is shown in figure 6.

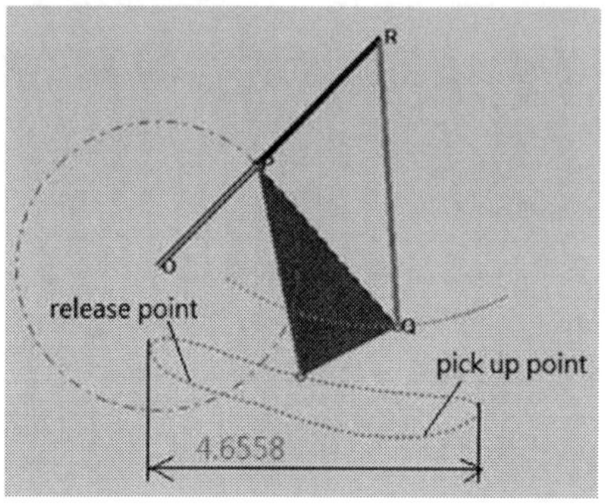

Figure 6. Trajectories for the four-bar linkage with length ratio 4.5: 2: 3: 4: 3 (θ_1=45°)

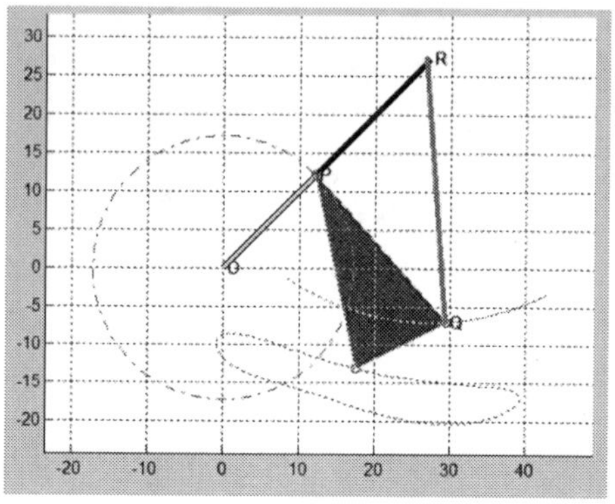

Figure 7. The design trajectories as request

The width of the trajectory in X-axis direction can be obtained by calculate the difference of minimal and maximal position of the clip. In this case, the horizontal width is 4.6558 as shown in figure 6. If the width is required to be 40 cm, then the lengths of the rod 1, rod 2, rod 3, rod 4 and coupler are 38.66 cm, 17.18 cm, 25.77 cm, 34.37 cm and 25.77 cm, respectively. The result is shown in figure 7. It can be seen from the diagram, the grab mechanism requires working space about 60 cm wide and 50 cm high. In this way we can design the grab mechanism with many different link lengths to serve grapping requirement of different glove lengths.

The purpose of this study is to design a new grab mechanism of a plastic product to replace the conventional machine. The basic four-bar position analysis can be used to calculate the clip path quickly under different design requirements. In the direction of the adjustment, the width of the clip path can be obtained by the program analysis. By means of the scale enlargement, the actual size of the new grab mechanism can be determined.

5 Conclusion

This study shows that a new design of grab mechanism in automatic plastic bag machine can be carried out. The development of a new design can provide high production capacity and high rigidity. The number of clips on the connecting link can be increase to provide more production capacity. The mechine can be easily set up for different glove length production by adjusting the length of the four links. The development of technology can make the machine more competitive in Taiwan, Europe and the United States markets.

References

1. M.T. Ho, Wu-Feng University Technical Reports, *Design of Molding Mechanism of PE Plastic Glove Machine*, 2-5 (2015)
2. C.C. Tsai, Thesis of Department of Mechanical and Automation Engineering National Kaohsiung First University of Science and Technology, *The Optimal Dimensional Synthesis of Four-Bar Linkage for Open Path Generation*, Kaohsiung, Taiwan ROC (2012)
3. Y.H. Su, Thesis of Department of Mechanical Engineering Kun Shan University, *Design, Analysis, and Fabrication of a Composite Four-Bar Linkage for Pick and Place*, Tainan, Taiwan ROC (2012)
4. S.Y. Zong, Thesis of Graduate School of Opto-Mechatronics and Meterials Wu-Feng University, *Design of Grab Mechanism of Plastic Glove Machine* (2016)
5. Y.H. Kang, *Mechanism* (Gau Lih Book Co., Ltd., 1996)
6. Y.H. Fan, Y.K. Zeng and Z.S. Chen, *Mechanism* (Gau Lih Book Co., Ltd., 2014)

Numerical and experimental diagnosis of complex rotor system by time-frequency techniques

Bernard Xavier Tchomeni[a] and Alfayo Alugongo

Department of Mechanical Engineering, Vaal University of Technology, 1911, South-Africa.

Abstract. This paper describes the application of Discrete Wavelet Transform (DWT) to identify various types of nonlinear damage caused by, unbalance, rotor-stator contact and a breathing crack in rotating machinery. Multiple faults have been investigated based on numerical and experimental signal analysis using Fast Fourier Transform (FFT) and DWT. A four degree of freedom fully coupled model of the rotor-stator system that includes the nonlinear damage in the rotor vibrations was established using Energy principles. Existence of high system nonlinearity could not allow exhaustive discrimination of rub and crack by classical FFT. Therefore, the DWT was employed. The results provide detailed feature analysis of the fault signals. Practical vibration measurements through a data acquisition system interfaced with Rotor Kit-4 and crack simulator provided the test data. Experimental Time-Frequency analysis gave more realistic faults responses with variable faults features. Irregularity of orbit, harmonic peaks in the presence of rub and crack were unique and distinguished periodic motion from other types of motion. The presence of a crack shifted the critical speed location and exhibited sub-harmonic components, which were more prominent with rub in vibration response. The detailed decomposition signal by DWT method established inherent feature patterns that effectively discriminated the multiple faults.

1 Introduction

Rotating shafts play an important role in many sectors of industry, and their dynamic behaviour and fault diagnosis have a significant economic importance. Thus, the risk of unbalance, rub-impact and cracks is an important form of rotor damage that can lead to disastrous failures if undetected in time. Therefore, it is imperative for fault analysis of the rotating system to extract significant features from time-domain vibration response of various faults combinations. Rotors are relatively more flexible and operate under harsh and hostile environmental conditions. Thus, condition monitoring of rotor system becomes more complicated, especially when faults existing in the system are more than one.

Rotor systems may harbour unbalance, undergo rub contact with a stationary part under tighter clearances or could develop fatigue crack under severe thermal and mechanical stresses [1], amounting to a coexistence of multiple faults. Several researchers have used various techniques to diagnose a crack and rotor-stator rub under exciting centrifugal force.

The cracked rotor has been analysed experimentally to investigate nonlinear dynamic characters of rotor system containing a transverse fatigue crack [2]. Results from this analysis revealed from the

[a] Corresponding author : bernardt@vut.ac.za

orbit and spectrum of the rotor's centre that, the effect of the crack becomes obvious only if the crack is deep enough. Based on the model-based diagnostic approach, [3] studied multiple faults identification of rotor system by a least-squares fitting approach in the frequency domain.

The signal-based approach for vibration analysis of the rotor system with combined crack and asymmetry faults was performed in [4]. Both the Jeffcott rotor model and Timoshenko beam element were used to numerically and experimentally investigate the rotor vibration characteristics of the unbalance, crack, and rotor-stator rub faults [5]. Recently, several authors have shown interest in new techniques of data analysis such as the Wigner-Ville distribution for analysis of linear but non-stationary signal [6, 19, 20]. Furthermore, various nonlinear time-frequency transformations methods such as Wavelet Transform (WT) were designed for non-stationary and nonlinear systems. The principle of WT and their numerical applications in signal processing have been widely used to simulate the transient response of an accelerated cracked Jeffcott rotor with a switching crack and investigated the detection and monitoring of cracks in a rotor-bearing system [7-10].

Literature review indicates that WT technique has been successfully used to simulate various faults and to perform damage localization. [11] presented an experimental study of the dynamic response of an overhang rotor with a propagating transverse crack using the Discrete Wavelet Transform (DWT). [12] showed the efficacy of WT in detecting transient waves in a dispersive medium by applying WT to the experimental analysis of the transient vibrations of an impact excited uniform beam.

Mostly, previous literature in multiple fault analysis is purely simulation and the need of validating experimentally the different numerical proposed method is requested.

Therefore, this paper presents both a numerical and experimental study on the dynamic response of a rotor with multiple faults namely unbalance, rub-impact and transverse crack using Fast Fourier Transform and the Discrete Wavelet Transform (DWT). Start-up and steady-state vibration signatures have been investigated using Daubechies (Db*n*) mother wavelet. Results are discussed and a number of observations on the rotor-stator system are highlighted in this paper. The basic system layout used is shown in figs 1 through 2.

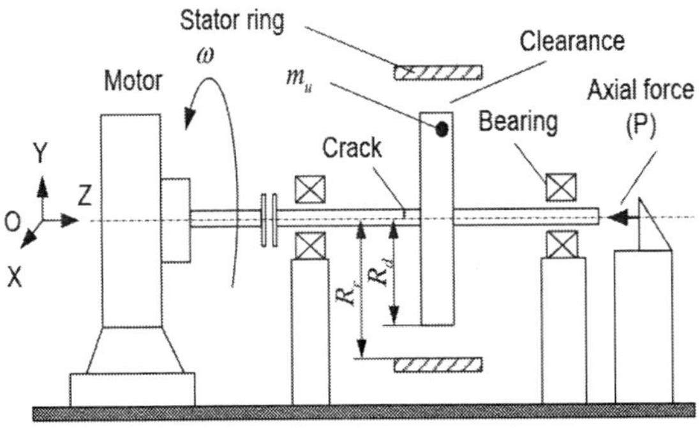

Figure 1. Motor with shaft-disc-stator-transverse crack

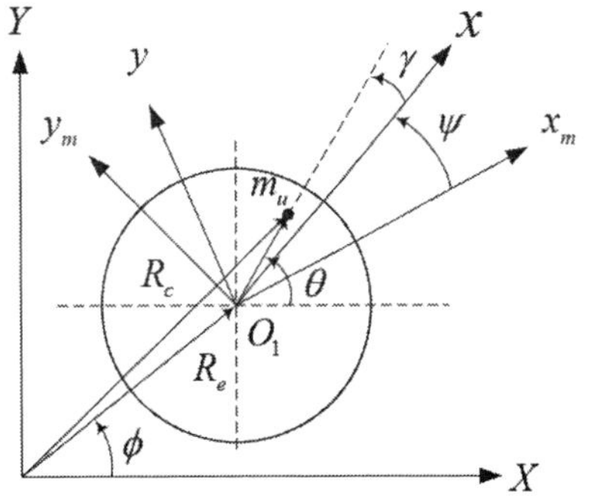

Figure 2. Section view of the disc the shaft

2 Model of Study and Numerical Simulation

Based on the geometry of Figs 1 to 2, and considering modal damping, the kinetic energy, potential energy, and Rayleigh's dissipation function of the motor-disc-shaft system are gave below [13]:

$$U_D = \frac{1}{2}M\left(\dot{X}^2 + \dot{Y}^2\right) + \frac{1}{2}J_D\left(\dot{\theta}_1 + \dot{\psi}_1\right)^2 + \frac{1}{2}J_M\dot{\theta}^2 + \frac{1}{2}m_u\dot{R}_e^T\dot{R}_e$$
$$V = \frac{1}{2}K_{XX}X^2 + \frac{1}{2}K_{YY}Y^2 + \frac{1}{2}K_{\psi\psi}\psi^2 \quad (1)$$

$$D = \frac{1}{2} C_{XX} \dot{X}^2 + \frac{1}{2} C_{YY} \dot{Y}^2 + \frac{1}{2} C_{\psi\psi} \dot{\psi}_1^2$$

where, $m_u, J_D, J_M, \theta, \psi$ and $\dot{R}_e$ respectively are, unbalance mass, disc-mass moment of inertia, rotor-mass moment of inertia, rigid-body rotation, twist about Z-axis and velocity vector of m_u, K_{XX}, K_{YY}, and $K_{\psi\psi}$ are the stiffness coefficients associated with the system degrees of freedom, and C_{XX}, C_{YY}, and $C_{\psi\psi}$ are the respective damping coefficients. A high rotor speed excites axial and lateral vibrations which in turn excites an axial force and torque that affects bending stiffness as follows [14]:

$$K_{XX} = K_{YY} = K_b - \frac{P\pi^2}{2L} - \frac{F_\psi \pi^3}{2L^2} \tag{2}$$

where, K_b, P and F_ψ correspondingly are, modal stiffness for first mode bending, fluctuating axial force and axial torque on the rotor. In this paper, P and F_ψ are defined as follows:

$$P = P_0 + P_f \sin\theta, \quad F_\psi = T_0 + T_f \sin\theta \tag{3}$$

where, P_f, T_f and θ are amplitudes and rigid-body rotation angle respectively. In this paper P and F_ψ are prescribed a priori as in [13]:

$$P = 0 \;,\; F_\psi = 0 \tag{4}$$

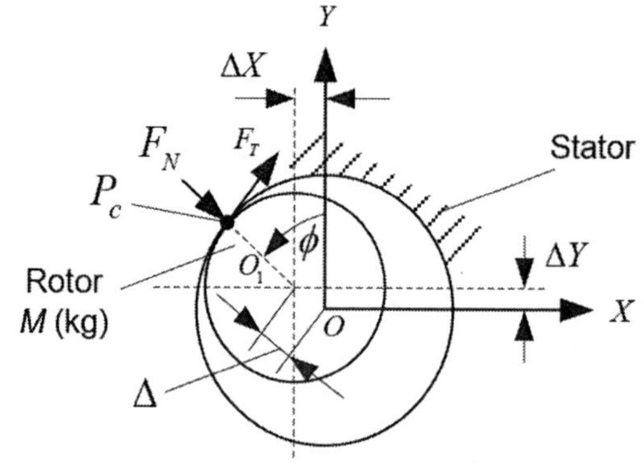

Figure 3. Geometry of rotor-stator forces and a clearance.

Applying the approach in [13], to Fig. 3, the position of the shaft center and the rub-impact forces in Cartesian coordinate frame are:

$$\begin{aligned} F_X &= -F_N \cos\phi + F_T \sin\phi \\ F_X &= -F_X \sin\phi - F_T \cos\phi \end{aligned} \tag{5}$$

where, F_N, F_T respectively are, radial component of rub force, tangential component of rub force. The lateral forces in X and Y directions are [13]:

$$\begin{aligned} F_X &= -K_S(1 - \frac{\Delta}{R})X + K_S(1 - \frac{\Delta}{R})\mu Y \\ F_Y &= -K_s(1 - \frac{\Delta}{R})\mu X - K_s(1 - \frac{\Delta}{R})Y \end{aligned} \tag{6}$$

where, μ and K_S respectively are coefficient of dry friction and stator radial stiffness.

Subsequently, a hinge mechanism in [10] was adopted in the shaft to provide for local flexibility effect of a breathing crack. Ignoring shear stresses, observing Saint-Venant's principle and following the approach in [16], the cracked rotor stiffness is given as,

$$K_0+\Delta k(\theta)=\begin{bmatrix} k_{xx} & 0 \\ 0 & k_{yy} \end{bmatrix}-\frac{1}{2}f(\theta)\Delta k_{xx}\begin{bmatrix} 1-\cos 2\theta & \sin 2\theta \\ \sin 2\theta & 1+\cos 2\theta \end{bmatrix} \tag{7}$$

where, Δk_{xx}, K_0 and $f(\theta)$ respectively, are change in stiffness of the shaft in the weaker axis due to the crack, uncracked shaft stiffness and steering function $f(\theta)$ as indicated in [16].
Considering Eqs. (6) and (7) the dynamic equation of the unbalanced-cracked-rub rotor-stator system is:

$$\begin{bmatrix} m_{\theta\theta} & m_{\theta\psi} & m_{\theta X} & m_{\theta Y} \\ m_{\psi\theta} & m_{\psi\psi} & m_{\psi X} & m_{\psi Y} \\ m_{X\theta} & m_{X\psi} & m_{XX} & 0 \\ m_{Y\theta} & m_{Y\psi} & 0 & m_{YY} \end{bmatrix}\begin{Bmatrix} \ddot{\theta} \\ \ddot{\psi} \\ \ddot{X} \\ \ddot{Y} \end{Bmatrix}+\begin{bmatrix} 0 & 0 & 0 & 0 \\ 0 & C_{\psi\psi} & 0 & 0 \\ 0 & 0 & C_{XX} & C_{XY} \\ 0 & 0 & C_{YX} & C_{YY} \end{bmatrix}\begin{Bmatrix} \dot{\theta} \\ \dot{\psi} \\ \dot{X} \\ \dot{Y} \end{Bmatrix}+\begin{bmatrix} 0 & 0 & 0 & 0 \\ 0 & K_{\psi\psi} & 0 & 0 \\ 0 & 0 & K_{XX} & K_{XY} \\ 0 & 0 & K_{YX} & K_{YY} \end{bmatrix}$$
$$\begin{Bmatrix} \theta \\ \psi \\ X \\ Y \end{Bmatrix}+\begin{Bmatrix} Q_\theta \\ 0 \\ Q_X \\ Q_Y \end{Bmatrix}+\frac{1}{2}f(\theta)\Delta k_{xx}\begin{bmatrix} 0 & 0 & 0 & 0 \\ 0 & 0 & 0 & 0 \\ 0 & 0 & 1-\cos 2\theta & \sin 2\theta \\ 0 & 0 & \sin 2\theta & 1+\cos 2\theta \end{bmatrix}\begin{Bmatrix} 0 \\ 0 \\ 0 \\ Y_S \end{Bmatrix}=\begin{Bmatrix} F_\theta \\ F_\psi \\ F_X \\ F_Y \end{Bmatrix} \tag{8}$$

3 A brief description on WT and DWT

WT is a signal processing tool and decomposes a nonstationary signal into scales (also known as levels) with different time and frequency resolutions. It is probably the most recent solution to overcome the shortcomings of the FFT. As opposed to FFT which breaks up a signal into sine waves of different frequencies, WT breaks up a signal into shifted and scaled versions of the original (mother) wavelet [17]. For nonstationary signals, the Fourier analysis is not effective since it transforms the signal into the frequency domain and the time information gets lost. WT analysis, allows a windowing technique with variable-sized regions, and permits the use of a time adaptive interval to analyse high and low frequency components of a signal [17]. Furthermore, researches have indicated that, as compared to FFT, WT produces results from which faults are clear to distinguish and easy to interpret [17]. WT can be categorized as Discrete WT or Continuous WT. The CWT time-frequency transform of the initial signal $s(t)$ is expressed as:

$$CWT_s(a,b)=\int s(t)\cdot\psi^*_{a,b}(t)dt \tag{9}$$

where a and b are the, scale parameter and time translation factor respectively. The term, $\psi^*_{a,b}(t)$ denotes the complex conjugate of $\psi_{a,b}(t)$. By varying a and b, the WT coefficients $W_s(a,b)$ can offer the representation of the signal $s(t)$ at different levels of resolution and time shift. In the process, thus, the WT enables one to extract features of the signal $s(t)$. The DWT is derived from the discretization of CWT. It serves as a tool that separates data into various frequency components, and enable to us to study each component with resolution matched to its scale. DWT is computed with a cascade of filtering followed by a factor 2 subsampling [6], and is given by:

$$DWT_s(j,k) = 1/\sqrt{2^j}\int_{-\infty}^{+\infty} s(t)\psi^*((t-2^j k)/2^j)dt \tag{10}$$

where j and k are integers, a and b are replaced by 2^j and $2^j k$. By using the DWT the signal $s(t)$ can be analysed at various levels of resolution, known as "multiresolution signal decomposition technique" [17]. In this work DWT has been computed by a successive low and high-pass filtering of the discrete signal $s(k)$, so that the original signal is broken down into many lower resolution components.

4 Expérimental Setup

Fig. 4 shows the adopted setup of the RK-4 test rig. The rig is constituted of a motor speed control, four proximity probes and a set of data acquisition interface unit controlled by Ascent toolbox. The piece of equipment designed for a test was set up on a rigid base to avoid disturbances that could influence the measurements. The test rig carried an elastic shaft and one rigid disc driven by a motor shaft through a flexible coupling. The rotor shaft was supported on two roller bearings at both ends and sustained in position by two fixed steel supports (designated as supports A1 and A*2* in Fig. 4)

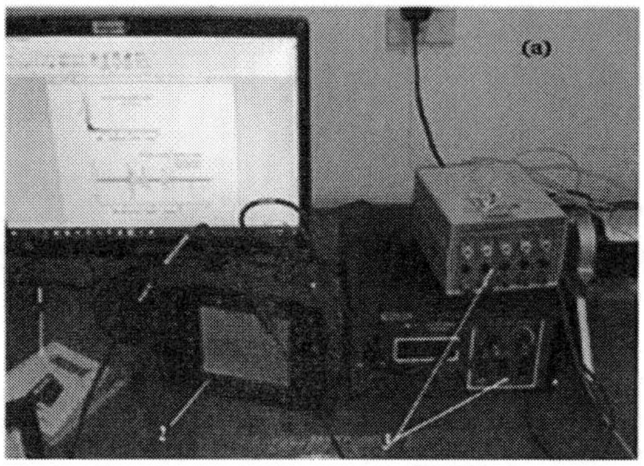

(a) Data acquisition. 1- Ascent; 2-Scout 100; 3-Data Acquisition

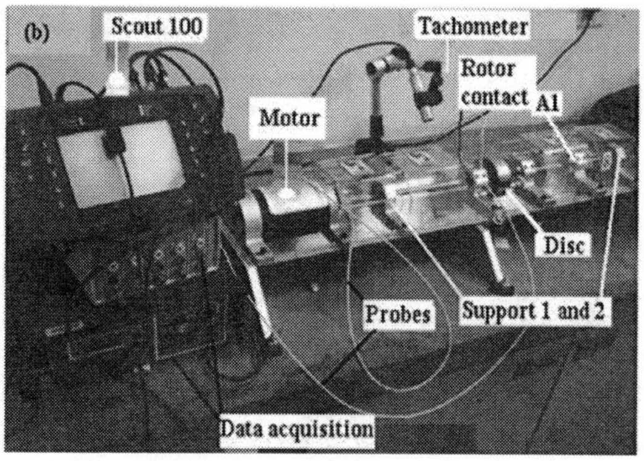

(b) Experimental set up

(c) R-K- 4 faults. 4-Crack simulator; 5- Tight screw for rub; 6-Unbalance mass

Figure 4. R-K-4 faults detection test rig with data acquisition

The shaft's acceleration was captured by a laser tachometer. Probes were placed at distances appropriate to detect and measure rotor shaft vibration in various positions. Caution was taken to avoid prospective contact between the probe and a section of the revolving assembly under extreme conditions of vibration. All the outputs were integrated with a filter and displayed on a screen of the oscilloscope for further analysis.

Table 1 gives the physical parameters of the RK-4 rig. The signals were recorded in radial direction at various rotating speeds by four proximity probes staggered in the probe block as shown in Fig. 4. The rotor was driven by an AC motor. The maximum rotating speed was set at 5000 rpm for unloading case. For a loading case the rotating speed was fixed at 3000 rpm. Two mutually perpendicular proximity probes recorded the relative displacement of the shaft. Two other probes were used for phase reference pulse generation and motor speed control. Rotor-stator rub was realized at point (A1) by introducing a small tight screw at a favourable distance between the rotor and the stator. Motor speed control was set to ensure that, the rotor was driven only in one direction in order to avoid reversal of the friction forces induced by rub.

The data analysis software used for signal's visualization, combined Ascent 2013 linked with a Scout 100 data acquisition board. Scout100 offered a processing unit that incorporates the capabilities of oscilloscopes, spectrum analysers, filters, recording instruments and directly interfaced with a computer. The acquired data was then subjected to digital processing by FFT and DWT.

Table 1. Rotor system RK- 4 parameters

Shaft	**Value**	**Crack**	**Value**
Length of the rotor (L)	570 mm	From the left bearing	240 mm
Shaft Diameter	10 mm	Depth	5 mm
Outer Diameter (D)	75 mm	Unbalance eccentricity	10 mm
Inner Diameter (d)	10 mm	**Motor**	
Density of the material	7800 kgm^{-3}	Frequency	500 Hz
Modulus of elasticity	$211x10^{9}$ Pa	Volt	415 V
Damping coefficient	$0.8x10^{-6}$ s^{-1}	Ampere	1.05 Amp
Discs	**Value**	**Discs**	**Value**
Mass	0.845 kg	Eccentricity mass	0.018 kg

5 Simulation, experimental results and discussion

To investigate vibrations induced by the external forces, the system parameters in [12] have been adopted for simulation. Two investigations were conducted, the first being the numerical simulation of model Eq. (8) on diverse hypothesis by a Modified Runge-Kutta algorithm in MATLAB, giving X, Y, θ at a time step $\Delta t = 0.001\,s$ in the interval $t = [0 \;\; 20]s$. The response at rotor start-up with and without external forces, have been presented and discussed. Subsequently, in a set of experiments, analysis was carried out first in absence of external excitation forces. Vibrations in axial direction did not provide information on rub and crack and therefore its results will not be presented hereafter.

5.1 Effect of lateral motion and axial force

In this simulation the torsional deflections ψ was considered to be negligible in comparison to other deflections. The variable stiffness was taken to be equal to the dynamic stiffness and the cracked shaft stiffness considered. Cross-coupling stiffness and damping were neglected *(C_{XY}=C_{YY}=K_{XY}= K_{YY}=0)*. Subsequently, a joint examination of frequency spectrum, rotor trajectory, orbit patterns, and DWT, enabled the features of unbalance, rub, and crack combined with rub to be discretized from model and experimental vibration test data. Unbalanced rotor responses at a high speed were obtained numerically by integrating Eq. (8). Figs. 5 and 6 illustrate the obtained results.

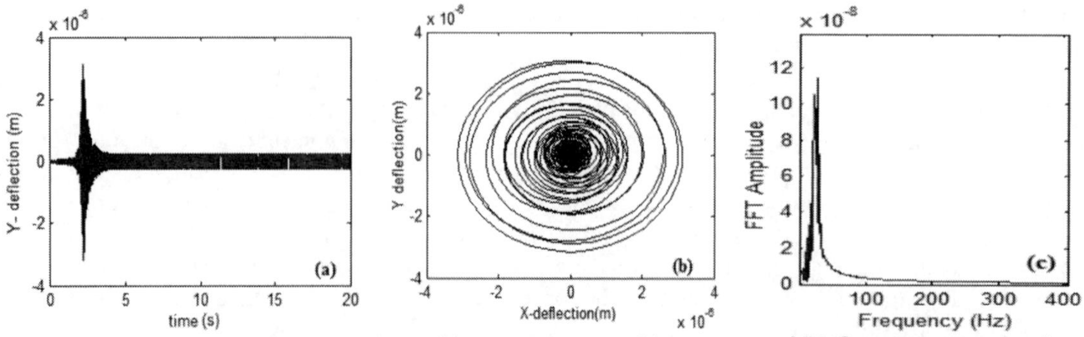

a) Vertical shaft deflection b) Rotor response orbit c) Y- frequency rotor signal

Figure 5. Simulated dynamic response of unbalanced shaft, with axial force but no rub, crack, $\Delta K/K_b = 0$, .

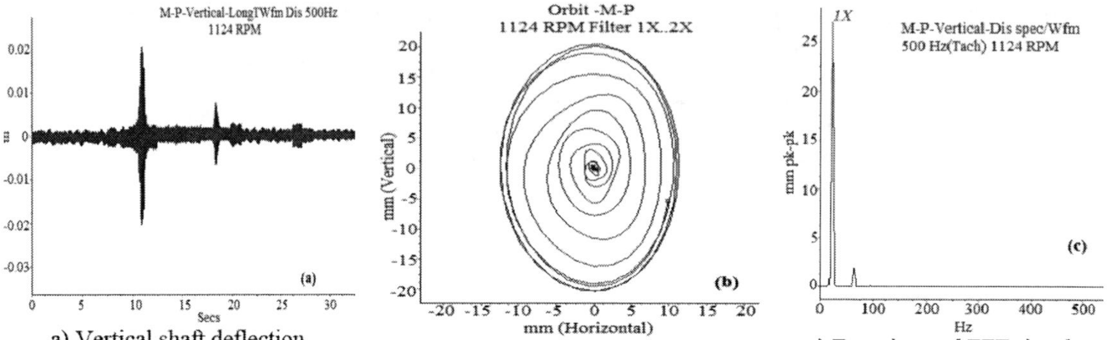

a) Vertical shaft deflection b) Unbalanced rotor response orbit c) Experimental FFT signal

Figure 6. Experimental dynamic rotor system response with unbalance at high speed 2570 rpm

Obtained results of unbalance in the absence of a crack and rub are shown in Figs. 5 and 6. Comparing the orbits in Figs. 5 and 6, it was apparent that, the unbalance was a source of extra evolutions in the response, especially when the rotating speed fell near half of the system critical speed. It is noted in Figs. 5 and 6 that, unbalance induced a closed-loop containing a circular loop. As seen in Figs. 5 (b) and 6 (b), the orbit response initially varied from a minimum at the start-up to a maximum at $R = 23.3\ mm$ at resonance and ceded into a regular closed inside loop. The maximum orbit amplitudes in Fig. 5(b) were modified from $R_1=3.167\times 10^{-6}\ m$ to $R_2=2.3\times 10^{-6}\ m$ in Fig. 6(b) when the critical speed changed correspondingly from, $\omega 1$ =1132.27 *rpm* in Fig. 5 (c) to $\omega 2$=124.43 *rpm* in Fig. 6 (c). In Fig. 6 (c), the frequency amplitudes of the experimental data was greater than that of the model data under a small axial force in Fig. 5 (c). The FFT in Fig. 6 (c) reveal that, a high vibrational level was attained by order *1X* harmonic at 23.46 *Hz* . Figs 5 (a) and 6 (a) indicate that, resonance occurred at different times corresponding to the critical speeds, $\omega 1$ and $\omega 2$. Correlation between model and experimental results was perceived in deflection shape, frequencies and orbits.

Further, Daubechies wavelet were adopted for diagnosis of unbalance, rub and crack respectively. Using DWT, a multiresolution analysis of the vibration response of the rotor with and without faults was performed and results illustrated in Figs.7and 8.

To explore the constituent components of the original signal at each level, DWT was performed up to the decomposition level 3 (D3) and approximation coefficients 3 (A3) using Daubechies db_i, $i = [1....7]$ on experimental data. The response was further reconstructed to examine the constituent component of the original signal at each detail level.

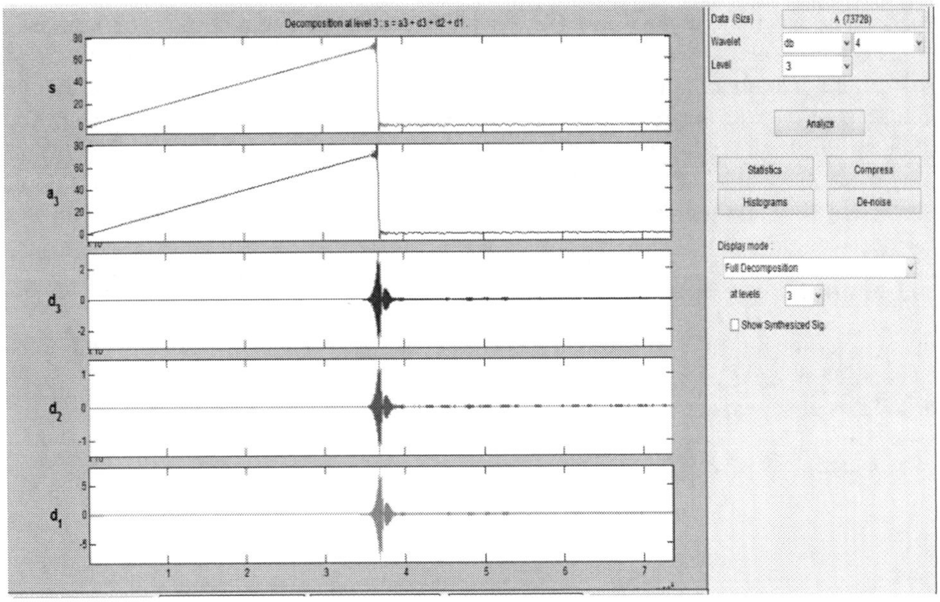

Figure 7. High-level Wavelet signals results from the DWT of experimental Balanced RK-4 rotor at a higher speed of 2570 rpm by a db4 Wavelet.

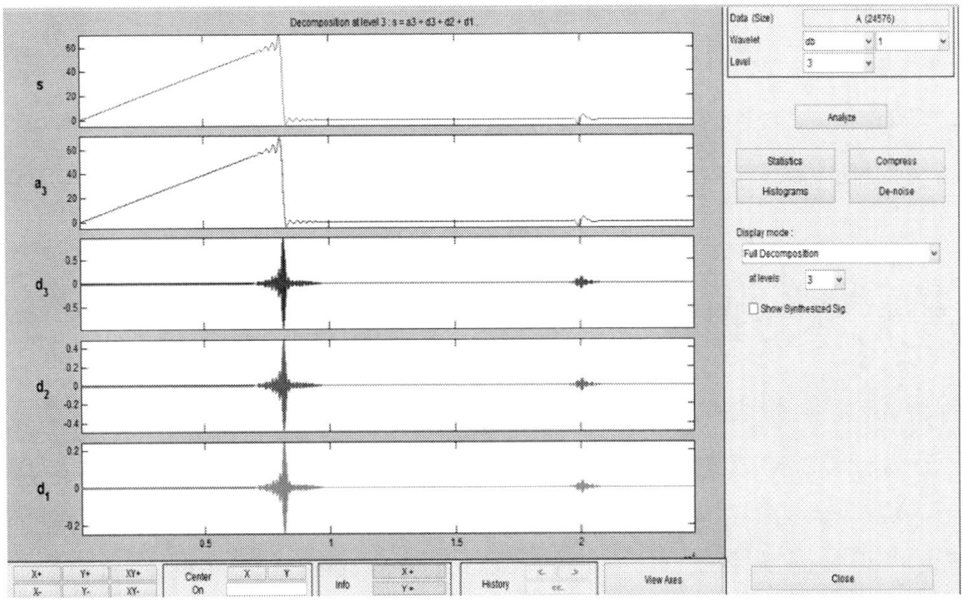

Figure 8. Signal Approximation and Decomposition of unbalanced RK-4 rotor at a speed of 2570 rpm by a db1 Wavelet at level(s) 1, 2, 3.

Using DWT, a multiresolution analysis of the vibration signals of the healthy rotor was also performed. The measured vibration signal was decomposed into three levels using Daubechies family as the mother wavelet. The vibration signal of the healthy rotor and the detail components of the 3 level decomposition are depicted in Fig. 7 and for unbalance effect in Fig. 8. The scaled signal values of detail coefficient *d3* obtained from the balanced rotor response at 2570 rpm are shown in Figs. 7 and 8. The results indicate that, the level-3 detail coefficient *d*3 were more sensitive to the effect of faults. It was noted that, the balance alters the transient responses in the baseline, which can be detected by carefully comparing Figs. 7 and 8.

Comparison between the plots of DWT in Figs. 7 and 8 shows slight changes in response. In Fig.8, the main resonance peak is represented by a straight-line change in shape and magnitude (stretching the signal at the maximum and minimum frequency sub-bands), standard deviations were still observable at level *a3* with insignificant vibration which could be attributed to the centrifugal force. The 3rd detail-level plot showed a little difference between the two cases, the difference was significant to identify unbalance features. The 3rd level approximation *(a3)* response contains the dilated signal at a very low frequency and diagnoses rotor unbalance.

5.2 Effect of unbalance and rub

Evolution of this phenomenon at various motor speeds has been investigated in this section. The clearance was set at $\Delta = 1.65\times 10^{-6}\, m$ and experimental data on RK-4 was taken and analysis using FFT and DWT as observed and analysed using Figs. 9, 10, and 11.

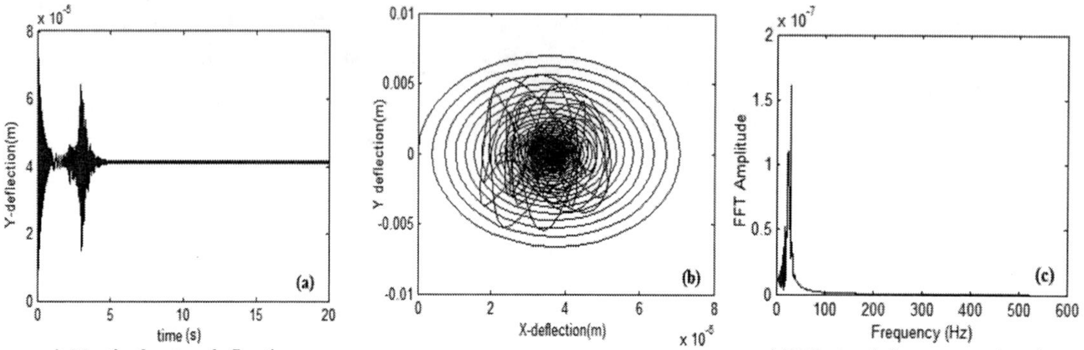

a) Vertical rotor deflection b) Rotor response orbit c) Y-Lateral frequency signal

Figure 9. Simulated rotor response of unbalanced rotor, with rub, axial force: $\Delta K/K_b = 0,\ \psi = 0$

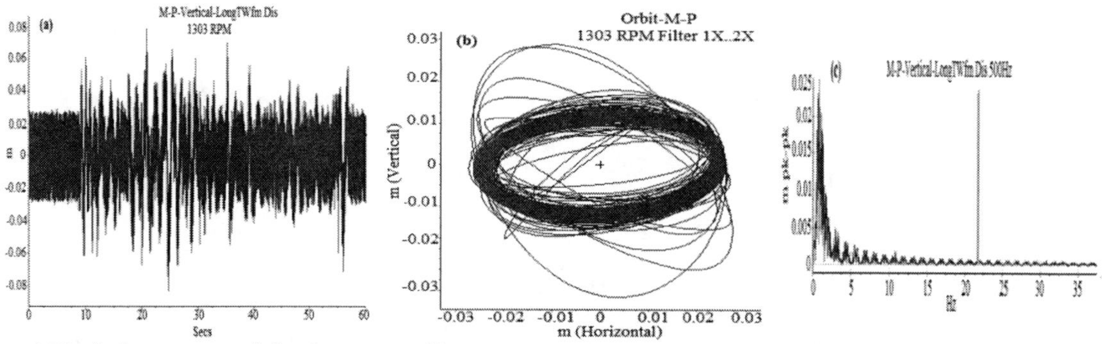

a) Vertical rotor-stator deflection b) Rotor-stator response orbit c) Shaft Frequency domain

Figure 10. Experimental unbalanced rotor system response of the RK.4 model at high speed 2570 rpm, with rotor-to-stator rub.

Fig. 9(a) displays the time-domain synchronous lateral response of the rotor. Critical speed of the rotor was found to be *799.6140 rpm* at *t=3s*, a point at which the peak amplitude features in Fig. 9(a). The unbalance and reduction in clearance generated a large frictional force at each rotor-stator contact. Therefore, in Fig. 9, the amplitude of the peaks caused by unbalance and rub increased with the number of contacts. Alternatively, this observation could be explained by a decrease in clearance and the increased percentage of solid contact between the rotor and the stator. The experimental orbit in Fig. 10(b) shows a complex shape due to multiple rotor-stator rebounds at points of impact. A significant distortion in orbits due to rub in the system was obvious and correlates with the simulated orbit in Fig. 9(b). The inside circular-loop orbit observed for a case of unbalance was drastically modified to an outside elliptic shaped characteristic of severe rub in Fig. 10(b). Existence of considerable disparities in time-deflections under rub was noted with an increased level of vibration contacts in Fig. 10 (a) where the first critical speed was observed at *t=32s* experimentally and at *t=3s* by simulation in Fig. 9(a).

Experimentally it was noted that, rotor-stator rub increased the rotor critical speed to 1303 rpm. This is because rotor-stator contact increased the rotor dynamic stiffness, which in turn increased the rotor critical speed from $\omega\, 1=1124.27$ *rpm* in Fig. 5 (c) to $\omega\, 1=1102.43$ *rpm* in Fig. 10 (c). The experimental frequency spectrum in Fig. 10 (c) indicates an increased magnitude of frequency components as observed previously in Fig. 9 (c) by the appearance of squashed sub-harmonic peaks before the main peaks. It should be noted that, the tightening of 1/2X frequency peaks (sub-harmonics) in Fig. 10 (c) was mainly due to rub. In Fig. 10 (c), the main peak coincided with the rotor frequency at *22.5 Hz*. The gap between two consecutive peaks was in the range of *22.5 Hz*, which indeed was the shaft's rotation frequency.

Severity of rub is therefore a function of the number and the size of the various peaks observed in the scalogram. Based on the size of the inside loop, the abrupt change in frequency, and the variation in deflection waveform as the rotor traverse the first critical speed can be used as an indicator of rub and the bounces inside the orbit loops can be assigned as rub signature. The assigned rub signature has been experimentally used in the Discrete Wavelet analysis of wavelet coefficients to detect the occurrence of rotor-stator rub as illustrated in Fig. 11.

The results, presented in Fig. 11 revealed a series of shocks characterized by the lateral bands, spaced from the rotational frequency *(about 7.88 Hz)* around a frequency of 588Hz observed on the Scout 100 data acquisition screen. Further, unbalance features were observable in the approximated signal *a2* before contact occurred. The signal in Fig. 11 presented the characteristics of an unbalance, observed before the main peak of resonance represented by a straight line with greater magnitude and deviations observable at level *a2*, and followed by several harmonics at lower amplitudes.

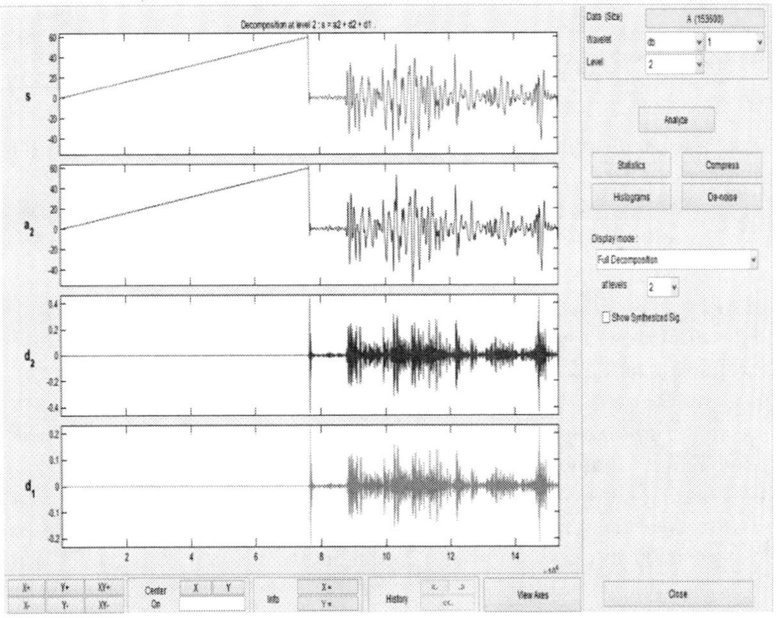

Figure 11. Two-level detail and approximate wavelet coefficients of unbalanced RK-4 rotor with rub at 2570 rpm by a db1 wavelet at level(s) 1, 2.

5.3 Excitation of unbalanced rotor by a crack

The analysis was performed on an unbalanced rotor with a transverse crack. The responses were processed by FFT and DWT. The system response with a crack gave the orbit and the amplitude spectrum in Figs. 12, 13 and 14.

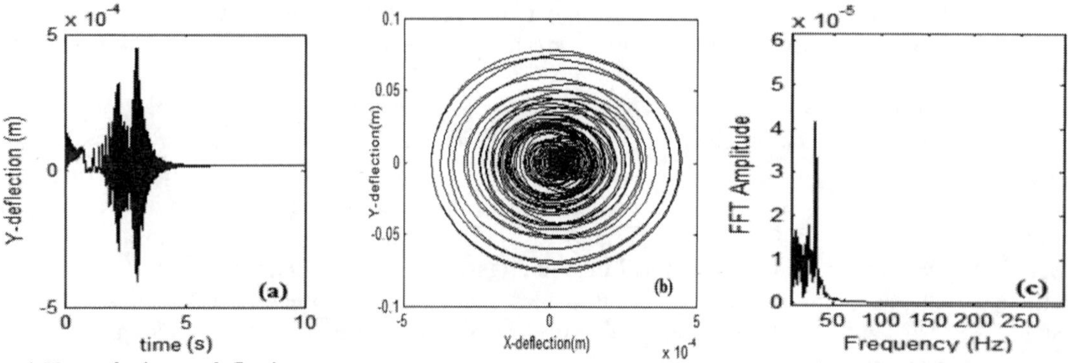

a) Y- cracked rotor deflection b) Cracked rotor response orbit c) Y-Lateral frequency

Figure 12. Simulated dynamic response of cracked rotor via the critical speed, with axial force, and no rub.

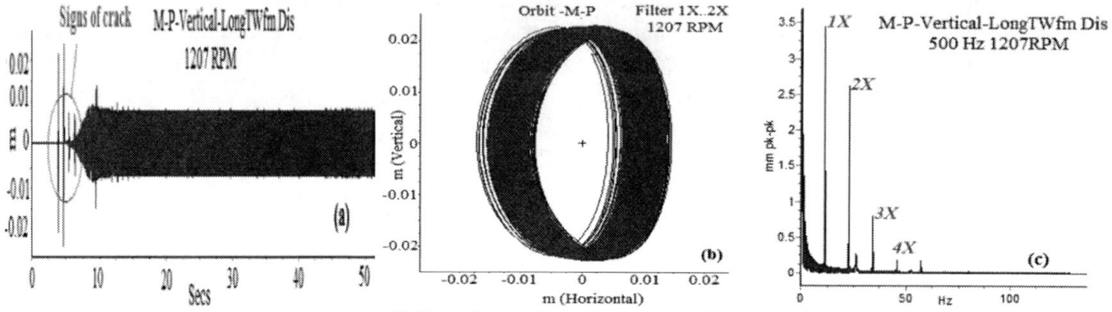

a) Vertical cracked rotor deflection b) Experimental cracked rotor orbit c) Frequency signal domain

Figure 13. Experimental rotor system response of the RK.4 model at 2570 rpm, with unbalance and crack ($\Delta K/K_b = 0.0542$).

Comparing the simulated orbits in Figs. 12 (b) and 5(b), it was apparent that, the breathing crack caused extra evolutions in the vibration response near the system critical speed. The maximum orbit amplitudes in Fig. 5 (b) were modified from $R_1=3.167\times 10^{-6}\,m$ to $R_2=5.03\times 10^{-4}\,m$ when the speed varied respectively, from $\omega\,1=1132.27\,rpm$ in Fig. 5 (c) to $\omega\,1=1855.43\,rpm$ in Fig. 12 (c), i.e, from an uncracked to a cracked shaft.

The presence of a crack indicated strong tightening loops in Fig. 12 and 13 (b). In Fig. 13 (c), a high vibrational level was attained by order *1X* harmonic at *23.46 Hz*. At this point, the crack generated a tremendous increase in vibration amplitudes, characterized by the *2X* and *3X* super harmonic peaks. From the deflection in Figs. 12 (a) and 13 (a), crack features were manifested by small tight peaks at the rotor start-up, unlike the observations in the previous analysis. In spite of crack features, unbalance characteristic could not be well discriminated. Therefore, DWT was performed and the corresponding responses are shown in Fig. 13.

An analysis of the decomposed and reconstructed response indicated that, the vibration amplitudes were significant in the three lowest-frequency bands, *d1*, *d2*, and *d3*. The presence of the peak linked to the second sub-critical speed appears more evident in the *d3* analysis than the approximated signal *a3*. The spectrums of a crack fault obtained by using db23 reveal the presence of two prominent peaks corresponding to *1X* and *2X* harmonics, at two different times respectively, (*t=9500s* and *t=13880s*) in Fig. 14. The effects of a crack generated a huge increase in vibration amplitude as previously observed in *2-D* deflection in Figs. 12 (a) and 13 (a). The vibration amplitude of *(1X)* harmonic at *t=1000s* was slightly higher than that of *(2X)* harmonic at *t=14000s*, until passing the critical speed after which, the amplitude became less than the *1X* harmonic. These observations are a unique indicator and are in agreement with the results obtained in [10].

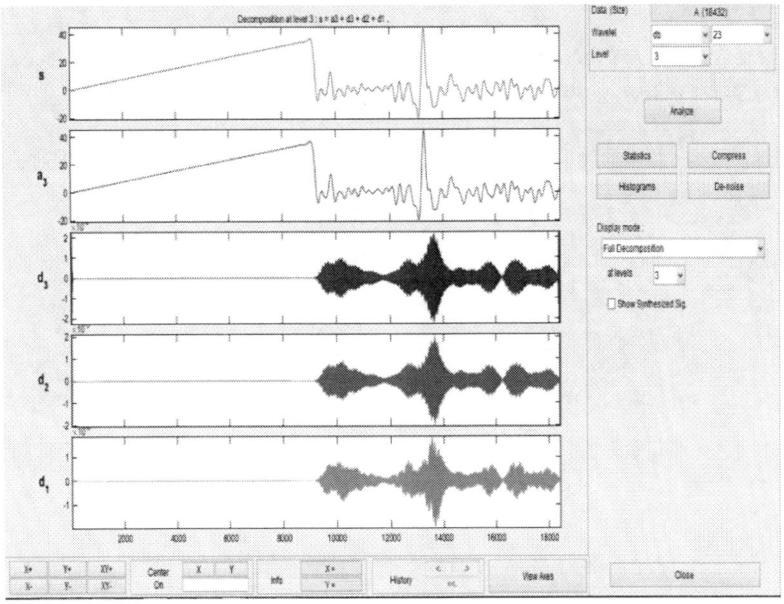

Figure 14. Signal Approximation and Decomposition of unbalanced RK-4 rotor with crack at 2570 rpm by a db23 Wavelet at level(s) 1, 2, 3.

5.4 Excitation of unbalanced rotor by a crack and rub

In this section, the vibration of the unbalanced rotor-stator system at a high speed under rub and a breathing crack was analyzed by simulation and experiment. The system responses are presented in the Figs. 15, 16 and 17.

In Figs. 15 (a) and 15 (b), the crack and rub combined generated high vibration as indicated by vertical deflection and orbit patterns. Similar to the results in the previous section, unbalance and rub increased the rotor dynamic stiffness, which in turn increased the rotor critical speed. However, the crack increased shaft flexibility and decreased the critical speed of the rotor as observed in Fig. 15 (a). The harmonic resonances *1X, 2X, 3X* and *1/2X* were observed in the FFT in Fig. 15 (c), *1X* were excited by unbalance, whereas the *2X, 3X* (super-harmonic) occurred due to the crack, and the *1/2X* were excited by rub, with a maximum limit on frequency at 500 Hz. The RK-4 speed became extremely unstable and in the range 1207–2570 rpm, it was not feasible to experimentally acquire measurements. In Fig. 16, the response was characterized by a main peak at the synchronous frequency of 22.5 Hz. Abnormal movements of the shaft in the bearing can be detected using orbit plots, complicated motions, such as quasi-periodic even chaotic vibrations, that are observed under multi-faults operating conditions in Fig. 16(b). It was noted in Fig. 16 (c) that, for a given crack depth, unbalance excited the *1X* sup-harmonic resonance, whereas rub excited the *1/2X* and *1/3X* sub-harmonic resonance. Finally, it can be inferred that, the emergence of the sub-harmonics, *2X, 3X* provide evidence on existence of a breathing crack. These can be easily distinguished near the first critical speed. This inference concurs with the results in [18].

In Fig. 16 (c), the amplitudes of the sub-harmonic and sup-harmonic resonance increased due to the interaction of rub, a breathing crack and unbalance at a high speed. The crack and noise induced the sup-harmonics, *2X* and *3X* in Fig. 16(c), which were of smaller magnitude compared to the *1X* harmonic. The trend in Fig. 16 (c) suggests that, the harmonics were sensitive to the rotor rotation speed. The results in Fig. 16 indicate that, the effect of the crack on the rotor response was considerably larger than that of rub as the amplitude of the later is considerably attenuated in comparison to that of the former. Once again, the experimental results correlated well with the simulated results in Fig. 15. Analysis of experimental test data revealed that, an unbalanced rotor with rub and a breathing crack could be identified by observing the evolution of the sub-harmonic and sup-harmonic frequencies; of *1/2X* and *2X*, *3X* near the critical speed region of *1X*. However, due to high non-linearity the evolution of the orbit patterns and the associated noise peaks became considerably complex for combined fault identification.

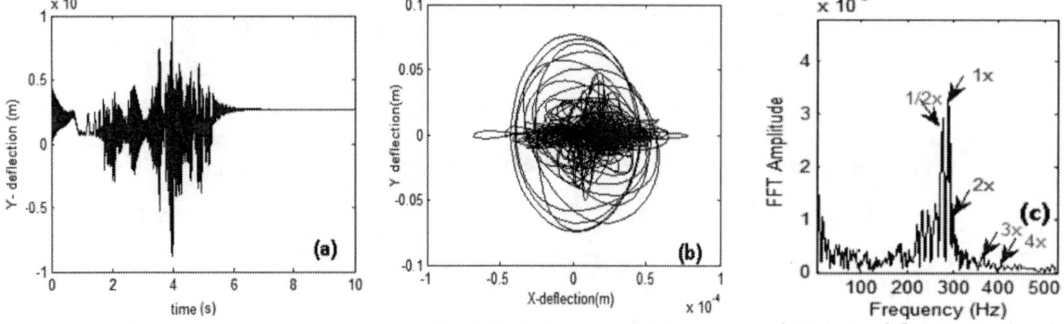

a) Multi-faults rotor deflection b) Multi-faults rotor orbit c) Y-Lateral frequency signal

Figure 15. Dynamic response of simulated unbalanced rotor model with rub, and crack, $\Delta K / K_b = 0.0542$

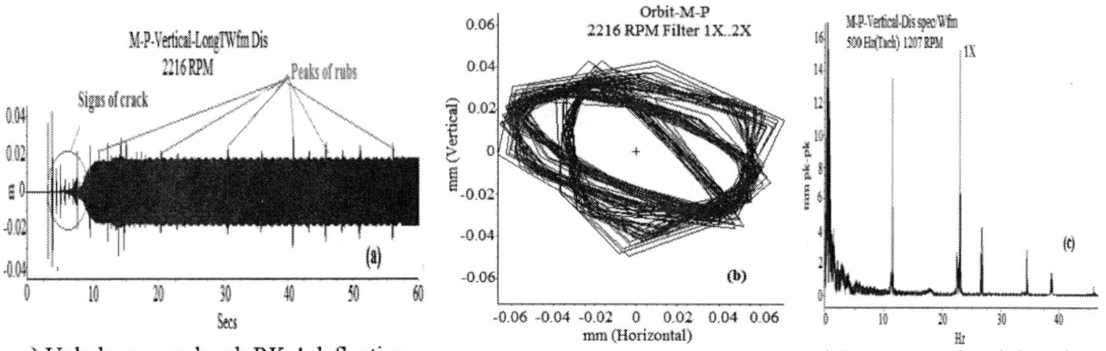

a) Unbalance-crack-rub RK-4 deflection b) Experimental shaft orbit c) Frequency signal domain

Figure 16. Experimental rotor system response of the unbalanced RK.4 model at higher speed 2570 rpm with rub and crack ($\Delta K / K_b = 0.0542$).

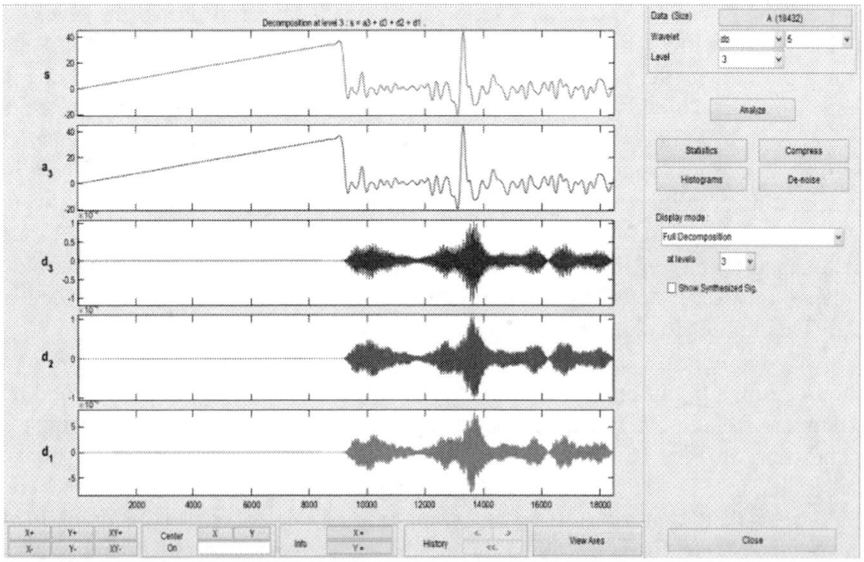

Figure 17. Signal Approximation and Decomposition of unbalanced RK-4 rotor with rub and crack at a speed 2570 rpm by a db5 Wavelet at level(s) 1, 2, 3.

For further exploration, the system was accelerated to a higher speed of *2570 rpm,* and held constant. The response near the rotor critical speed analyzed using DWT is illustrated in Fig. 17. The multiresolution analysis applied to the signal using the db5 at level 3 decomposition extracted the detail and the approximation coefficient of vibrated RK-4 system. A qualitative difference in this case is apparent in the Fig.17. The result revealed as found previously, the features of the faults namely, unbalance, a breathing crack, and rub. The consistency of the experimental results presented so far, using DWT at various levels was manifested again in the Fig. 17, with the sole difference being that, the signal in Fig. 17 presented the noise feature due to contact.

From the results in Fig.17, it was observed that the measured signals were corrupted by multiple peak well observed in *(d3)* which is attributed to rub-impact. These vibration signals contained useful information in detecting the multiple faults. The unbalance and crack features are dominant in Fig. 17 at level *a3* approximation and the rub-impact characteristics are identifiable in *d3* decomposition. In summary, it was logical to conclude that sensitivity in the vibration tests could be enhanced by using Wavelet analysis. The latter provides a considerable diagnosis and an efficient method for multiple faults analysis.

6 Conclusion

This paper has presented a method for online multi-faults diagnosis; rub, a transverse breathing crack on unbalanced rotor-stator system. An analytical model of an extended Jeffcott rotor based on Energy Principles was used. Dynamics of a "breathing" crack was integrated into the model using Mayes steering function and rub was generated numerically using elastic impact-contact at the rub contact point. A unique pattern of vibration signals acquired from multiple probes mounted at various positions of the shaft allowed practical extraction and discrimination of features of mixed faults from the system response. Simulation strategies were applied to predict and discriminate the coexistence of multi-fault characteristics for validation. It was observed that, at a high rotating speed when a crack and rub coexist, the response becomes very complex to analyse using FFT. Multiresolution decomposition by DWT was performed in the form of discrete-time approximate and detail wavelet coefficients, respectively to break the measured unfaulty and multi-faulted system signals, into detailed versions. The DWT analysis using the Daubechies family demonstrated that, the magnitude of level three detail wavelet coefficient was sensitive to faults diagnostics and provides good discretisation signature from the original rotor signal. The effectiveness of the adopted method for multiple fault diagnosis has been demonstrated by the analysis of approximated and decomposed response signal. Modal parameters such as shaft deflection, orbit patterns and natural frequencies obtained from simulation results were sensitive to rub-impact and crack. The differences in simulated shaft deflection, orbit, by FFT and DWT analysis at higher speed reveal unique patterns of the various faults. The simulated and experimental results are well correlated. The simulated and experimental analysis demonstrated the consistency of the proposed method in the multi-faults identification by DWT in a real system. The DWT accurately discriminated unbalance, rub and crack symptoms in the rotating system.

Acknowledgment

Support by the Department of Mechanical Engineering, VUT towards this work is highly acknowledged by the authors.

References

1. T.H. Patel and A.K. Darpe, Journal of Sound and Vibration, **326**(3), 740-752 (2009)
2. T. Zhou, J. Xu and W. Han, Journal of dynamic systems, measurement, and control, **127**(3), 313-320 (2005)

3. N. Bachschmid and P. Pennacchi, *In IMechE 7th Int. Conf. on Vibrations in Rotating Machinery* (571-580, 2000)
4. A.K. Darpe, Dynamics of cracked rotor, PhD Thesis, IIT Delhi (2000)
5. T.H. Patel and A.K. Darpe, Journal of Sound and Vibration, **317**(3), 841-865 (2008)
6. P. Flandrin, Time-frequency/Time-scale analysis, Academic Press (1999)
7. H. Kantz and T. Schreiber, *Nonlinear time series analysis* (Vol. 7). Cambridge university press (2004)
8. F. Al-badour, M. Sunar and L. Cheded, Mechanical Systems and Signal Processing, **25**, 2083-2101 (2011)
9. C.S. Burrus, R.A. Gopinath and H. Guo, *Introduction to wavelets and wavelet transforms: a primer* (Vol. 1, New Jersey: Prentice hall, 1998)
10. S. Prabhakar, A.S. Sekhar and A.R. Mohanty, Mechanical Systems and Signal Processing, **15**(2), 447-450 (2001)
11. S.A. Adewusi and B.O. Al-Bedoor, Journal of sound and vibration, **246**(5), 777-793 (2001)
12. T. Önsay and A.G. Haddow, The Journal of the Acoustical Society of America, **95**(3), 1441-1449 (1994)
13. B.O. Al-Bedoor, Journal of Sound and vibration, **229**(3), 627-645 (2000)
14. R. Sukkar and A.S. Yigit, Kuwait J. Sci. Eng, **35**(2B), 143-170 (2008)
15. A.S. Yigit and A.P. Christoforou, Journal of sound and vibration, **234**(1), 67-83 (2000)
16. A.D. Dimarogonas, Engineering fracture mechanics, **55**(5), 831-857 (1996)
17. M. Misiti, Y. Misiti, G. Oppenheim and J. Poggi, *Wavelet Toolbox for Use with MATLAB*: User's Guide, The MathWorks, Natick, NA (2000)
18. J.J. Sinou, A. Loyer, O. Chiello, G. Mogenier, X. Lorang, F. Cocheteux and S. Bellaj, Journal of Sound and Vibration, **332**(20), 5068-5085 (2013)
19. Z. Feng, M. Liang and F. Chu, Mechanical Systems and Signal Processing, **38**(1), 165-205 (2013)
20. J. Zou and J. Chen, Journal of Sound and Vibration **276**(1-2), 1-11 (2004)

Separation criteria of nanoscale water droplets from a nozzle plate surface

Van Quang Nguyen[1], Xuan Son Phung[1] and Jau-Wen Lin[2,a]

[1]*Faculty of Mechanical Engineering, Ha Noi University of Industry, Cau Dien Street, Bac Tu Liem District, Hanoi*
[2]*Department of Mechanical Engineering, National Kaohsiung University of Applied Sciences, Taiwan*

Abstract. This paper studies the water nanojet ejection process using molecular dynamics simulation. The results show that nanoscale water droplets cannot separate from the nozzle plate surface when the nozzle aperture has a diameter of 27.5 Å or smaller. The maximum height of the produced water nanojet is reduced after reaching to its highest position when the jet does not separate from the plate surface. Separation phenomenon between the water nanojet and the nozzle plate surface is the most obvious with the 27.5-Å-diameter nozzle aperture for this simulation setup. In addition, initial findings on the characteristics of nanoscale water droplets (width and contact angle) impinging onto a fixed plate surface are revealed in preparation for future investigation.

1 Introduction

Nanofluid ejection technology is employed in many industrial applications such as high precision print heads in the printing industry, cooling systems in metallic processing, printed circuit board manufacturing industry, and many more. In the nanofluidic ejection process, the formation of the fluid nanojet and the separation of the nanoscale fluid droplet have attracted research attention for the design and the accurate operation of micro and nanofluidic devices.

The molecular dynamics (MD) simulation method has been adopted to investigate the physics of nanoscale fluid ejection systems. The MD method employed to perform a number of simulations for propane nanojet ejection with a range of exit nozzle orifice diameters between 2 and 6 nm. In their work, various nozzle size conditions were applied to characterize the necking of the fluid nanojet. The effects of viscosity, wetting, and nozzle configuration factors on nanojet formation were discussed [1]. The MD simulation method also used to simulate the ejection process of water nanojets and the formation of nanoscale water droplets. The effects of nozzle aperture diameters, pressing force, system temperature and fluid flow dynamics on the formation and shape of the nanoscale water droplet were investigated [2, 3].

The formation of fluid droplets during the nanojet ejection process is affected by the nozzle size, temperature system, pressing force, and the properties of the fluid. The MD simulation is applied to study the role of thermal fluctuations on the formation and stability of nanoscale fluid droplets in drug delivery, printing, nanoscale manufacturing and conduct nanojet ejection simulation using non-equilibrium molecular dynamics in a high pressure environment [4, 5].

[a] Corresponding author : daniel@kuas.edu.tw

Some characteristics of the breakup and formation of fluid nanojets are mentioned in previous researches. The model to simulate the thermal induced breakup of liquid nanojets [6], phenomena of liquid thread breakup and nanoscale droplet formation in the nanoscale ejection process for various pushing periods of oscillation. The simulation results revealed that the sizes of nanoscale droplets are linear depending on the pushing periods [7].

Experimental method has also been used to empirically study the formation of nanoscale fluid droplets. The formation of a nanoscale liquid droplets from a wettable nozzle, the effects of nozzle size and fluid flow rate on the various behaviors of the droplets were investigated and discussed [8]. For inkjet printer heads [9], a moving piezo element which covered with thin foil is adopted to deform the ink channel and eject droplets from the nozzle opening.

Another method of nanofluidic ejection is via bubble expansion. A numerical method is used to describe the three-dimensional Navier-Stockes equation for ink motion both inside and outside of the nozzle by growing bubbles on the chamber wall [10].

In order to observe the phenomenon of the impingement process between the nanoscale fluid and solid molecules. The nanojet ejection process is investigated by spreading droplets on a solid surface using molecular dynamics simulation. The results showed that the spreading width increased as time increased, and the droplet's contact angle varied within a range when the atoms in the ejection process reached a relatively steady state [11]. In another study, the MD simulation based on Lennard-Jones potential is employed to study the nanoscale droplet ejection process and the impingement of the fluid nanojet onto a solid surface [12, 13]. The collisions of two opposed liquid argon nanojets is described. Nanojet exit velocities are the highest for non-wetting surfaces and decrease as solid-liquid interactions are enhanced [14].

Most of the above mentioned research works concern the simulation and experimental studies of the fluid nanojet ejection process, fluid nanojet breakup, and the formation of the nanoscale fluid droplets. However, the effects of the nozzle aperture diameter, various pressing forces on the ejection of the water molecules and separation of the nanoscale water droplets have rarely been investigated. This paper is a continuation of a previous study [15] using MD simulation methods to investigate the separation of nanoscale water droplets from the nozzle plate surface and the causes of separation of the nanoscale water droplet from the nozzle plate surface were also investigated and discussed. Furthermore, cursory observations on the impingement of nanoscale water droplets onto a fixed metallic plate are described.

2 Methodology

2.1 Details of simulation model

The configuration of the simulation model was arranged as illustrated in Fig. 1. Periodic boundary conditions of the simulation box are applied along the x-, and y-directions. Water molecules are confined in the space between the nozzle and back plates. The distance between the nozzle and back plates is 60.872 Å. The dimensions of the box are 101.9 Å along both the x-, and y-directions. The model contains 20 691 water molecules having density of 0.999972 g/cm^3 at 310 K. The oxygen atoms of the water molecules are arranged into simple cubic crystal lattices with lattice constant a = 3.104 Å.

The nozzle and back plates are composed of gold (Au) atoms and arranged into face-centered cubic (fcc) crystal lattice structures with the lattice constant a = 4.076 Å. The nozzle plate is fixed while the back plate can move vertically when subjected to a pressing force in the +z-direction. The thickness of the nozzle and back plates is 8.152 Å, each having 5 layers of gold atoms. The number of atoms of the back plate is 6 503. The nozzle plate has an aperture in the center. The number of atoms for the nozzle plates are 6 200, 6 144, 6 060, 5 924, and 5 772 gold atoms for the 25-, 27.5-, 30-, 35-, and 40-Å-diameter nozzle apertures, respectively.

The simulation of impingement of the nanoscale water droplets onto a fixed metallic plate is also performed. The dimensions of the fixed metallic plate are 150.812 Å along both the x and y-directions.

The thickness of the fixed metallic plate is 6.114 Å, consisting of 11 250 gold atoms arranged into a FCC structure with 4 layers of gold atoms.

LAMMPS software and C++ code were used as the tools for the simulation work and the data analysis.

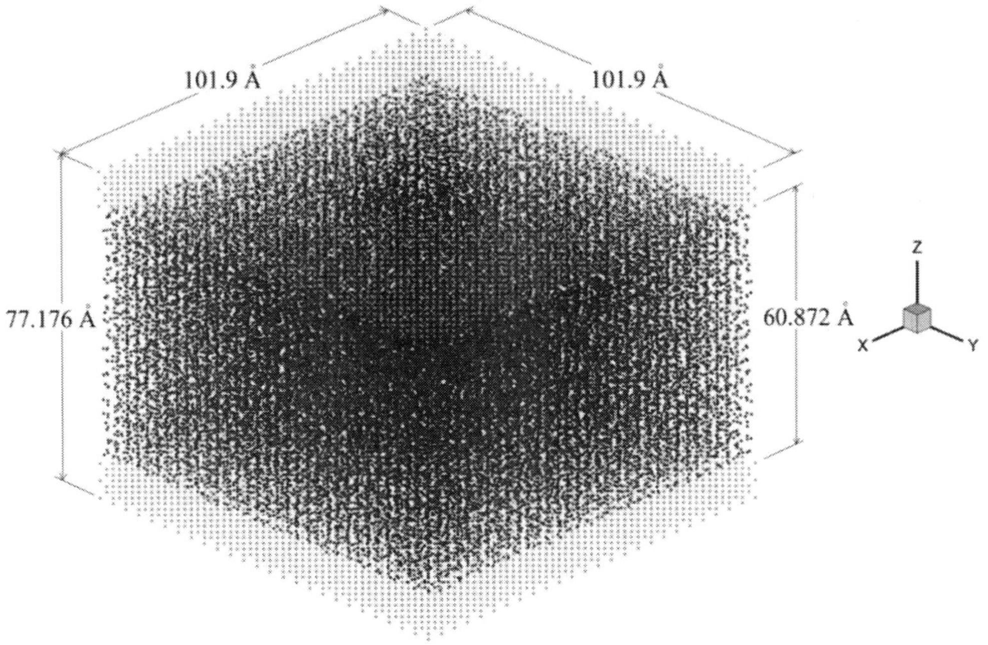

Figure 1. Molecular dynamics simulation model.

2.2 Interaction between atoms

The MD simulation method was used for analyzing the structure of the flexible three-centered (F_3C) water molecules, and provided insight into the features of the nanoscale structure and the dynamics of the fluid nanojet flow. Specific potential model was also employed to represent the interaction among the atoms in the water molecules. In this paper, the shifted Lennard-Jones 12-6 potential was applied to the model of the intermolecular interactions for H_2O - H_2O and H_2O - Au components of the system. Harmonic potential function was used to model the vibrations of the liquid water molecules. The equilibrium values of the oxygen-hydrogen bond length and the H-O-H bond angle were 1.0 Å and 104.5 degrees, respectively.

In the F_3C water molecular model [16], the complete form of the energy function used here to represent the water molecules is given as the sum of the bonding, the bending, the van der Waals and the Coulomb terms and can be expressed as:

$$U = U_{bond} + U_{bend} + U_{vdw} + U_{els}$$

$$= \sum K_b^{OH}(r_i - r_{eq})^2 + \sum K_\theta^{HOH}(\theta_i - \theta_{eq})^2 + \sum [A_{SC}\varepsilon_{ij}\left(\frac{r_0^{ij}}{r_{ij}}\right)^{12} - 2\varepsilon_{ij}\left(\frac{r_0^{ij}}{r_{ij}}\right)^6 - S_{vdw}(r_{ij})] + \sum [\frac{q^i q^j}{r_{ij}} - S_{els}(r_{ij})], \quad (1)$$

where $S_{vdw}(r_{ij})$ is the truncation shift function of the van der Waals force, which is expressed as:

$$S_{vdw}(r_{ij}) = \begin{cases} U_r - U_{r_c} - (r - r_c)\left(\dfrac{dU}{dr}\right)_{r_c} & for\ r < r_c \\ 0 & for\ r \geq r_c \end{cases}, \tag{2}$$

and $S_{els}(r_{ij})$ is the shift function for the Coulomb's force and is expressed as:

$$S_{els}(r_{ij}) = (r - r_c)\left[\frac{q^i q^j}{r_c^2}\right], \tag{3}$$

where A_{sc}, K_b^{OH}, K_θ^{OH}, r_i, r_{eq}, θ_i, θ_{eq}, r_{ij} and r_c are the coefficients that adjust the short distance force, the harmonic force constant, the angle bending force constant, the ith O-H bond length, the equilibrium length of the O-H bond, the ith H-O-H bond angle, the equilibrium angle of the H-O-H bond angle, the interatomic distance, and the cutoff distance (10 Å), respectively. Terms q_i and q_j are the partial charges of O or H.

The interactions between the water molecules and the gold atoms are represented by the Spohr potential functions of

$$U_{Au-H_2O} = U_{Au-O}(r_{Au-O}) + U_{Au-H_1}(r_{Au-H_1}) + U_{Au-H_2}(r_{Au-H_2}), \tag{4}$$

with

$$U_{Au-O}(r) = S_2(r) D_0 [\exp(-2\alpha_0(r - r_{e1})) - 2\exp(-\alpha_0(r - r_{e1}))] \tag{5}$$

and

$$U_{Au-H}(r) = \gamma D_0 \exp(-2\alpha_H(r - r_{e2})). \tag{6}$$

The switch function $S_2(r)$ has the form of

$$S_2(r) = \begin{cases} 1 & for\ r \leq r_{on} \\ \dfrac{(r_{off}^2 - r^2)^2 (r_{off}^2 + 2r^2 - 3r_{on}^2)}{(r_{off}^2 - r_{on}^2)^3} & for\ r_{on} < r < r_{off} \end{cases}, \tag{7}$$

where r_{on} and r_{off} are the start and end distances of the switch function and are respectively set to 7.0 and 11.0 Å.

2.3 Simulation scheme

In this paper, 11 simulation cases with 5 nozzle aperture diameters were studied. A random initial velocity was assigned to each atom of each water molecule. The atomic velocities were scaled such that the system temperature was 310 K. The pre-run time for each simulation was 10 000 time steps for the water molecules to reach an equilibrium state, with one time step being 1 femtosecond (fs). After the pre-run, the back plate is subjected various upward pressing forces to eject the water molecules out of the nozzle aperture. The magnitudes of the pressing force were 9.25 $\times 10^{-10}$, 10.0 $\times 10^{-10}$, 11.0 $\times 10^{-10}$, and 12.0 $\times 10^{-10}$ N for 25-Å-diameter nozzle aperture. The force magnitudes were 9.0 $\times 10^{-10}$, 10.0 $\times 10^{-10}$, 11.0 $\times 10^{-10}$, and 12.0 $\times 10^{-10}$ N for the 27.5-Å-diameter nozzle aperture. The magnitudes of the minimum pressing forces were 7.5 $\times 10^{-10}$, 6.5 $\times 10^{-10}$, and 5.75 $\times 10^{-10}$ N for the 30-, 35-, and 40-Å-diameter nozzle apertures, respectively.

The simulation scheme is listed in Table 1. The simulation length in each case is 150 000 time steps. The data output frequency is once per 2000 time steps.

Table 1. Simulation scheme.

Case	Nozzle aperture diameter (Å	Temperature (K)	Pressing force (N)
1	25	310	9.25×10^{-10}
2	25	310	10.0×10^{-10}
3	25	310	11.0×10^{-10}
4	25	310	12.0×10^{-10}
5	27.5	310	9.0×10^{-10}
6	27.5	310	10.0×10^{-10}
7	27.5	310	11.0×10^{-10}
8	27.5	310	12.0×10^{-10}
9	30	310	7.5×10^{-10}
10	35	310	6.5×10^{-10}
11	40	310	5.75×10^{-10}

3 Results and discussion

3.1 Water nanojet ejection process results of various nozzle aperture diameters

Snapshots of the oxygen atoms of the water nanojets corresponding to various pressing forces and nozzle aperture diameters at 150 000 fs are shown in Figure. 2. Figures 2(a)-(d) and (e)-(h) respectively show the snapshots of the water nanojets ejected through 25- and 27.5-Å-diameter nozzle apertures with the different pressing forces. In these simulations, the minimum pressing forces to eject all the water molecules though the 25- and 27.5-Å-diameter nozzle apertures were 9.25×10^{-10} and 9.0×10^{-10} N, respectively, as shown in Figures. 2(a) and (e). The water nanojets after ejection through the 25- and 27.5-Å-diameter nozzle apertures did not separate from the nozzle plate surface and form a nanoscale water droplet, regardless of pressing force. Figures 2(i)-(k) show the snapshots of the nanoscale water droplets ejected through the 30-, 35-, and 40-Å-diameter nozzle apertures with the minimum pressing forces of 7.5×10^{-10}, 6.5×10^{-10}, and 5.75×10^{-10} N, respectively. The nanoscale water droplets separated from the nozzle plate surface into jets in these cases.

Table 2 shows circumference-length-to-area ratios for the 25-, 27.5-, 30-, 35-, and 40-Å-diameter nozzle apertures in cross-section. The respective minimum pressing forces were 0.04, 0.036, 0.033, 0.029, and 0.025. The water molecules ejected through the nozzle apertures form water nanojets at these ratios. Based on a combination of the circumference-length-to-area ratios and the simulation results, criteria of the circumference-length-to-area ratios for the separation of the nanoscale water droplets from the nozzle plate surface in the water nanojets can be established as follows:

First, when the circumference-length-to-area ratio (L/A) is in a range from 0.036 to 0.040, the water molecules are ejected out through the nozzle aperture. The water nanojets are formed on top of the nozzle plate surfaces. However, the water nanojet does not separate out from the nozzle plate surface to produce droplets.

Second, if the L/A is less than or equal to 0.033, the formation of the water nanojets is similar to the first scenario however, the nanoscale water droplets are formed when the nanojet separates from the nozzle plate surface.

The ejection process of the water nanojets is affected by the diameter of the nozzle aperture. In order to determine the causes of the phenomena, the typical simulation cases of water nanojets ejected through the 27.5- and 40-Å-diameter nozzle apertures with respective 9.0×10^{-10} and 5.75×10^{-10} N pressing forces are analyzed in this study. The 27.5-Å-diameter nozzle aperture is chosen because the separation phenomenon of water nanojet from the nozzle plate surface is the most obvious, though nanoscale water droplet did not separate out from the nozzle plate surface. Furthermore, this case is also employed to compare with the case of the 40-Å-diameter nozzle aperture where the water droplet can separate from the nozzle plate surface. The following analysis focuses on these simulation cases.

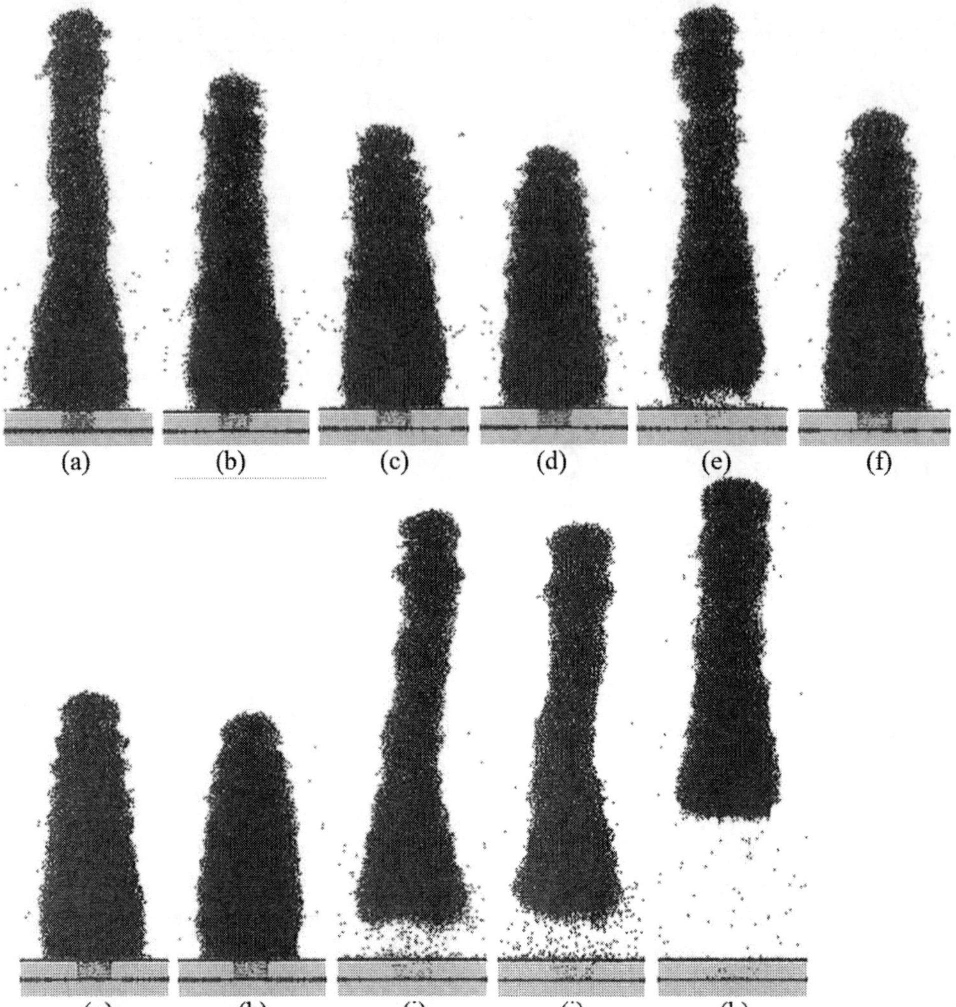

Figure 2. Snapshots of the water nanojet at 150 000 fs. (a)-(d) Water nanojets of the 25-Å-diameter nozzle aperture for various pressing forces (e)-(h) Water nanojets of the 27.5-Å-diameter nozzle aperture for various pressing forces (i) 30-Å-diameter nozzle aperture water nanojet (j) 35-Å-diameter nozzle aperture water nanojet (k) 40-Å-diameter nozzle aperture water nanojet.

Table 2. Features and circumference-length-to-area ratios of the nozzle apertures.

Nozzle aperture diameter (Å)	Minimum pressing force (N)	Circumference length of nozzle aperture in cross-section (L)	Area of nozzle aperture in cross-section (A)	Ratios of circumference Length to area (L/A)
25	9.25×10^{-10}	78.540	1963.494	0.040
27.5	9.0×10^{-10}	86.394	2375.827	0.036
30	7.5×10^{-10}	94.248	2827.431	0.033
35	6.5×10^{-10}	109.956	3848.448	0.029
40	5.75×10^{-10}	125.664	5026.544	0.025

3.2 The separation or non-separation of the nanoscale water droplet

Snapshots of the oxygen atoms of the water nanojets through the 27.5-Å-diameter nozzle aperture with a 9.0×10^{-10} N pressing force at 60 000 and 100 000 fs are shown in Figure. 3. The trajectories of the molecules in the water nanojet ejection process are used to analyze the formation of the water nanojet and explain why the nanoscale water droplet cannot separate from the nozzle plate surface and the increase of the water nanojet base diameter. For clarification, only oxygen atoms are shown.

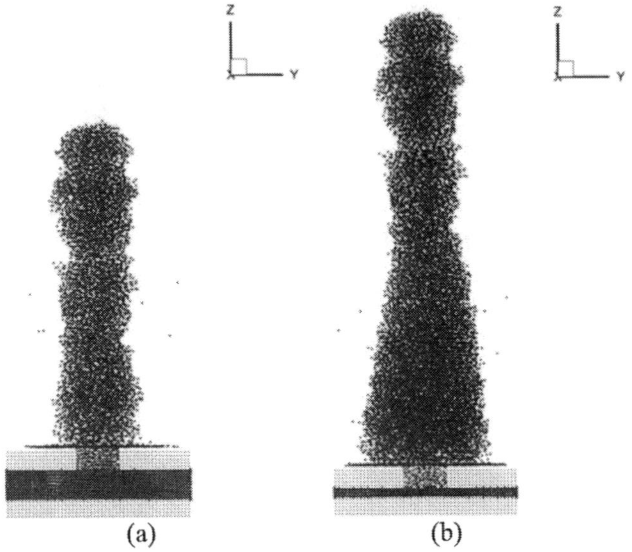

Figure 3. Snapshots of the water nanojet through the 27.5-Å-diameter nozzle aperture at (a) 60 000 fs and (b) 100 000 fs.

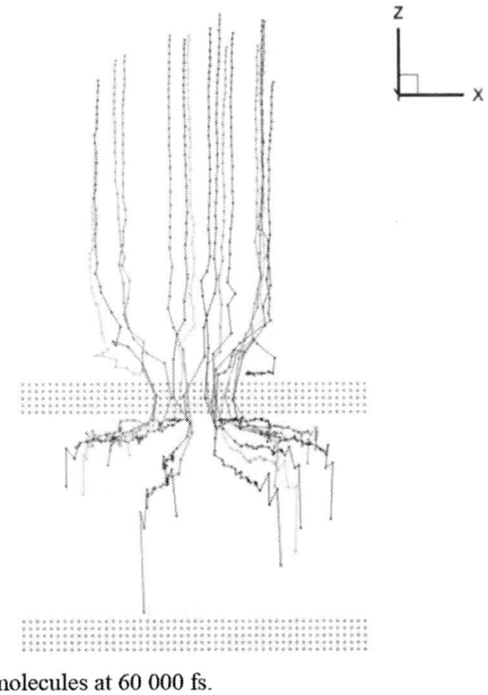

Figure 4. Trajectories of water molecules at 60 000 fs.

Figure 4 shows the trajectories of a few oxygen atoms at 60 000 fs. The water molecules close to the nozzle aperture moved with a zigzag trajectory. Their trajectories become straight after ejecting out of the nozzle aperture. Most of these molecules are ejected at the beginning and then move away from the nozzle plate surface form the upper half of the water nanojet. The distance between any two adjacent dots on each trace expresses the velocity of the particle. The velocities of the molecules are reduced along the top of the water nanojet. In addition, a few molecules with low velocity after ejection through the nozzle aperture moved sinuously near the nozzle plate surface, and then fall down to form a wetting molecular layer on the nozzle plate surface.

The trajectories of a few oxygen atoms at 100 000 fs are shown in Figure. 5. Some water molecules adjacent to the wall of the box and far away from the nozzle aperture are ejected out at this time. When subjected to a pressing force, most of the molecules move up with high velocity. Shortly after, these velocities rapidly low and move along the box wall. After having collisions with the bottom of the nozzle plate, the direction of the molecular trajectories is changed from the vertical direction to the horizontal direction, and then move towards the nozzle aperture before ejecting out with low velocity. Hence, a large number of molecules sinuously move near the nozzle plate surface. The attraction force among the molecules and the cohesion in water contribute to pull the water nanojet downward. Therefore, the water nanojet cannot separate from the nozzle plate surface because many molecules with low velocity move equivocally near the nozzle plate surface in a short period and thereafter, drop on the wetting molecule layer which was formed by the previous stage.

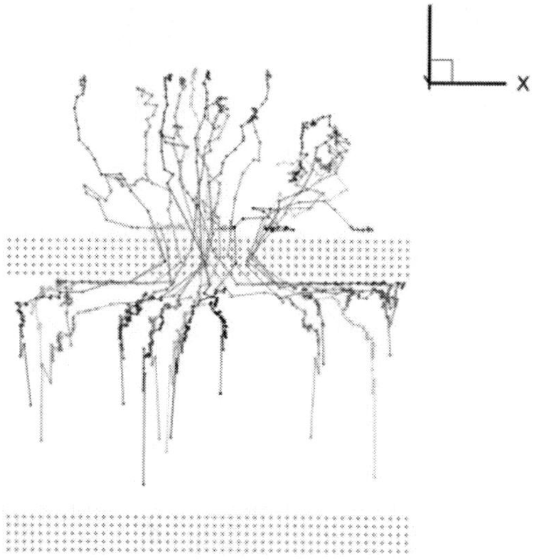

Figure 5. Trajectories of water molecules at 100 000 fs.

Figure 6 presents the molecular ratios between the two water molecular groups inside the nozzle aperture at each 10 000 fs instant. The molecules are divided into two groups along the z-direction as shown in Figure. 7. The first group is an annular volume close to the wall of nozzle aperture with a 2 Å annular radius. The second group is the rest. The curves in Figure. 6 marked by triangles and circles respectively show the molecular ratios for the 27.5- and 40-Å-diameter nozzle apertures. There are two observations of the molecular ratio during the ejection process of molecules through the nozzle aperture. In the first stage, these ratios change in a narrow band and increase at the end of this stage. The ratio value of 27.5-Å-diameter nozzle aperture is greater than that of 40-Å-diameter nozzle aperture. The curves show a high density of molecules in the annular volume resulting in many collisions on the wall of the 27.5-Å-diameter nozzle aperture. This affects the velocity and moving direction after ejection out of the nozzle aperture. This also induces an increasing diameter at the bottom of the water nanojet and affects the separation ability of the water nanojet from the nozzle plate surface. In the second stage, almost all of the molecules are ejected out of the nozzle aperture already. At this point, the low velocity molecules close to the nozzle aperture dropped back into the nozzle aperture. The attraction force of the gold atoms constituting the nozzle plate pulls most of these molecules towards the wall of the aperture, thus the molecular ratio significantly increases at this stage.

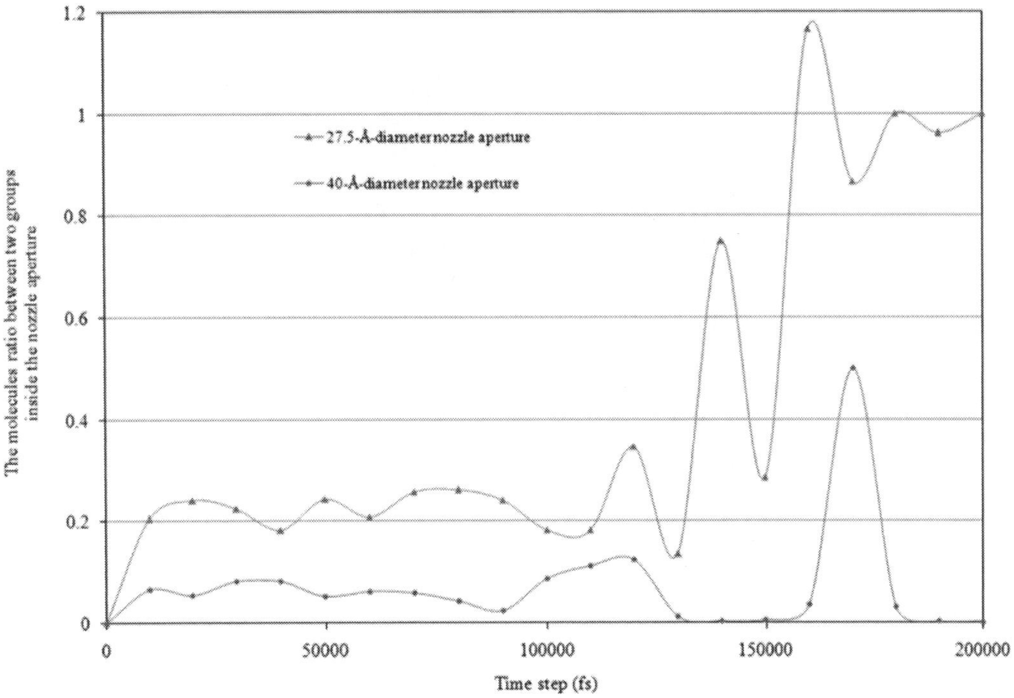

Figure 6. Plot of the molecular ratio between two sections inside the nozzle aperture.

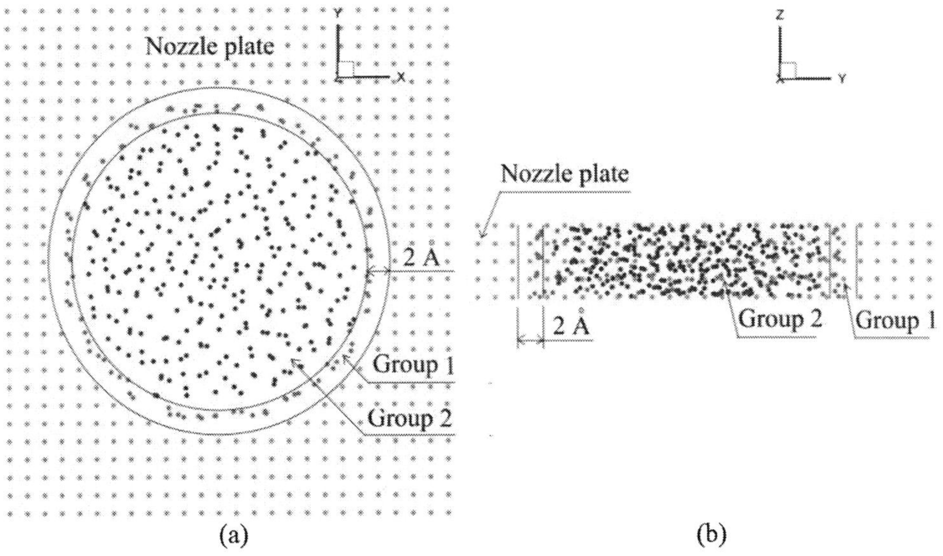

Figure 7. The number of molecules of two groups inside the nozzle aperture volume: (a) view in projection from top to bottom; (b) view in vertical projection.

Figures 8 (a)-(b) show the snapshots of the water nanojets at 200 000 fs through the 27.5- and 40-Å-diameter nozzle apertures, respectively. In comparison to the snapshot of the 27.5-Å-diameter nozzle aperture at 150 000 fs (Figure. 2(e)), the snapshot at 200 000 fs (Figure. 8(a)) shows a downwards movement, an increasing diameter and a formation of necking of the water nanojet. The molecular velocity in horizontal direction inside the nozzle aperture and the molecular density inside the simulation box are used to explain these phenomena. In contrast, Figure 8 (b) shows the continued upward movement of the water nanojet.

https://doi.org/10.1051/matecconf/201816901016

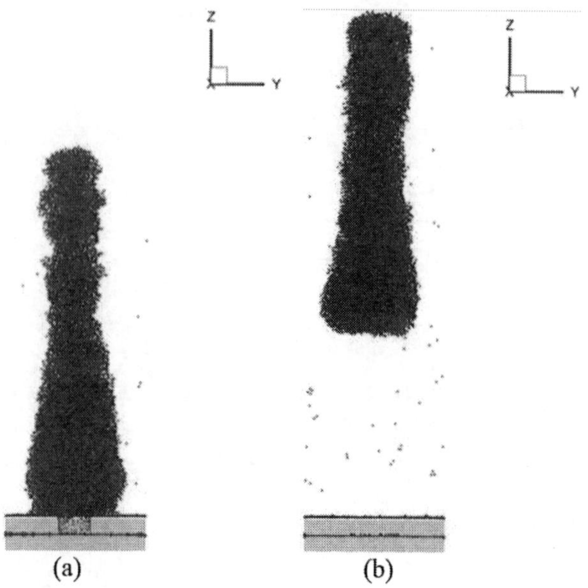

Figure 8. Snapshots of the water nanojet at 200 000 fs: (a) 27.5-Å-diameter nozzle aperture; (b) 40-Å-diameter nozzle.

The sum of the molecular velocities in the x- and y-horizontal directions near the wall of the nozzle aperture is shown in Figure. 9. The velocities rapidly increase at the initial stage of the molecular ejection process and reach to an equilibrium value at 120 000 and 60 000 fs for the 27.5-, and 40-Å-diameter nozzle apertures, respectively. Smaller nozzle aperture diameter yields higher velocity. Thus, direction of molecules is changed after ejection out of the nozzle aperture due to many molecules colliding on the wall. Most of these molecules move perpendicular to the axis of the water nanojet. This is a main reason for an increase of diameter of the water nanojet at the base. After 170 000 fs, the molecular velocity associated with the smaller nozzle apertures increases. This results in the downward movement of the water nanojet after reaching the maximum height but not separating from the nozzle plate surface. This is because the molecules are pulled by the attraction force of the gold atoms and the water molecules on the aperture wall.

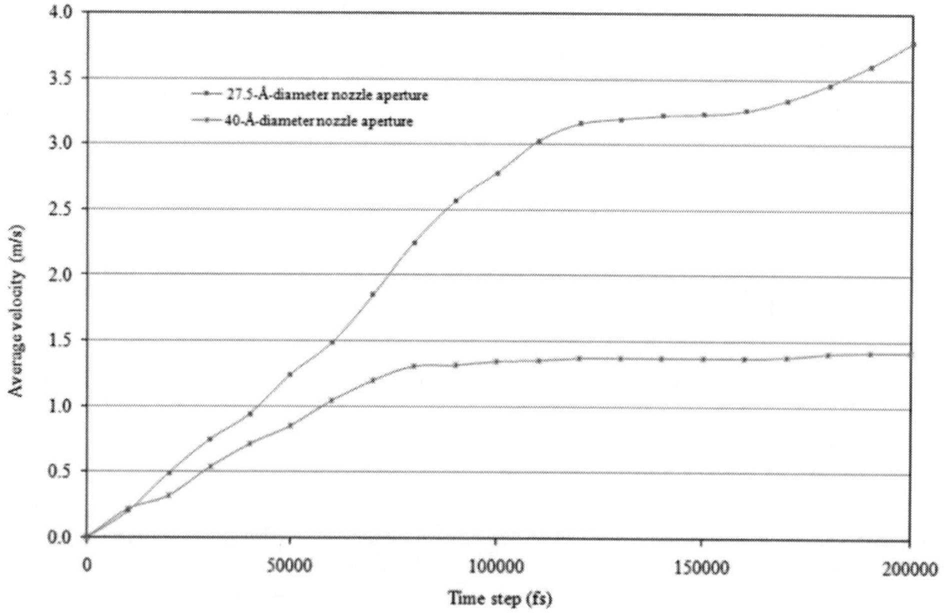

Figure 9. Plot of molecular velocity in x- and y-directions inside the nozzle aperture.

Figure 10 shows molecular density inside the water reservoir at each 2 000 fs instant. The initial density value is 0.378 g/cm^3. When the back plate moves upward and acts on the water molecules, the molecular density rapidly increases in the beginning and the density of molecules inside the reservoir fluctuates in a wide band with the 27.5-Å-diameter nozzle aperture. These fluctuations affect the continued ejection stream and cause the necking of the water column. The collisions of molecules in the water column increase and thus affect the upward z-direction velocity of the water nanojet. Shortly after, these densities gradually reach the steady values of 0.748 and 0.692 for the 27.5-, and 40-Å-diameter nozzle apertures, respectively. Intuitively, the ejection time is longer when the nozzle aperture diameter is smaller. The density of the molecules inside the reservoir fall back to near the initial value at the end of the molecular ejection process.

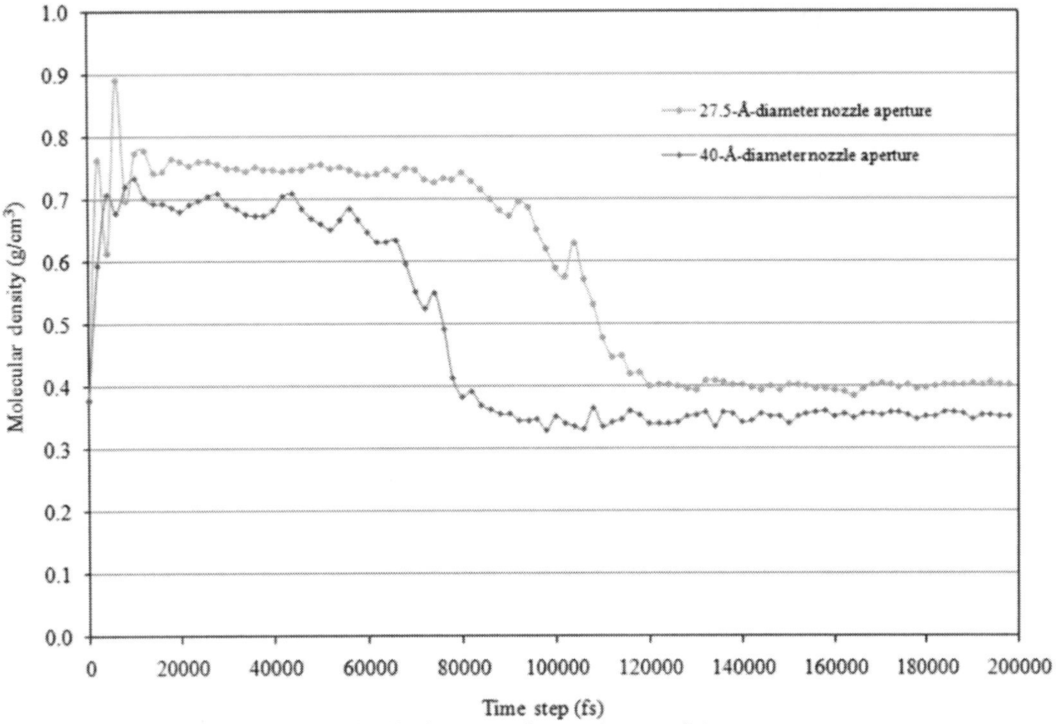

Figure 10. Plot of molecules densities inside the reservoir at each 2 000 fs instant.

3.3 Studying the impingement of the nanoscale water droplet onto the fixed plate surface

Impingement of a nanoscale water droplet onto a fixed plate surface was modeled by MD simulation using a 50-Å-diameter nozzle aperture with pressing forces of 8.0×10^{-10}, 10.0×10^{-10}, and 12.0×10^{-10} N under a system temperature of 310 K. Figures 11 (a)-(c) show snapshots of the nanoscale water droplet impingement onto a fixed plate surface at 1 200 000 fs with pressing forces of 8.0×10^{-10}, 10.0×10^{-10}, and 12.0×10^{-10} N, respectively. It can be seen that the number of water molecules on the solid plate surface and the width of the spreading droplet increase as the impingement time increases. Comparing the snapshots of Fig. 11, the results show that the contact angle of the water droplet with the fixed plate surface and the height of the nanoscale water droplet reduce as time and pressing force increase.

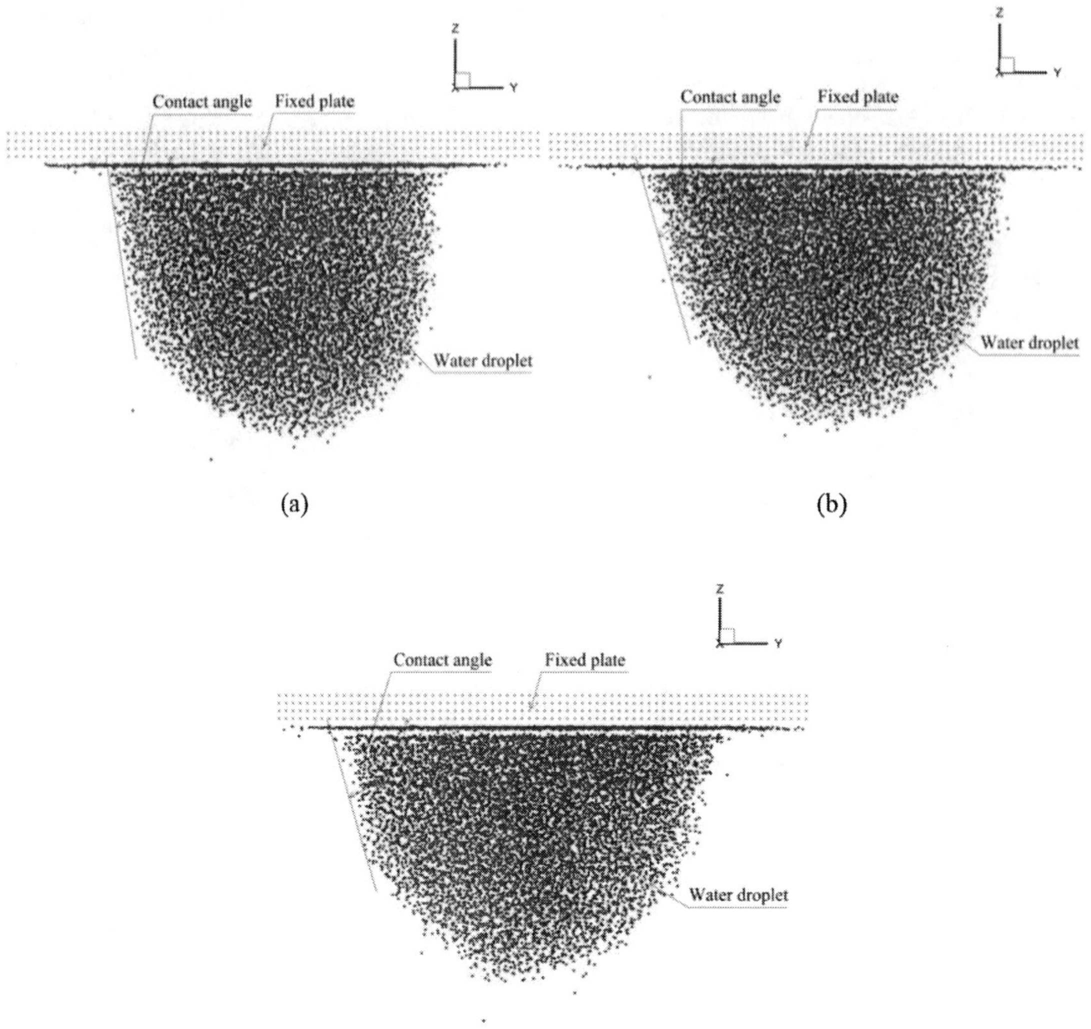

Figure 11. Snapshots of the nanoscale water droplets onto a fixed plate surface at 1 200 000 fs.

4 Conclusions

The molecular dynamics method is used in this study to simulate the formation of the water nanojet and the separation or non-separation of the water nanojet from the nozzle plate surface. Furthermore, initial observations on the impingement of a nanoscale water droplet onto a fixed solid surface are given. Based on the simulation results, the following five conclusions were drawn:

1. When the circumference-length-to-area ratio (L/A) is in the range from 0.036 to 0.040, all the water molecules in the reservoir are ejected through the nozzle apertures. However, the water nanojets do not separate from the surface of the nozzle plate. If the L/A is less than or equal to 0.033, the nanoscale water droplets are formed from the separation of the jet from the surface of the nozzle plate.

2. For small nozzle aperture diameters, the attraction force among the molecules and the cohesion in water are the main reasons which pull the water nanojet downward. This is enough to present the separation of the water nanojet from the nozzle plate surface.

3. The distribution and velocity in x-, and y-direction of molecules close to the nozzle wall are large for the smaller aperture diameters. These molecules have many collisions with the wall, so direction of molecules is changed after ejection from the nozzle aperture. Most of these molecules move in direction perpendicular to the axis of the water nanojet. This induces an increasing diameter

at the base of the water nanojet and affects the separation ability of the water nanojet from the nozzle plate surface.

4. The fluctuation of molecular density inside the reservoir is a wide band at the beginning of the ejection process for the 27.5-Å-diameter nozzle aperture. This affects the ejection stream and causes the necking of the water column.

5. The width of the spreading droplet on the fixed plate surface increases as the impingement time increases. The contact angle of the water droplet in relation to the fixed plate surface is reduced when the pressing force increases.

References

1. M. Moseler, U. Landman, Science, **289**, 1165-1170 (2000)
2. J.W. Lin and S.X. Chu, Proceedings of MNHT2008 Micro/Nanoscale heat transfer international conference, Taiwan, Taiwan, No. MNHT2008-52044, 519-524 (2008)
3. Y. Li, J. Xu and D. Li, Microfluid. Nanofluid., **9**, 1011-1031 (2010)
4. N. Gopan and S.P. Sathian, Colloids Surf. A Physicochem Eng. Asp., **432**, 19-28 (2013)
5. D.Y. Shin, P. Grassia and B. Derby, J. Fluids Eng., **127**, 98-109 (2005)
6. A. Tiwari, J. Abraham, Microfluid. Nanofluid., **4**, 227-235 (2008)
7. C.F. Dai and R.Y. Chang, Mol. Simul., **36**, 847-855 (2010)
8. K.T. Chang and C.I. Weng, Comput. Mater. Sci., **43**, 1048-1055 (2008)
9. B. Beulen, J.D. Jong, H. Reinten, M.V.D. Berg, H. Wijshoff and R.V. Dongen, Exp. Fluids, **42**, 217-224 (2007)
10. A. Asai, J. Fluids Eng., **114**, 638-641 (1992)
11. T.H. Fang, W.J. Chang and S.L. Lin, Appl. Surf. Sci., **253**, 1649-1654 (2006)
12. T.H. Fang, W.J. Chang and S.C. Liao, J. Phys.: Condens. Matter, **15**, 8263-8270 (2003)
13. S. Murad and I.K. Puri, Phys. Fluids, **19**, 128102 (2007)
14. G. Balasubramanian, I.K. Puri and S.A. Ragad, Chem. Phys. Lett., **491**, 177-182 (2010)
15. J.W. Lin, Microfluid. Nanofluid., **13**, pp. 37-48 (2012)
16. M. Levitt, M. Hirshberg, R. Sharon, K.E. Laidig and V. Daggett, J. Phys. Chem. B, **101**, 5051-5061 (1997)

Thermal hazard assessment of TMCH mixed with inorganic acids

Chi-Tang Yeh, Wei-Chun Chen and Chi-Min Shu[a]

Doctoral Program, Graduate School of Engineering Science and Technology, National Yunlin University of Science and Technology, Taiwan

Abstract. 1,1-Bis(*tert*-butylperoxy)-3,3,5-trimethylcyclohexane (TMCH) is a typical peroxide with two peroxy groups that may runaway and/or explode due to mixing with inorganic acids, such as HCl, HNO_3, H_2SO_4, or H_3PO_4. In this study, reactivities of TMCH mixed with the above inorganic acids were assessed by differential scanning calorimetry (DSC). Furthermore, data obtained by DSC, such as exothermic onset temperature (T_0), maximum temperature (T_{max}), and heat of decomposition (ΔH_d) could be employed to acquire thermal safety parameters. Moreover, thermal activity monitor III (TAM III) was employed to investigate the thermal hazards while storing or transporting TMCH and TMCH mixed with four types of commonly used inorganic acids, here as HCl, HNO_3, H_2SO_4, or H_3PO_4 under isothermal conditions. Mixing TMCH with those inorganic acids resulted in higherΔH_d except H_3PO_4, and mixing TMCH with HCl clearly decreased T_0. Therefore, the phenomena of mixing those incompatible materials with TMCH can be concluded as the worst cases in terms of contamination hazards during storage and transportation of TMCH.

1 Introduction

Most organic peroxides (OPs) can decompose readily even with inadvertent physical impact, while mixing with incompatible materials or being under extreme conditions, such as external fire and the high temperature environment. Furthermore, they will inevitably release a large amount of energies resulting in fire, explosion, or chemical release accidents [1–3]. The peroxy functional group (–O–O–) in OPs' structure, which is intrinsically unstable and thermally reactive, and it readily causes thermal runaway accidents because the peroxy functional group will break while reacting with incompatible materials, for example, acids, bases, metals, and ions. In view of process safety, as well as environment protection, some liquid organic peroxides are intrigue while handling. For instance, it is hardly predict to 1,1-bis(tert-butylperoxy)-3,3,5-trimethylcyclohexane (TMCH), a highly sensitive, exothermic, and unstable material. Accordingly, it must conserve well and proactively design safety protocol during preparation, manufacturing, application, transportation, and storage [4–8]. In addition, TMCH is used in the manufacturing process of styrene or polyethylene and is treated as an initiator of polymerization or cross-linking agent. Furthermore, TMCH can be employed as an activator in the process industries. Consequently, TMCH must be well mastered and stored with caution in order to eliminate any potential thermal hazard which caused by mixing TMCH with other incompatible

[a] Corresponding author : shucm@yuntech.edu.tw

materials during the whole gamut of its process span [9–13]. Historically, there have been incidents related to organic waste solvents that have caused thermal hazard accidents, because the mixtures of organic waste solvent may have further induced a decomposition reaction due to improper disposal and inferior safety management. Decomposition hazards can be coped with through adequate process safety. Process safety not only protects the property from accidents, but also reduces environmental impact because of runaway reaction. OPs will decompose and may contribute to process hazards, thermal runaways, and environmental impacts.

The aim of this study was to appraise the environmental concerns with thermal hazard of TMCH and the thermal decomposition reaction of TMCH mixing with various incompatible substances, such as hydrochloric acid (HCl), nitric acid (HNO_3), sulfuric acid (H_2SO_4), or phosphoric acid (H_3PO_4). The thermal stability parameters including apparent onset temperature (T_0), heat of decomposition (ΔH_d), and maximum temperature (T_{max}) were reckoned by differential scanning calorimetry (DSC). Furthermore, we used the thermal activity monitor III (TAM III) to calculate time to maximum rate under isothermal conditions (TMR_{iso}), highest or peak heat flow, and reaction time. The apparent activation energy (E_a) for TMCH under different isothermal temperatures (60.0, 70.0, 80.0, and 90.0 °C) were tested by TAM III. Therefore, we can establish inherently safer designs in the storage and transport of TMCH to avoid jeopardizing the system under thermal upsets.

2 Experiments and methods

2.1 Samples

To assess reaction and determine the related parameters of TMCH with four inorganic acids (HCl, HNO_3, H_2SO_4, H_3PO_4), 88.0 mass% of TMCH was procured from the ACE Chemical Corp. in Taoyuan, Taiwan and then stored in a refrigerator at 4.0 °C. Mixing TMCH with the four types of commonly used inorganic acids, HCl, HNO_3, H_2SO_4, and H_3PO_4, whose concentrations of HCl, HNO_3, H_2SO_4, and H_3PO_4 were 37.0, 69.0, 95.0, and 85.0 mass%, respectively. HCl and H_2SO_4 were purchased from the Merck Display Technologies Ltd. in USA and HNO_3 and H_3PO_4 were purchased from the Panreac in USA.

2.2 Differential scanning calorimetry (DSC)

Temperature-programmed screening experiments were performed on a Mettler TA8000 system coupled with a DSC 821^e measuring cell [14]. The DSC results show the profiles of heat flow versus temperature that the reference sample and the target sample are heated at the same time. The temperature between the reference sample and the target sample are different because the target sample releases energy during thermal decomposition and it, in turn, resulted in heat generation in the cell. Therefore, the temperature in reference sample's cell and target sample's cell did not sustain equally after heating test. The system of programmable temperature control applied in DSC, which is used to acquire the various heating and cooling rates from 0.01 to 100.0 °C and measure the range of temperature of target sample from –150.0 to 700.0 °C. For a better equilibrium, the heating rate in this study was set at 4.0 °C min^{-1}, and the rising range of temperature was chosen from 30 to 300.0 °C.

2.3 Thermal activity monitor III (TAM III)

Isothermal calorimetry, TAM III, is a delicate thermal sensitivity and elaborate stability detecting equipment. It is easy to operate and only requires few sample quantity during detecting process. Accordingly, TAM III is broadly applied in many areas, such as material science, pharmaceutical science, chemistry, physical chemistry, and life science [15–17]. The liquid in TAM III's thermostat was mineral oil, which is in a reactor with a total volume of 22 liter (L), and the maximum heating rate of TAM III was ±2.0 K h^{-1} for usage of chemical and physical equilibrium. Moreover, the temperatures were set at 15.0–150.0 °C, and the decomposition of sample was, in turn, detected by TAM III under isothermal conditions. The software of TAM III was furthermore employed to control the thermostat and compute some thermokinetic parameters [18, 19]. The aim of this study was to use TAM III to determine the thermal decomposition reactions of 88.0 mass% TMCH and 88.0 mass% TMCH individually mixed with 6.0 N HCl, 6.0 N HNO_3, 6.0 N H_2SO_4, or 6.0 N H_3PO_4. Here, the amount of the sample in a cell was 100.0 mg TMCH mixing with 10.0 mg HCl, HNO_3, H_2SO_4, or H_3PO_4 at 60.0, 70.0, 80.0, and 90.0 °C.

3 Results and discussion

3.1 Thermal analysis by DSC

The DSC test results were detected by DSC 821[e] with thermal spectrum analysis including heating and temperature conditions, showing heat release or heat decomposition, starting temperature, or the dynamics calculations. Figure 1 shows the thermal curve of TMCH with or without inorganic acids, and it is one exothermic peak while TMCH decomposing.

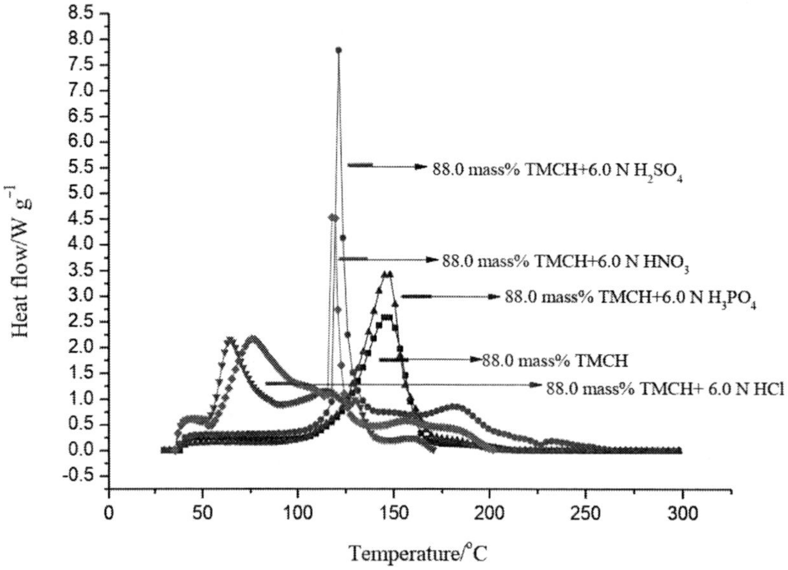

Figure 1. DSC thermal curves of heat flow versus temperature for TMCH mixing with HCl, HNO_3, H_2SO_4, and H_3PO_4 decomposition at scanning rate of 4.0 °C min^{-1}.

Table 1. Calorimetric data from the dynamic scanning experiments of TMCH 88.0 mass% and TMCH 88.0 mass% mixed with 6.0 N HCl, HNO_3, H_2SO_4, and H_3PO_4 by DSC.

Samples	Mass/mg	T_0/°C	T_{max}/°C	ΔH_d/kJ mol^{-1}	Total heat of reaction/W g^{-1}
88.0 mass% TMCH	2.9	120.2	146.6	1112.8	3227.1
88.0 mass% TMCH+ HCl (6.0 N, 1.0 mg)	3.1	52.7	64.3	1646.3	5103.6
88.0 mass% TMCH+HNO_3 (6.0 N, 1.0 mg)	3.1	115.7	118.4	1720.4	6895.8
88.0 mass% TMCH+H_2SO_4 (6.0 N, 1.0 mg)	2.8	119.0	120.8	1783.1	4992.7
88.0 mass% TMCH+H_3PO_4 (6.0 N, 1.1 mg)	2.9	123.4	146.6	1398.9	4056.9

The results of TMCH with inorganic acids which were used as the contaminants are displayed in Table 1. It caused the T_0 emerged earlier while TMCH mixing with HCl, especially, and not only the T_0 occurred earlier but also it formed a fierce thermal reaction and then causing high temperature when TMCH was mixed with H_2SO_4. In addition, it revealed that the similar reaction phenomenon between TMCH and TMCH mixed with H_3PO_4, such as the similar T_0, but the ΔH_d of TMCH mixed with H_3PO_4 increased significantly. The T_0 was induced to occur earlier as HNO_3 was mixed with TMCH, and it formed a relatively high heat of reaction after the mini-exothermic reaction of the first peak. In conclusion, all of the inorganic acids could alter the original decomposition reaction of TMCH while TMCH is mixing with the inorganic acids.

3.2 Isothermal analysis by TAM III

TAM III was used to evaluate the decomposition reaction of TMCH and the reaction of TMCH mixing with various inorganic acids. Figures 2 to 6 delineate the decomposition of pure TMCH and TMCH mixed with incompatible substances under different isothermal temperatures by TAM III. Figure 2 shows the decomposition of 88.0 mass% TMCH at various temperatures, in which the heat flow was low and not visible when the temperature was at 60.0 °C, whereas the heat flow was high and substantial at 90.0 °C

Table 2. Scanning data of the thermal runaway decomposition at 60.0 °C by TAM III.

Samples	Mass/mg	TMR_{iso}/h	Highest heat flow/W g^{-1}
88.0 mass% TMCH	100.1	94.71252	0.00029
88.0 mass% TMCH+ HCl (6.0 N, 10.2 mg)	99.5	1.27623	0.23796
88.0 mass% TMCH+HNO_3 (6.0 N, 11.8 mg)	100.6	0.41744	0.11802
88.0 mass% TMCH+H_2SO_4 (6.0 N, 10.6 mg)	100.6	18.93276	0.24074
88.0 mass% TMCH+H_3PO_4 (6.0 N, 12.0 mg)	100.2	93.37667	0.00038

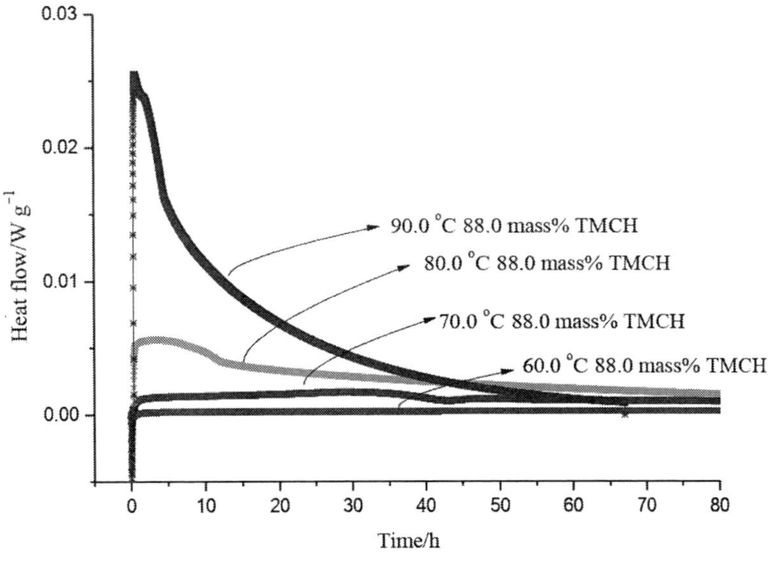

Figure 2. TAM III thermal curves of heat flow versus time for 88.0 mass% TMCH decomposition at 60.0, 70.0, 80.0, and 90.0 °C.

When TMCH mixed with 6.0 N HCl, the heat flow was prominent at 60.0 °C, and they were similar at decomposition at 80.0 or 90.0 °C as shown in Fig. 3. We found that there are two exothermic peaks when TMCH mixed with 6.0 N HNO_3. In addition, Fig. 4 illustrates that the energy generated from first exothermic peak, which forced the second exothermic reaction to appear earlier. Furthermore, the first exothermic peak almost occurred at the same time around 60.0–90.0 °C, but the most high second exothermic peak was at 90.0 °C. Therefore, TMCH mixed with HNO_3 not only resulted in two exothermic peaks, but also released enormous amount of heats, which may lead to serious thermal hazard. The heat flow was higher at 90.0 °C when TMCH mixed with 6.0 N HNO_3, 6.0 N H_2SO_4, or 6.0 N H_3PO_4 as revealed in Figs. 3, 4, 5, and 6. The heat flow of TMCH mixed with H_2SO_4 was higher than TMCH mixed with other inorganic acids. In Fig. 6, there are also two exothermic peaks when TMCH mixed with H_3PO_4 at 80.0 and 90.0 °C. The exothermic peak was similar, but ΔH_d was higher at 90.0 °C than at 80.0 °C. TMCH mixed with H_3PO_4 could increase the probability of the second thermal hazard to be happened. The thermal stability parameters, such as TMR_{iso} and the maximum heat flow, can be calculated by TAM III. TMR_{iso} is employed to assess the emergency response, and we therefore applied TMR_{iso} in prevention of thermal runaway reactions.

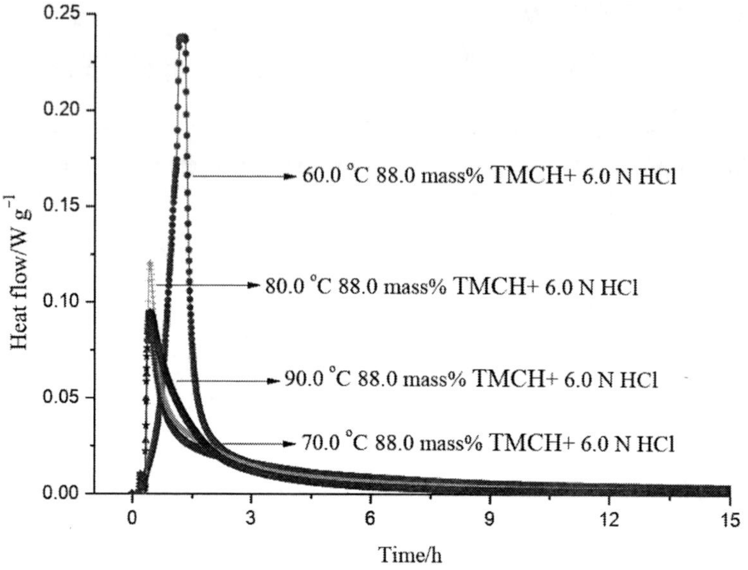

Figure 3. TAM III thermal curves of heat flow versus time for TMCH mixing with 6.0 N HCl decomposition at 60.0, 70.0, 80.0, and 90.0 °C.

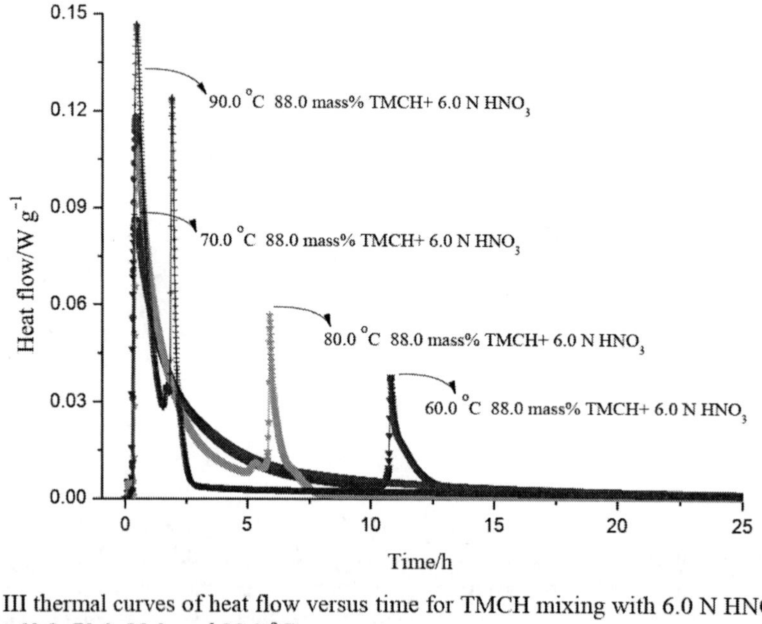

Figure 4. TAM III thermal curves of heat flow versus time for TMCH mixing with 6.0 N HNO_3 decomposition at 60.0, 70.0, 80.0, and 90.0 °C.

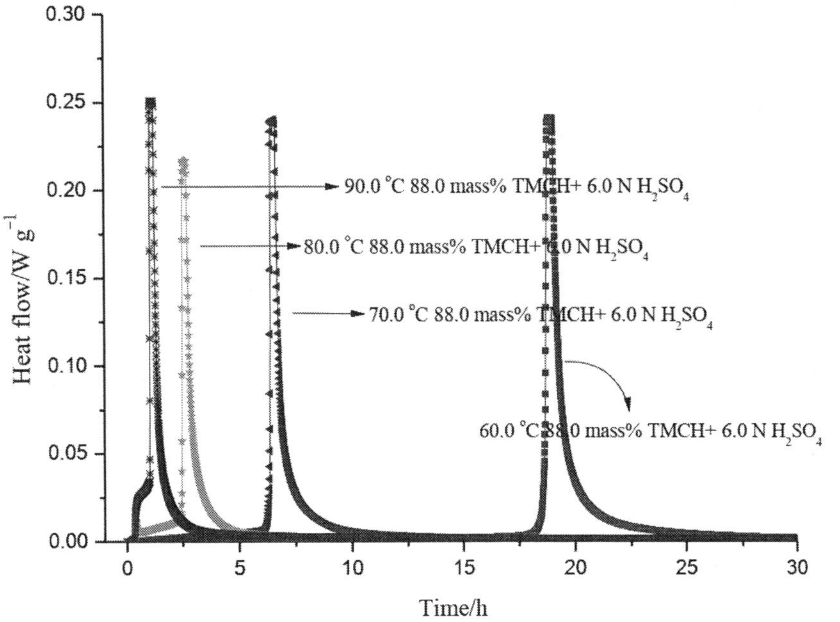

Figure 5. TAM III thermal curves of heat flow versus time for TMCH mixing with 6.0 N H_2SO_4 decomposition at 60.0, 70.0, 80.0, and 90.0 °C.

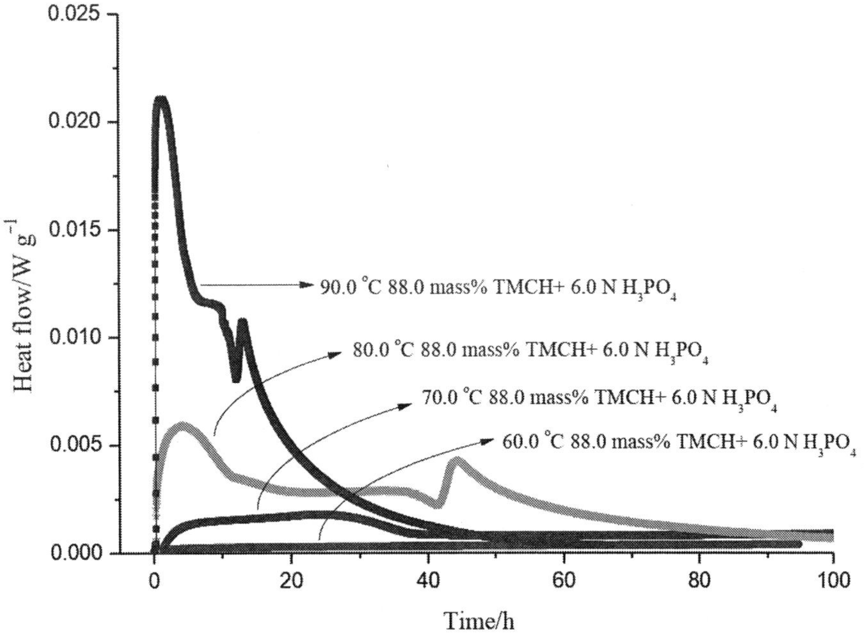

Figure 6. TAM III thermal curves of heat flow versus time for TMCH mixing with 6.0 N H_3PO_4 decomposition at 60.0, 70.0, 80.0, and 90.0 °C.

In Tables 2 to 5, the results presented that the TMR_{iso} of TMCH was 94.71252, 28.49707, 3.53645, and 0.58584 h and the TMR_{iso} of TMCH mixing with HCl, at 60.0, 70.0, 80.0, and 90.0 °C were reduced to 1.27623, 0.42682, 0.43239, and 0.46303 h, correspondingly. The results indicated that the TMR_{iso} of TMCH mixing with HNO_3 was 0.41744, 0.42490, 0.50855, 0.46303 h and the TMR_{iso} of TMCH mixing with H_2SO_4 at 60.0–90.0 °C were reduced to 18.93276, 6.60274, 2.66964, 1.23899 h, respectively.

Table 3. Scanning data of the thermal runaway decomposition at 70.0 °C by TAM III.

Samples	Mass/mg	TMR_{iso}/h	Highest heat flow/W g^{-1}
88.0 mass% TMCH	100.3	28.49707	0.00163
88.0 mass% TMCH+ HCl (6.0 N, 9.5 mg)	99.1	0.42682	0.09202
88.0 mass% TMCH+HNO_3 (6.0 N, 10.9 mg)	99.7	0.42490	0.08624
88.0 mass% TMCH+H_2SO_4 (6.0 N, 9.5 mg)	100.2	6.60274	0.26238
88.0 mass% TMCH+ H_3PO_4 (6.0 N, 9.0 mg)	99.3	24.29267	0.00174

Table 4. Scanning data of the thermal runaway decomposition at 80.0 °C by TAM III.

Samples	Mass/mg	TMR_{iso}/h	Highest heat flow/W g^{-1}
88.0 mass% TMCH	101.0	3.53645	0.00557
88.0 mass% TMCH+ HCl (6.0 N, 9.7 mg)	101.2	0.43239	0.12052
88.0 mass% TMCH+HNO_3 (6.0 N, 10.0 mg)	100.1	0.50855	0.10800
88.0 mass% TMCH+H_2SO_4 (6.0 N, 11.8 mg)	100.2	2.66964	0.21654
88.0 mass% TMCH+ H_3PO_4 (6.0 N, 11.5 mg)	100.6	4.30909	0.00588

Furthermore, the TMR_{iso} of TMCH mixing with H_3PO_4 at 60.0, 70.0, 80.0, and 90.0 °C were reduced to 93.37667, 24.29267, 4.30909, and 1.32272 h. In contrast, as TMCH was mixed with H_2SO_4, whose TMR_{iso} was not visible at 80.0 and 90.0 °C. However, the TMR_{iso} of TMCH was smaller when mixed with HCl, HNO_3, and H_3PO_4. Moreover, the maximum heat flow of TMCH mixed with above inorganic acids was also increased significantly. However, the TMR_{iso} of TMCH mixing with H_3PO_4 was increased due to decomposition mechanisms. The higher maximum heat flow and temperature could cause prominent runaway excursions, and HCl, HNO_3, and H_2SO_4 could be classified as incompatible substance because TMCH mixed with those inorganic acids may result in the diminished TMR_{iso}.

Table 5. Scanning data of the thermal runaway decomposition at 90.0 °C by TAM III.

Samples	Mass/mg	TMR_{iso}/h	Highest heat flow/W g^{-1}
88.0 mass% TMCH	100.0	0.58584	0.02551
88.0 mass% TMCH+ HCl (6.0 N, 10.1 mg)	99.0	0.46303	0.09479
88.0 mass% TMCH+HNO_3 (6.0 N, 11.7 mg)	99.6	0.46303	0.14684
88.0 mass% TMCH+H_2SO_4 (6.0 N, 11.5 mg)	99.4	1.23899	0.25069
88.0 mass% TMCH+H_3PO_4 (6.0 N, 11.6 mg)	99.3	1.32272	0.02104

Table 6 lists the estimated results of apparent activation energy of TMCH mixed with H_3PO_4 by using Arrhenius equation, as shown in Eq. (1):

$$k_i(T) = Aexp\left(-\frac{E_a}{RT}\right) \quad (1)$$

where A, E_a, and R represent the frequency factor, the apparent activation energy of the stage, and the gas constant, individually. The E_a of pure TMCH and TMCH mixed with H_3PO_4 was 147.37 and 133.34 kJ mol^{-1}, respectively. When TMCH was mixed H_3PO_4, its E_a is lower than pure TMCH, indicating that the reaction could occur easily.

Table 6. Comparison of TMCH and TMCH+6.0 N H_3PO_4 of apparent activation energy analysis graph for Arrhenius plot with different isothermal temperatures at 60.0, 70.0, 80.0, and 90.0 °C.

Samples	E_a /kJ mol^{-1}	A/s^{-1}	R^2
88 mass% TMCH	147.37	3.99436E+19	0.99569
88 mass% TMCH+ H_3PO_4	133.34	3.26106E+17	0.99855

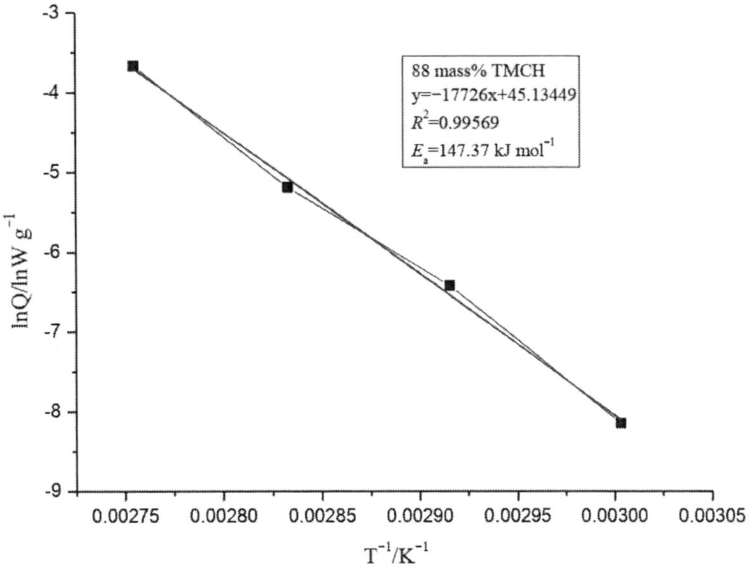

Figure 7. TMCH of activation energy analysis graph for Arrhenius plot with different isothermal temperatures at 60.0, 70.0, 80.0, and 90.0 °C.

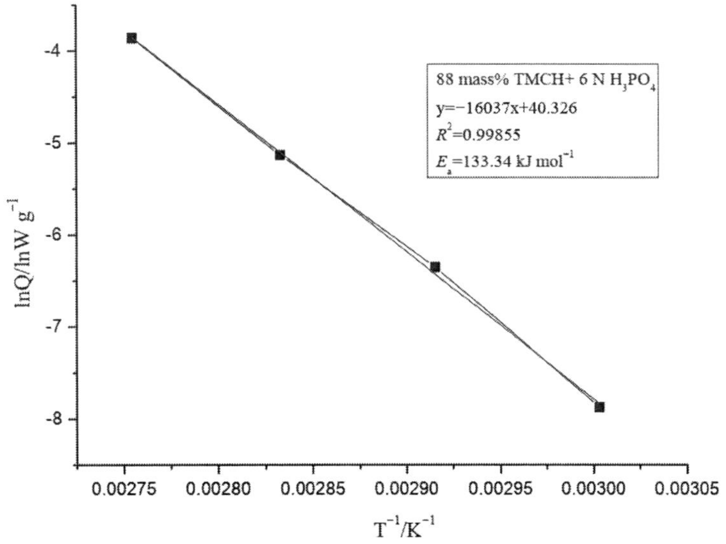

Figure 8. TMCH mixing with 6.0 N H_3PO_4 of apparent activation energy analysis graph for Arrhenius plot with different isothermal temperatures at 60.0, 70.0, 80.0, and 90.0 °C.

4 Conclusion

In view of environment protection, the calorimetric technique is very important in chemical industry for assessing the thermal hazard influence during the whole gamut of the manufacturing process. The inherent thermal hazard characteristics of TMCH and TMCH mixed with HCl, HNO_3, H_2SO_4 or H_3PO_4, were completely described in the results and discussion which depicted the thermal stability parameters of TMCH and TMCH with inorganic acids. It significantly changed not only the original reaction but also the TMR_{iso} of TMCH while TMCH was mixed with HCl, HNO_3, H_2SO_4, or H_3PO_4. When TMCH mixed with HCl, it may generate high heat flow and has second potential hazard because it formed two prominent peaks during decomposition reaction. Therefore, those related process safety parameters were assessed by DSC and TAM III that could apply in the prevention of runaway reaction of TMCH. The incompatible behavior of the TMCH incompatible hazards should be considered carefully in industrial processes and the environment. Therefore, thermal decomposition analysis plays a vital role in environmental engineering. Furthermore, we can employ chromatography instruments, such as gas chromatography/mass spectrometer (GC/MS), thermogravimetry/GC–mass spectrometer (TG/GC/MS) coupled with Fourier transform infrared (FT/IR) spectrometer to determine the decomposition pathway of TMCH as well as mixing with incompatibles, here HCl, HNO_3, H_2SO_4, or H_3PO_4.

References

1. Y.L. Chen, Y.P. Chou, H.Y. Hou and C.M. Shu, J. Therm. Anal. Calorim., **95**, 535-539 (2009)
2. S.H. Wu, Y.W. Wang, T.C. Wu, W.N. Hu and C.M. Shu, J. Therm. Anal. Calorim., **93**, 189-194 (2008)
3. K.Y. Chen, S.H. Wu, Y.W. Wang and C.M. Shu, J. Loss Prev. Process Ind., **21**, 101-109 (2008)
4. S.H. Liu, C.P. Lin and C.M. Shu, J. Therm. Anal. Calorim., **106**, 165-172 (2011)
5. J.R. Chen, S.H. Wu, S.Y. Lin, H.Y. Hou and C.M. Shu, J. Therm. Anal. Calorim., **93**, 127-133 (2008)
6. L.K. Wu, K.Y. Chen, S.Y. Cheng, B.S. Lee and C.M. Shu, J. Therm. Anal. Calorim., **93**, 115-120 (2008)
7. H.Y. Hou, Y.S. Duh, W.H. Lin and C.M. Shu, J. Therm. Anal. Calorim., **85**, 145-150 (2006)
8. Y.S. Duh, X.H. Wu and C.S. Kao, Process Saf. Prog., **27**, 89-99 (2008)
9. A.C. Li, Y.T. Tsai, S.H. Wu, C.W. Chiu, S.J. Shen, R.H. Chang and C.M. Shu, J. Therm. Anal. Calorim., **113**, 1611-1618 (2013)
10. Y.W. Wang, C.M. Shu, Y.S. Duh and C.S. Kao, Ind. Eng. Chem. Res., **40**, 1125-1132 (2001)
11. S.H. Liu, H.Y. Hou and C.M. Shu, Thermochimica Acta, **556**, 226-232 (2013)
12. Y.T. Tsai, M.L. You, X.M. Qian and C.M. Shu, Industrial & Engineering Chemistry Research, **52**, 8206-8215 (2013)
13. C.Y. Jhu, Y.W. Wang, C.Y. Wen and C.M. Shu, Applied Energy, **100**, 127-131 (2012)
14. M. Toledo, Stare Software with Solaris Operating System, Operating Instructions, Switzerland (2004)
15. S.Y. Lin, J.M. Tseng, Y.F. Lin, W.T. Huang and C.M. Shu, Journal of Thermal Analysis Calorimetry, **93**, 257-267 (2008)
16. W.T. Chen, C.C. Ma, M.H. Lee, Y.C. Chu, L.C. Tsai and C.M. Shu, Appl. Energy, **100**, 187-192 (2012)
17. H.X. Chen and N.A. Liu, J. Therm. Anal. Calorim., **90**, 449-452 (2007)
18. J.M. Tseng, M.Y. Liu, S.L. Chen, W.T. Hwang, J.P. Gupta and C.M. Shu, J. Therm. Anal. Calorim., **96**, 789-793 (2009)
19. Y.C. Chu, F.C. Tsai, F.R. Lin, T.C. Chiang and C.M. Shu, Energy Education Science and Technology Part A, **30**, 977-982 (2013)

Molecular characterization, overexpression and comparison of esterases-encoding LipRT, Lip4 and Lip20 from moderately thermophilic and mesophilic bacteria

Yu-Pei Chen[1], Shu-Shan Liang[2], Ing-Er Hwang[2], Mei-Chih Chien[2], Guey-Horng Wang[1] and Li-Ling Liaw[2,a]

[1]*Application Technique Engineering Center of Natural Cosmeceuticals, College of Fuijan Province, Xiamen Medical College, China*
[2]*Bioresource Collection and Research Center, Food Industry Research and Development Institute, Taiwan*

Abstract. Thermostable enzymes have the potential as the biocatalyst for industrial applications. To compare the relationship of enzymatic thermostability, the moderately thermophilic and mesophilic bacteria were utilized to explore the properties of esterases. By using the shotgun libraries of mesophilic *Thalassomonas agarivorans*, and *Aeromonas* sp., and moderately thermophilic *Ralstonia* sp., esterases-encoding Lip20, Lip4 and LipRT for α/β-hydrolase fold were cloned, sequenced, and characterized. According to the recombinant proteins overexpressed by *Escherichia coli*, these results indicated that Lip20, Lip4 and LipRT preferred to hydrolyze short-length p-nitrophenyl (*p*-NP) esters. The optimal temperature required for the activity of Lip20, Lip4 and LipRT was 30, 40 and 60°C, respectively, corresponding to the trend of bacterial growth temperature. Even at low temperature, cold-adapted Lip4 from *Aeromonas* sp. revealed well enzymatic activity. In addition, after 60 min incubation between 40-60°C, over 92% residual activity can be retained by the thermostable analysis of LipRT from *Ralstonia* sp.. Inspecting the predicted structures and amino acid composition, we found that the high helix content was exhibited in LipRT. Also, high frequency residues of Val, Phe and Arg for increasing hydrophobic and salt-bridge interactions were observed. These factors could improve LipRT thermal stabilization and lead to become more rigid.

1 Introduction

The worldwide demand for enzymes is increasing, and expenditure on enzymes is expected to reach $6.9 billion by 2017. Lipolytic enzymes account for 21% of the worldwide industrial enzyme market and are the third largest commercialized enzymes after proteases and carbohydrases [1]. They are widely present in animals, plants, and microorganisms. According to the capacity for hydrolyzing carboxyl esters of different acylglycerol lengths, lipolytic enzymes can be divided into two major groups: triacylglycerol acylhydrolases (lipases) (EC 3.1.1.3) and carboxylic-ester hydrolases (esterases) (EC 3.1.1.1).

[a] Corresponding author : lll@firdi.org.tw

Lipolytic enzymes are used in various industrial areas, such as food, bioremediation, pharmaceuticals, detergents, biodiesel production, and waste treatment [2]. Therefore, the development of high-yield production and thermostable lipolytic has attracted considerable attention. Over the past years, a considerable number of studies have been made on the thermostability by structural comparison of enzymes between the different optimal temperatures [3]. In addition, according to the different amino acid composition between thermophilic and mesophilic proteins, thermostable features can provide guidelines for the design of thermo-tolerance proteins [4].

The hot-adapted esterases can be explored from thermophilic bacteria which represent a valuable source towards numerous applications. In this study, the esterases were carried out to explore the thermostable relationship between moderately thermophilic and mesophilic bacteria. Hence, the novel esterases, Lip20, Lip4 and LipRT from mesophilic *T. agarivorans* and *Aeromonas* sp., and moderately thermophilic *Ralstonia* sp., were successfully cloned and overexpressed in *E. coli* for characterization and comparison.

2 Materials and methods

2.1 Shotgun library construction, sequencing and phylogenetic analysis

The three bacterial strains isolated and purified from Hsinchu, Taiwan were capable of hydrolyzing tributyrin by agar plate determination. They were identified as *Aeromonas* sp., *Ralstonia* sp. and *T. agarivorans*. *T. agarivorans* was incubated in Polypepton-Yeast (PY) medium at 25°C for 3 d. The *Aeromonas* sp. and *Ralstonia* sp. were incubated in Tryptic Soy Broth (TSB) medium at 30°C and 40°C for 3d, respectively. The bacterial pellet was harvested for genomic extraction. The shotgun library and DNA sequencing were performed as described in Liang et al. [5]. Putative open reading frames were further searched and examined using the protein BLAST program available on the NCBI website (https://blast.ncbi.nlm.nih.gov/Blast.cgi). The phylogenetic tree of the esterase gene was constructed using the neighbor-joining method by using MEGA 6.0 software with 1000 bootstrap replicates. Protein structure prediction was carried out using SWISS-MODEL [6] and the results were implemented in PyMol (www.pymol.org).

2.2 Esterase expression plasmid construction and overexpression in *E. coli*

To amplify the esterase genes from *T. agarivorans*, *Aeromonas* sp. and *Ralstonia* sp., the forward primers 5′-CACCATGCGAGGTATAGCGAGC-3′, 5′-CACCATGAGCATATCGGTAGAG-3′ and 5′-CACCATGGACAGGCAAGCCA CG-3′ (underlined sequence was designed for ligation with the pET151/D-TOPO plasmid) and the reverse primers 5′-TTATTTGTTTACGAATCTGGC-3′, 5′-TTAGCCAACAAAATCTCGA-3′ and 5′-TCAACGCCCGCCCGCACC-3′ were used for polymerase chain reaction (PCR). A Champion pET151 Directional TOPO Expression Kit (Invitrogen, Carlsbad, CA, USA) was used to obtain the esterase overexpression plasmids. Three esterase overexpression plasmids were transformed into competent cells of *E. coli* C41(DE3), *E. coli* C43(DE3), and *E. coli* BL21(DE3). For purifying the esterase protein, a Ni-NTA purification system (Invitrogen) was used. Protein content was determined using a Bradford protein assay kit (Bio-Rad, Hercules, CA) and monitored using SDS-PAGE and colloidal Coomassie blue staining.

2.3 Enzymatic activity assay of the esterase protein

Enzyme activity was determined using a spectrophotometric assay with *p*-nitrophenyl (*p*-NP) esters as substrates. Various *p*-NP esters were estimated, including *p*-NP butyrate (C4), *p*-NP valerate (C5), *p*-NP octanoate (C8), *p*-NP decanoate (C10), *p*-NP dodecanoate (C12), *p*-NP myristate (C14), and *p*-NP palmitate (C16). The mixture used in the enzyme activity assay contained 0.25 mM *p*-NP esters (C4–C10), 2.5% ethanol, 50 mM Tris-HCl (pH 7.4), and a final volume of 800 µL of esterase. For determining long-chain *p*-NP esters (C12–C16), 0.2 mM *p*-NP esters, 2.5% ethanol, 0.1% gum arabic, 0.1% deoxycholate, 50 mM Tris-HCl (pH 7.4), and esterase were added. The reaction mixtures were incubated at different temperatures or pH for 10 min and interrupted by incubation at 100°C. The mixture was then measured at OD405 by using a spectrophotometer.

3 Results and discussion

3.1 Genomic library of Lip20, Lip4 and LipRT, and identification of the esterase genes

Three genomic libraries of *T. agarivorans*, *Aeromonas* sp. and *Ralstonia* sp. containing over 15,000 clones with 2–4-kb DNA inserts was constructed. Three clones from *T. agarivorans*, *Aeromonas* sp. and *Ralstonia* sp. libraries produced significantly clear zones, indicating that they were capable of producing esterases, and were selected for shotgun sequencing. The sequencing reads were assembled and insertions showed the full open reading frames for BLAST analysis. The results revealed three genes with 927-bp, 765-bp and 1263-bp esterases which were named as *lip20*, *lip4* and *lipRT* (GenBank accession nos. KR401219, KX756573 and KX756574). The deduced sequences of Lip20, Lip4 and LipRT had 308, 254 and 420 amino acids, suggesting the calculated molecular weights of 34.1, 28.2 and 45.5 kDa. They respectively shared 53%, 97% and 99% identities with the sequences of *Shewanella frigidimarina* lipase, *Aeromonas veronii* pimeloyl-[acyl-carrier protein] methyl ester esterase and *Ralstonia mannitolilytica* serine hydrolase (GenBank accession no. WP_011638606, WP_005340633.1 and WP_045218811.1), which remain uncharacterized and unpublished.

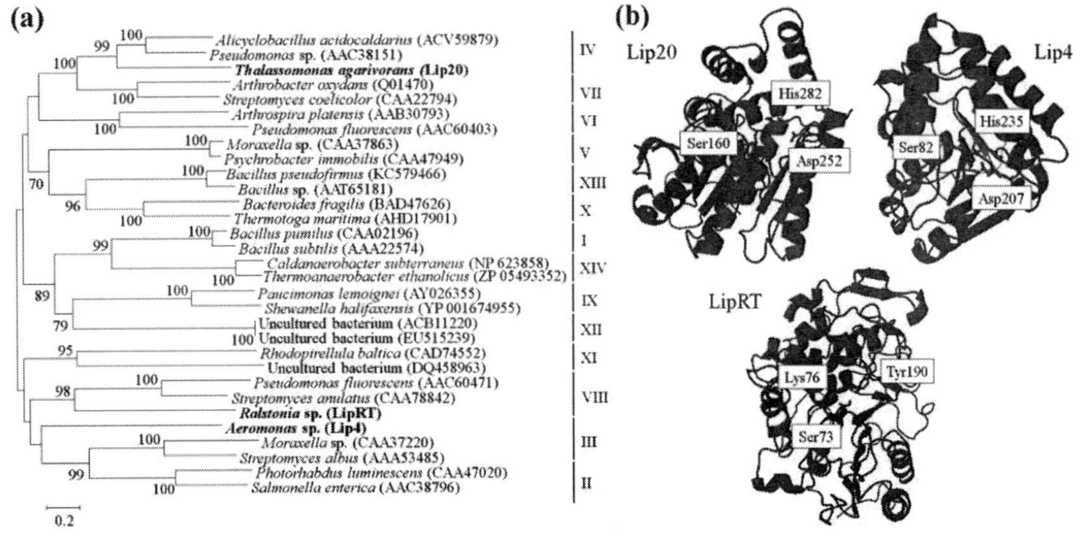

Figure 1. Phylogenetic tree and structure prediction of the Lip20, Lip4 and LipRT. (a) Phylogeny based on the complete open reading frame of esterase sequences of 14 families from *T. agarivorans*, *Aeromonas* sp., *Ralstonia* sp. and various bacteria. (b) The protein structures were predicted by SWISS-MODEL and implemented in PyMol.

The 14 families of lipolytic enzymes identified by Rao et al. [7] were used for constructing a phylogenetic tree. Figure 1a shows that Lip20, Lip4 and LipRT were classified into the Family IV, III and VIII clades. Moreover, considerable research has been conducted on the three-dimensional structures of esterases, which show the conserved α/β-hydrolase fold [2]. The protein structures of Lip20, Lip4 and LipRT were predicted by using SWISS-MODEL [6] and the results are shown in Figure 1b. The predicted catalytic triads were observed at Ser160, Asp252, and His282 for Lip20 and

Ser82, Asp207, and His235 for Lip4. Moreover, Lip20 and Lip4 shared the consensus sequences of glycine–Xaa–serine–Xaa–glycine at residues 158 to 162 and 80 to 84, respectively [2]. The Family VIII is different from Family III and IV without G-X-S-X-G motif [8]. The predicted catalytic triad of LipRT was observed at Ser73, Lys76 and Tyr190. Furthermore, LipRT shared the consensus sequences of serine–Xaa–Xaa–lysine between 73 and 76 residual, like the class C β-lactamase.

3.2 Construction and characterization of recombinant esterase

The expression plasmids harboring *lip20*, *lip4* and *lipRT* were transferred into *E. coli* BL21(DE3), *E. coli* C41(DE3), and *E. coli* C43(DE3). The results indicated that the clones with *E. coli* C41(DE3) and *E. coli* C43(DE3) formed a more marked clear zone than did those with *E. coli* BL21(DE3)(Fig. 2). Low esterase activity in *E. coli* BL21(DE3) may be due to hydrophobicity, resulting in protein aggregation [9]. Our result can provide the advice to the recombinant proteins leading to the production of insoluble inclusion bodies in *E. coli*. The selection of appropriate host can reduce the complex refolding process of recombinant proteins and has the potential to apply in industry. Therefore, *E. coli* C43(DE3) was used for protein overexpression and esterase activity analysis. After protein purification was performed using a nickel column and the three proteins were analyzed using SDS-PAGE, the three major bands were observed (Fig. 2). The expected molecular weights of the Lip20, Lip4 and LipRT-fused polyhistidine-tag were 37.9, 32 and 49.3 kDa.

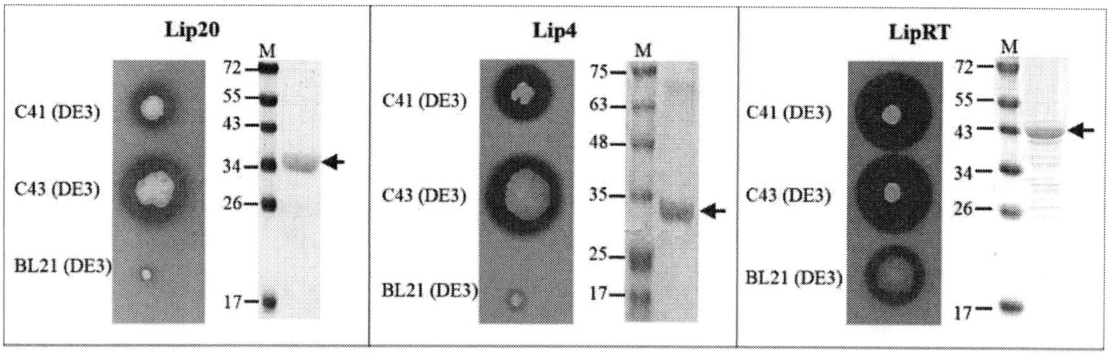

Figure 2. Overexpression of *lip20*, *lip4* and *lipRT* in *E. coli*. For observing esterase activities, transformants with *E. coli* C41(DE3), *E. coli* C43(DE3), and *E. coli* BL21(DE3) harboring Lip20, Lip4 and LipRT were spread on an LB agar plate with 0.01% arabinose, 0.4 mM IPTG, 1% tributyrin, and 12.5 μg/mL chloramphenicol. Recombinant Lip20, Lip4 and LipRT proteins in *E. coli* C43(DE3) were purified and performed by SDS-PAGE analysis.

3.3 Esterase activity assay

For demonstrating the esterase activity of recombinant Lip20, Lip4 and LipRT, various *p*-NP esters were used as substrates. The results showed the hydrolytic activities of recombinant Lip20, Lip4 and LipRT toward p-NP butyrate (C4) and *p*-NP valerate (C5) of short-length *p*-NP esters (Fig. 3). When the *p*-NP ester chain was over *p*-NP octanoate (C8), the activity of Lip4 was lower than 7.9% and no activity was detected with Lip20. However, the activity of LipRT was still maintained with 53.2% using *p*-NP octanoate (C8). Little activity of LipRT was determined as over *p*-NP dodecanoate (C12). The temperature and pH profiles of the recombinant Lip20, Lip4 and LipRT were further analyzed. The optimal temperatures of Lip20, Lip4 and LipRT were 30, 40 and 60°C, respectively. Nevertheless, the residual activity of Lip20 remained 39% at 40°C and was completely lost at 60°C. The Lip4 activity still had 32.2% at 60°C. Even at low temperature of 10 and 20°C, 58.7% and 77.1% of relative activity can be achieved. This implied that the Lip4 was regarded to be the cold-adapted enzyme. It can offer the novel opportunities as the biocatalyst exploitation in food industry, medical, domestic and environmental application [10]. Interestingly, the LipRT possessed the tolerance of high temperature and its activity maintained 82% at 65°C. Accordingly, a correlation existed between bacterial growth temperature and optimal temperature of enzymatic activity [11]. The pH profiles of Lip20, Lip4 and LipRT activity indicated that residual activities were 56.3%–100%, 39.4%–93.1% and 70.9%–100% at pH 7–8.6 with Tris–HCl buffer. With phosphate buffer, esterase activities of Lip20, Lip4 and LipRT were 8.5%–83%, 7.3%–100% and 18.7%–83.9% at pH 7–6. Furthermore, the relative activities of Lip20 and Lip4 were 33.5%–33% and 71.6%–40.9% at pH 8.6 and pH 9.0, respectively, with boric acid/NaOH buffer. However, no activity was detected with LipRT at pH 8.6 and 9.0. This revealed that the three enzymes preferred to potassium phosphate rather than Tris-HCl at pH 7. On the contrary, they were suitable for Tris-HCl instead of boric acid/NaOH at pH 8.6. The different buffer system acquires the different activities of esterases. Consequently, the specific activity of esterase was detected using the preferred substrate, optimal temperature and pH. The results of Lip20, Lip4 and LipRT specific activities were 42.4, 11.7 and 15.2 U/mg, respectively.

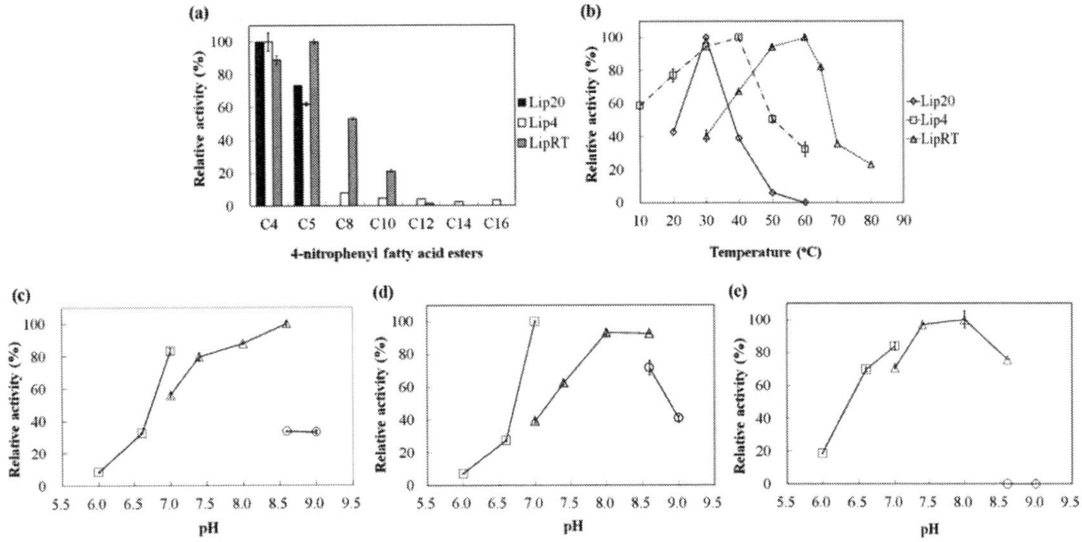

Figure 3. The optimal substrate, temperature and pH analyses of recombinant Lip20, Lip4 and LipRT activities. (a) Lip20, Lip4 and LipRT activities toward various p-nitrophenyl esters were assessed. Reaction mixtures were incubated at 30°C and pH 7.4 with Tris–HCl buffer for 10 min and interrupted by incubation at 100°C. (b) The temperature profiles of Lip20, Lip4 and LipRT were measured at 10°C–80°C and pH 7.4 with Tris–HCl buffer for 10 min using *p*-NP butyrate (C4). The pH profiles of (c) Lip20, (d) Lip4 and (e) LipRT were determined at 30, 40 and 60°C with potassium phosphate buffer (pH 6–7), Tris–HCl buffer (pH 7–8.6), and boric acid/NaOH buffer (pH 8.6–9) for 10 min using *p*-NP butyrate (C4).

3.4 Effect of thermostability on Lip20, Lip4 and LipRT

The various temperatures were conducted for thermostable analysis of Lip20, Lip4 and LipRT from 0 to 60 min (Fig. 4). The result revealed that the LipRT was more thermostable than Lip20 and Lip4 in correspondence with the optimal temperature of enzymatic activities. The Lip20 and Lip4 activities were dramatically drop to 45% and 1.6% after 15 min incubation at 50°C. The cold-adapted enzymes generally have a low thermal stability [10]. The result of Lip4 was corresponding to this finding. Additionally, over 92% relative activities of LipRT were retained for 60 min between 40°C and 60°C. The half-life, the time over which the 50% activity is lost, of LipRT was approximately maintained for 26 min at 70°C. This result suggested that the LipRT was a thermostable protein and superior to previous study, an alkaline lipase from *Ralstonia* [12].

Several structural features have been explained that could contribute to thermostability, including an increased protein hydrophobicity, additional salt bridges, a decreased number of cavities, an increased helix stability, a decreased chain flexibility and so forth [3]. In this study, the helix content of thermostable LipRT was higher than those of Lip20 and Lip4, and was an important role for thermal resistance (Table 1) [3]. Even if the helix content of MtEst45 from *Microbulbifer thermotolerans* was not as high as other thermally tolerant esterases, increasing β-sheet contents may contribute to the enzymatic stabilization [13]. In addition, some high frequency residues for facilitating thermostability, such as Val, Phe and Arg, were recognized in LipRT. These residues may lead to reinforce the hydrophobic and salt-bridge interactions.

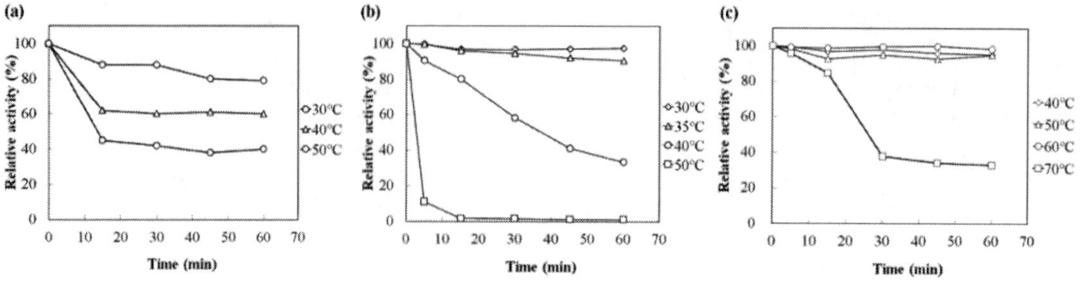

Figure 4. The thermostable analyses of Lip20, Lip4 and LipRT relative activities. The (a) Lip20 and (b) Lip4 were pre-incubated from 30-50°C by the time course from 0 to 60 min. The (c) LipRT was pre-incubated from 40-70°C by the time course from 0 to 60 min. Then, the remaining Lip20, Lip4 and LipRT enzymatic activities were measured at 30, 40 and 60°C and pH 8.6, 7 and 8 for 10 min, respectively.

Table 1. Properties of the esterases from Lip20, Lip4, LipRT, various bacteria and metagenome.

Esterase family	Esterase	Bacterial growth temperature	Esterases properties				Reference
			Optimal temperature activity	Residual activity of thermostability[a]	α-helix[b]	β-sheet	
III	**Lip4**	30°C	40°C	1%	11	8	This study
	EstA	—	45°C	0%	11	10	[14]
	MtEst45	37°C	46°C	30%	10	20	[15]
IV	**Lip20**	25°C	30°C	40%	9	10	This study
	LipP	28°C	45°C	47%	9	10	[16]
	REst1	37°C	50°C	~75%	12	9	[17]
VIII	**LipRT**	40°C	60°C	95%	13	9	This study
	PBS-2	25°C	50°C	80%[c]	11	10	[18]
	EstC	25°C	30°C	—	12	9	[19]

[a] The residual activities of esterases were detected after 50°C incubation for 1 h.
[b] The secondary structures of α-helix and β-sheet was predicted and referenced by YASPIN and SWISS-MODEL [6,20].
[c] The PBS-2 residual activity was determined after 50°C incubation for 0.5 h [18].

4 Conclusions

Lip20, Lip4 and LipRT encoding esterases were cloned from mesophilic *T. agarivorans* and *Aeromonas* sp., and moderately thermophilic *Ralstonia* sp., and successfully overexpressed in *E. coli*. These enzymes were α/β-hydrolase structures and were capable of hydrolyzing short-length *p*-nitrophenyl esters. Moreover, the optimal temperature of esterase activities was in agreement with the bacterial growth temperature. Better thermostability of enzymatic activity was observed by LipRT from *Ralstonia* sp.. According to the comparison of predicted structures, the helix contents may play a critical factor to enhance the enzymatic thermostability. On the other hand, the cold-adapted Lip4 from *Aeromonas* sp. also revealed well enzymatic activity at low temperature, whereas the weak thermostable ability was perceived. Further study of these esterases, including the X-ray structure and site-directed mutagenesis analysis, will clarify the conformational difference and ion interactions for thermostability.

Acknowledgments

This work was supported by the Bureau of Energy, Ministry of Economic Affairs, Taiwan under Grant No. 106-D0111.

References

1. F. Hasan, A.A. Shah and A. Hameed, Enzyme. Microb. Tech., **39**, 235-251 (2006)
2. L. Casas-Godoy, S. Duquesne, F. Bordes, G. Sandoval and A. Marty, Methods Mol. Biol., **861**, 3-30 (2012)
3. X.X. Zhou, Y.B. Wang, Y.J. Pan and W.F. Li, Amino Acids, **34**, 25-33 (2008)
4. H. Yu and H. Huang, Biotechnol. Adv., **32**, 308-315 (2014)
5. S.S. Liang, Y.P. Chen, Y.H. Chen, S.H. Chiu and L.L. Liaw, J. Appl. Microbiol., **116**, 563-572 (2014)
6. M. Biasini, S. Bienert, A. Waterhouse, K. Arnold, G. Studer, T. Schmidt, F. Kiefer, T.G. Cassarino, M. Bertoni, L. Bordoli and T. Schwede, Nucleic Acids Res., **42**, W252-258 (2014)
7. L. Rao, Y. Xue, Y. Zheng, J.R. Lu and Y. Ma, PLoS One, **8**, e60645 (2013)
8. J.L. Arpigny and K.E. Jaeger, Biochem. J., **343**, 177-183 (1999)
9. X.Z. Lin, S.S. Cui, G.Y. Xu, S.A. Wang, N. Du and J.H. Shen, Polar Res., **29**, 421-429 (2010)
10. B. Joseph, P.W. Ramteke and G. Thomas, Biotechnol. Adv., **26**, 457-470 (2008)
11. P.J. Haney, M. Stees and J. Konisky, J. Biol. Chem., **274**, 28453-28458 (1999)
12. H.Y. Yoo, J.R. Simkhada, S.S. Cho, D.H. Park, S.W. Kim, C.N. Seong and J.C. Yoo, Bioresour. Technol., **102**, 6104-6111 (2011)
13. C. Niu, L. Zhu, P. Zhu and Q. Li, J. Agric. Food Chem., **63**, 5249-5256 (2015)
14. X. Chu, H. He, C. Guo and B. Sun, Appl. Microbiol. Biotechnol., **80**, 615-625 (2008)
15. Y.S. Lee, Front. Microbiol., **7**, 218 (2016)
16. D.W. Choo, T. Kurihara, T. Suzuki, K. Soda and N. Esaki, Appl. Environ. Microbiol., **64**, 486-491 (1998)
17. A.P. Virk, P. Sharma and N. Capalash, J. Microbiol. Biotechnol., **21**, 667-674 (2011)
18. Y.O. Kim, I.S. Park, B.H. Nam, D.G. Kim, Y.J. Jee, S.J. Lee and C.M. An, J. Microbiol. Biotechnol., **24**, 1260-1268 (2014)
19. Y.S. Kim, H.B. Lee, K.D. Choi, S. Park and O.J. Yoo, Biosci. Biotechnol. Biochem., **58**, 111-116 (1994)
20. K. Lin, V.A. Simossis, W.R. Taylor and J. Heringa, Bioinformatics, **21**, 152-159 (2005)

Organic food consumption in China: the moderating role of inertia

Tsai-Fa Yen[a]

School of Economics, Sichuan University of Science and Engineering, China

Abstract. Despite the progressive development of the organic food sector across Taiwan Strait, little is known about how consumers' self congruity will influence organic food decision through various degrees of attitude and whether or not consumers with various degrees of inertia will vary in their intention to buy organic foods. The current study aims to examine the effect of consumption self congruity on behavioral intention related to organic food consumption under the mediating role of attitude as well as the moderating role of inertia. Research data were collected from organic food consumers across Taiwan Strait via a questionnaire survey, eventually obtaining 500 valid questionnaires for analysis. This study tested the overall model fit and hypotheses through structural equation modeling method (SEM). The results show that consumer attitude significantly mediates the effects of self congruity on organic food purchase intention. Moreover, the moderating effect of inertia is statistical significance, indicating that the relationship between attitude and purchase intention becomes weaker in the condition of consumers with higher degree of inertia. Several implications and suggestions are also discussed for organic food providers and marketers.

1 Introduction

In the past two decades, organic food has developed impressively, from a neglected niche market to the food market mainstream [1]. More and more researchers and practitioners are attracted and involved. As [2] ascertained from several studies, the attitude of consumers towards organic food is in general positive with typically associated benefits being superior taste, more environmental-friendliness, improved health, safer food, and more animal welfare. The foci of prior research into organic foods consumption included identifying motivations behind their purchase to understand the "whys" of consumers' choices. The key drivers previously identified include health benefits and environmental concerns (e.g., [3, 4]), as safety concerns regarding conventional foods (e.g., [3,5]), as well psychological motives (e.g., [6]).

However, despite the valuable insights provided by previous studies, at least two research gaps exist, which must be addressed. That is, previous studies have neglected psychological motives (i.e., self-congruity) and an important role of organic foods, gender. In particular, initial psychological motives, such as self-congruity, deserve attention given their impact on consumers' product choices. Consumers purchase goods/services because they are driven by a motivation to express their own self-images [7-9]. These self-images to a product or brand was defined, sustained, and enhanced by

[a] Corresponding author : 1722997311r@qq.com

consumers. When the images of a product or brand have matched their own self-images, the congruence as known self-congruity was met and they are more likely to be selected and purchased [10]. Consequently, the possibility of psychological motives (i.e., self-congruity) might play an important role in organic product consumption.

Moreover, attitude toward a product/brand has been viewed as an antecedent of behavioral intention and evidence from leisure farming has illustrated it could impact behavioral intention of a consumer [9, 11, 12]. Clarifying its impacts on behavioral intention in the context of organic foods would extend the knowledge for practitioners and academia.

Furthermore, researches in marketing and consumer behavior agree that inertia is a facilitator of behavioral loyalty [13]. It's been described as a condition where customers' repeat purchases occur because of their laziness, inactiveness, or passiveness [13]. A strong inertia customers perceived means their disinterest in actively processing relevant information to make the best possible choice[13, 14]. Thus, inertia leads them to repeat the same purchase pattern, similar to a habit [15], concentrating on familiar few organic/non-organic foods. However, fewer studies have been done to the buying behavioral of organic food. Thus, examination of the role of inertia in intention formation contributes to the existing organic foods literature.

Therefore, this study examines the role of a psychological motive (i.e., self-congruity) and inertia that have been identified in previous organic food studies or marketing literature. The current research demonstrates the relative importance of this variable in stimulating consumers' consumption of organic foods. To our knowledge, this study is likely to be the first study to clarify the importance of self-congruity in predicting consumers' intention to purchase organic foods. Therefore, the results enrich our understanding of key drivers of consumers' organic food purchase.

2 Hypothesis development

2.1 The concept of self-congruity

The concept of self-congruity comes from self-concept or self-image. Self-congruity is referred to a matching process which a shopper/buyer make a decision to patronize a store, he/she is likely to compare the store patron image with his/her own self-concept [16]. Another also stated the congruity of self-concept with product or brand mental representation refers to the match between those two images (i.e. products/service and his/her own) and plays an important role in consumer behavior since it constitutes a basic mechanism on which individuals base their brand preferences [17]. Research has demonstrated that the need for individual reassurance from others is one of the basic principles of self-concept that drive consumers [18]. This means that the consumers towards a product/service were likely to driven because their needs for reassurance. They symbolically relate their belongings in such a way that, on analysis, some significance can be read into their aspirations to success, social acceptance, ostentation, or prestige through the self-concept/product image congruity mechanism [17].

The self-congruity hypothesis is rooted in the assumption that individuals strive to maintain cognitive consistency in their beliefs and behaviors. Thus, consumers prefer those products or brands whose image is congruent with their own self-image [18]. Empirical study have evidenced that a higher level of self-congruity is experienced when a shopper perceives a greater match between the typical mall shopper image and his/her own self-image [16]. And this higher level of self-congruity can lead to a higher level of behavioral intention. Likewise, the stronger the congruence between the perceived personality of the brand and the consumer's own personality, the more likely the consumer is to buy this brand [19]. This implies that an organic food/product- self congruity is likely to enhance consumer's behavioral intention. Hence, building on these theoretical arguments and empirical findings, a hypothesis is proposed as follows:

H1: Self-congruity consumer perceived has positive influence on behavioral intention toward organic food.

Self-congruity was proved to be an antecedent of customer attitude toward products and a predictor of their shopping behavior [20, 21, 22, 23]. According to functional attitude theory, one of the important motives for customers in their purchase behavior is to solidify and express those values which are important to them [24-25]. When customers identify that the store image is congruent with their self-image, they agree that the store can represent their identity, and thus, favorable attitudes will be formed [18]. This implies that after the consumers identify the organic food image is congruent with their self-image; the level of their attitudes is likely to be higher. Based on the discussion above, a hypothesis is proposed as follows:

H2: Self-congruity consumer perceived has positive influence on attitudes toward organic food.

2.2 The concept of attitudes

Attitude is "a person's consistently favorable or unfavorable evaluation, feelings, and tendencies toward an object or idea"[26]. It can also be described as "the degree to which a person has a favorable or unfavorable evaluation or appraisal of the behavior in question" [27]. Attitudes toward a behavior is believed to be a function of one's salient beliefs (i.e., behavioral beliefs (BB)), which represent the perceived consequences of the behavior and his/her evaluation of the significance of the consequences (i.e., outcome evaluation (OE)) [28]. It includes judgment on whether the behavior under consideration is good or bad, and whether the actor wants to do the behavior [29].

Research has also demonstrated it is a function of behavioural beliefs; if an individual believes the performance of a specific behaviour will lead to a positive outcome then he/she will develop a favourable attitude towards that behaviour [30]. It influence the intentions held and the more favourable the attitude, the greater the intention to perform the behaviour will be. When determining whether to perform a specific behavior, a person is likely to assess the benefits and the costs resulting from the behavior [31]. For example, in our case, consumers could perceive eating organic foods as experiencing a healthy environmentally friendly activities, eating fresh and healthy food, and being more socially responsible, etc.

Additionally, attitudes are predictors of purchase intentions and consequently purchase behaviour [27]. Research has demonstrated that an individual tends to possess a favorable attitude when the outcomes are positively evaluated and, thus, he/she is likely to engage in that specific behavior [27, 31, 32, 33]. In other words, an individual's positive attitude toward a certain behavior strengthens his/her intention to perform the behavior [27]. In our case, consumers who possess favorable attitudes towards organic foods are likely to engage in that specific behavior in terms of the positive intentions of words of mouth, recommend to others, and purchase behaviour.

Attitudes are the main important predictor of behavioral intention and are the psychological emotion routed through consumers' evaluations and, if positive, BI tend to be more positive [34]. More specifically, in the context of green products, a positive relationship between attitude and behavioral intention has been established across many cultures [35]. Based on the discussion above, a hypothesis is proposed as follows:

H3: Attitudes consumer perceived have positive influence on behavioral intention toward organic food.

2.3 The effects of inertia on variables' relationships

Inertia is perceived as customers not switching because it is 'too much bother in terms of time and effort' [36]. Yanamandram & White (2006) described inertia as a condition where customers' repeat purchases occur because of their laziness, inactiveness, or passiveness. In other words, such customers' repeat purchases result simply from a lack of goal-directed behaviors [37], from a lack of conscious decision to change [38], or from being conditioned by habit [15].

In this situation, the customer repeats the same buying behaviour to limit the information search process and the cost of thinking [39]. The repeat purchase may even be in spite of the customer reflects a non-conscious process [38]. Repeat purchase as a result of inertia is unstable, reflecting little or no brand commitment [40] and merely represents acceptance [41]. Thus, high-inertia customers are passive in contemplating switching, thereby resulting in them remaining status quo with their service providers. Low-inertia customers may be actively searching for and comparing deals, consequently leading to their defections [36, 37]. Therefore, inertia is likely to play a moderated role in decision-making process. Hence, the hypotheses are proposed as follows:

H4a: Inertia significantly moderates the self-congruity-attitudes relationship.

H4b: Inertia significantly moderates the self-congruity-behavioral intention relationship.

H4c: Inertia significantly moderates the attitudes-behavioral intention relationship.

3 Methodology

3.1 Questionnaire design

Having considered the data collection requirements of this study such as a need of large sample of organic food consumers in Fujian province, it would be appropriate to employ the field survey with a self- administered questionnaire as the primary data collection technique. The population was set on those organic food consumers who consume organic food during the past six month. Organic firms were conducted for collecting data. After giving them phone call/interview by the researcher, finally four organic firms agreed to give us help in collecting data. They were found more than 5 years, stood on a standard of operation, and could be the representation of organic food firms in Fujian province. Their main products of those firms were fresh organic vegetables, organic tea, and others.

The field study method was chosen in order to gain information directly from organic food consumers who consume organic food during the past six month. As such, their perceptions about the setting with respect to self-congruity (SC), attitude(AT), and behavioral intention (BI) are likely to be clearly in mind. The survey instruments were borrowed from previous studies and adapted for this study. Self congruity was defined as how a consumer hopes others see her/him. A consumer's self congruity comprises their actual self-congruity, ideal self-congruity, social self-congruity, and ideal social self-congruity [18]. Four items for self congruity are distinguished, based on the studies of Ryu & Lee (2013) [42]and Yen (2014) [43]. Attitudes toward to organic foods were defined as a consumer's evaluation of the significance of the consequences/outcomes[27]. The three items adopted for attitude are based on the studies of Ajzen (1991) [27]and Kang et al. (2012) [21]. Behavioral intention was defined as a consumer's evaluation of the intents to word-of-mouth, recommendation, and repurchase[44].Three items for behavioral intentions are drawn from Lee & Yun (2015) [44]and Yen (2013) [45].

3.2 Sampling and Respondent profile

Quota sampling was adopted for collecting data based upon firms' scale. Regarding to the feasibility and accessibility, four firms were adopted for collecting data after conducting to legal organic firms in Fujian province. They have offered the member list for sending questionnaires or space for surveyors to interview the organic food consumers. Survey was down during May to June in 2016. Finally a total of 305 samples were received. After non-qualified and incomplete surveys were excluded, the remaining 295 respondents were valid reporting a responds rate of valid samples was 96.7%.

3.3 Respondent profile

Of 295 questionnaires obtained, about 40.7% were male and 55.3% were from female respondents. At about 40.3% of respondents were 20-29 years of age, 23.7% of respondents were 30-39 years of age, 16.6% of respondents were 40-49 years of age, 8.1% of respondents were 50-59 years of age, and 3.1% exceed 60 years of age. Regarding to married status, 51.2% of respondents was single, 11.5% of respondents were married with no children while 31.5% of respondents were married with children. Approximately 3.8% of respondents were graduated from senior high school, 10.2% of respondents received a bachelor degree and the majority of respondents got a master degree (61.4%). With regard to the occupation, 14.2% of respondents work at government related sectors, 5.1% of respondents belong to industrial sector, 37.3% of respondents were businessman, 13.2% of respondents were farmers, 22% of respondents were students, and 5.4% of respondents were retired/else. Approximately 42.4% of respondent's monthly income was below 3000RMB, 32.9% of respondent's monthly income was 3001-6000 RMB, and 13.2% of respondent's monthly income was 6001-8000 RMB while 5.1% of respondent's monthly income exceeds 8000 RMB. With regard to the frequency consumed within six month, 45.8% of respondents were less than three times, 28.8% of respondents were 4-6 times, and 21% of respondents were more than seven times. Table 1 reports the demographic characteristic.

Table 1. Demographic statistics (n=295)

Items		Freq.	%	Items		Freq.	%
Gender	Male	120	40.7	Occupation	Public sector	42	14.2
	Female	163	55.3		Industrial sector	15	5.1
	NA	12	4.1		Business	110	37.3
Age	20-29	119	40.3		Agriculture	39	13.2
	30-39	70	23.7		Student	65	22.0
	40-49	49	16.6		Retired (else)	16	5.4
	50-59	24	8.1		NA	8	2.7
	>60	9	3.1				
	NA	24	8.1	Monthly Income	<3000	125	42.4
Married Status	Single	151	51.2		3001-6000	97	32.9
	Married with No Children	34	11.5		6001-8000	39	13.2
	Married with Children	93	31.5		>8001	15	5.1
	Separate	7	2.3		NA	19	6.4
	NA	10	3.4				
Education	Primary	2	.7	Frequency consumed within six month	<3	135	45.8
	High	9	3.1		4-6	85	28.8
	College	30	10.2		>7	62	21.0
	Bachelor	52	17.6		NA	13	4.4
	Master	181	61.4				
	NA	21	7.2				

NA: not available

4 Result

4.1 Common method variance assessment

The potential occurrence of common method variance (CMV) was tested by employing a Harman's one-factor test. An exploratory factor analysis containing all seven reflective indicators was conducted. Consistent with the measurement model, the results of the principal components factor analysis reveal that there are three factors that account for 73% of the total variance. CMV does not appear to be a problem, since the first factor did not account for the majority of the variance (only 23.5%). Moreover, a CFA approach to Harman's one-factor test was used to assess whether a single latent factor accounts for all manifest variables. The one-factor model yielded a $\chi 2$ of 4207.81 (df=702) compared with a χ^2 of 1322.80 (df=647) for the three-factor measurement model in which manifest variables were assigned to load onto their theoretical constructs. The fit is considerably worse for the unidimensional model than for the measurement model ($\Delta\chi^2$=1885.01, Δdf=55, pb.01) suggesting that CMV is not a serious threat in the study. Finally, the major focus of this study is on interaction effects; however, CMV usually works against differential validity.

4.2 Measure assessment for formative scales

A confirmatory factor analysis using AMOS 17.0 is conducted to test the measurement model. Table 2 reports the results. The chi-squares (82.43) are significant ($p < 0.05$; [46]). The ratios of chi-square to degrees of freedom (df= 31) are 2.65 for measurement model within the acceptable range of 2 to 5 [47]. The values for GFI (0.949), AGFI (0.909), CFI (0.975), and RMSEA (0.075) are acceptably close to the standards suggested by Hu & Bentler (1999) [48] greater than 0.9 for GFI and AGFI, greater than 0.95 for CFI and less than 0.08 for RMSEA. Given that these batteries of overall goodness-of-fit (GFI) indices were accurate and that the model was developed on theoretical bases, and given the high level of consistency samples, no respecifications of the model were made. This enables authors to proceed in evaluating the reliability and validity.

Table 2. Reliability and validity

Conpt.	Items	λ	t-values	SMC	CR	AVE
SC	SC1	.755	-	.570	0.791	0.486
	SC2	.839	14.785	.704		
	SC3	.888	15.675	.789		
	SC4	.826	14.537	.683		
AT	AT1	.754	-	.568		
	AT2	.899	15.683	.808		
	AT3	.847	14.864	.718		
BI	BI1	.871	-	.759	0.864	0.683
	BI2	.907	20.663	.822		
	BI3	.737	15.070	.543		

SC: Self congruity; BI: Behavioral intention; λ: Standardized factor loading; CR: Composite reliability; AVE: Average variance extracted; χ^2=82.43，df=31(p=.000)，χ^2/df=2.65，GFI=.949，AGFI=.909，CFI=.975，RMSEA=.075

Assessment of a formative index requires the examination of five critical aspects: (1) content specification, (2) indicator specification, (3) indicator collinearity, (4) individual indicator validity, and (5) construct validity. The definition stems from a detailed and systematic literature review supplemented with field-based interviews, a process that resulted in the generation of four indicators specifying the one dimension comprising the index. The domain of the dimension has been captured because all the items characterizing each dimension have been included in the corresponding index. Similarly, an inclusive set of items capturing the domains of one dimension was generated. Consequently, the definition and measurement of the index capture the breadth of the conceptual domain, thereby satisfying the content and indicator specification criteria.

To assess collinearity, we regressed indicators on the sum score of the self congruity index; the highest variance inflation factor (VIF) was 3.55, which is considerably below the threshold of 10, and thus, multicollinearity is not an issue. Individual indicator validity assessment was conducted by fitting a multiple indicators and multiple causes (MIMIC) model. The MIMIC model was identified by specifying two structural relations from the formative index to two reflectively measured constructs (attitudes and behavior intention). The results of the MIMIC model indicate that the overall model has a very good fit ($\chi 2$= 93.04, df= 69, p-value= .028; SRMR= .050; RMSEA= .045; CFI= .98; NNFI= .98; and GFI= .93) and all four γ-parameters exerted a significant effect on sales strategy.

Construct validity was assessed in three ways. First, the CFA model and the regression analyses (to be discussed subsequently) show that self congruity is significantly related to attitude and behavioral intention measures. Second, the error variance of the self congruity construct as estimated in the CFA model is .24, which indicates that it is a valid construct. Finally, we calculated the inter correlations among the self congruity as well as the Cronbach's alpha reliability coefficient for the construct. Again, the results speak against the possibility of a reflective structure for the self congruity measure because the inter correlations ranged from .06 (pN.10) to .32 (pb.01) with an average value of .17, whereas the reliability coefficient was .896.

4.3 Measure assessment for reflective scales

This study evaluated reflective scales by means of one covariance-based CFA. The CFA was run on attitude and behavioral intention. The overall fit of the first measurement model was good: root mean square error of approximation (RMSEA)= .075; comparative fit index (CFI)= .975; goodness-of-fit index (GFI)= .949; and adjusted goodness-of-fit index (AGFI)= .909. In addition, all indicators loaded significantly on their hypothesized latent construct, which demonstrates convergent validity. Table 2 reveals that the average variance extracted (AVE) for each of the measures is greater than the shared variance between any pair of measures, thereby satisfying the criterion of discriminant validity [49]. Finally, the constructs' composite reliabilities (CRs) indicate that each exceeded the accepted reliability threshold of .70.

4.4 Hypotheses testing

The overall model fit indices show that the chi-squares (82.43) are significant ($p < 0.05$); the ratios of chi-square to degrees of freedom (df= 31) are 2.65 for structural model within the acceptable range of 2 to 5; the values for GFI (0.949), AGFI (0.909), CFI (0.975), and RMSEA (0.075) are acceptably close to the standards suggested by Hu & Bentler (1999) greater than 0.9 for GFI and AGFI, greater than 0.95 for CFI and less than 0.08 for RMSEA. This suggests that the hypothesized model fits the empirical data well. Table 3 shows the estimated model with the standardized path coefficients. SC can directly and significantly impact AT (t=9.069, β=0.635). Thus H1 was supported. Moreover, SC can also directly and significantly impact BI (t=3.828, β=0.225). Thus H2 was supported. Finally BI can be influenced by AT (t=9.869, β=0.703). Thus H3 was supported. The R^2 is for AT .404 and .745 for BI. Table 3 reports the results of structural model.

Table 3. Results of SEM

Path and items	β	t	SMC
SC-AT	.635	9.069	.404
SC-BI	.225	3.828	.745
AT-BI	.703	9.869	

χ^2=82.43，df=31(p=.000)，χ^2/df=2.65，GFI=.949，AGFI=.909，CFI=.975，RMSEA=.075

4.5 The moderated effects of inertia

The mean-centered of relevant variables were done before creating the interaction terms. Then two interactional variables were added into the estimated model. Table 4 reports the results of moderated effects. Again, the overall model fit indices show that the chi-squares (98.27) are significant ($p < 0.05$); the ratios of chi-square to degrees of freedom (df= 46) are 2.136 within the acceptable range of 2 to 5; the values for GFI (0.949), AGFI (0.914), CFI (0.977), and RMSEA (0.062) are acceptably close to the standards suggested by Hu & Bentler (1999) greater than 0.9 for GFI and AGFI, greater than 0.95 for CFI and less than 0.08 for RMSEA. This suggests that the hypothesized model fits the empirical data well. The results show that SC can directly and significantly impact AT (t=8.625, β=0.612). Thus H1 was supported. Moreover, SC can also directly and significantly impact BI (t=3.875, β=0.231). Thus H2 was supported. Finally BI can be influenced by AT (t=9.884, β=0.705). Thus H3 was supported. The R^2 is for AT .411 and .751 for BI. Moreover, the moderator, inertia has moderated effects on SC-BI (t=1.974, β=0.104) and AT-BI (t=-1.816, β=-0.096) relationships. Thus H5 and H6 were supported while H4 was not supported.

Table 4. Results of Moderated Effects

Path and items	β	t	SMC
SC-AT	.612	8.625	.411
SC-BI	.231	3.875	.751
AT-BI	.705	9.884	

INT*SC-AT	-.080	-1.494
INT*SC-BI	.107	1.974
INT*AT-BI	-.096	-1.816

χ^2=98.27, df=46(p=.000), χ^2/df=2.136, GFI=.949, AGFI=.914, CFI=.977, RMSEA=.062

5 Conclusion and Suggestion

The present study tested a model to determine whether behavioral intention is predicted by self-congruity and attitudes, and how the predictive relationships are moderated by inertia. The results found: (1) behavioral intention was predicted by self-congruity and attitudes; (2) however, the predictive power of attitudes was stronger and more direct than self-congruity; (3) attitudes toward organic food were predicted by self-congruity; (4) the high inertia group exhibited stronger associations between self-congruity and behavioral intention than the low one; and (5) the high inertia group exhibited weaker associations between attitudes and behavioral intention than the low one. Implications are suggested based on these findings.

As previous studies explained, the simultaneous of self-congruity have been understood, while its effects have not been examined in the theoretical context of inertia. Thus, the results indicate several theoretical contributions.

First, social self- congruity was found to be the most important one in influencing attitudes and behavioral intention, and followed by ideal self- congruity, ideal social self- congruity, and actual self-congruity when purchasing organic food. This finding was different from previous studies. In the self-congruity literature, whether actual- and ideal self- congruities are similar or dissimilar are conflicting. Given these divergent findings, it's reasonable to speculate that the conflicting results are partially due to the range of products investigated [50]. Specifically, it might be that actual self- congruity and social self- congruity are not the case in organic food contexts because such purchases do not fall into the category of mundane and routine occurrences. This pattern is quite different from other food purchases due to its high price and special meanings for health benefits and environmental concerns.

Furthermore, the finding that the predictive power of attitudes was stronger and more direct than self-congruity contributes to the attitudes literature. Since self-congruity represents the value-expressive function, this finding suggests that organic foods are dissimilar to conventional foods. Their attitudes toward organic foods are likely to determine by multiple ways in terms of the level of monthly income, educational background etc. Specifically, attitudes toward organic foods are also a mediator between self-congruity and behavioral intention. This implies it is a key factor when forming behavioral intention to consumers in organic food contexts. Therefore, both self-congruity and attitudes are appropriate theories for identifying dominant of organic foods.

Moreover, the finding that attitudes toward organic food were predicted by self-congruity implies self-congruity can be one of the antecedents of attitudes. In line with previous studies, Attitude is "a person's consistently favorable or unfavorable evaluation, feelings, and tendencies toward an object or idea"[26]. These evaluation, feelings, and tendencies should comprise the comparison of products image and consumers' self-image. The attitudes are likely to be favorable while the evaluation toward the products can match his/her self-image. Therefore, these self-congruities are likely to enhance his/her attitudes toward

Specifically, the finding that the high inertia group exhibited stronger associations between self-congruity and behavioral intention than the low one; and the high inertia group exhibited weaker associations between attitudes and behavioral intention than the low one contributes to the existed inertia literature. That is previous studies state that high levels of inertia weaken the association between post purchase decision/behavior and its predictors [33]. However this study suggests that both positive and negative effects of inertia on variables can be met. Such individual-level factors as inertia moderates the influence of self-congruity and behavioral intention. In particular, individuals' self-congruity has greater effects on behavioral intention at high values of inertia than at low values of inertia. This means that as the comparisons of products image and his/her image are congruence, the high intents to word-of-mouth, recommendation, and repurchase are met. Because high-inertia customers are passive in contemplating switching, thereby resulting in them remaining status quo with their service providers.

In contrast, individuals' attitudes have greater effects on behavioral intention at low values of inertia than at high values of inertia. It indicates such contingency variable will destroy the attitudes-intention relationship. Because low-inertia customers may be actively searching for and comparing deals, consequently leading to their defections, the effects of attitudes on behavioral intention are thus decline.

References

1. J. van Doorn and P.C. Verhoef, Journal of Retailing, **91**(3), 436-450 (2015)
2. L. Marian, P. Chrysochou, A. Krystallis and J. Thogersen, Food Quality and Preference, **37**, 52-60 (2014)
3. N. Michaelidou and L.M. Hassan, International Journal of Consumer Studies, **32**(2), 163-170 (2008)
4. A. Milman and A. Pizam, Annals of tourism research, **15**(2), 191-204 (1988)
5. M. Wandel and A. Bugge, Food Quality and Preference, **8**(1), 19-26 (1997)
6. S. Baker, Thompson, K.J. Engelken and K. Huntley, European journal of marketing, **38**(8), 995-1025 (2004)
7. J. Hwang, Journal of Retailing and Consumer Services, **28**, 281-287 (2016)
8. J.L. Aaker, Journal of marketing research, **34**, 347-356 (1997)
9. F. Kressmann, M. Sirgy, A. Herrmann, F. Huber, S. Huber and D.D. Lee, Journal of Business Research, **59**(9), 955-964 (2006)
10. P.H. Chiu, T.F. Yen and K.K. Chu, Journal of Food, Agriculture & Environment, **13**(3), 168-174 (2015)
11. T. Graeff, Journal of consumer marketing, **13**(3), 4-18 (1996)
12. T. Ramayah, J. Lee and O. Mohamad, Green product purchase intention: some insights from a developing country. Resources, Conservation and Recycling, **54**(12), 1419-1427 (2010)
13. M. Kotchen and S. Reiling, Ecological Economics, **32**(1), 93-107 (2000)
14. V. Yanamandram and L. White, International Journal of Service Industry Management, **17**(2), 158-192 (2006)
15. W. Kim, C. Ok and D.D. Canter, International Journal of Hospitality Management, **29**(1), 136-147 (2010)
16. C. Bozzo, Journal of Customer Behaviour, **1**(3), 335-355 (2002)
17. K.El. Hedhli, H. Zourrig and J.C. Chebat, Journal of Retailing and Consumer Services, **31**, 1-13 (2016)
18. A. Beerli, G.D. Meneses and S.M. Gil, Annals of Tourism Research, **34**(3), 571-587 (2007)
19. M.J. Sirgy, Journal of consumer research, **9**, 287-300 (1982)
20. A.Usakli and S. Baloglu, Tourism management, **32**, 114-127 (2011)
21. D. Lee and M.R. Hyman,. Journal of Marketing Theory and Practice, **16** (3), 219-232 (2008)
22. J. Kang, L. Tang, J.Y. Lee and R.H. Bosselman, International Journal of Hospitality Management, **31**(3), 809-818 (2012)
23. T.F. Yen, The effects of gender on self congruity- behavioral intention relationships: An empirical study of organic food consumers in Fujian province, China. J. Lei., Tou., Spo., & Hea. **7**, x-xx (2017, in press)
24. M.J. Sirgy and A. Samli, Journal of the Academy of Marketing Science, **13**(3), 265-291 (1985)
25. M.J. Sirgy, D. Grewal and T.F. Mangleburg, Journal of Business research, **49**, 127-138 (2000)
26. H.H. Chang and Y.M. Liu, The Service Industries Journal, **29**(12), 1687-1706 (2009)
27. I. Ajzen, Organizational Behavior and Human Decision Processes, **50**(2), 179-211 (1991)
28. A.H. Eagry and S. Chailen, *Psychology of attitudes*. Fort Worth, TX: Harcourt Brace Jovanovich. (1993)
29. M. Leonard, S. Graham and D. Bonacum, Quality and Safety in Health Care, **13**, 85-90 (2004)

30. I. Ajzen and M. Fishbein, *Understanding attitude and predicting social behavior*, Englewood Cliffs, NJ: Prentice-Hall, Inc (1980)
31. S. Cheng, T. Lam and C.H.C. Hsu, International Journal of Hospitality Management, **24**(4), 475-492 (2005)
32. M.J. Lee, Effects of attitude and destination image on association members' meeting participation intentions: development of meeting participation model. Unpublished doctoral dissertation, Manhattan, KS: Kansas State University. (2005)
33. H. Han, Y. Kim and E. Kim, International Journal of Hospitality Management, **30**(2), 1008-1019 (2011)
34. M.F. Chen and P.J. Tung, International journal of hospitality management, **36**, 221-230 (2014)
35. M.M. Mostafa, Psychological Marketing, **24**(5), 445-473 (2007)
36. M. Colgate, B. Lang, Journal of consumer marketing, **18**(4), 332-347 (2001)
37. M. Zeelenberg and R. Pieters, Journal of business Research, **57**(4), 445-455 (2004)
38. M. Huang and S. Yu, Psychology and Marketing, **16**(6), 523-544 (1999)
39. K. Bawa, Marketing Science, **9**(3), 263-278 (1990)
40. M.R. Solomon, G. Bamossy and S. Askegaard, *Consumer behaviour: A European Perspective*, 2nd edition, Pearson Education, Harlow (2002)
41. H. Assael, Consumer behaviour and marketing action, Southwestern College. Publishing, Cincinnati, OH (1998)
42. K. Ryu and J.S. Lee, International Journal of Hospitality Management, **33**, 29-40 (2013)
43. T.F. Yen, J. Spo. Lei. Hos. Res. **9**(1), 27-55 (2014)
44. H.J. Lee, Z.S. Yun, Food Quality and Preference, **39**, 259-267 (2015)
45. T.F. Yen, Leisure Industry Research, **11**(3), 1-10 (2013)
46. K.A. Bollen, *Structural Equations with Latent Variables*, New York, John Wiley and Sons. (1989)
47. H.W. Marsh and D. Hovecar, Psychological Bulletin, **97**(3), 562-582 (1985)
48. L. Hu and P.M. Bentler, Structural Equation Modeling: A Multidisciplinary Journal, **6**(1), 1-55 (2009)
49. C. Fornell and D.F. Larcker, Journal of marketing research, **18**, 382-388 (1981)
50. H. Choi, L.N. Reid, Journal of Business Research, **69**(9), 3430-3438 (2016)

MATEC Web of Conferences **169**, 01020 (2018) https://doi.org/10.1051/matecconf/201816901020
IMETI 2017

Cognitive diagnostic assessment based on knowledge structure

Sue-Fen Huang[a]

Department of Digital Game and Animation Design, Taipei University of Marine Technology, Taiwan

Abstract. The purpose of this study is to provide an integrated method of fuzzy theory basis for individualized concept structure analysis. In order to insight the misconception of learning basic mathematics and progress teaching. This method integrates Fuzzy Logic Model of Perception (FLMP) and Interpretive Structural Modelling (ISM). The combined algorithm could analyze individualized concepts structure based on the comparisons with concept structure of expert. In this paper, some well-known knowledge structure assessment methods will be discussed. For item connection, Bart et al ordering theory and Takeya's item relational structure provided ordering coefficient to construct item relationships and hierarchies. For concepts or skills connection, Warfield's ISM and Lin et al **Concept Advanced Interpretive Structural Modelling** (CAISM) provided to construct graphic relationship among elements and display the individualized concept hierarchy structure by numeric and picture. Samples contain 427 which come from Min-Hwei Junior College. Subjects were analyzed by CAISM. It shows the traditional assessment is not the only criteria; it must be combined with other assessment tools. The result shows that CAISM gives meaningful learning and lacks of learners.

1 Introduction

Achievement assessment is a series of operations or examinations of learners' ability, knowledge, or learning outcome, consisting of questions for them to answer or activities for them to carry out. Achievement assessment perhaps is a most part of criteria for evaluating learners' learning performance, the people who has the same score will be classified into the same achievement group. However, there might exists distinct misconceptions from learners who have the same score, hence achievement assessment is no more a major criterion for evaluating learners' performance. Cognitive diagnostic assessment (CDA) is designed to obtain specific knowledge structures, processing skills and causes in terms of learners' understanding. Interpretation and explanation are required to providing feedback information for educator and learner which consisting of learners' lack of understanding or the primary cause of their performance [1][2].**Cognition diagnosis can help represent knowledge structure** [3] [4] [5]. It is a common viewpoint that human knowledge is stored in the form of structural relationship among concepts and their subordinate relationship is fuzzy, not crisp. There are some methodologies for concept structure analysis but little is known about methodologies of individualized concept structure [6] [7] [8] [9]. Therefore, the development for methodology of individualized concept structure is an important issue and it is essential for cognition

[a] Corresponding author : yih@mail.ntcu.edu.tw

diagnosis and pedagogy [10]. In this study, the integrated method of individualized concept structure based on fuzzy logic model of perception (FLMP) and interpretive structural modelling (ISM) will be developed [11] [12] [13]. An example of empirical test data of linear algebra concept for students of learning deficiencies will also be analyzed and discussed. **For the feasibility of remedial instruction based on the cognition diagnosis**, clustering method is needed so that students within the same cluster own similar knowledge structures and students among different clusters have the most variance on knowledge structures [14].

2 Literature review

The utility of interpretive structural modelling (ISM) is to construct graphic relationship among elements within a complex system which was proposed by Warfield, J. N. [15] in 1982. For a subordinate matrix of a system, ISM will arrange elements in the form of hierarchical structure. Lin, Y. H., Hung, W. L., & Huang, K. J. (2006) [16] extended Warfield's interpretive structural modelling and proposed concept advanced interpretive structural modelling. The object of concept advanced interpretive structural modelling (CAISM) is to provide individualized hierarchy structure of knowledge of examinees based on response patterns of tests. CAISM incorporates principles of concept vector matching and α-cut of fuzzy set theory into ISM to display the individualized concept hierarchy structure by numeric and picture.

2.1 Concept advanced interpretive structural modelling

The definition of data structures are described as follows.

$X=[x_{nm}]_{N\times M}$ is the response matrix of all examinees on the test. There exist M items in a test and N examinees. We know that $x_{nm}=1$ when examinee n gives correct answer on item m; otherwise $x_{nm}=0$ means examinee n gives wrong answer on item m. $Y=[y_{ma}]_{M\times A}$ is the item-concept matrix, which containing A concepts. If item m measures concept a, then y_{ma} is denoted by 1; otherwise y_{ma} is denoted by 0. $Z=[z_{ia}]_{I\times A}$ is the ideal concept matrix, where $z_i=(z_{i1},z_{i2},\ldots,z_{iA})$, $i=1,2,\ldots,I$, $I=2^A$. if ideal concept vector z_i contains concept a then $z_{ia}=1$; otherwise $z_{ia}=0$. $R=[r_{im}]_{I\times M}$ is the ideal response matrix and $r_i=(r_{i1},r_{i2},\ldots,r_{iM})$ is the ideal response vector. Ideal response vector is the response pattern on all items based on corresponding ideal concept vector z_i. It is $r_{im}=1$ if ideal concept vector z_i could provide correct answer on item m; otherwise $r_{im}=0$. $SC=[sc_{ni}]_{N\times I}$ is the standardized closeness matrix which measures the closeness between response vector $x_n=(x_{n1},x_{n2},\ldots,x_{nM})$ and ideal response vector $r_i=(r_{i1},r_{i2},\ldots,r_{iM})$. Greater sc_{ni} means more similarity between x_n and r_i. The algorithm of CAISM is described as follows:

The ideal response matrix is defined

$$r_{im}=\begin{cases}1 & ,z_{ia}\cdot y_{ma}=y_{ma}\\ 0 & otherwise\end{cases},\ \forall a=1,2,\ldots,A \tag{1}$$

$$c_{ni}=\frac{\sum_{m=1}^{M}x_{nm}\circ r_{im}}{M} \tag{2}$$

where

$$x_{nm} \circ r_{im} = \begin{cases} 1 & , x_{nm} = r_{im} \\ 0 & , x_{nm} \neq r_{im} \end{cases} \quad (3)$$

c_{ni} is the closeness between the x_n and r_i , then the standardized closeness sc_{ni}

(a) It is crisp recognition if $K(K \geq 1)$ different c_{ni} values satisfy $c_{ni} = 1$, and

$$sc_{ni} = \begin{cases} 1 & , \forall c_{ni} = 1 \\ 0 & , otherwise \end{cases} \quad (4)$$

(b) It is fuzzy recognition if $c_{ni} \neq 1$

$$sc_{ni} = \frac{c_{ni}}{\sum_{i=1}^{I} c_{ni}}, \quad \forall i = 1,2,...,I \quad (5)$$

2.2 ISM approach on concept structure analysis

J. N. Warfield provided the foundation of ISM [17]. The integrated algorithms consist of three steps of algorithms, AMC, ASC and AFISM.

3 Knowledge structure analysis

The test includes 19 items with 427 task-takers of College students. The data set comes from the Min-Hwei Junior College. used in the empirical study with learning Abstract Algebra. There are 7 concept attributes within each item and they are depicted in Table1.

Table 1. The Details of the Used Datasets.

Classes	Characters about Concepts
1	Number
2	Coordinates
3	Geometry
4	Function
5	Distance
6	Statistics
7	Equation

As shown from Figure 1, To student 139, All concepts are equal. All concepts are so easy to the student. The index of diffcult is closed to zero. The concept of structure is very like expert.

Figure 1. Knowledge Structure of Student 139.

As shown from Figure 2, To student 378, mastery of concept 5 and 6 are 0.50. Concept 5 and 6 are the basis for concept 2,7. Concept 1 is difficult for student 378. It is also clearly understood that concept structures of these two students vary a lot.

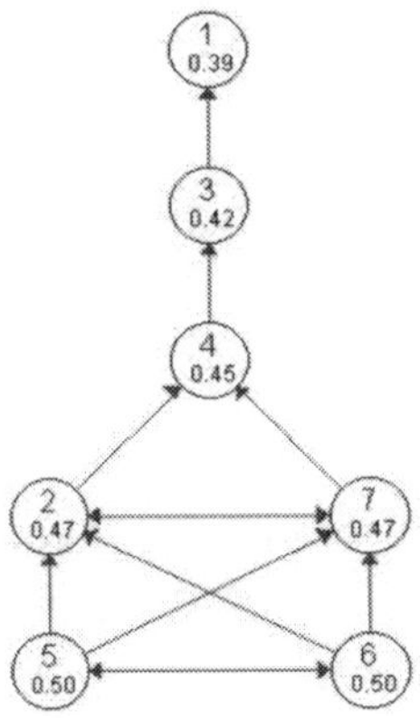

Figure 2. Knowledge Structure of Student 378.

4 Conclusion

Each cluster of data can easily describe features of knowledge structures. Manage the knowledge structures of Mathematics Concepts to construct the model of features in the pattern recognition completely. To sum up, this integrated algorithm could improve the assessment methodology of cognition diagnosis and manage the knowledge structures of Mathematics Concepts easily. It shows the traditional assessment is not the only criteria; it must be combined with other assessment tools. The result shows that CAISM gives meaningful learning and lacks of learners.

References

1. J.P. Leighton and M.J. Gierl, Cambridge University Press, 3-18 (2007)
2. S.P. Norris, J.S. Macnab and L.M. Phillips, Cognitive diagnostic assessment for education: Theory and practices, Cambridge University Press, 61-84 (2007)
3. M. Smithson and J. Verkuilen, Fuzzy Set Theory: Applications in the Social Sciences, Sage Publications (2006)
4. R. Coppi, P. Giordani and P. D'Urso, Psychometrika, **71**, 733-761 (2006)
5. T. Sato, The S-P Chart and The Caution Index, NEC Educational Information Bulletin 80-1, C&C Systems Research Laboratories. Nippon Electic Co., Ltd,. Tokyo, Japan (1980)
6. J.P. Doignon and J.C. Falmagne, Knowledge Space. Springer-Verlag (1999)
7. K. VanLehn, Journal of the Learning Sciences, **8**, 71 (1999)
8. R.W. Schvaneveldt, Pathfinder Associative Networks. Ablex (1991)
9. W.P. Jr. Fisher, Rasch Measurement Transactions, **9**, 442 (1995)
10. R.J. Mislevy and N. Verhelst, Psychometrika, **55**, 195 (1990)
11. J.N. Warfield, Societal Systems Planning. Policy and Complexity, Wiley (1976)
12. J.N. Warfield, Interpretive Structural Modeling (ISM). In S. A. Olsen (Eds.), Group Planning & Problem Solving Methods in Engineering, 155-201, Wiley (1982)
13. L.A. Zadeh, Information and Control, **8**, 338 (1965)
14. Y.H. Lin, M.W. Bart and K.J. Huang, Generalized Polytomous Ordering Theory. [manual and software], National Taichung University, Taiwan (2006)
15. J.N. Warfield, Interpretive structural modeling. In S. A. Olsen (Ed.), Group Planning and Problem-solving Methods in Engineering, 155-201, New York: Wiley (1982)
16. Y.H. Lin, W.L. Hung and K.J. Huang, CAISM software [Manual and Software for CAISM]. Taiwan, Taichung City: National Taichung University,(2006)
17. J.N. Warfield, Crossing Theory and Hierarchy Mapping, **7**, 505 (1977)

Study on the relationship between project management and organizational efficacy in nonprofit organizations

I-Chan Kao[a]

General Education Center, Open University of Kaohsiung, Taiwan

Abstract. This study treats the members in nonprofit organizations (NPOs) as subjects, and explores the origination, planning, control, and completion of project management in NPOs, as well as the general performance of organizational efficacy, such as environmental satisfaction, organizational atmosphere, operational performance, job engagement, and work quality. It also probes into the relationship and effect. By various research methods, such as literature review and questionnaire survey, this study attempts to determine if project management in NPOs can significantly enhance organizational efficacy. This study finds that different NPOs have significant differences in the general performance of project management and organizational efficacy. When the performance of project management in NPOs is more significant, organizational efficacy is higher. Project management in NPOs has a significant path relationship to organizational efficacy; therefore, reinforcement of vocational training in the project management of NPOs could improve performance, which would have significant effect on enhancing organizational efficacy.

1 Introduction

1.1 Research motives

Nonprofit Organizations (NPO) are organizations or groups that do not aim to make profits[1]. The principal objective is to support or cope with issues or events of concern by individuals or the public; therefore, the related fields are broad, including art, charity, education, politics, public policy, religion, academic, and environmental protection. They also supplement the gap between social demand and governmental supply. As enterprises, the operations of NPO aim to produce profits; however, the difference is that NPOs make profits by the service subjects and service content initiated by the organizations, which tend to be considered as the main characteristic of this type of organization.

However, some experts have argued that the major difference between NPO and enterprises is that NPO are restricted by law or morality, and cannot distribute profits to owners or shareholders. Therefore, they are independent, public, and civil. Nowadays, in society, NPOs are sometimes called "the third sector" and in comparison to the governmental sector (the first sector) and private sector of enterprises (the second sector), they are the third force to influence society. Although NPOs state "nonprofit", they must produce the profits of organizational operations as capital for activities. However, as the revenues and expenditures are limited, they obtain funds through public fund raising

[a] Corresponding author : kic0929@ouk.edu.tw

or donations from public and private sectors. In addition, they are often exempt from taxation. Sometimes, private donations to NPOs can be benefited by tax deductions. In addition, charity groups are one kind of NPO and non-governmental organizations can also be called NPOs.

The activities of NPO project management include the following 6 fields[2]: (1) environmental protection, ecology, and resource protection; (2) poverty relief and promotion of economic and social development of developing countries; (3) rights protection for minorities; (4) community services for townships; (5) economic intermediary services, including various types of economic intermediary organizations, such as civil groups and industrial groups consisting of managers, professional lawyer offices, accountants, and auditors; (6) charity and relief. Project management emphasize the limitations of time, cost, quality, and scope, and must consider execution, monitoring risks, human resources, stakeholders, communication, and procurement in order to accomplish their goals. Therefore, in terms of organizational type and the operational items of NPO's, project management is the core of the said management and operations.

Hsu (2000) argued that NPO's play the roles of development and innovation, reform and initiation, value maintenance, service offering, expansion of social participation, etc. However, when implementing projects, NPO encounter the instability of funds, labor, and materials, and lack project management competence. Thus, the performances of NPO role functions can be extremely different. Situ (1999), Hsu (2000), Hsiao (2000), and Chang et al. (2005) stated that the problems encountered by Nonprofit Organizations include incomplete guidance and management of governmental regulations, unstable positions, financial difficulties, overly complicated political background, lack of specific efficacy, and failure to implement institutionalized management. Thus, how do NPO's enhance organizational effectiveness and accomplish their role functions through effective project management performance? This is the motive of this study.

1.2 Research purposes

Based on the previous statements of research motives, the research purposes are shown, as follows:

1. To recognize the implications of NPO project management and organizational efficiency.
2. To explore general performance of project management and organizational efficiency of NPO with different backgrounds.
3. To analyze the path relationship between NPO project management competence and organizational effectiveness.
4. According to the research findings, this study proposes suggestions to enhance NPO project management competence performance and organizational effectiveness as reference for NPO's to strengthen their project management competence and organizational effectiveness.

1.3 Research method

By literature review, this study collects books, journals, magazines, and papers associated with NPO's project management, and organizational effectiveness, and analyses them to obtain the basic concepts. Secondly, it generalizes the related literature and suggestions of scholars and experts of research and practice on NPOs, project management, and organizational effectiveness, and edits the *questionnaire on general NPO project management and organizational efficiency*. This study adopts questionnaire survey to investigate the on-the-job adult students of the Open University of Kaohsiung that have participated in NPO activities. Through questionnaire collection and statistical analysis, this study explores the subjects' perceptions of the current performances of NPO project management and organizational efficiency. The questionnaires of this study are completed as both written and online questionnaires.

2 Implications of NPO project management and organizational efficiency

2.1 Implications of NPO project management

Most NPOs have active, innovative, flexible, and public organizational characteristics, and provide multiple channels for the public to consider and participate in public policies. By the force of groups, they directly or indirectly initiate related value and actions, which they promote to society. They represent the essence of sovereignty for the people, and are a critical factor of progress and stability in a democratic society[3]. Thus, NPOs play the role of service Vanguard or Service Pioneer, Value Guardian, Improver or Advocator, Educational Provider, Service Provider, policy assistant, reminder and counseling, supervision, and balance; such activities enhance social welfare, initiate reform, maintain value, improve a negative system, provide diverse negotiation and communication channels for society, accomplish highly-efficiency objectives through flexible organization, increase exchanges between international societies, train international professional talents, and offer opportunities for international recognition [4-6].

In order to effectively demonstrate organizational functions to accomplish organizational visions and missions, NPOs must execute and cope with human resource management, marketing management, fund raising management, performance management, accountability, social business, etc. [7]. Nevertheless, NPOs encounter difficulties and challenges, such as organizational mission, public persistence, establishment of goals, effective power, organizational legitimacy, finance and resource shortages, cost measurement and control, commercialization and industrialization, stimulation of work teams, service performance evaluation, autonomic governance, multiple demands of service subjects, organizational structure and service expansion, professional research, international development, and sustainable operation [3-11]. Peter. F. Drucker (1990) argued that the priority of the operations of NPOs is to transform donators into contributors; secondly, it proposes objectives for communities and the entire organization, validates missions and positions, leader roles, defines performance, develops marketing strategies, reforms, and funds, reinforces human resources and international relationships (including employees, board of directors, volunteers, and community), and develops self-development opportunities and potential from the perspectives of individuals, executors, and leaders, in order to break through the difficulties and blind spots of NPOs in operation and management.

The operational patterns of NPOs are very similar to the operational processes of project management, and include the following project management characteristics: specific management objectives, extremely complicated and numerous uncertain task factors, cross-departmental or cross-field cooperation for tasks, funds and operational resources based on external donations, and stakeholders include industry, government, academic fields, research, and community, thus, while they are numerous and have complicated demands, they value the social benefit of tasks instead of economic benefit [2, 12]. According to standards of project management competence (IPMA Competency Baseline, ICB 3.0), as established by the International Project Management Association (IPMA), "project" denotes a certain period of time, an organization's temporary involvement, and efforts to accomplish certain goals [13]; in addition, the Project Management Institute (PMI) states that "project" is unique product, service, or result created by temporary efforts [14].

Based on the definitions of previous international professional organizations (IPMA & PMI) and project management scholars [15-30] of project management, this study suggests that "NPO project management" means that, with limited resources and fixed and unique visions, NPOs apply effective management methods and systematic management processes to integrate labor and resources to accomplish tasks.

According to the definitions of scholars [14], [30-36], [62] of competence and project management competence, this study argues that "NPO project management competence" means NPO personnel's project management knowledge and explicit competence of skills. Kao (2017) treated project management initiation, planning, execution, control, and closing as the framework for project management competence, and cognitive and skills performance of the five processes of project management as the items of project management competence. Upon attitude and moral regulations of project management, the said researcher regarded the project management executors of industry, government, academic, and research circles in Taiwan as the subjects. By empirical research, it constructed Taiwan's project management competence scale of 80 items, including 30 items of project management cognitive competence performance, 30 items of project management technical competence performance, 10 items of project management attitude competence performance, and 10 items of project ethical competence performance.

This study adopts the technical competence performance of project management, as established by Kao (2017), and 5-point Likert scale as the design criteria to construct a *scale of general NPO project management*, which includes project initiation (5 items), project planning (10 items), project control (10 items), and project closing (5 items). When the scores of items and dimensions are higher, NPO project management performance is more significant. When the scores are lower, NPO project performance is less satisfying. The content is shown, as follows [30]:

2.1.1 Technical competence performance of project initiation

01. I will initiate projects with benefits.
02. I will set project goals.
03. I will propose a project concept book (draft).
04. I will operate the tools and techniques used for project feasibility analysis.
05. I will propose a project authorization letter.

2.1.2 Technical competence performance of project planning

01. I will propose a project plan.
02. I will operate the tools and techniques used for project schedule planning.
03. I will operate the tools and techniques used for project budget planning.
04. I will operate the tools and techniques used for project quality planning.
05. I will operate the tools and techniques used for project scope planning.
06. I will operate the tools and techniques used for project risk planning.
07. I will operate the tools and techniques used for project human resource planning.
08. I will operate the tools and techniques used for project equipment and materials planning.
09. I will operate the tools and techniques used for project procurement and bid planning.
10. I will operate the tools and techniques used for communication planning with a project's interested parties.

2.1.3 Technical competence performance of project controlling and management

01. I will control and manage the project schedule.
02. I will control and manage the project budget.
03. I will control and manage project quality.
04. I will control and manage project scope.
05. I will control and manage project risk.
06. I will control and manage the project's human resources.
07. I will control and manage the project's equipment and materials.
08. I will control and manage the procurement bidding of a program.
09. I will effectively control and manage the interested parties of the project.
10. I will control and manage the related laws and contracts of the project.

2.1.4 Technical competence performance of project closing

01. I will operate the tools and techniques of project closing.
02. I will operate the tools and techniques of closing the project's contracts.
03. I will operate the tools and techniques of closing the project's administrative affairs.
04. I will hold project review meetings.
05. I will manage the knowledge and experience documents of the project.

2.2 Implication of organizational effectiveness

While the implications of effectiveness are broad, they can be generally classified into "quality" and "quantity"; however, these two views are not totally different or confronting, but are part of complicated organizations; therefore, they enhance complexity. Profit organizations mostly rely on financial indicators as the criteria of measurement. NPO schools, public sectors, hospitals, and various types of civil groups tend to treat organizational innovative performance, learning and growth, operational processes, the accomplishment of goals, job engagement, organizational climate, and job quality as criteria of measurement. Thus, the two show certain degrees of differences [37-41].

Based on the definitions and views of scholars [37-44] regarding organizational effectiveness, this study defines NPO efficacy, as follows: NPO members of different levels and categories fulfill their duties, accomplish the organizational objectives, and make efforts for the plans set by the organization to result in the positive performance of organizational management and operation. NPO efficacy in this study refers to the following five items:

(1) Environment satisfaction: NPO members' satisfaction with business execution, colleagues' relationship, salary, high-rank supervision, and promotion of the current organization.
(2) Organizational climate: NPO members' perceptions of colleagues' relationships.
(3) Operational performance: NPO members' satisfaction with treatment and outcome of affairs related to the people and things in the organization.
(4) Job engagement: NPO members' involvement in organizational affairs.
(5) Job quality: NPO members' perceptions of the progress of professional competence, mastery of affair treatments, proper application of various resources to enhance job efficiency, presentation of business outcomes, etc.

This study generalizes literature related to organizational effectiveness as the *scale of general NPO efficacy* and adopts a five-point scale; where content refers to environment satisfaction (6 items), organizational climate (5 items), operational performance (5 items), job engagement (5 items), and job quality (6 items), for a total of 27 items. When the scores of items and dimensions are higher, NPO efficacy is higher. When the scores are lower, NPO efficacy is lower. The content is shown, as follows:

2.2.1 Environment satisfaction

01. I am properly respected in the organization.
02. I am satisfied with the award system in the organization.
03. I am satisfied with the work environment in the organization.
04. I am satisfied with the work assigned by the organization.
05. I am satisfied with colleagues' coping with affairs in the organization.
06. I am satisfied with the welfare in the organization.

2.2.2 Organizational climate

01. Colleagues' relationships are harmonious.
02. Colleagues get along joyfully at work.
03. My organization is a vigorous team.
04. Interaction between the cadre and members is positive.
05. When colleagues have different opinions, they can communicate and negotiate with each other.

2.2.3 Operational performance

01. Colleagues can make efforts to solve the problems encountered by the organization.
02. Different units of the organization establish concrete duties of work and operations are successful.

03. Colleagues of units in the organization can punctually accomplish the resolutions of meeting.
04. Work performances of colleagues are acknowledged by authorities and different circles of society.
05. The organization can immediately cope with the problems encountered.

2.2.4 Job engagement

01. Colleagues always make efforts to the job.
02. I tend to treat organizational goals as personal work objectives.
03. Colleagues tend to share the opinions of reform in various kinds of meeting in the organization.
04. I usually can experience pleasure at work.
05. I have intention to actively cooperate with all measures in the organization.

2.2.5 Job quality

01. Colleagues tend to discuss the problems of work and share their experiences during spare time at work.
02. Colleagues can properly use labor, materials, and financial resources in and out of the organization to enhance job outcome.
03. Colleagues all pay attention to the executive strategy and method design of work tasks.
04. The organization assigns jobs according to colleagues' specialties.
05. The organization frequently holds forums for work experience exchange.
06. The organization frequently holds visits in model units with outstanding performance.

3 Research framework and steps

Through the previous literature review and analysis, this study generalizes and edits the *first draft of the questionnaire on the general situation of NPO project management and organizational efficiency,* and invites NPO workers and university scholars and experts in the field of project management as the experts to examine the expert validity of the questionnaire. Regarding NPO workers, this study invites 1 labor group supervisor, 1 supervisor of the chamber of commerce, and 1 supervisor of the community development association as experts to test the draft of the questionnaire scale of this study; secondly, regarding university professors or scholars in the field of project management, by the assistance of colleagues in the Taiwan Project Management Association, this study invites 1 scholar or expert from National Pingtung University, 1 from National Pingtung University of Science and Technology, 1 from Lunghwa University of Science and Technology, 1 from National Kaohsiung First University, 1 from Aletheia University, 1 from the Republic of China Military Academy, and 1 from the Open University of Kaohsiung, to examine the expert validity of this questionnaire.

This study focuses on the on-the-job adult learners from different counties and cities around Taiwan who have participated NPO affairs in the Open University of Kaohsiung, where the sampling unit is the class, and adopts random sampling as the method of the questionnaire survey to explore their different personal background variables, such as gender, age, educational background, job title, organizational history, organizational scale, educational and training, general performance of project management business of the project initiation, project planning, project control, project closing in the NPO, as well as the general perceptions of the organizational effectiveness of organizational environment satisfaction, organizational climate, operational performance, job engagement, and job quality. It also probes into the effective measures to reinforce the organizational efficacy of NPO project management performance, and the performance differences of the different dimensions of project management and organizational effectiveness, as reference for future improvement of educational training systems and reinforcement of project management labor and organizational effectiveness.

Operations of the research framework (see Figure 1) are based on statistical analysis, such as t test and one-way ANOVA, to conduct difference comparison analysis of subjects' personal background variables on project management variables and the outcome variables of organizational effectiveness; secondly, by path analysis of Structural Equation Modeling (SEM), it probes into the path relation between subjects' operational variables of project management and the outcome variables of organizational effectiveness; finally, based on the research findings, this study states that NPOs properly apply project management to effectively enhance efficacy in order to accomplish organizational objectives and visions.

The framework of this study is shown, as follows:

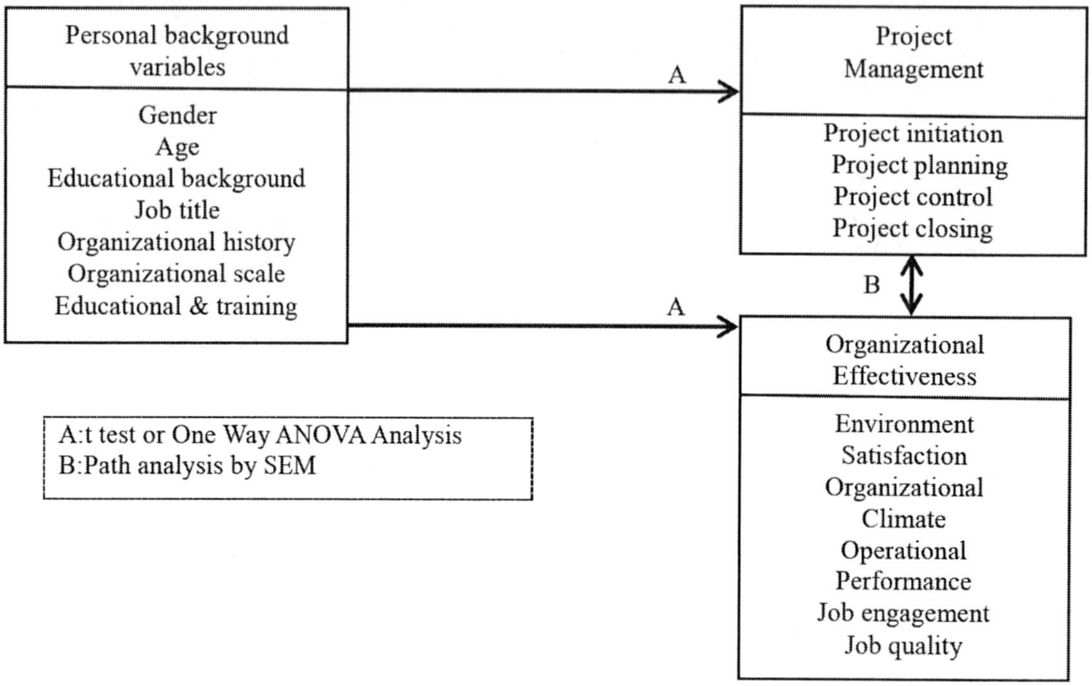

Figure 1. Research framework.

The steps of the research process are shown, as follows:

Step 1: Related literature collection and generalization of research variables.
Step 2: Selection of research samples.
Step 3: Editing of the questionnaire.
Step 4: Practice the questionnaire survey.
Step 5: Questionnaire retrieval and statistical analysis.
Step 6: Generation of statistical results and suggestions.
Step 7: Completion and submission of the research findings report.

4 Research results

4.1 Descriptive statistics result

The descriptive statistical results of the samples are shown in Table 1. Regarding gender, there are 150 males, accounting for 54.0% of the total subjects, and 128 females, accounting for 46.0%. Regarding age, "21-40 years old" is the most, and there are 189 subjects, accounting for 68.0%; second is "41-60 years old, there are 79 subjects, accounting for 28.4%; the least is above 61 years old, there are 10 subjects, accounting for 3.6%. Regarding educational distribution, "Below (including) Senior high school and vocational school" is the most, there are 146 subjects, accounting for 52.5%; the second is "college or university", there are 116 subjects, accounting for 41.7%; the least is "graduate school", there are 16 subjects, accounting for 5.8%. Regarding subjects' job title, "Member" is the most, there are 153 subjects, accounting for 55.0%; the second is "Officer", there are 104 subjects, accounting for 37.4%; the least is "Responsible person", there are 21 subjects, accounting for 7.6%. Regarding subjects' history in the organization, "More than 21 years" is the most, there are 98 subjects, accounting for 35.3%; the second is "2--10 years" and "Less than 1 year", there are 68 and 60 subjects, accounting for 24.5% and 21.6%, respectively; the least is "11--20 years", there are 52 subjects, accounting for 18.7%. Regarding subjects' organizational scale, "Under 50 people" is the most, there are 119 subjects, accounting for 42.8%; second is "Over 201 people" and "51-100 people", there are 72 and 68 subjects, accounting for 25.9% and 24.5%, respectively, the least is "101-200 people", there are 19 subjects, accounting for 6.8%. Regarding subjects' educational training, 151 subjects never received project management related training courses, accounting for 54.3%; 127 subjects have received project management related training courses, accounting for 45.7%.

Table 1. Questionnaire survey on general NPO project management and organizational efficiency.

Variables	Group	Number of subjects	Percentage	Cumulative Percentage
Gender	Male	150	54.0	54.0
	Female	128	46.0	100.0
Age	21--40 years old	189	68.0	68.0
	41--60 years old	79	28.4	96.4
	Above 61 years old	10	3.6	100.0
Educational background	Below (including) Senior high school and vocational school	146	52.5	52.5
	College or university	116	41.7	94.2
	Graduate school	16	5.8	100.0
Job title	Member	153	55.0	55.0
	Officer	104	37.4	92.4
	Responsible person	21	7.6	100.0
Organizational history	Less than 1 year	60	21.6	21.6
	2--10 years	68	24.5	46.0
	11--20 years	52	18.7	64.7
	More than 21 years	98	35.3	100.0
Organizational scale	Under 50 people	119	42.8	42.8
	51—100 people	68	24.5	67.3
	101—200 people	19	6.8	74.1
	Over 201 people	72	25.9	100.0
Educational & training	Never received project management related training courses	151	54.3	54.3
	Have received project management related training courses	127	45.7	100.0

N=278

4.2 Confirmatory Factor Analysis result of questionnaire scale

4.2.1 Validation of convergent validity

Thomopson (2004) argued that a measurement model can accurately reflect the dimensions. Thus, before SEM analysis, measurement model analysis must be conducted. According to the two-stage model analysis of Kline (2005), this study examines the SEM model for model fit. When it is acceptable, it subsequently evaluates the total SEM model [45]. According to the suggestions of Hair, Anderson, Tatham, and Black (2009) and Fornell and Larcker (1981), the CFA measurement standard is shown, as follows: (1) factor loading is higher than .50; (2) composite reliability is higher than .60; (3) average variance extracted is higher than .50. CFA analytical results of this study show that the factor loading of dimensions is .0.76~.98; composite reliability is .94~.99, and average variance extracted is .75~.88 (see Table 2), which shows that dimensions of this study reveal convergent validity.

Table 2. Summary of the data for convergence validity verification.

Parameter estimates of model						**Convergent validity**			
Latent variables	Manifest variables	Non-standardized factor loading	Standard error S.E.	t-value	P	Standardized factor loading	SMC	C.R	AVE
Project initiation	Aa01	1				0.88	0.77	0.96	0.81
	Aa02	1.02	0.05	22.65	***	0.91	0.83		
	Aa03	1.07	0.05	22.49	***	0.91	0.82		
	Aa04	1.07	0.05	23.09	***	0.92	0.85		
	Aa05	1.1	0.05	21.11	***	0.88	0.78		
Project planning	Ab01	1				0.85	0.73	0.98	0.83
	Ab02	1.12	0.05	23.32	***	0.94	0.89		
	Ab03	1.15	0.05	23.54	***	0.95	0.90		
	Ab04	1.12	0.05	23.88	***	0.95	0.91		
	Ab05	1.11	0.05	23.25	***	0.94	0.89		
	Ab06	1.06	0.05	21.71	***	0.91	0.83		
	Ab07	1.08	0.05	22.02	***	0.92	0.84		
	Ab08	1.04	0.05	21.50	***	0.91	0.82		
	Ab09	1.07	0.05	19.93	***	0.87	0.76		
	Ab10	0.96	0.05	19.08	***	0.85	0.73		
Project control	Ac01	1				0.96	0.93	0.99	0.88
	Ac02	1.00	0.02	45.49	***	0.98	0.95		
	Ac03	0.95	0.03	38.42	***	0.95	0.90		
	Ac04	0.98	0.02	41.33	***	0.96	0.93		
	Ac05	1.00	0.03	39.97	***	0.96	0.92		
	Ac06	1.00	0.03	36.66	***	0.94	0.89		
	Ac07	1.00	0.03	36.31	***	0.94	0.89		
	Ac08	0.82	0.04	21.41	***	0.81	0.66		
	Ac09	0.97	0.03	33.84	***	0.93	0.86		
	Ac10	1.05	0.03	32.56	***	0.92	0.85		
Project closing	Ad01	1				0.94	0.88	0.96	0.83
	Ad02	1.00	0.03	33.56	***	0.96	0.91		
	Ad03	0.99	0.03	32.86	***	0.95	0.90		
	Ad04	0.88	0.04	20.93	***	0.82	0.68		

Parameter estimates of model						Convergent validity			
Latent variables	Manifest variables	Non-standardized factor loading	Standard error S.E.	t-value	P	Standardized factor loading	SMC	C.R	AVE
	Ad05	0.88	0.04	24.69	***	0.88	0.77		
Organizational environment satisfaction	Ba01	1				0.79	0.63	0.96	0.78
	Ba02	1.22	0.07	17.46	***	0.89	0.79		
	Ba03	1.24	0.07	17.63	***	0.89	0.80		
	Ba04	1.28	0.07	18.17	***	0.91	0.83		
	Ba05	1.26	0.07	18.56	***	0.93	0.86		
	Ba06	1.24	0.07	17.51	***	0.89	0.79		
Organizational colleagues' interpersonal relationships	Bb01	1				0.91	0.82	0.95	0.80
	Bb02	1.03	0.04	27.06	***	0.94	0.89		
	Bb03	1.07	0.04	24.38	***	0.91	0.82		
	Bb04	1.05	0.04	24.25	***	0.91	0.82		
	Bb05	0.96	0.05	18.78	***	0.81	0.66		
Organizational administration and management performance	Bc01	1				0.86	0.73	0.96	0.83
	Bc02	1.06	0.05	21.90	***	0.92	0.85		
	Bc03	1.07	0.05	22.14	***	0.93	0.86		
	Bc04	1.07	0.05	21.92	***	0.92	0.85		
	Bc05	1.04	0.05	21.62	***	0.92	0.84		
Organizational colleagues' job engagement	Bd01	1				0.85	0.72	0.94	0.75
	Bd02	0.98	0.05	20.13	***	0.90	0.81		
	Bd03	0.93	0.05	18.30	***	0.85	0.73		
	Bd04	1.00	0.05	19.30	***	0.88	0.77		
	Bd05	0.88	0.05	17.96	***	0.85	0.71		
Organizational colleagues' job quality	Be01	1				0.91	0.83	0.95	0.77
	Be02	1.09	0.04	27.59	***	0.94	0.88		
	Be03	1.05	0.04	25.34	***	0.91	0.83		
	Be04	1.03	0.05	21.68	***	0.86	0.77		
	Be05	1.07	0.05	22.85	***	0.88	0.77		
	Be06	0.86	0.05	16.79	***	0.76	0.58		

***P<.001

4.2.2 Validation of discriminant validity

Table 3. Bootstrap correlation coefficient 95% confidence interval.

Parameters			Estimates	Bias-corrected		Percentile method	
				Lower Bound	Upper Bound	Lower Bound	Upper Bound
A0 Project management	<-->	B0 Organizational efficacy	0.27	0.20	0.35	0.20	0.34

AMOS provides two types of estimations for the confidence interval: one is the Bias-corrected Percentile Method and the other is the Percentile Method [46]. This study applies the bootstrap confidence interval to test the discriminant validity of the dimensions. With a 95% confidence level, it repeats 1000 times of estimations. According to the calculation results, when the correlation coefficient of the dimensions is lower than 1 with a 95% confidence interval, it means there is discrimin ant validity among the dimensions [46-48]. The estimated results of this study are shown in Table 3. The confidence intervals of the dimensions are lower than 1, thus, there is discriminant validity between the two dimensions.

4.2.3 Verification of goodness-of-fit

In this study, fit measures are based on the suggestions of Wu (2007), Hsu (2010), Bagozzi and Yi (1988), and Hair et al. (1998). It adopts $\chi 2$ to test ratio of $\chi 2$ and freedom. GFI, AGFI, Root Mean Square Error of Approximation (RMSEA), Comparative Fit Index (CFI), Incremental Fit Index (IFI), Non-Normed Fit Index (NNFI), and NFI are the fit measures of the total model. When SEM samples are more than 200, the chi-square value ($\chi 2$=(n-1)Fmin) can be too large; in other words, when samples are oversize, the chi-square value can be too large, and the P value tends to be rejected [46, 49], thus, Bollen and Stine (1992) proposed model revision by Bootstrap. After adjustment, as the chi-square value becomes lower, all fit measures should be recalculated. Regarding calculation results are shown in Table 4; all model fit measures match the standard.

Table 4. Model goodness-of-fit indices.

Goodness-of-fit Indexes	Standard Value		Model Goodness-of-fit
χ^2	The less the better	2824.99	Passed
χ^2/df	<3	1.85	Passed
GFI	>.9	0.89	Accepted
AGFI	>.9	0.88	Accepted
RMSEA	<.08	0.06	Passed
TLI (NNFI)	>.9	0.95	Passed
CFI	>.9	0.95	Passed
NFI	>.9	0.89	Accepted

4.3 Difference analysis of the variables of personal background variables and NPO project management and organizational efficiency

Independent sample t testing aims to calculate the mean of two independent samples, and generalizes it to the mean of the original two population, in order to determine if there is significant difference [50]. This study adopts independent sample t testing to conduct difference analysis of different genders and educational training on the variables of the correlation model of NPO project management and organizational efficiency. One-way ANOVA aims to determine if the means of the manifest variables show significant difference upon the effect of different levels[50]. When the test result shows that the F value is significant, the means shows significant difference between at least one group. Therefore, "multiple Post Hoc" analysis should be applied to further verify the difference [51].

4.3.1 Difference analysis of gender on variables of correlation model of NPO project management and organizational efficiency

Regarding the difference comparative analysis of subjects' genders on the dimensions, according to the result of independent sample t testing, different genders on project control (t=-.09, p=.93), organizational administration and management performance (t=-1.87, p=.06) do not show significant difference of means; means of project initiation (t= -3.74*, p=.00), project planning (t=-4.39*, p=.00), project closing (t=-4.52*, p=.00), organizational environment satisfaction (t=-2.89*, p=.00), organizational colleagues' interpersonal relationship (t=-2.50*, p=.01) organizational colleagues' job engagement (t=-2.72*, p=.01), and organizational colleagues' job quality (t=-3.23*, p=.00) show significant differences. Females are more significant than males.

4.3.2 Difference analysis of the variables of the correlation model of Education and Training on NPO project management and organizational efficiency

Regarding the difference analysis of subjects' educational training on the dimensions, according to the result of independent sample t testing, the means of different educational training on project control (t=1.21, p=.23) do not show significant difference; the means of project initiation (t= -7.10*, p=.00), project planning (t=-6.54*, p=.00), project closing (t=-6.66*, p=.00), organizational environment satisfaction (t=-3.97*, p=.00), organizational colleagues' interpersonal relationship (t=-4.57*, p=.00), organizational administration and management performance (t=-4.46, p=.00), organizational colleagues' job engagement (t=-4.96*, p=.00), and organizational colleagues' job quality (t=-4.18*, p=.00) show significant differences. Moreover, subjects who have received project management related training courses are more significant than those who never received project management related training courses.

4.3.3 Difference analysis of age on the variables of the correlation model of NPO project management and organizational efficiency

Regarding the difference analysis of subjects' ages on the dimensions of project management and organizational effectiveness, according to the result of one-way ANOVA, the means of different ages on project initiation (F= 1.99, p=.14), project planning (F=.90, p=.41), project control (F=1.90, p=.15), project closing (F=1.10, p=.34), organizational environment satisfaction (F=2.13, p=.12), organizational colleagues' interpersonal relationship (F=1.23, p=.30), organizational administration and management performance (F=.96, p=.38), organizational colleagues' job engagement (F=2.15, p=.12), and organizational colleagues' job quality (F=.39, p=.68) do not show significant difference. In other words, different ages on the variables of the correlation model of NPO project management and organizational efficiency do not show significant difference.

4.3.4 One-way ANOVA of educational background on the variables of project management and organizational effectiveness

Regarding the difference analysis of subjects' educational background on the dimensions of project management and organizational effectiveness, according to the result of one-way ANOVA, the means of different educational background on project initiation (F=2.59, p=.08), project control (F=1.16, p=.32) and project closing (F=2.65, p=.07) do not show significant difference; the mean of project planning (F=3.11*, p=.05) reveal significant difference. After Scheffe Post Hoc Testing, "Graduate school" is more significant than "College or university"; the mean of organizational colleagues' job engagement (F=9.07*, p=.00) reveal significant difference. After Scheffe Post Hoc Testing, "Graduate school" and "Below (including) Senior high school and vocational school" are higher than "College or university". "Graduate school" and "senior high school and vocational school (including)" do not reveal significant difference; the means of organizational environment satisfaction (F=6.53*, p=.00), organizational colleagues' interpersonal relationship (F=10.89*, p=.00), organizational administration and management performance (F=7.69*, p=.00), and organizational colleagues' job quality (F=7.56*, p=.00) show significant difference. After Scheffe Post Hoc Testing, "Graduate school" is higher than "Below (including) Senior high school and vocational school" and "College or university". "Graduate school" and "Below (including) Senior high school and vocational school" do not show significant difference.

4.3.5 One-way ANOVA of Job title on the variables of project management and

organizational effectiveness

Regarding the difference analysis of subjects' job titles on the dimensions of project management and organizational effectiveness, according to the result of one-way ANOVA, the means of different job titles on project initiation (F= 4.30*, p=.01) and organizational environment satisfaction (F=3.43*, p=.03) show significant difference. After Scheffe Post Hoc Testing, "Member" is higher than "Responsible person"; the means of different job titles on project planning (F=2.80, p=.06), project control (F=.29, p=.75), project closing (F=2.17, p=.12), organizational colleagues' interpersonal relationship (F=1.24, p=.29), organizational administration and management performance (F=1.77, p=.17), organizational colleagues' job engagement (F=1.33, p=.27), and organizational colleagues' job quality (F=.86, p=.43) do not show significant difference. In other words, different job titles on the variables of the correlation model of NPO project management and organizational efficiency do not show significant difference.

4.3.6 One-way ANOVA of organizational history on the variables of project management and organizational effectiveness

Regarding the difference analysis of subjects' organizational history on the dimensions of project management and organizational effectiveness, according to the result of one-way ANOVA, the means of different organizational history on organizational colleagues' job quality (F=3.63*, p=.01) show significant difference. After Scheffe Post Hoc Testing, "2--10 years" is higher than "Less than 1 year"; the means of different organizational history on project initiation (F=.83, p=.48), project planning (F=1.94, p=.12), project control (F=.75, p=.52), project closing (F=1.15, p=.33), organizational environment satisfaction (F=.61, p=.61), organizational colleagues' interpersonal relationship (F=.81, p=.49), organizational administration and management performance (F=.50, p=.69), and organizational colleagues' job engagement (F=.18, p=.91) do not show significant difference. In other words, different organizational history on the variables of the correlation model of NPO project management and organizational efficiency does not reveal significant difference.

4.3.7 One-way ANOVA of organizational scale on the variables of project management and organizational effectiveness

Regarding the difference analysis of subjects' organizational scale on the dimensions of project management and organizational effectiveness, according to the one-way ANOVA result, the means of different organizational scales on project initiation (F=1.06, p=.37), project planning (F=1.64, p=.18), project control (F=2.29, p=.08), project closing (F=1.52, p=.21), organizational environment satisfaction (F=1.13, p=.37), organizational colleagues' interpersonal relationship (F=.65, p=.59), organizational administration and management performance (F=.31, p=.82), organizational colleagues' job engagement (F=.24, p=.87), and organizational colleagues' job quality (F=3.33, p=.80) do not show significant difference. In other words, different organizational scales on the variables of the correlation model of NPO project management and organizational efficiency do not reveal significant difference.

4.4 Path analysis of NPO project management and organizational efficiency

Table 5. Path analysis of the relation between NPO project management and organizational efficiency.

Path relation			Path value	C.R.	*p*
Organizational efficacy	<->	Project management	.58	10.56*	***

*P<.05 ***P<.001

The path coefficient of NPO project management and organizational efficiency is 0.58, which is significant. The results are shown in Table 5.

4.5 Statistical model and path relation between NPO project management and organizational efficiency

This study integrates the stratified factors of NPO project management, project initiation, project planning, project control and project closing, and stratified factors of environment satisfaction, organizational climate, operational performance, job engagement, and job quality of NPO efficacy for Confirmatory Factor Analysis of Structural Equation Modeling and path relation analysis. The analytical results are shown in Figure 2.

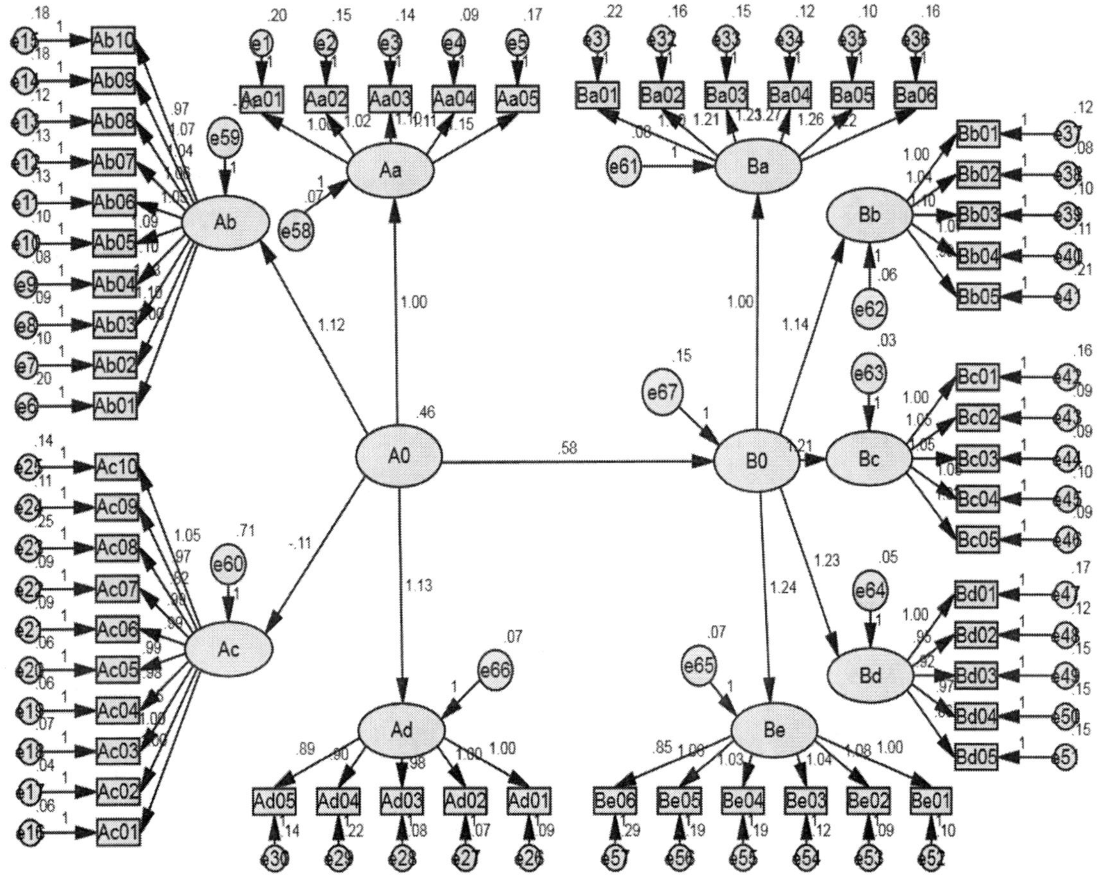

Figure 2. Statistical model and path relation between NPO project management and organizational efficiency.

5 Conclusion and suggestions

Based on the research findings, this study generalizes the following conclusions. First, regarding NPO members' cognitive differences of personal background variables, NPO female members' project management performance and organizational effectiveness perception are superior to male members. NPO members who have received project management training are superior to those who never received project management training in terms of project management performance and organizational effectiveness perception; NPO members graduated from graduate school are superior to those graduated from university, college, senior high school, or vocational school in terms of project management performance and organizational effectiveness perception; subjects in NPO with the job title of "Member" are superior to those with the job title of "Responsible person" in terms of project initiation performance and perception of satisfaction with organizational environment; NPOs with organizational history within 10 years is superior to that founded within 1 year in terms of perception of organizational job quality. Secondly, there is a path relation between NPO project management and organizational effectiveness.

Based on previous conclusions, this study proposes suggestions for NPO operations and management, as follows. First, it should actively plan project management competence training for NPO male members in order to reinforce the executive outcome of project management and strengthen NPO operational efficacy. NPOs should introduce project management competence training as the occupational training course for organizational members' professional advanced study. It should encourage NPO members to actively study courses in graduate school programs, in order to enhance their professional knowledge, particularly the advanced study of professional degrees in project management and organizational effectiveness, in order to enhance the performance of NPO project management and organizational efficiency; NPO owners should actively participate in project management competence training in order to fulfill organizational missions and positions, vision construction, strategic planning, goal setting, program design, process improvement, value creation, performance reinforcement, and sustainable operation, and lead members to demonstrate NPO functions, such as development and innovation, reform and promotion, value maintenance, service offering, and expansion of social participation; newly founded NPOs should enhance organizational members' project management competence, mastery to cope with affairs, proper use of various resources to increase job efficiency, and active and effective holding of the presentations of business outcomes, in order to upgrade the efficacy of job quality. Secondly, NPOs should actively reinforce organizational members' project management competence, such as project initiation, planning, control, and closing to strengthen the organizational effectiveness of environment satisfaction, organizational climate, operational performance, job engagement, and job quality, in order to effectively accomplish organizational missions and visions.

References

1. R.D. Anthony and D.W. Young, *Management control in Nonprofit organization* (Irwin Series in Graduate Accounting, Irwin, 1993)
2. L.L. Chin, Journal of Beijing Technology and Business University (version of social science), **20**(6), 63-67(2005)
3. P.H. Kao, *NPO operational strategy and management* (Hwa Li Publishing, Taiwan, 2012)
4. P.S. Yu(trans), Peter.F. Drucker, *Operation of nonprofit organizations* (bookzone, Taiwan, 1994)
5. D.H. Situ, *Operation and management of nonprofit organizations* (bookzone, Taiwan, 1999)
6. T.Y. Chang, T&D Fashion, **36**, 6-24 (2005)
7. S.H. Lin, *NPO management (additional version)* (Sanmin, Taiwan, 2015)
8. M.H. Chiang, *NPO management* (Taipei, Taiwan, 2002)
9. C.T. Chiu(ed), *Nonprofit department study: governance, departmental interaction and social innovation* (Bestwise, Taiwan, 2007)
10. H.H. Hsiao, Y.Y. Kuan and W.P. Lu(ed), *Nonprofit department: organization and operation (3rd ed.)* (Blueocean, Taiwan, 2000)
11. H.H. Hsiao(ed), *Nonprofit department: organization and operation* (Blueocean, Taiwan, 2000)
12. M. Wang(ed), *Introduction of NPO management* (Beijing, China, 2002)
13. Taiwan Project Management Association, *Selection of Cases of Management Quality Certification of International Items and C Class Interview* (Beijing, China, 2012)
14. PMI, *A Guide to the Project Management Body of Knowledge (5th ed.)* (Pennsylvania, USA, 2012)
15. H. Kerzner, *Project Management: A System Approach to Planning, Scheduling, and Controlling (9th ed.)* (New York, USA, 2006)
16. J. Lewis, *Fundamentals of Project Management (2nd ed.)* (New York, USA, 2002)

17. J. Lewis, *Mastering Project Management* (New York, USA, 2007)
18. C.C. Wei, *General project management knowledge system* (Taipei, Taiwan, 2013)
19. M.J. Kuan, *Teaching materials of APMP project manager training* (Taoyuan, Taiwan, 2009)
20. K.M. Lu, *Project management of software* (Taipei, Taiwan, 2009)
21. H.Y. Hsu, P.L. Hsiung, Y.F. Chu, Y.H. Chang, M. Fan, C.M. Huang, H.D. Chou and Z.P. Lu, *Basic knowledge and application practice of project management (5th ed.)* (New Taipei, Taiwan, 2014)
22. H.H. Lin, M.C. Huangand and W.L. Wang, *Project management of software (2nd ed.)* (Taipei, Taiwan, 2005)
23. W.D.Yu, *Project Management in Practice* (Taipei, Taiwan, 2015)
24. W.W. Chuang, *11 lessons to learn project management (revision)* (Taipei, Taiwan, 2014)
25. I.C. Kao, Civic knowledge management: feasible strategy to construct project knowledge management system of school education. Published in Y.M. Wu, T.C. Tsai, ed.: *Civic Urbanology* (18-35) (Kaohsiung, Taiwan, 2009)
26. I.C. Kao, Storage Management Solutions, **1**, 154-173 (2013)
27. I.C. Kao, Storage Management Solutions, **1**, January, 198-229 (2013)
28. I.C. Kao, Storage Management Solutions, **2**, 42-65 (2014)
29. Taiwan Project Management Association, *International project management knowledge system* (Kaohsiung, Taiwan, 2014)
30. I.C. Kao, MATEC Web of Conferences, **119**, 01037 (2017)
31. ICB: IPMA Competence Baseline, IPMA, http://ipma.ch/resources/ipma-publications/ ipma-competence-baseline (2009)
32. S.C. Chen, P.C. Lin and Y.C. Chen, Psychological Testing, **56**(3), 397-430 (2009)
33. D.C. McClelland, American Psychologist, **28**(1), 1-24 (1973)
34. L.M. Spencer and S.M. Spencer, *Competence at work: Models for superior performance* (New York, USA, 1993)
35. C.L. Kang, Employment and training, **15**(6), 3-9 (1997)
36. S.C. Wu, Psychological Testing, **60**(2), 397-428(2013)
37. W.K. Hoy and C.G. Miskel, *Educational administration: Theory, research, and practice, 9th edition* (New York, USA, 2013)
38. K.F. Huang, An Interdisciplinary Journal, **64**, 51-78(2013)
39. M.D. iPaola and W.K. Hoy, *Improving Instruction Through Supervision, Evaluation, and Professional Development* (Charlotte, USA, 2014)
40. C.S. Wu, *Research on school effectiveness* (Psychological Publishing, Taipei, 1992)
41. I.C. Kao, *Research on relationship between school teachers' organizational operation strategy and school effectiveness* (Taipei, Taiwan, 2003)
42. C.S. Chang, *The Effects of Transformational and Transactional Leadership of Elementary School Principals on the Culture Characteristics and Effectiveness of School Organization* (Kaohsiung, Taiwan, 2001)
43. M.C. Chin, C.F. Wu and C. Hien, Education and Psychological Study, **37**(1), 1-35 (2014)
44. H.Y. Kuo and L.Wu, Journal of Cheng-Shiu University, **19**, 279-300 (2006)
45. S.Y. Chen, M.C. Hsu and F. E. Li, Journal of Leisure and Recreation Industry Management, **5**(3), 1-21 (2012).
46. W.H. Chang, *Techniques of thesis writing* (Tingmao, Taipei, 2011)
47. M.C. Hsu, F.C. Pan and C.M. Huang, Journal of Taiwan Society for Sport Management, **11**(2), 85-107 (2011)
48. G. Torkzadeh, X. Koufteros and K. Pflughoeft, Structural Equation Modeling, **10**(2), 263-275 (2003)
49. M.C. Hsu and H.P. Yen, International Journal of LISREL, **6**(1), 24-56 (2013)
50. K.Y. Chen and C.H. Wang, *Thesis statistical analysis practice SPSS and operation of AMOS* (Wu Nan Books, Taipei, 2011)
51. M.L. Wu and C.T. Tu, *SPSS and statistical application analysis* (Wu Nan Books, Taipei, 2005)

52. P.F. Drucker, *Managing the Non-profit Organization: Principles and Practice* (New York, USA, 1990)
53. M.L. Wu, *Structural Equation Modeling: AMOS operation and application* (Wu Nan Books, Taipei, 2007)
54. M.C. Hsu, Journal of Sport and Recreation Management, **7**(1), 174-18 (2010)
55. C.Fornell and D.F. Larcker, Journal of Marketing Research, **18**, 39-50 (1981)
56. J.F. Jr.Hair, R.E. Anderson, R.L. Tatham and W.C. Black, *Multivariate data analysis (7th ed.)* (Prentice Hall, New Jersey, 2009)
57. R.B Kline, *Principles and practice of structural equation modeling (2nd ed.)* (Guilford Press, New York, 2005)
58. B. Thomopson, *Exploratory and confirmatory factor analysis: Understanding concepts and applications* (American Psychological Association, Washington DC, 2004)
59. R.P. Bagozzi and Y. Yi, Journal of the Academy of Marketing Science, **16**, 74-94(1988)
60. J.F.Jr. Hair, R.E. Anderson, R.L. Tatham and W.C. Black, *Multivariate data analysis (5th ed.)* (Prentice Hall, New Jersey, 1998)
61. S.Y. Hsu, *Nonprofit organizations and public administration. Published in M.H. Chiang (ed): The third sector: operational strategy and social participation* (Taipei, Taiwan, 2000)
62. H.L. Cheng, Research on competence construction of nonprofit organization project management personnel(master's thesis), Graduate Institute of Management, Ming Chuan University (to be published)

MATEC Web of Conferences **169**, 01022 (2018) https://doi.org/10.1051/matecconf/201816901022
IMETI 2017

A research on innovative management of teaching assessment- using I.T. to implement students' on-line peer review of learning outcomes through rubric

Huei-Zhen Gu[1,a] and Chien-Wei Ku[2]

[1]*Department of Information Management, Lunghwa University of Science and Technology, Taiwan*
[2]*Department of Engineering Science, National Cheng Kung University, Taiwan*

Abstract. The assessment of a student's learning outcome in a particular course is aimed not only to evaluate the learning effects but also to actively provide the student with immediate feedback to improve performance. However, the conventional methods for assessment are mostly concerned with the learning effects. In this study, the free software of General Public License is adopted to develop a database platform to let students of different learning teams conduct a peer review online, both quantitatively and qualitatively. With the system, inter and intra team peer reviews are also implemented. The findings help facilitate the conduct of online rubric quantitative and qualitative reviews, regardless of self or mutual ones. A multidimensional evaluation of learning outcome is therefore achieved. More crucially, the finding offer students an opportunity to drill down their learning effects, to understand current learning status, and to be acquainted with the feedback or comments made by peers. It also helps the learners modify their pace as well as strategies to eventually better their learning outcome.

1 Introduction

The quality of university students are commonly suspected in modern society due to the rapid increase in the number of higher education institution; therefore, the learning outcome and the assessment of students' core abilities have become the main concerns worldwide [1]. To improve the learning outcome through various kinds of strategies and skills has also become a trend for colleges and universities in the management of teaching quality. Education theories have suggested that the assessment of a student's learning outcome(SLO) is aimed not only to give comments on a student's learning progress but also acts as a mechanism to advise how to rectify learning strategies and activities in order to improve the current learning situation.

Conventionally, in school education, teachers are usually the ones who evaluate SLO through a summative assessment. This method is seen too subjective because of lack of unbiased criteria. Also, the teachers might be influenced by personal liking or disliking towards particular students. Furthermore, it ignores the subjectivity and the initiative of the students, which will finally write off their characteristics and creativity in learning [2].

In addition, the traditional method of assessment, in a summative form, gives no feedback to the students for improving the quality of learning. This seems having greatly degraded the main functions

[a] Corresponding author : jasongu@mail.lhu.edu.tw

and the perspectives of the assessment. Based on the reasons mentioned, in consideration of increasing self-improvement and learning effects of the students, this research develops an online database platform with the aims not only for teachers to easily conduct an assessment that emphasizes equally on both formative and summative evaluations but also for students to carry out intra and inter team peer to peer reviews. Through the process and the results of such an assessment, it is hoped that the students are able to understand their weaknesses in learning and eventually make improvement.

2 Literature review

2.1 Assessment of students' learning outcomes

From a macro viewpoint to look at the learning outcomes of the students, scholars suggested in 2010 that SLO refers to the knowledge learned in school and the ability to apply these knowledge to various situations after completion of the course or graduation [3]. On the other hand, in any kind of courses, learning outcome mainly focuses on the performance of the students, especially the objectives of the learning and the core abilities achieved after completion of a particular course [4]. To further explain this, SLO can be properly examined after each session of learning. The results are then used both by the teacher to adjust teaching plan and design, and by the students to review their learning strategies and paces.

The assessment of learning outcome is usually classified into direct and indirect ones. The former one reviews a student's knowledge as well as mental state according to his or her learning status, skills, attitude, and behaviour through a set of test design, psychological scale, or questionnaire. Certificate test, in class examination, doing reports on special topics, and capstone course are common tools to assess. The indirect one, on the other hand, utilizes questionnaire, alumni follow-up, and employer's satisfaction with the alumni to evaluate learning outcomes in a long term. It is noticeably that the assessment of learning outcome should not focus merely on the evaluation itself. The ways of teaching, course syllabus, and the methods conducting the assessment should be taken into consideration as well [3]. Due to the discrepancy in the levels of different courses, the assessment of learning outcome should be designed based on the level of the course to make it more specific and targeted in order to avoid evaluation errors. In other words, courses might be categorized into three levels - preliminary, intermediate, and advanced. For preliminary students, their abilities are shown in describing, memorizing, and didactic. The intermediate students should focus their abilities in application, presentation, explanation, analysis, making comments, comparison, and problem-solving. The advanced course, meanwhile, would concentrate on cultivating students' initiative abilities in creativity, planning, administration, and construction.

2.2 Rubric

The term 'rubric', a Latin word, originally derived in the 13th century or even earlier days, which refers to the headlines or the highlights of important meanings in religion marked by red ochre or chalk in manuscripts. At that time, red colour pens were only used to make remarks in religious books or the authoritative publications. The term was later defined as a brief and authoritative criterion. Nowadays, 'rubric' is commonly defined as the red mark, the title, the prompt, or the instruction in books or on exam papers. It is also utilized in the process of teaching assessment.

In teaching activities, 'rubric' is a method of assessment that enhances learning activity [5]. It is also a structured evaluating standard with the characteristics of flexibility, strong operability, and oriented process [2]. As rubric was gradually developed into an evaluation tool, it has become a terminology for education assessment, standing for the scoring guide to evaluate students' learning quality in learning process [6]. It usually gives the definition of learning quality and achievement, and specifies the strategies and the standard of the assessment. It is usually shown in a form of a table, enabling teachers to mark and students to plan for future learning [7].

2.3 Implementation of paper-based rubric on peer review

To take the researcher of this paper as example, a paper-based rubric was ever conducted for peer to peer reviews. This experience gave an important reference to the development of the database platform online. The details are shown as follows.

2.3.1 Grouping

Prior to the commencement of the class, the students were grouped into different learning teams to prepare together for the assigned report and presentation. The number of each team varied, according to the total number of the participants. The students of the same team were seated adjacently. In this study, seven teams were formed, with four to seven members in each one.

2.3.2 Preparation of paper-based rubric scale

Generally speaking, to conduct a typical rubric assessment, a form of peer to peer review is needed. In the case of the study, three forms were designed, including the ones for 'inter team peer review', 'self-review', and the 'assessment of the exercise.' The inter team peer review form was used to evaluate the performance of the students on stage. The second one enabled the one in the reporting team to review the performances of him/herself and his or her teammates. The form of 'assessment of the assigned exercise' was used by students of the reporting team to evaluate the achievement of the students out of their team, based on the scores gained in the assigned exercise.

The intra team peer review form is composed of a quantitative scoring assessment and a section for making qualitative comments. The scale for the quantitative scoring assessment is a cross matrix. Within each cell of the matrix, the lowest standard of the score obtained by the subject is recorded, namely 'rubric'. The teacher can design the assessment items in accordance with the goals of each course and create a scale of four or five levels. In the case having been discussed so far, six items of assessment were made, including the content of the report, the application and the example(s) used, the presentation skills, the design of the presentation, the composure on stage, interactivity, and collaboration. Each item was graded based on a five-level scale- '5' for 'excellent', '4' for 'good', '3' for 'fair', '2' for 'needing improvement', and '1' for 'poor'. As for the part of the qualitative comment, each reviewer was required to write down his or her subjective opinion(s) and feeling of the report or suggestions to the subject with at least 20 words.

The self-review form also includes sections of quantitative scoring and qualitative comment. The quantitative scoring form is the same as that of the intra team peer review form. However, the comment part is different. While the reviewer and the subject are different individuals, the comment is made by the reviewer to give the subject pieces of advice. On the contrary, the comment will act like a learning feedback if the reviewer and the subject are the same one. In the case discussed, students were required to write down the self-learning feedback, which should comprise the items such as the favourite part of the course, the most beneficial part received from the course, and the part(s) of the course that need(s) to improve.

The evaluation of students' grades for doing the assigned exercise should be principally included in the content of the 'assessment of the assigned exercise.' In the case, the items on the assessment list included 'the timing used for answering the questions and the correct rate of the answers. Each item was also graded according to the five-level policy.

2.3.3 To convert scores on five-level scale to centesimal grades

In general, centesimal grading policy is widely adopted in Taiwan's education system to evaluate SLO. Therefore, to convert the scores on the five-level scale to centesimal grades is essential. From level one to four, in the case of the study carried out by the researcher, each level was made equivalent to 12 grades in the hundred mark system. Based on this principle, a score between 5.0 and 4.1 is converted to one between 100 and 89.2; between 4.0 and 3.1 calculated as one between 88.0 and 77.2; between 3.0 and 2.1 converted to a score between 76 and 65.2; between 2.0 to 1.1 scored between 64 and 53.2. The score between 1.0 and 0 was converted to the one between 52 and zero.

2.3.4 Calculation of the grade of general SLO

Principally, the ratio of the scores of peer to peer review, self-review, and the exercise work in final grade can be determined either by the teacher or based on the discussion of the teacher with the students. In the previous study, it was firstly proposed by the teacher, and the decision was finally made after discussion with the students. According to the consensus reached, peer to peer review, self-review, and the score for the exercise work counted for 40%, 20%, and 40%, respectively.

3 Research architecture

Fig. 1 illustrates the research architecture of this work. In the research, SLO, the assessment standard of rubric, and literature reviews of several case studies related to the implementation of rubric were firstly examined. Next, a class in which the researcher had ever conducted flipped classroom and rubric assessment was chosen as the object of study. Then, a discussion with the students on how to improve the items on the assessment forms took place after the assessment forms used for the previous rubric were collected and reviewed. After this, a platform and tools for system development were chosen to conduct demand analysis, function design, program coding for the prototype system. The process of testing was finally conducted.

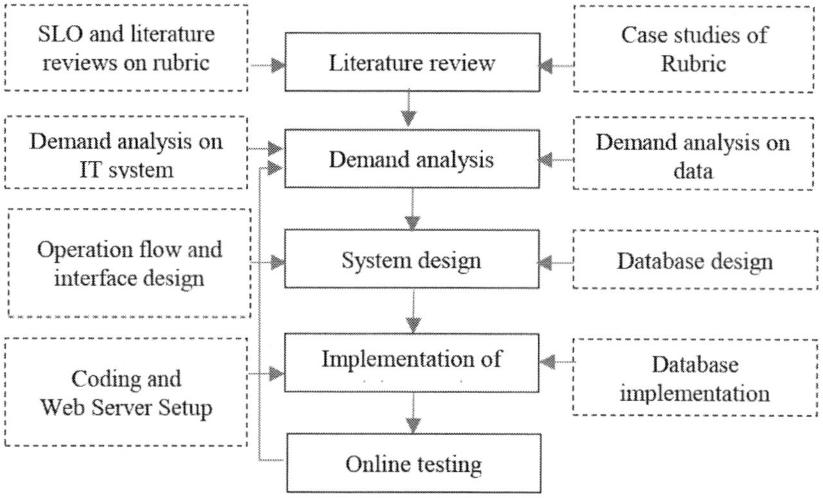

Figure 1. Research architecture of the research

4 System development

4.1 The platform

In consideration of the development costs as well as the features of the internet, i.e. being able to be used for multiple work by many people, the free software of General Public License (GPL) is adopted to develop a system platform, using the tools of Ubuntu Linux as operation system, MySQL as database server, Apache 2 as web server, and PHP7, Java script, and HTML5 for program coding and the design of operation interface. The structure of the system platform is shown in Fig. 2.

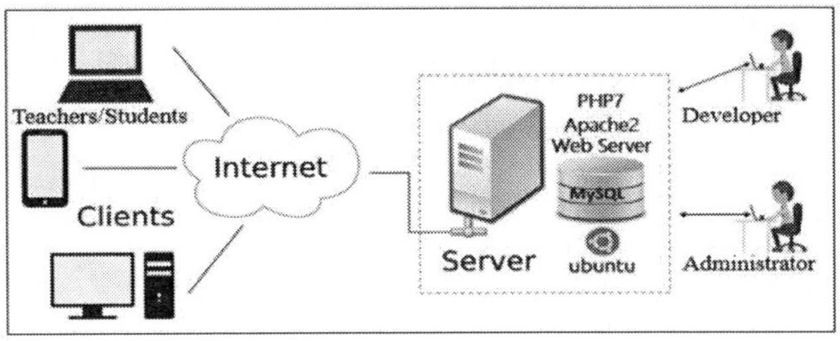

Figure 2. Structure of the system platform

4.2 Design of the functions and activities

4.2.1 Functions of the system

Fig. 3 represents the design of the system functions with UML use case diagram. The functions of the platform developed can be divided into two kinds- one for the teacher and the other for the students. As for the teacher's one, it helps the users to assign report team, adjust learning groups, inquire about the team members, the learning quantitative feedback or check the qualitative feedback of peer to peer reviews, and enquire about the grades of the students. The teacher can also calculate the general grades of the students. The later one has the functions of join learning team, peer to peer review(inter review) online, self-review (intra review) online, peer to peer review of the exercise work online, real time inquiry about individual learning outcome, and inquiry about the qualitative feedback from other members.

Figure 3. Design of system functions

Through this platform, teachers can easily implement rubric online for a real time multiple assessment to examine SLO in both qualitative and quantitative ways. On the other hand, following the completion of group report, students can not only carry out empathic learning and inter group peer reviews but also inquire about the grades of their report or the assigned exercise work. They can also make qualitative comment to themselves and other students.

4.2.2 Procedures and activities in the initial state

The platform can be administered by the computer centre to be used by faculties and students of the whole institution. It can be also kept by any department for its members or by individual teacher for his or her own use. Prior to a new semester, the data of the previous semester will be exported by the administrator from the rear end of the database by using MySQL Workbench for archive. The data in the database will be deleted, too, in order for the teachers to use the system in the coming new semester.

After the names of the teacher and the students are confirmed, the administrator will import the information about the teachers, the courses, and the students into the database in the format of CSV file. The professor's information should include the professor ID, name, the department the professor belongs to, and the password. The file of the courses information includes the titles of the class and course, the ID of professor who offer the course, the course ID, the year and semester the course belongs to. The file of the student information includes the class and course title, the ID, name and password of student who join the course, and the group number jointed. The file formats are shown in following table.

Table 1. Examples of the information in CVS file format to be imported

Professor information format: ID, name, the department belonged, and password
Course information format: Class and course titles, professor's ID, course ID, year and the semester
Student information format: Class and course titles, the ID, name, password of student, and the group number

4.2.3 Procedure and activities in general status of use

The procedure and activities of using this system in classroom for teacher are described as follows. First, personal account and password have to be entered to log in. Then, using the drop down menu to choose the course title. At the beginning of class, the teacher must require three to five students to form a learning team and login to the system for cooperative study. After the student completes the study team, the teacher assigns the study topic to each study group. After a period of teaching, students need to submit a study experience report. In class, the teacher must specify the day's report team using the 'assign report team' functional menu, then the report team members will be able to come to the stage to make report, while other students will be able to conduct an online peer review for the reporting student based on the evaluation criteria which the teacher previously defined in the system. After the report team completes the report, each student in the report group have to conduct a self-assessment. In addition, the teacher can regroup the students upon request by using the function menu of 'team adjustment', and the function menu of 'team member inquiry', 'review inquiry', 'score inquiry', and 'score summary' are used for checking the member list of each team, the mutual reviews, scores of peer review and exercise work, and the grades of all students for each learning item, respectively.

For the students, initially, they need to be grouped and adjust the seating in classroom in accordance with the teacher's order. Next, they have to be connected with the internet by using PC or mobile device to choose the course participated in. After this, students have to log in the system using their account and passwords. The function of 'join learning team' must be used for the students who use the system for the first time to join any particular team. In classroom, the learning group is divided into two sub groups- the one to report and the one not doing so. The function of 'inter review' is then used by the group not delivering report on stage to grade the students delivering their presentation. A subjective review or comment in at least 20 words will also be given afterwards. On the other hand, in the wake of the presentation and the teacher's comment on the report, the team members of the report group use the 'intra review' function to assess oneself as well as his or her teammates, based on the five-grade scoring criterion.

As for the part of self-review, in addition to a quantitative score, a learning feedback and a subjective comment on the teammates, both written in more than 20 words, should be made. Besides, after the exercise questions are revealed and explained, the report team members will assess the work done by the students not reporting by using the function of 'exercise review'. Moreover, through the connection with the system online, students are able to enquire about the grades of peer to peer review, self-review, and the exercise work, as well as the up-to-date summary of their scores. The qualitative comment of oneself is also acquirable.

4.3 Design of the database

4.3.1 Data demand analysis

In this system, a professor can instruct several courses in a semester, and the data of the professor is stated in the table of 'professors,' specifying the information such as 'professor_id' (in the form of character with a maximum length of 20 letters), 'name' (character form, less than 12 letters), the 'department' (character form, less than 24 letters), and 'password' of a maximum length of 15 letters.

The course data is stored in the table of 'courses'. Each course must be taught by at least one professor. The 'courses' table should state 'class_course_name' (in characters, not exceeding 45 letters), 'course_id' (in characters with a maximum length of 20 letters), and 'professor_id' (in characters, not exceeding 20 letters). The 'academic_year' is recorded in the format of integer with three digits. The Arabic numeral 1 refers to the first semester and 2 stands for the second semester.

The data of the students is stored in the table of 'students'. Every student must take part in at least one course, and one can keep several grades of intra review, inter review, and the assigned exercise. Specifying the information such as 'student_id' (in characters with a maximum length of 12 letters), 'class_course_name' (in characters, less than 45 letters), 'student_name' (in characters, not exceeding 15 letters), 'password' (in characters, not exceeding 15 letters), and 'group_id' is stated in integer format, with two digits at most.

The score of inter review is stored in the table of 'inter_review_score'. Each score belongs to a particular student, too. Information including 'class_course_name' (in characters with a maximum length of 45 letters), 'intra_review_id' (in integer form with a maximum length of two digits), 'owner_id' (in characters with a maximum length of 12 letters), 'giver_id' (in characters with 12 letters at most), score 'item_1' to 'item_6' (in integers, not exceeding two digits), 'comments' (in characters not exceeding 1,000 words), and 'review_date' (in date format).

The score of student's exercise work is stored in the table of 'exercise_score', and each grade belongs to one student. The data recorded includes 'class_course_name' (in characters with a maximum length of 45 letters), 'exercise_id' (in the form of integer, not exceeding two digits), 'owner_id' (in characters with a maximum length of 12 letters), and 'giver_id' (in characters, with a maximum length of 12 letters). Score 'item 1' and 'item2' are both recorded in integer, with a maximum length of three digits. Finally, 'review_date' is stated in date format.

The score of student's intra review is stored in the table of 'intra_review_score'. Each item must belong to a particular student only. Information including 'class_course_name' (in characters with a maximum length of 45 letters), 'intra_review_id' (in the form of integer with two digits), 'owner_id' (in characters, not exceeding 12 letters), 'giver_id' (in characters with 12 characters at most), score 'item_1' to 'item_6' (in integers, not exceeding two digits), and the learning feedback are recorded in 'comments' field(in characters with a maximum length of 1,000 words), 'review_date' stated with date format.

4.3.2 Database ERD design

Based on the demand analysis of the data, MySQL Workbench is used for the system database ERD design, which is shown below in Fig. 4.

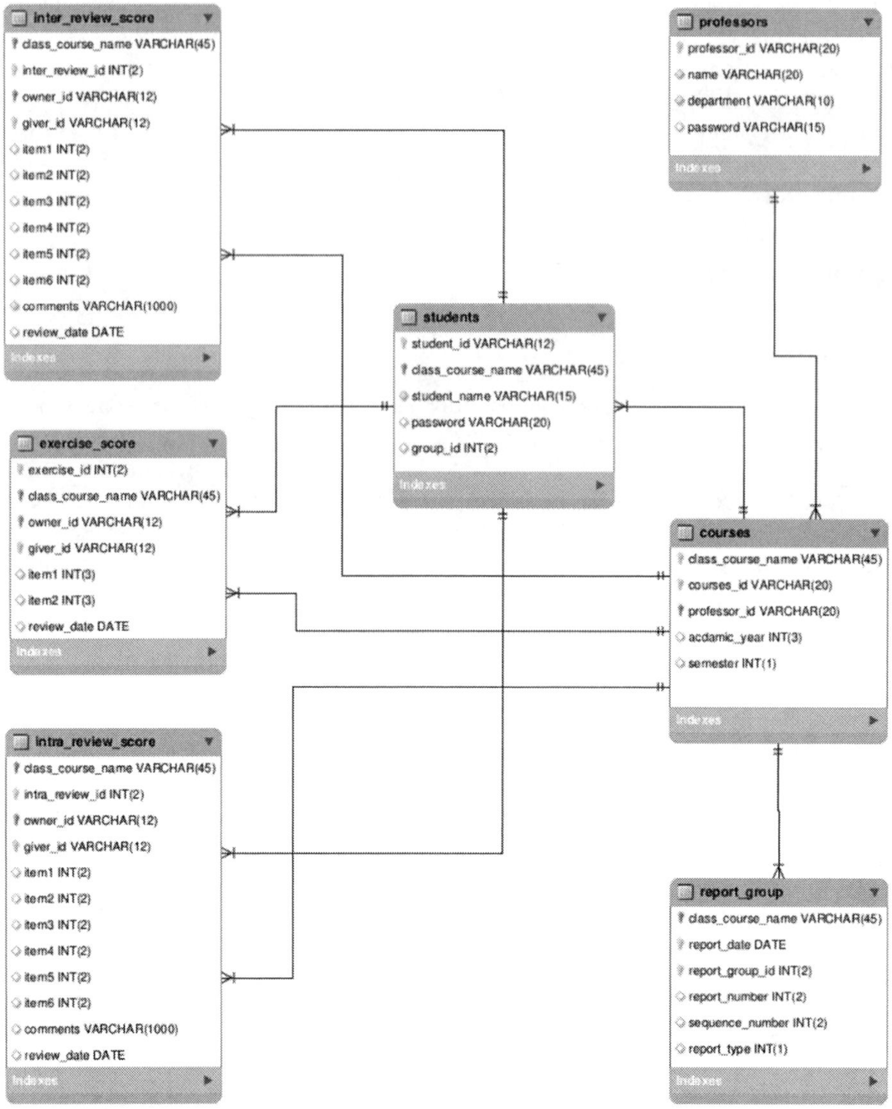

Figure 4. ERD of the database on this system

5 Online test

Following the completion of the prototype system, the researcher, according to the default work flow and the rules of the system, conducted an online platform testing program in a post graduate class. Due to the limited space of the paper, this section illustrates briefly the program modules relations, self and mutual reviews, and the inquiry about scores. Students' feedback is also shown as follows.

5.1 Program module relations and function menu

The relations between the program modules operated by the teacher and its linking program modules are illustrated in Fig. 5. Fig. 6 describes the one used by the students. In the centre of the figure is the name of the program module. The ones on the left are the modules linked to the main module, and the right ones are the modules that can be linked to from the operation item.

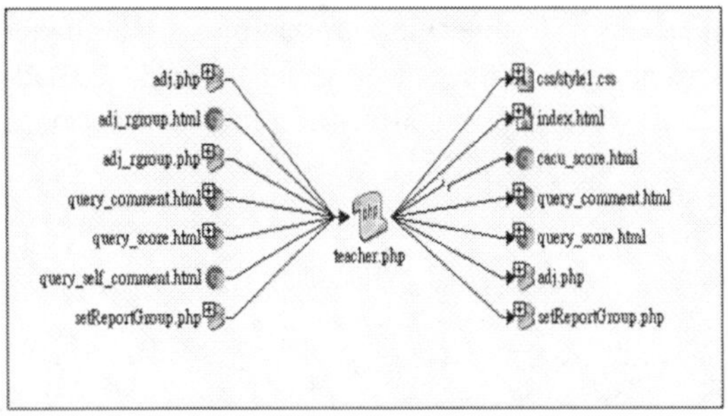

Figure 5. Teacher's related modules implementation

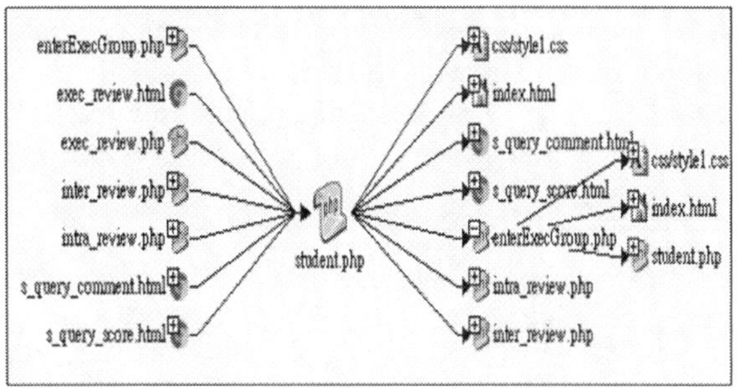

Figure 6. Student's related modules implementation

Fig. 7 is a screenshot on the operation menu for the student after successful logging in, and the user's name is shown on the screen to create a friendly environment. On the platform, the function menu is classified into several items, such as 'Join Learning Team', 'Inter Review', 'Intra Review', 'Exercise Review', 'Reviews Inquiry' and 'Scores Inquiry'. Fig. 8 is a screenshot on the operation menu for the teacher. Similarly, the teacher's name is shown above the logo of the platform after logging in. The function menu for the teacher is categorized into 'Assign Report Team', 'Team Adjusting', inquiry about 'team members', 'student peer reviews', 'individual student scores' as well as 'Score Summary' of all students.

Figure 7. Screenshot example on student's operation menu

Figure 8. Screenshot example on teacher's operation menu

5.2 Self and mutual reviews of the students

After the teacher has assigned the current report team, the students are able to conduct mutual and self-reviews on the reports on the platform online. Fig. 9 shows the operation interface for peer to peer mutual review. At the beginning, the reviewer clicks the name of the subject, and the information of the subject, the reviewer's name, and the date will be subsequently shown on the screen. By using the radio button, the reviewer grades the subjects on a five-level scale- excellent, good, fair, needing improvement, and poor(the default grade is fair), based on the content of the report, the application and the example(s) used, the imposing manner and style on stage, the design of the presentation, interaction with audience, and collaboration with team members. Finally, a comprehensive comment at least 20 words is made. Similar to Fig. 9, Fig. 10 is the operation interface for self-review. The major difference is that on this interface learning feedback of at least 20 words are required to be keyed in.

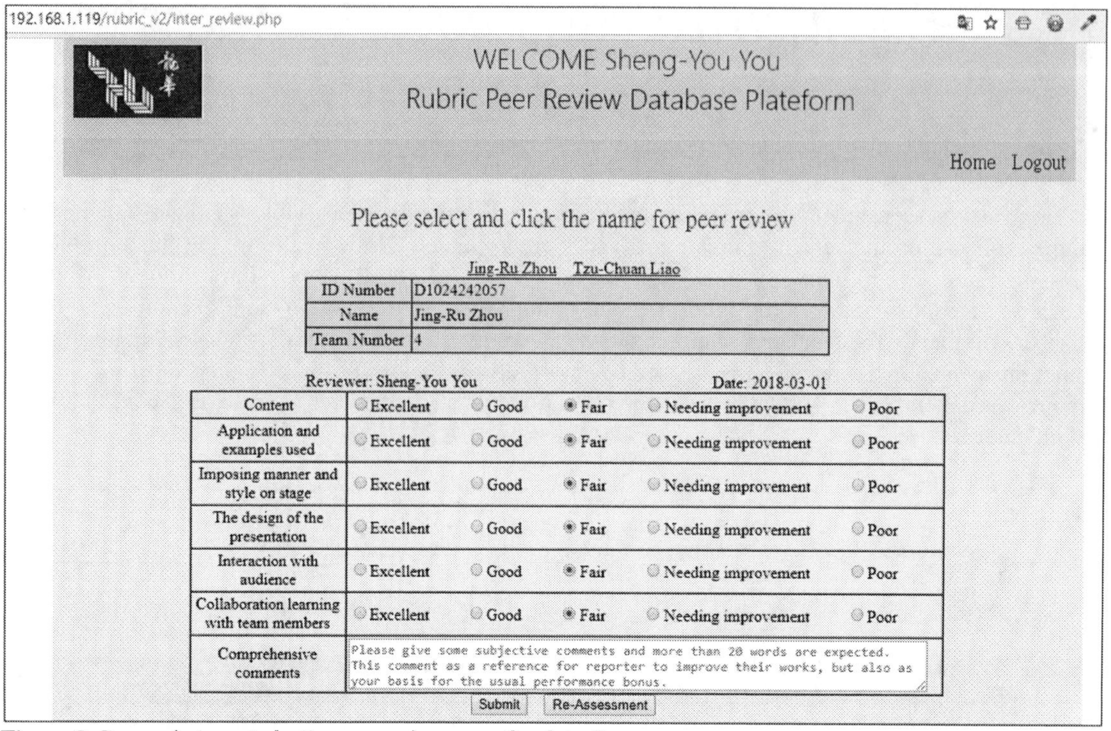

Figure 9. Screenshot on student's peer-review operation interface

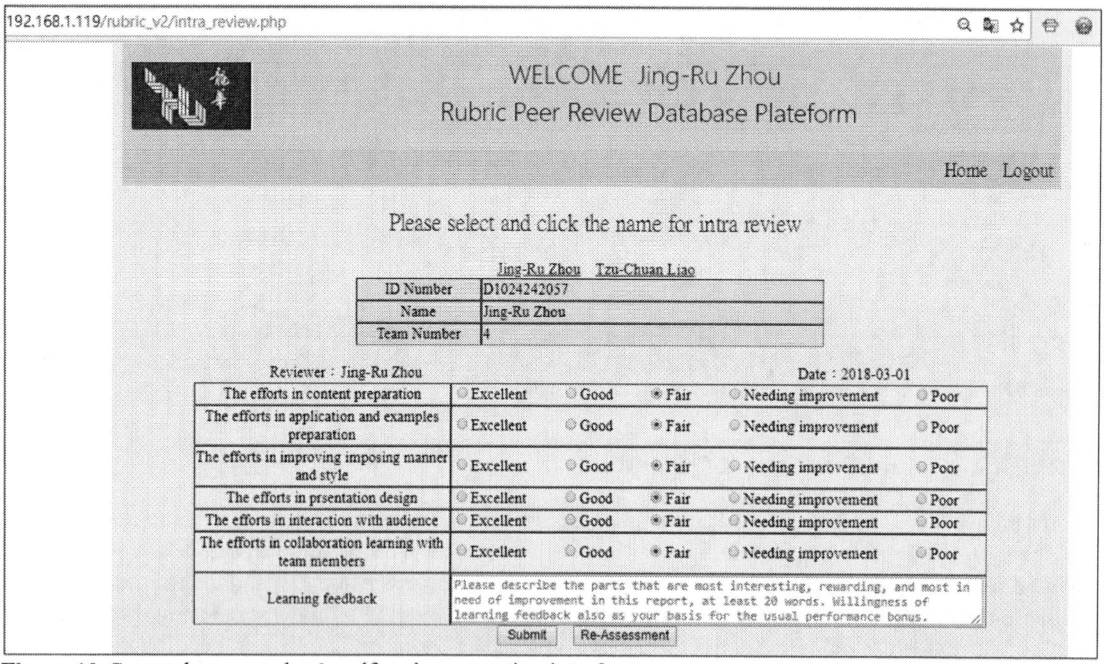

Figure 10. Screenshot on student's self-review operation interface

5.3 Inquiry examples

The inquiry function of this platform is divided into two groups- the one for teacher and the other one for student. The teacher can inquire individual student's information in different learning team, his or her score received and its scorer, the reviews and the reviewer, and the total scores obtained by the whole students. The students, on the other hand, can find out the reviews and the average score given, and then drill down the original score of each item. Fig.11 is screenshot example to show the results of teacher's inquiring the grades of all students, and Fig.12 is screenshot example of student's inquiring peer reviews.

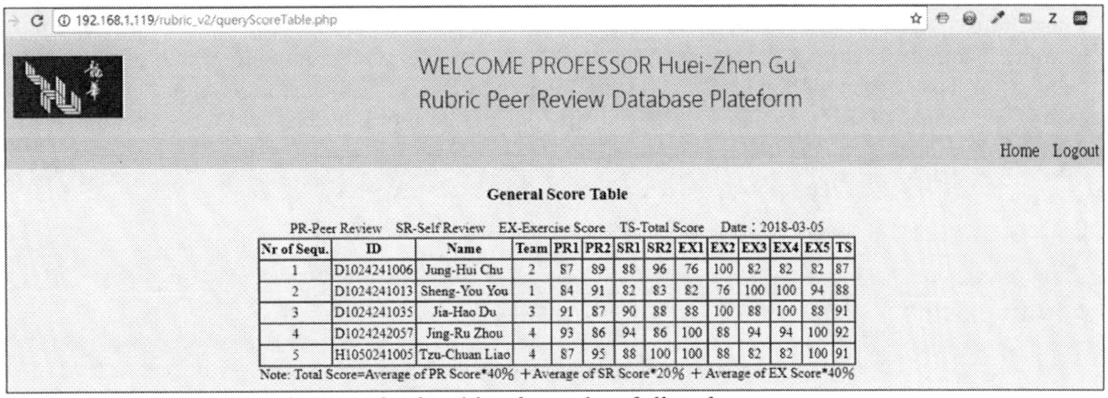

General Score Table

PR-Peer Review SR-Self Review EX-Exercise Score TS-Total Score Date : 2018-03-05

Nr of Sequ.	ID	Name	Team	PR1	PR2	SR1	SR2	EX1	EX2	EX3	EX4	EX5	TS
1	D1024241006	Jung-Hui Chu	2	87	89	88	96	76	100	82	82	82	87
2	D1024241013	Sheng-You You	1	84	91	82	83	82	76	100	100	94	88
3	D1024241035	Jia-Hao Du	3	91	87	90	88	88	100	88	100	88	91
4	D1024242057	Jing-Ru Zhou	4	93	86	94	86	100	88	94	94	100	92
5	H1050241005	Tzu-Chuan Liao	4	87	95	88	100	100	88	82	82	100	91

Note: Total Score=Average of PR Score*40% +Average of SR Score*20% + Average of EX Score*40%

Figure 11. Screenshot example on teacher inquiring the grades of all students

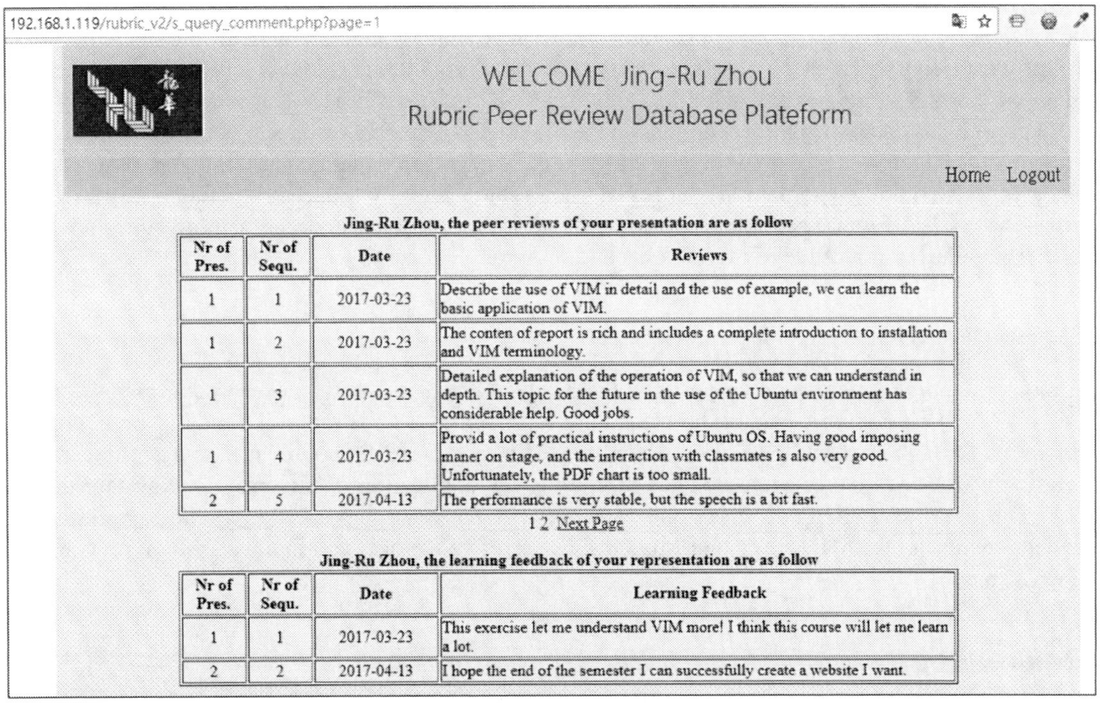

Jing-Ru Zhou, the peer reviews of your presentation are as follow

Nr of Pres.	Nr of Sequ.	Date	Reviews
1	1	2017-03-23	Describe the use of VIM in detail and the use of example, we can learn the basic application of VIM.
1	2	2017-03-23	The conten of report is rich and includes a complete introduction to installation and VIM terminology.
1	3	2017-03-23	Detailed explanation of the operation of VIM, so that we can understand in depth. This topic for the future in the use of the Ubuntu environment has considerable help. Good jobs.
1	4	2017-03-23	Provid a lot of practical instructions of Ubuntu OS. Having good imposing maner on stage, and the interaction with classmates is also very good. Unfortunately, the PDF chart is too small.
2	5	2017-04-13	The performance is very stable, but the speech is a bit fast.

1 2 Next Page

Jing-Ru Zhou, the learning feedback of your representation are as follow

Nr of Pres.	Nr of Sequ.	Date	Learning Feedback
1	1	2017-03-23	This exercise let me understand VIM more! I think this course will let me learn a lot.
2	2	2017-04-13	I hope the end of the semester I can successfully create a website I want.

Figure 12. Screenshot example on student inquiring peer reviews

5.4 Feedback opinions

This section makes excerpts from the opinions of three students who ever used the online platform, hoping their ideas serve as references for examining the research purpose and further improvement of the system.

Student 1: After delivery of my report each time, I could immediately understand my weakness through the reviews given by my classmates. On the other hand, I could write some questions for my classmates to answer. Through this process, I could have a better understanding of my project and enhance my learning effects as well.

Student 2: Using this platform, I could know my scores and the reviews given by other students right away. This helps me do self-review as well. If the number of the participants is small, the effect of this system may weaken. It is easy for us to know who scored us. To benefit each other, we might give a good mark, and the score received/given might be unreal.

Student 3: This is a very good evaluation system. We can not only mark and make comments anonymously on the learning outcomes of the students of other teams but also conduct self and intra-learning team peer reviews. As for giving reviews, it requires careful listening to each report, or it must be hard for us to write anything about it. From this point of view, I suggest the system be widely used by undergraduate students, which will help them concentrate more in class and let them perceive the views or opinions of others on their works.

6 Conclusions and future works

6.1 Conclusion

Based on this research, it is suggested that learning methods are facilitated through the real time function of intra and inter peer reviews on this platform. By using PC or mobile device to be connected online, they can conduct quantitative intra and inter peer reviews, make qualitative comments and learning feedbacks, inquire about personal score of the report, check the quantitative grade of the exercise, drill down the original score of each report, and acquire the qualitative advice on learning made by others. All of these serve as feedback to self-learning status and provide the students with valuable references to adjust learning strategies and activities, and eventually improve current learning situation.

Through this database platform online, on the other hand, teachers can easily conduct multiple rubric assessments, regardless of intra or inter reviews of the students in either general or computer classroom. They can also quickly examine the scores of each report done by a particular student, drill down in discovery of the original score, check the homework and its score and count the result, and finally understand the learning state of the students and make adjustment to the teaching plan.

6.2 Future works

Although the prototype system proposed in this study can achieve online peer to peer review and self-assessment in classroom teaching, however, the function of the system still needs to be continually improved through constant feedback from users, so as to achieve a smooth state more. Therefore, the follow-up work of this research includes promoting the application of this prototype system through different levels of teaching seminars and working with teachers who use the system to continuously collect feedback from teachers and students as a reference for continuous system revision and improvement.

References

1. CHEA, Washington, D.C.: CHEA. (2003)
2. M.X. Xu, Dist. Edu. J., **179**(2), 60-64 (2007) (In Chinese)
3. S.L. Huang and J.J. Chi, Eval. Bimo., **28** (2010), Access from http://epaper.heeact.edu.tw/archive/2010/10/27/3683.aspx (In Chinese)
4. J.L. Wang, Eval. Bimo., **28** (2010), Access from http://epaper.heeact.edu.tw/archive/2010/11/01/3698.aspx (In Chinese)
5. J. Dong, Fund. Engl. Edu., 14-15 (2015)

6. Popham and James, Edu. Lead., **55**(2), 72-75 (1997)
7. Dawson, Phillip, Asse. & Eval. H. Edu. doi:10.1080/02602938.2015.1111294 (2015)

Evaluating key factors affecting knowledge exchange in social media community

Li-Wen Chuang[1] and Shu-Ping Chiu[2,a]

[1]*Department of Film and TV Technology, Fuzhou University of International Studies and Trade, China*
[2]*Department of Fine Art and Design, Fuzhou University of International Studies and Trade, China*

Abstract. Today, social media opens up multiple options to add a new channel to learn and obtain knowledge. In particular, social media allows users to learn formal and informal social settings. Users can find like-minded people or community and organize knowledge exchange for educational or other purposes. This paper takes the theory of social network and social capital as the core factors to explore social media community members how to use social media platform to exchange knowledge. The paper uses social network and social capital as predictors of knowledge exchange so that we can use social media as a way to advance knowledge exchange ceaselessly. Therefore, there were 263 members of well-known social media community of knowledge exchange filling out the questionnaires completely, and then evaluated with structural equation modeling, and confirmatory factor analysis was also applied, using SmartPLS 2.0, to test if the empirical data conform to the proposed model. However, results also imply that social capital and social network distinctly play the important part of affecting knowledge exchange and communication about social issues on social media community goes hand in hand with knowledge exchange. Finally, this article proposes implications for theory and practice on the current and future research.

1 Introduction

Knowledge exchange and knowledge flows all have become an important issue in affecting the development of the social media communities. Because of the mutual communication and collaboration in activities ranging from the economic and marketing to social and educational field, an invisible relationship and other factor have been formed among members about knowledge exchange of social media community. Therefore, the good characteristics of the social media communities can be of some help in understanding where knowledge flows, how knowledge and information is exchanged among members. It can promote the communication and collaboration among members and the formation and development of social media communities. As a result, knowledge exchange is very important issue in the fields of design, information and knowledge management, and considered as a significant research topic by current academic and business circles.

This article is structured as follows: First, the paper discusses the theoretical foundations to review relevant literature and research model related to this study to present the proposed hypotheses along with the research model. Next, the third section, we describe the research methodology and this survey

[a] Corresponding author : sphill200406@163.com

data was used to test these hypotheses in this study. Following from the methodology, the empirical results of this article are reported in section 4.

2 Theoretical foundations and research model

In section 2, Figure 1 describes the proposed research model. The dependent variable, knowledge exchange, refers to conceptualize a co-production of new knowledge (e.g. information, skills, or ability) through the social interactions of academics and non-academics, individuals and communities(e.g. members, friends, families, groups, organizations) , which is of gain or profit to both parties and is explicit from the one way spread of formal and informal research findings[1]. This study, following Wasko and Faraj (2005) [2], contends that social network and social capital are all the critical determinants of knowledge exchange in social media community.

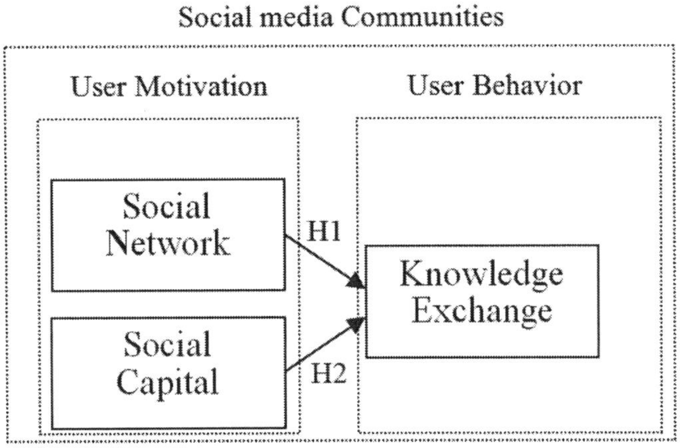

Figure 1. Research Model.

2.1 Determinants of knowledge exchange

2.1.1 Social network and knowledge exchange

Various knowledge management sources (such as members, friends, families, communities, organizations, customers, common interest groups and government agencies) are valuable for users to control knowledge and information in effect. Individuals can easily obtain unique knowledge through interactions and socials with members within or outside of their social media communities. Even more influential than knowledge/information knows where this knowledge/information is located and how it can be applied. Therefore, to exchange, acquire and share necessary knowledge and information, persistent effort must be made to interact formally and informally and to connect with members and join social networks associated with proper knowledge databases to promote the performance of social media community [3]. And then, some studies have shown that social and human networks are important channels for exchanging, acquiring, transferring and sharing knowledge [2, 4, 5, 6, 7]. Accordingly, following hypothesis is proposed in this study:

H1 、Social network is positively related to knowledge exchange.

2.1.2 Social capital and knowledge exchange

Following Chiu et al.(2006)[9], social capital is the network of relationships had by an individual or a social network and the set of resources embedded within it, strongly affect the extent to which interpersonal knowledge exchange and knowledge sharing occurs. Thus, another study also argues that individuals' behavior is a product of their social network. Through close social capital, members are able to increase the depth, breadth, and efficiency of mutual knowledge exchange [9]. Therefore, Wasko and Faraj (2005) also indicated that the significant levels of social capital and knowledge exchange will develop and contribute knowledge exchange to others in electronic networks of practice [2]. Therefore, the next hypothesis predicts the following:

H2 、Social capital is positively related to knowledge exchange.

3 research methodology

3.1 Measurement development

The measurement items of our research in the instrument were adapted from the literature. For all the measures, a five-point Likert scale was used with anchors ranging from strongly disagree (1) to strongly agree (5). Items for knowledge exchange were adapted from Yang et al. (2009) [10]. Social identity was measured with items adapted from Chiu et al. (2006) [9]. Social network was measured with items adapted from Bruque et al. (2008) [11].

3.2 Data analysis and results

The proposed model was tested using data collected from members of famous social media community, one of the largest and widely social media communities of knowledge exchange and to have observed a solid rise in its adoption and use within the last 2 to 6 years in the real world. By the time this survey was concluded, 263 complete and usable questionnaires had been collected for data analysis.

We analyzed the proposed model in this research, based on the standpoints of Ahuja and Thatcher (2005) and Hair et al. (2010) [12, 13]. This study used Structural Equation Modeling (SEM) with Partial Least Squares (PLS) skill to analyze our proposed hypotheses, applied the following procedures. First, by using SmartPLS, the research carried out reliability and validity tests were performed to validate the measurement model. Next, we tested the structural relationships among latent constructs. Data analyses in detail for both the measurement and structural models were performed by using PLS. The remaining of the section describes the investigation procedures.

3.3 Reliability and validity tests

The research model using SmartPLS version 2.0 M3 from Ringle et al. (2005) [14], this research individually assessed the proposed model for the full sample and each factor to assure the reliability and validity.

Table 1. Descriptive statistics

Constuct	Item	Factor loading	Cronbach's α	Composite reliability	AVE
Social Capital	SC1	0.91	0.84	0.91	0.76
	SC2	0.90			
	SC3	0.81			
Social Network	SN 1	0.96	0.95	0.97	0.90
	SN 2	0.96			
	SN 3	0.93			
Knowledge Exchange	KE1	0.93	0.94	0.96	0.89
	KE2	0.95			
	KE3	0.95			

As for reliability, most scholars use 0.7 as the recommended threshold for reliable dimensions [15, 16]. As shown in Table 1, all composite reliabilities and Cronbach's α of the constructs had a value greater than 0.7, expressing the acceptable reliability of the constructs.

As for validity, following Fornell and Larcker (1981) and Anderson and Gerbing (1988) [16, 17], convergent validity is verified when all factor loadings are significant and exceed 0.6, construct reliabilities exceed 0.8, and average variance extracted (AVE) by each construct should exceed the variance due to measurement error for that construct (i.e. AVE should exceed 0.50). For the full sample, all factor loadings were above the 0.7 threshold (see Table 1), the composite reliabilities of the constructs ranged between 0.97 and 0.91 and AVE ranged from 0.76 to 0.90 (see Table 1). Moreover, discriminant validity is concerned, if the square root of AVE of each construct is larger than its correlations with the other construct, discriminant validity is proved [16]. Table 2 shows all the diagonal values exceed the interconstruct correlations coefficient for the full sample. Therefore, this sample indicates an adequate validity value.

Table 2. Discriminant validity

Construct	KE	SN	SC
KE	**0.94**		
SN	0.74	**0.94**	
SC	0.76	0.82	**0.87**

notes: 1.Diagonal:Square Root of AVEs reported along diagonal in bold.
Off-diagonals: Correlation between latent variables .
2. KE: Knowledge Exchange;
SN: Social Network;
SC: Social Capital.

3.4 Hypothesis testing

To examine the statistical significance of the parameter estimates, the research implemented bootstrapping procedure with replacement using 500 subsamples [8, 15]. A summary of these results is presented in Fig 2. As shown in Fig 2, the variables explain considerable proportions of the variance— 62% for knowledge exchange. As expected, Social network had the significant effect on knowledge exchange (β = 0.343, t = 4.129), thus supporting hypothesis H1. Social capital was positively associated with knowledge exchange (β = 0.481; t = 6.179), supporting hypotheses H2.

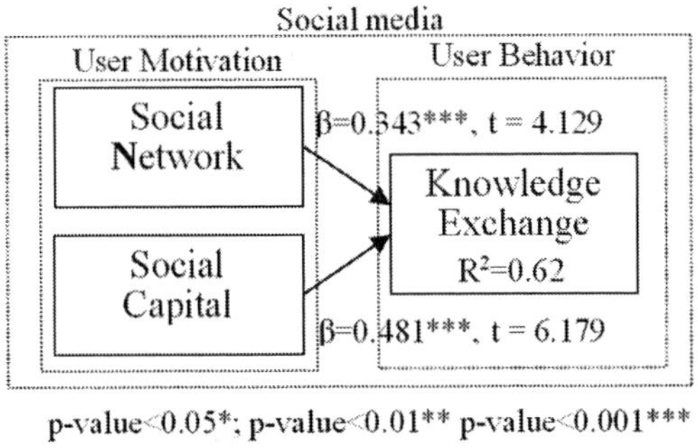

Figure 2. Model Analysis Results

4 Discussion and conclusion

The research indicates a proposed model including distinct kinds of knowledge exchange motivations (social network and social capital). This research model demonstrates that for an individual who is dominated by the source of motivation, users can be motivated by the proposed motivation. We, however, do not suppose that users can not be motivated by other motivations. According to Chiu et al. (2006) indicated that individual can be characterized by motivational profiles which reflect the relative strength of each of the sources and every individual is dominant by a source of motivation [9]. This implies that people can be motivated by a second source of motivation if the dominated one is absence. This research model supports results of hypotheses H1 and H2 proposed in the research. Investigation of the results shows social network and social capital all have positive and significant influence on knowledge exchange. The unique finding agrees with scholars' viewpoints. Furthermore, it is illustrated that social network and social capital are the key motivation for knowledge exchange of social media community. After the motivation is satisfied by communication and interaction on social network and social capital, user's later knowledge exchange behavior will happen with more efficiency.

To the end, two practice implications can also be applied. Firstly, why it is important to understand the main motivation of knowledge exchange? Because knowledge exchange is not a human natural instinct of exchange/sharing. Knowledge exchange will make members loss value and benefits brought by unique knowledge in community. Actually, in social media community, all members take pleasure from knowledge exchange. The root reason deserves further study. Utilizing social network and social capital, this study finds relevant users may actively choose a social media community due to their demands on communication and interaction. After users' demands are met, their subsequent use behaviors will be influenced. Such finding conforms to scholars' viewpoints. Moreover, it is demonstrated that social network and social capital are the main motivation for knowledge exchange of social media community. After the motivation is satisfied by communication and interaction on social network and social capital, user's subsequent knowledge exchange behavior will happen with equal efficiency. Secondly, when knowledge published by a member is trusted by other members or members all have positive interactions in knowledge exchange and communication in social media community, members will be more willing to exchange knowledge, and quality of knowledge and desire of continuous knowledge exchange behavior will be enhanced.

References

1. C. Fremantle, M.N. Gulari, S.M. Fairburn, L.A. Hepburn, G. Valentine and L. Meagher, *Proceedings of the 20th DMI: Academic Design Management Conference (DMI:ADMC)* (2016)
2. M.M. Wasko and S. Faraj, MIS Quarterly, **29**(1), 35-57 (2005)
3. H.S. Lee, S.H. Lee and J.T. Han, International Journal of Digital Content Technology and its Applications, **7**(12), 307-312 (2013)
4. G. Lei and G. Xin, Journal of System and Management Sciences, **1**(3), 79-89 (2011)
5. J.A. Baum, T. Calabrese and B.S. Silverman, Strategic Management Journal, **21**, 267-294 (2000)
6. J.H. Dyer and K. Nobeoka, Strategic Management Journal, **21**, 345-367 (2000)
7. M.L. Wasko, S. Faraj and R. Teigland, Journal of the Association for Information Systems, **5** (11):2, 493-513 (2004)
8. A. Majchrzak, C.M. Beath, R.A. Lim and W.W. Chin, MIS Quarterly, **29**(4), 653-672 (2005)
9. C.M. Chiu, M.H. Hsu and E.T. Wang, Decision Support Systems, **42**(3), 1872-1888 (2006)
10. H.L. Yang and C.Y. Lai, Behaviour & Information Technology, **30**(1), 131-142 (2009)
11. S. Bruque, J. Moyano and J. Eisenberg, Journal of Management Information Systems, **25**(3), 177-206 (2008)
12. M.K. Ahuja and J.B. Thatcher, MIS Quarterly, **29**(3), 427-459 (2005)
13. J.F. Hair, W.C. Black, B.J. Babin, R.E. Anderson and R.L. Tatham, *Multivariate Data Analysis* (NJ, America, 2010)
14. C.M. Ringle, S. Wende and A. Will, SmartPLS 2.0.M3 (2005).
15. W.W. Chin, *Modern methods for business research - The partial least squares approach for structural equation modeling* (NJ, America ,1998)
16. C. Fornell and D.F. Larcker, Journal of Marketing Research, **18**, 39-50 (1981)
17. J.C. Anderson and D.W. Gerbing, Psychological Bulletin, **103**, 411 (1988)

Opportunity recognition for small businesses in medical tourism in Taiwan

Ying-Hsun Hung, Jerome Chih-Lung Chou, Jung Ma, and Ching-pei Lin[a]

Department of Information Management, Hwa-Hsia Institute of Technology, Taiwan, ROC.

Abstract. Medical tourism has become an investment focus for many Asian countries. For example, India, Singapore, Thailand, South Korea, and Malaysia have governmental projects to foster this emerging industry, and successfully attract millions of medical tourists each year. Medical tourism consists of diverse industries that can provide abundant opportunities of new businesses. Currently Taiwan is trying to catch up in this trend by making national development policies, setting up institutions, modifying laws and regulations, and encouraging investment of private sector. The purpose of this study is to analyze the status quo of international medical tourism in Taiwan, and identify several opportunities for small businesses to catch and to participate in forming a solid medical tourism industry for Taiwan.

1 Introduction

Medical tourism has materialized from a broader concept of health tourism as one's organized travel outside their local environment for the maintenance, enhancement, or restoration of one's well-being in mind and body. Particularly, in medical tourism the travelers seek medical intervention such as cosmetic enhancements, dental treatments, or major or minor required surgery. Medical tourism is partly resulted from globalization of healthcare and tourism itself constituting huge economic potential (Bookman & Bookman, 2007).

Medical tourism is also a popular cultural phenomenon when people make a long journey in order to obtain medical services while vacationing (Connell, 2006). Medical travelers are looking for specialty treatments, privacy, expedited medical attention, or, frequently, lower costs than available at home country (Woodman, 2008). The market drivers of medical tourism are cost savings, comparable or better quality, and quicker access to the care (Keckley, 2008). Medical or health tourism has become one of the fastest growing sectors that many countries are embracing for planning their economic expansion because it drives the development of a high-value and technology oriented service industry. The growth of medical tourism has been reportedly high for each year recently.

2 Medical tourism in Taiwan

Taiwan has full spectrums of medical services, including preventive medicine, health screening, diagnosis, and treatment of disease. All services are accompanied with advanced equipment and customized health examination, or more diversified herbal medicine and diagnosis. The five

[a] Corresponding author : jerome@cc.hwh.edu.tw

competitive items of medical tourism of Taiwan are craniofacial reconstructive surgery, living donor liver transportation, artificial reproductive technology, cardiovascular therapy, and joint replacement surgery (Wu, 2009). Price competitive items in the international markets of Taiwan are health examination, surgery of prostate, hysterectomy and face lifts.

Taiwanese physicians must finish 7 years of medicine school education: 4 years of basic medicine courses, 2 years of specialized medicine courses, and 1 year of internship training. Besides, they must be certified by a national examination. A qualified doctor-in-charge must go through extra years of specialist training, do research and publish proper papers. Averagely, it takes 12-13 years for a student of school of medicine to become a doctor-in-charge. The high quality of medicine school education and practical training enable Taiwan to provide high quality of medical services.

Currently, there exist several referral organizations helping recruit medical travelers to Taiwan, such as Zion Health Control and Youth Rejuvenation Center, Formosa Medical Travel, and World Taiwanese Chambers of Commerce.

However, the development of medical tourism of Taiwan is facing some problems that limit its potential. Firstly, the distribution of nationality of medical travelers to Taiwan is largely skewed to mainland China. Due to recent improvement of political relationship with China, the number of Chinese tourist to Taiwan has been increasing, and some of them went to Taiwan with medical visas in tours. These tours are recruited by travel agents of any kind in China, and the medical visa is issued easily with the help of 39 selected Taiwanese hospitals. Because the demand of tourism to Taiwan is high in China and there exists quota for Chinese travelers to Taiwan, medical tours became one of the special channels for Chinese travelers whose need are not really medical.

Secondly, the distribution of medical services consumed is skewed to low priced items. Since many of travelers with medical visas go to Taiwan on purposes other than obtaining medical treatments, they tend to spend less in hospital. Recently, Taiwanese government posed a new regulation on the minimum requirement on medical expenses of travelers who apply medical visa through hospitals. The new regulation mandates hospitals that accommodate medical travelers must present receipt exceeding 10,000 NTD (about 340 USD) in advance when applying medical visa for the patient.

Thirdly, famous and talented Taiwanese physicians are attracted to mainland China to provide high-end medical services. Chinese hospitals offer advanced equipment, convenient and condensed schedule, and high payments to Taiwanese physicians in fields like anti-aging, dental implant and cosmetic surgery, which are not covered by Taiwan National Health Insurance. Such arrangement is good for physician's private benefit, but does not help the development of medical tourism industry in Taiwan.

3 Issues of incentive and information

The problems above root in incentive and information gaps among the immature industry. Because of the broad coverage of National Health Insurance system of Taiwan, part of the role of large hospitals in Taiwan is social welfare provider. Large hospitals receive large government subsidy and serve large amount of citizen patients. This leaves little capacity and poor incentive for large hospitals to serve international patients as long as the income from National Health Insurance is big. However, famous physicians and advanced equipment resides in large hospitals. Recently, due to financial shortage of National Health Insurance, some large hospitals has stared rearrange organization and capacity for more international patients.

Traditional travel agents have incentive problems, too. Due to the difficulty to serve diverse medical needs of travelers, serving clients of medical tourism are not easy to provide high profit margin to travel agency. Therefore travel agents lack of incentive to promote medical tourism through their cooperative overseas partners, which is the most effective channel to disperse tourism information overseas, and consequently, the information about the benefit and advantage of Taiwan medical services are poorly communicated over this traditional channel.

Local hotels and local health promotion service providers such as hot spring resort, oxygen bar, functional food store and medical equipment venders are well motivated to welcome international medical travelers, but they are not proper to take the leading role to coordinate the joint effort of international marketing and medical service providing.

4 Role of medical travel facilitator

To solve the incentive and information problems, fostering the development of medical travel facilitator is one solution. Growth of medical tourism has been facilitated by the emergence of healthcare intermediaries or medical tourism facilitators between international patients and hospital networks (Connell, 2006). Medical travelers could seek help from medical travel facilitators in finding reliable healthcare providers, ensuring trouble-free travel arrangements and having a comprehensive understanding of any pre and post operation cautions. Facilitator could close the incentive and information gaps by integrating knowledge of medical services, health services, tourism facilitation and concierge services, which are currently poorly performed by hospitals, traditional travel agents, and local hotels.

Services provided by medical travel facilitators can be stand alone or in a form of partnership integration. Partnering with industry key players can provide better start-to-finish experience to travelers and thus create more value for the industry. These services can include air travel and visa arrangement, long-distance medical screening and consultation, communication between physicians of parent and host countries, transportation of medical records, arrangement of personal nurse, arrangement of accommodation, and arrangement of follow-up care.

5 Opportunities for small business

Applying TRIZ theory, we identify several opportunities that are suitable for small and medium enterprises (SME) to catch.

To close the incentive gap of large hospitals in serving international patients, SMEs with medical technology background could act as overseas liaison for large hospitals if they have overseas subsidiaries or strong online presence. As a liaison, the SME could access the service capacity information of hospital and only use the hospital allocated medical service items and capacity to help foreign medical travelers plan their journeys to Taiwan. With the endorsement of large hospitals from the parent country, the SME could improve its brand image and enhance profitability. With liaisons, large hospitals could close the gap of information about high-end and advantageous medical services in the global market. The candidate SMEs for hospital liaisons are medicine, functional food and medical equipment providers, many of which could be carefully selected from the hospital's current suppliers.

To close the incentive gap of traditional travel agents, SMEs with staff of local tourist guides could act as outside suppliers of medical concierge services of travel agents. Traditional travel agents could advertise their allying hospital networks with medical concierge services to their clients of free and independent travel (FIT). And as the service is included in the FIT package, the SMEs could expect more quantity of genuine medical travelers and provide them customized local services with high value, such as translation and personal nursing care, in coordination with local hospitals, hotels and health promotion service providers.

6 Methodology

Medical tourism consists of diverse industries that can provide abundant opportunities of new businesses. However, few studies in tourism and medical cares have discussed the medical tourism services, or proposed the services quality evaluation strategies to assist in improving performance for long-term development. The purpose of this study is to address this problem using the novel method of hybrid MCDM (multiple criteria decision-making), including DEMATEL (decision-making trial and evaluation laboratory), DANP (the DEMATEL-based analytic network process) and VIKOR, to examine the influential relationships among dimensions and criteria of the empirical case and to ultimately present the best improvement schemes, which are valuable for both practitioners and researchers.

The decision-making process involves identifying problems, constructing preferences, evaluating alternatives, and determining the best alternative. However, when decision-makers evaluate the alternatives with multiple criteria, many problems, such as the weights of the criteria, preference dependence, and conflicts among criteria, seem to complicate the decision making process and should be resolved by more sophisticated methods. The DEMATEL (Decision MAking Trial and Evaluation Laboratory) method gathers collective knowledge to capture the causal relationships between strategic criteria. This paper applies the DEMATEL method in the strategic planning of medical services Gaps to help managers address the above situations and related questions. The procedures of the DEMATEL method and the evaluation methods can be summarized as follows:

Step 1: Producing the direct-relation matrix

Comparison scale of the DEMATEL method

Numeral	Definition
0	No influence
1	Low influence
2	Medium
3	High influence
4	Very high influence

Step 2: Normalizing the direct-relation matrix

On the basis of the direct-relation matrix $\boldsymbol{A}$, the normalized direct-relation matrix $\boldsymbol{X}$ can be obtained through formulas (3-1) and (3-2):

$$\boldsymbol{X} = k \cdot \boldsymbol{A} \tag{3-1}$$

$$k = \frac{1}{\max\limits_{1 \le i \le n} \sum_{j=1}^{n} a_{ij}} \quad {}_{i,j=1,2,\dots,n} \tag{3-2}$$

Step 3: obtaining the total-relation matrix

Once the normalized direct-relation matrix $\boldsymbol{X}$ has been obtained, the total-relation matrix $\boldsymbol{T}$ can be derived by using formula (3-3), in which the $\boldsymbol{I}$ denotes the identity matrix.

$$\begin{aligned} \boldsymbol{T} &= \boldsymbol{X} + \boldsymbol{X}^2 + \boldsymbol{X}^3 + \dots + \boldsymbol{X}^q = \boldsymbol{X}(\boldsymbol{I} + \boldsymbol{X} + \boldsymbol{X}^2 + \dots + \boldsymbol{X}^{q-1}) \\ &= \boldsymbol{X}(\boldsymbol{I} + \boldsymbol{X} + \boldsymbol{X}^2 + \dots + \boldsymbol{X}^{q-1})(\boldsymbol{I} - \boldsymbol{X})(\boldsymbol{I} - \boldsymbol{X})^{-1} \\ &= \boldsymbol{X}\left(\frac{\boldsymbol{I} - \boldsymbol{X}^n}{\boldsymbol{I} - \boldsymbol{X}}\right) \\ &= \boldsymbol{X}(\boldsymbol{I} - \boldsymbol{X})^{-1}{}_{,when\ q \to \infty, \boldsymbol{X}^q = 0} \\ \boldsymbol{T} &= \boldsymbol{X}(\boldsymbol{I} - \boldsymbol{X})^{-1} \end{aligned} \tag{3-3}$$

Step 4: Analyzing the results

In this step, the sum of rows and the sum of columns are used to derive vector $\boldsymbol{S}$ and vector $\boldsymbol{r}$ respectively, as shown in formulas (3-3), (3-4), and (3-5). Then, the horizontal axis vector ($\boldsymbol{S} + \boldsymbol{r}$), called "Prominence", is formed by adding $\boldsymbol{S}$ to $\boldsymbol{r}$, which indicates the level of importance of the criterion. Similarly, the vertical axis ($\boldsymbol{S} - \boldsymbol{r}$) called "Relation", is formed by subtracting $\boldsymbol{s}$ from $\boldsymbol{r}$, which may divide criteria into a cause group and an effect group. When ($\boldsymbol{s}-\boldsymbol{r}$) is positive, the criterion belongs to the cause group; otherwise, it belongs to the effect group. Therefore, the causal diagram can be derived by mapping the dataset of $(\boldsymbol{s}+\boldsymbol{r},\ \boldsymbol{s}-\boldsymbol{r})$, which provides valuable insights for making decisions.

$$\boldsymbol{T}=[t_{ij}]_{n\times n}\ ,i,j=1,2,\ldots,n \tag{3-4}$$

$$\boldsymbol{s}=\left[\sum_{j=1}^{n} t_{ij}\right]_{n\times 1}=[t_{i\cdot}]_{n\times 1} \tag{3-5}$$

$$\boldsymbol{r}=\left[\sum_{i=1}^{n} t_{ij}\right]_{1\times n}^{t}=[t_{\cdot j}]_{n\times 1} \tag{3-6}$$

where vector $\boldsymbol{s}$ and vector $\boldsymbol{r}$ respectively denote the sum of rows and the sum of columns from the total-relation matrix $\boldsymbol{T}=[t_{ij}]_{n\times n}$.

Opricovic (1998) and Opricovic and Tzeng (2002) developed VIKOR (the Serbian name, VlseKriterijumska Optimizacija I Kompromisno Resenje, means Multi-criteria Optimization and Compromise Solution). The basic concept of VIKOR lies in defining the positive and negative ideal solutions first. The positive ideal solution indicates the alternative with the highest value (score of 100), while the negative ideal solution indicates the alternative with the lowest value (score of 0). In our study, the highest performance value of medical services is 5, and the lowest performance value is 0. They were used to help DMs to represent the present status of services components for services assessment and adoption.

The VIKOR method was developed as a multi-criteria decision-making method to solve a discrete decision problem with non-commensurable and conflicting criteria (Opricovic and Tzeng 2004). The method ranks a set of alternatives, and selects the alternative with the highest score. It then suggests compromise solutions to a problem with conflicting criteria in order to help practitioners reach a final decision. Here, the compromise solution is the feasible solution that is the closest to the ideal, and a compromise means an agreement reached on the basis of mutual concessions.

7 Conclusion

Medical tourism sits at the growing intersections of medical technology, economy, cultural and global relations. However, Taiwan is sitting at the edge of market failure due to incentive and information gaps among key industry players. Medical travel facilitator could help close these gaps, and small businesses can find opportunities and play the facilitator role in shaping a healthy medical tourism industry for Taiwan. This study applies a modified VIKOR method to improve service quality among Medical Tourism Services in Taiwan. Our model allows decision-makers and policy-makers to understand the gaps between alternatives and aspired-levels in practice. We then applied the modified VIKOR method to establish the gaps in priorities between alternatives and aspired-levels. Finally, based on these gaps in priorities, we provide managerial implications to improve different carriers for satisfying the medical tourism customers' needs to achieve the aspired-level.

References

1. M.Z. Bookman, K.R. Bookman, *Medical tourism in developing countries* (NY: Palgrave Macmillan, 2007)
2. J. Connell, Tourism Management, **27**(6), 1093-1100 (2006)

3. P.H. Keckly, *Medical tourism*: Consumers in search of value. Washington DC: Deloitte Center for Health Solutions (2008)
4. J. Woodman, *Patients beyond borders*: Everybody's guide to affordable, world-class medical tourism. Chapel Hill, NC: Healthy Travel Media (2008)
5. H. Wu, *Database of Industry & Technology Intelligence Service*: Current status of medical tourism in Taiwan., Ministry of Economic Affairs of Taiwan (2009)
6. Y.H. Hung, T.L. Huang, J.C. Hsieh, H.J. Tsuei, C.C. Cheng and G.H. Tzeng, Knowledge-Based Systems, **35**, 87-93 (2012)
7. A. Pourahmad, A. Hosseini, A .Banaitis, H. Nasiri, N. Banaitienė and GH Tzeng, Technological and Economic Development of Economy, **21**(5), 773-796 (2015)
8. K.Y. Shen, G.H. Tzeng, Soft Computing, **19**(4), 859-874 (2015)

Summarizing the image adjectives for the construction of the picture database for lifestyle image board

Peng-Jyun Liu[1,2,a] and Ming-Chuen Chuang[1]

[1]*Institute of Applied Arts, National Chiao Tung University, Taiwan*
[2]*Department of Creative Product Design, Asia University, Taiwan*

Abstract. Design targets the promotion consumers' motivation to buy more products, whose sensual appeal has become the core of design. This research targets at the lifestyles of three major groups of people in Taiwan: "high-tech groups", "LOHAS groups", and "quality groups". Using literature review, questionnaires, and expert interviews, the frequently used images syntaxes used in three major dimensions: product design, designers, and lifestyle clusters are collected and summarized into 237 items. These image syntaxes are further categorized and selected, yielding 122 image syntaxes in six categories. At last, experts in different areas of design are requested to pick frequently used image syntaxes in the primary stage when designing for these three groups in order to construct the appropriate image syntaxes used in and their association with different areas of design and lifestyle clusters. The results can become the foundation of the next stage of this research in order to construct a lifestyle oriented image board database.

1 Introduction

Nowadays, consumers tend to buy merchandise with sensual appeal instead of functional purposes. Moreover, consumers are motivated to purchase merchandise that appeal to them sensually. In addition, the goal of design is to motivate consumers into consuming; therefore, the sensual appeal of products has become the core of design. In order to achieve this goal, designers introduce many different methods in the design process to inspire their design thinking. An image board is one of the methods frequently used by designers. According to related studies, consumers that have purchased products designed with image board show to have more pleasure and satisfaction from them [1]. Therefore, positive and active meanings in the overall design of products can be derived by introducing image board in the process of design. [2]

In the process of preparing and making image boards, the researchers encountered a few problems: 1. Most of the image graphics are collected in accordance with the designers' experience and personal feelings, showing to be subjective. 2. The association between image graphics and consumer groups are subjectively decided by the designers, and do not always comply with the intended objectives. 3. Image graphics must be collected separately for each case, lowering efficiency. Moreover, graphics are mostly printed materials, occupying much space. And it is difficult to discard old materials to renew the archive. 4. When engaging in interdisciplinary and joint design, it is difficult to integrate the

[a] Corresponding author : yu4038@asia.edu.tw

image boards prepared by different teams, exerting certain difficulties in cooperation or participation in exhibition or development. 5. The appropriateness of the images or themes communicated through the graphics is not examined.

Considering the problems, this research proposes to construct lifestyle image boards from a broader perspective with Baxter's (1995) method[3] to construct a graphic database with images associated to different consumers' lifestyles to solve the problems of time and energy exhausting in collecting pictures, and avoid positioning the pictures and grouping them with the designers' subjective views. In this stage, this research plans to find out the image adjectives corresponding to different lifestyles.[4]

2 Methodology

In order to further understand the association between lifestyle and product category, literature review, questionnaire, and expert interview are used as method in this research. It intends to finish the lifestyle image syntaxes used frequently in three different lifestyle clusters and products.

First, the researchers determined the three major lifestyle clusters with reference to E-ICP Taiwan Lifestyle Bible, 2014, [5] a marketing database that analyzes Asian consumers. Then, online questionnaire was conducted to collect the image syntaxes frequently used in product design. The questionnaire was completed voluntarily by net users, with 26 valid questionnaires. The age group falling between 19 and 30, 17 of them were working people, and 9 students. 12 of them had a design background, and 14 of them had no design background. Moreover, massive image syntaxes used frequently by designers were collected by reviewing related literature, newspaper, magazines, and more. Then, five professional designers with at least five years of experience in related fields were invited to attend the expert interview. They were requested to identify the syntaxes used frequently in design from the pool. At the same time, this research also collected and categorized image syntaxes related to three major lifestyles by reviewing related literature, newspapers, and magazines.

This research integrated the images syntaxes collected from three perspectives. Then five experts with at least five years of experience in design were invited to a focus group interview to categorize and delete the image syntaxes. The experts included four males and one female (as shown in Fig. 4), with ages between 30 and 50 years old. Two of them were design teachers in universities, and the other two were experts in the field of design. The last one was marketing expert, who contributed to point out the consumers' perspectives, with no design background.

The categorization of image syntaxes was conducted in three stages. In stage one, flash cards were made with all the integrated image syntaxes. They were categorized by experts with consensus derived from discussions in accordance with the similarities of the semantic meanings of the syntaxes. In stage two, the image syntaxes were further integrated in accordance with the similarities of images and the habitual similarities of the designers'' applications. In stage three, image naming was conducted with the materials categorized in stage two. In order to condense the large number of images derived after integration, the same group of experts was requested to choose the most appropriate syntax from each group of syntaxes.

This research invited six designers in different fields, including kitchenware, furniture, transportation, communication, and industrial machinery, to help fine-tune these syntaxes for convenient applicability in the designs of different products. These designers fell into the age group between 32 and 50, with at least 6 years of experience in design. During individual interviews, they were requested to pick the adjectives used in their areas and image syntaxes applicable in product design for the three major lifestyle clusters. During the interview, the features and pictures related to the three lifestyle clusters were introduced and provided for the designers' references. Then, the interviewees were requested to select the most frequently used image syntaxes in the three major lifestyle clusters.

3 Results and discussion

3.1 Investigation of lifestyle cluster

This research focus on the investigation of lifestyle clusters in Taiwan. At present, the database E-ICP Taiwan Lifestyle Bible (2014) constructed by Eastern Online Co Ltd provides very complete references to the lifestyle clusters and consumer behaviors in Taiwan. Therefore, this research consulted the databank and regarded the ten major lifestyle clusters as the sources of this project. From these ten lifestyle clusters, the research team has chosen three identifiable clusters with potential consuming ability as the targets. They is the "high-tech lifestyle cluster," "LOHAS lifestyle cluster," and "quality living lifestyle cluster," whose features are illustrated in Table 1:

Table 1. Ten major lifestyle clusters in Taiwan as shown in E-ICP 2014.

Lifestyle cluster	Personalities and features
High-tech lifestyle	Science and tech pro, high-tech lifestyle, innovative, anti-high-tech
LOHAS lifestyle	Self-regulated, self-complacent, indulging, wanderer, otaku
Quality living lifestyle	Active and adventurous, eager and enthusiastic, enjoy life, secluded and otaku, firm and steady

Remarks: The research design and statistics method of this table are results of the collaboration between Eastern Online and the Institute of Business Administration, NCCU.

3.2 Collection of adjectives

The adjectives were collected from three perspectives, including the adjectives used on wrappings, the adjective collections used frequently by designers, and adjectives of different lifestyle clusters.

3.2.1 Collection of adjectives used frequently on product design

Online questionnaire was conducted to collect the image syntaxes frequently used on product design. The questionnaire was conducted at random, opened to net users via the internet. The testees were requested to fill out the adjectives used frequently in design. The totals of 27 questionnaires were sent out, with 1 invalid questionnaire and 27 valid ones. Falling between the age of 19 and 45, 14 of the testees were working people, and nine of them were students. Eight of them were males, and 19 females, with 12 of them had backgrounds in design, and 14 of them had no design background. The total number of adjectives collected was 160, 31 of them had repeated twice or more. Nine of the testees expressed that they often used the term "practical;" six suggested the term "simple"; and five raised the term "handy." For the other repetitions, please consult Table 2. In Table 1, the figures listed behind the adjectives were the numbers of testees that used the terms frequents. For example, the number (9) means that nine of the testees use the term frequently.

Table 2. Results of the online investigation of the image syntaxes used frequently in design.

Practical (9)	Safe(4)	Streamlined(3	Humanistic(2)	Good-looking (2)	Chic (2)
Simple(6)	Magnificent(3)	Simplistic (3)	Refined (2)	Complicated (2)	Interesting (2)
Handy(5)	Durable (3)	Abstract (2)	Organic (2)	Fashionable (2)	Colorful (2)
Neat(4)	Creative(3)	Multifunctional (2)	Generous (2)	Pretty (2)	Amiable (2)
Luxurious(4)	Convenient (3)	Bright (2)	Cute (2)	Concise (2)	Amazing (2)
Hi-tech(4)					

3.2.2 Collection of adjectives used frequently by designers

From related literature [6-11], newspapers, magazines, and more, 188 adjectives used frequently by designers were collected. After reviewing, they were summarized into 144. Five experts were invited to choose the image syntaxes used frequently by them. Two or more of them picked 34 frequently used image syntaxes, as shown in Table 3.

Table 3. Collections of adjectives used frequently by designers.

Classical	Harmonious	Clean	Neat	Vintage	Light
Leisurely	Bright	Fast	Elegant	Magnificent	Sensual
Lively	Convenient	Quiet	Stable	Refined	Symmetric
High-tech	Avant-garde	Comfortable	Natural	Cute	Proportional
Character	Gentle	Manly	Cool	Stylish	Cozy
Romantic	Soft	Traditional	Classy		

3.2.3 Collection of adjectives used by lifestyle clusters

From related literature [12,5], newspaper, magazines, and more, 148 image syntaxes used by LOHAS lifestyle clusters were collected. After revision, there were 60 adjectives. From the high-tech lifestyle clusters, 81 image syntaxes were collected. After revision, there were 51 adjectives. From the quality living lifestyle cluster, 108 image syntaxes were collected. After revision, there were 78 adjectives.

3.2.4 Categorization of the adjectives

254 image syntaxes from the three perspectives mentioned above were collected. Deleting the repeats, 234 were derived, and they were further categorized by experts. In stage one, the experts divided the 234 image syntaxes into 15 categories. In stage two, the experts condensed the 15 categories into 6 types. In stage three, these six categories were named, "image of everyday life" with 41 image syntaxes; "traditional image, 21 syntax images", "image of style," 15; "image of luxury," 55; "image of sustaining," 26; and "image of high-tech," 76.

Then, the experts selected the most frequently used images from the above and yielded 21 images of everyday life, 12 images of tradition, 12 images of style, 32 images of luxury, 16 images of sustaining, and 29 images of high-tech with the total of 122, as shown in Table4. Image syntax (2) implies the syntaxes selected by two or more experts(table 4).

Table 4. Image syntaxes (after abbreviating).

Image categories	**Adjectives**
Images of everyday life	Life (5), home (5), cozy (4), fresh (3), relaxing (3), stress-free (3), humanistic (2), leisure (2), safe (2), free (2), harmonious (2), warm (2), tranquil (2), hospitable (2), simple, plain, clean, elegant, pleasing, comfortable, light
Images of tradition	Cultural (5), traditional (5), reserved (3), silent (3), typical (3), conservative (3), solemn (2), well-behaved (2), vintage (2), rustic (3), sincere, careful
Images of style	Ultra-simplistic (4), streamlined (4), classical (3), geometrical (3), neat (3), organic (3), three-dimensional (2), smooth (2), simplistic (2), simple, heavy, soft
Images of luxury	Refined (5), lofty (4), quality (4), luxurious (4), brand-name (3), classy (3), high-class (3), chic (3), deluxe (3), magnificent (3), pricy (2), material (2), eye-catching (2), good taste (2), gorgeous (2), fashionable (2), charming (2), complicated (2), vivid, artistic, aesthetics, graceful, aesthetics, modern, delicate, visual, sensual, meticulous, quality, fancy, romantic, bright
Images of sustaining	Sustaining (5), environmental friend (5), natural (4), ecological (4), healthy (4), energy saving (3), natural (3), valuable (3), handy (2), generic (2), convenient, durable, sturdy, manageable, practical, hand-made
Images of high-tech	Digital (4), efficiency (4), avant-garde (3), efficient (3), intelligent (3), professional (3), hard (3), high-tech (3), rational (3), designed (3), future (2), multifunctional (2), popular (2), accurate (2), fashionable (2), intelligent (2), creative (2), innovative (2), novel (2), pioneer (3), unique(2), Synthetic, cold, urban, special, urban, effective, prompt, systematic, conceptual

3.3 Integration of image syntaxes used by different lifestyle clusters and in different areas of design

3.3.1 Investigation results of applications of image syntaxes used in different areas of design

Designers from different areas picked the most frequently used images syntaxes from 122 items, listed above, applicable in three different lifestyle clusters – the high-tech lifestyle cluster, LOHAS lifestyle cluster, and quality living lifestyle cluster, as listed in Table 5

Table 5. Image syntaxes frequently used by designers in different areas of design.

Area Category		Kitchenware	Furniture	Vehicles	Computer appliances	Communication equipment	Machinery	Repetition rate
Everyday life	Humanistic	◎	◎		◎	◎	◎	83
	Pleasing	◎	◎	◎	◎			67
	Neat				◎	◎	◎	50
	....							
Tradition	Restrained	◎		◎	◎		◎	67
	Cultural		◎	◎	◎		◎	67
	Typical	◎	◎		◎		◎	67
	....							
Style	Classical	◎		◎	◎			50
	Three-dimensional			◎	◎			33
	Organic	◎		◎	◎			50
	....							
Luxurious	Expensive		◎	◎	◎	◎		67
	Quality		◎	◎	◎			50
	Brand name	◎	◎	◎	◎		◎	50
	....							
Sustaining	Natural		◎	◎	◎	◎	◎	83
	Hand-made				◎			17
	Sustaining		◎	◎	◎			50
	....							
High-tech	Artificial							0
	Future		◎	◎	◎	◎		67
	Conceptual		◎		◎			33
	....							

3.3.2 Investigation results of applications of image syntaxes used in the lifestyle clusters

The designers in different areas were requested to choose the image syntaxes used in different lifestyle clusters to obtain more accurate image syntaxes. The image syntaxes without targeting at any specific realm used by the designers in different areas were analyzed. The survey found that as shown in table 6.

According to Table 6, when constructing the construction of databank, the image syntaxes are arranged with the following guidelines:

- Level 1: Used frequently by designers in various fields.
- Level 2: Used frequently by designers in various fields and sometimes by three lifestyle clusters.
- Level 3: Used by both designers in various fields and three lifestyle clusters.
- Level 4: Area-dependent in each lifestyle cluster.

Table 6. Investigation of the image syntaxes used in the lifestyle clusters of luxury, sustaining, and high-tech.

Category	Design Category	Classification	Images of vocabulary
Life Images	Kitchen supplies	Both	Humanity、Pleasure Comfortable、Life、Relaxation、Comfortable
		3 ethnic groups have	Humanity
	Furniture Design	Both	Humanity、Pleasure、Comfortable、Harmony
		3 ethnic groups have	Humanity
	means of transport Design	Both	Pleasure、Easy、Comfortable、Life、Warm、Harmony
		3 ethnic groups have	Harmony
	Computer Equipment	Both	Humanity、Friendly、Pleasure、Austere、Easy、Homely、Life、Warm、Refreshing、Harmony、Leisure、Plain
		3 ethnic groups have	Humanity、Life
	Communications equipment	Both	Humanity
		3 ethnic groups have	
	Industrial machinery and equipment	Both	Humanity、Friendly、Austere、Easy、Security、Plain
		3 ethnic groups have	Humanity
Traditional Images	Kitchen supplies	Both	Introverted、Quiet、Rules
		3 ethnic groups have	……….

Source: This study investigated

Therefore, in this study the vocabulary of Images of extraction results shown in Table 7. The Construction database for design use, therefore, in order to extract Images of vocabulary designers use on all areas across two levels as the main basis for selection and coupled with individual lifestyle ethnic specificity Images of vocabulary. Overall, Life Images have 8 Images of vocabulary; Traditional Images have 6 Images of vocabulary; Style Images have 5 Images of vocabulary; Luxury Images have 11 Images of vocabulary; Technology Images have 12 Images of vocabulary. These images will be used as the second phase of stimulus vocabulary sample image of keyword selection. Therefore the image syntaxes summarized by this research are listed in Table 7.

Table 7. The image syntaxes summarized by this research are listed.

Level	Image of everyday life	Image of tradition	Image of style	Image of luxury	Image of sustaining	Image of high-tech
1	Humanistic Pleasing Simple Casual Comfortable Home Safe Everyday life Harmonious	Restrained Cultural Typical Well-behaved Vintage Traditional	Classical Organic Soft Smooth Streamlined Ultra-simplistic Concise Simple Neat	Expensive Quality Good taste Brand name Fashionable High-class Elegant Extravagant	Natural Sustaining Ecological Ease Durable Healthy Energy saving Practical Environmental friendly	Future Multi-functional Systematic Accurate Hi-tech Efficiency Special Hi-intelligent Professional Design Digital

2	Humanistic Pleasing Comfortable Everyday life	Restrained Typical Vintage Rustic	Streamlined Ultra-simplistic Neat Simple	Brand name Good taste Fashionable Quality Expensive Quality	Healthy Natural Sustaining Ecological Ease Energy saving Practical Environmental friendly	Hi-tech Efficiency Efficacy Hi-intelligent Professional Design Innovative Future Digital Conceptual
3	Humanistic Harmonious Everyday life	Vintage Restrained Cultural Rustic	Neat Three-dimensional Simple	Expensive Good taste Fashionable Quality Artistic High-class Fine	Natural Ecological Generic	Hi-intelligent Future Effective Accurate Urban Creative Innovative
4	Unrestrained Clean Graceful Warm	Careful		Fashionable Charming Refined Complicated Romantic	Handy Hand-made Value	Cold Avant-garde Popular

This research had collected 237 image syntaxes from three lifestyle clusters: the "high-tech cluster," "LOHAS cluster," and "quality living cluster," which are used frequently in product designs by designers and lifestyle clusters, through a literature review, questionnaire, and expert interview as its research method. Then focus group interviews with experts were conducted to select and categorize these syntaxes further to yield 122 image syntaxes in six categories. At last, experts in different fields picked the most frequently used image syntaxes from these 122 images to derive adjectives applicable in the three major lifestyle clusters to construct the relationships between different areas of design, different lifestyle clusters, and image syntaxes. The research results will be used in the studies in the next stage to construct the foundation lifestyle oriented image board database.

This research discovered that there is no significant different regarding the designers' cognition of the applicability of the image syntaxes in these three lifestyle clusters. It will be helpful in the collection of graphics, positioning, and grouping in the next stage. Eventually, there are still quite a number of terms derived from the shrinking of the 122 image syntaxes in three stages. These image syntaxes will be further selected in the next stage by the people in the three mentioned lifestyle clusters upon request to render these image syntaxes closer to the everyday life needs in each cluster.

References

1. M.W. Brown, *The teacher-tool relationship: Theorizing the design and use of curriculum materials*, Connecting curriculum materials and classroom instruction, 17-36 (New York, 2009)
2. P.J. Yi and Y.M. Zhang, A study of the process of mature period product differentiation-The proposal of multi-image features deduction method, Kung University master's thesis (2006)
3. M. Baxter, *Product design, practical methods for systematic development of new products*, Chapman & Hall (London, 1995)
4. G.A. Kelly, The psychology of personal constructs (New York, 1995)
5. Taiwan Lifestyle Bible, Eastern online (Taipei, Taiwan, 2014)

6. R.C. Chou, Create complex perceptual Image - to shape features based, Cheng Kung University Thesis (2000)
7. H.W. Lin, Research Product Images search mechanism used in the Internet, Cheng Kung University Thesis (2000)
8. M.C. Chuang and C.C. Chen, Journal of Design, **9**(3), 43-58 (2004)
9. Y.Y Chan, Studies hair like the imagery used in product shape-picture to picture motion imagery, for example, should be surgery, Chiao Tung University Institute of Thesis (2005)
10. S.H. Hung and M.C. Chuang, Taiwan Textile Research Journal, **18**(3), 42-52 (2008)
11. Y.R. Chun and Y.J. Ya, To investigate the behavior of daily life information from LOHAS lifestyle of view, Shih Hsin University Thesis (2007)
12. R.Y. Lee, The study of consumer goods consumers of cultural and creative motivation, lifestyle and consumer behavior of the Di Ye University Thesis (2013)

Comparative study on two civil engine thrust reversers for their maintainability analysis

Hua-Cong Xu[a], Yang Pan, Yue-Xi Xiong and Jing-Wu He

School of Aeronautic Science and Engineering, Beihang University, Beijing, China

Abstract. Trade-offs studies on two kinds of existing engine thrust reverser design were carried out, applying virtual maintenance method and the basic theory of ergonomics. Analysis on the maintainability of the two kinds of configurations, the O-duct thrust reverser owning the integrated propulsion system (IPS) and the traditional portal D-duct thrust reverser, were compared. According to the differences of O-duct and D-duct thrust reverser in the structure, working principle and mode of motion during the maintenance, structure modelling and kinematics simulation of two kinds of thrust reverser were finished in the same space constraints. Taking DELMIA software as the platform, the partial ergonomic research and evaluation including accessibility analysis, visibility analysis, specific disassembling and assembling time estimation and workspace analysis were put forward through the virtual maintenance simulation of two civil engine thrust reversers. Supplied technical reserves on design and selection, scheme analysis and technical evaluations for domestic engine thrust reverser in the future.

1 Introduction

In addition to undertake tasks like comprising core engine and accessories, engine air supply and exhaust, and aerodynamic shape adjustment to reduce drag, an important function of nacelle is to provide the reverse thrust of aircraft landing, which can be realized by thrust reverser system. Thrust reverser is used in speed reduction during landing and rejected takeoff by changing the direction of airstream in fan duct [1]. Different thrust reverser structures have effects on aerodynamic performance, structural weight and maintainability. Among them, maintainability is the important quality characteristic of aircraft in the stage of product design, which has a decisive effect on the aircraft reliability, efficiency and economy. Beyond that, the pros and cons of maintainability will also directly affect the aircraft ground turnaround time and attendance [2]. Therefore, to analyse its maintainability is of great significance when it comes to engine thrust reverser performance. The quality of maintainability must be evaluated and validated through the maintainability design analysis and tasks simulation. In this paper, taking a certain type of aircraft as an example, two kinds of maintainability models of thrust reverse configurations have been established in the same space constraints and digital simulation methods have been applied. Virtual maintenance method and basic theory of ergonomics are also used to make comparison in maintainability analysis and evaluation taking DELMIA software as a platform.

[a] Corresponding author : 1210552073@qq.com

2 Structure and maintainability analysis of two kinds of thrust reversers

Maintenance is the abbreviation of maintain and repair on the aviation engine. Aero engine maintainability refers to ease of engine maintenance [3]. This paper focuses on the influence of two different engine thrust reverse structures named D-Duct and O-Duct in engine maintenance, including inspection of core engine and thrust reverser, maintenance accessibility, time estimation and so on.

2.1 Introduction on traditional D-Duct type thrust reverser

2.1.1 Structure analysis on D-Duct

The traditional D-Duct type thrust reverser system is also called the grid-baffles thrust reverser, which is composed of left and right T/R halves attaching to the engine strut, including translating cowl, inner cowl, cascades, blocker doors, hydraulic actuators, torque box, upper and lower sliders and tracks [4]. The two sleeves are connected by tension latches and inner V-blades, working simultaneously but independent of each other. Figure 1 is the structure of D-Duct thrust reverser.

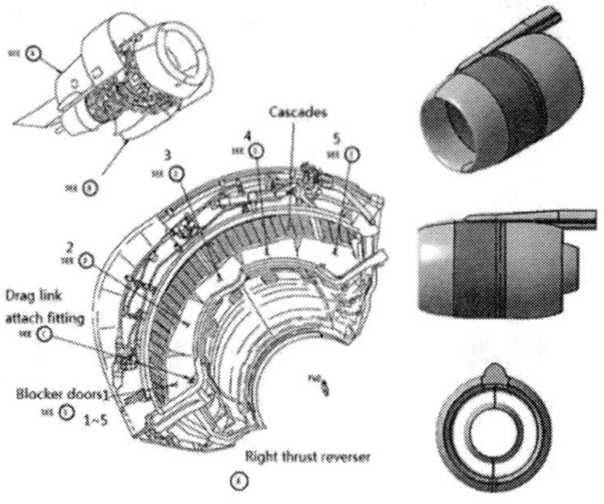

Figure 1. Structure of D-Duct thrust reverser.

2.1.2 Maintenance methods

When in maintenance, the thrust reverser is opened on the ground, and the fan cowl and translating cowl are installed on the pylon. The left and right halves are lifted from bottom to upper after releasing the latches under the sleeves, then the fan case, core engine, low-pressure turbines and other components are able to be inspected directly. The core engine including the inlet cowl can be removed or installed while the D-Duct cowl and fan cowl are left on the pylon when in removal/installation tasks. So the traditional D-Duct thrust reverser cowl has satisfied accessibility and visibility, and the operation is simple. Now we will do structure and maintainability analysis for new structure of O-Duct thrust reverser. The following in figure 2 are in maintenance position and latches structure.

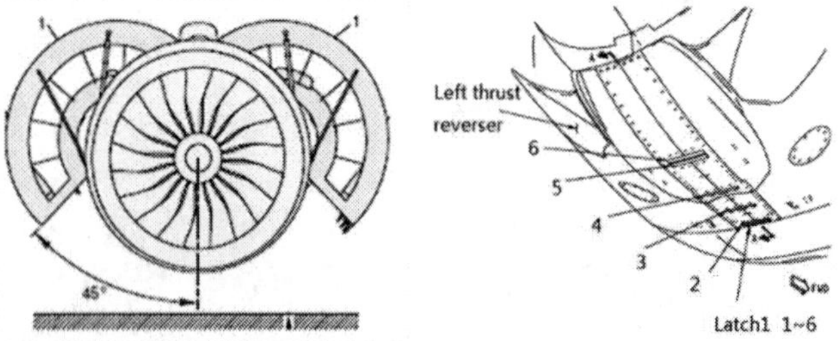

Figure 2. Maintenance position and tension latches of D-Duct.

2.2 Introduction on new O-Duct type thrust reverser

2.2.1 Structure analysis on O-Duct

The biggest difference of the two thrust reversers is that the new component has a next-generation integrated propulsion system(IPS) with a low drag front end with a directed flow nozzle deicing system; a fan cowl structurally integrated to the engine; an integrated mounting system for reduced engine distortion and enhanced on-wing performance; and an all-new translating O-Duct thrust reverser configuration. As figure 3 shows, the O-Duct system uses one-piece composite "O" duct, which replaces the two-piece "D" doors on traditional thrust reversers. When the thrust reverser is working, the entire O-Duct moves afterward to the reverse thrust position driven by electric thrust reverse actuated system(ETRAS), thereby eliminating the need for drag links in the secondary flow-path.

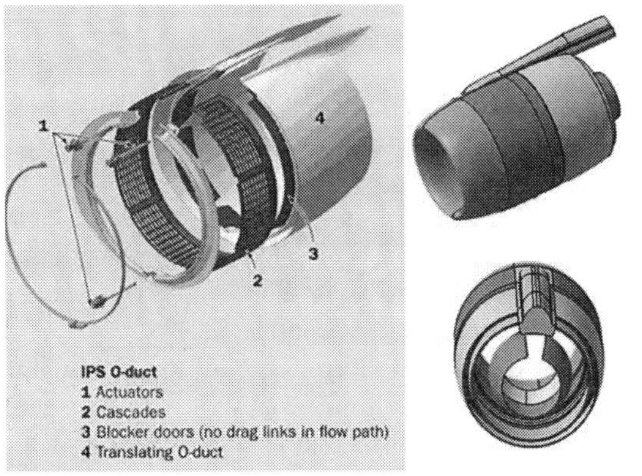

Figure 3. Structure of O-Duct thrust reverser.

2.2.2 Technique innovation on maintain

As showed in figure 3, a key feature of the O-Duct is its monolithic structures, 330-deg. carbon composite inner skin. This fact not only reduces the structural weight but also solve a persistent maintenance issue on D-Duct, for the O-Duct has no tension latch, no beam or lower bifurcation on the bottom of translating cowl, neither does lower track liner. Therefore thrust reverser with O-Duct of translating cowl may have easier operation and less maintain in routine maintenance.

Comparing with D-Duct, the integral translating cowl moves only along top slider automatically to intermediate maintenance positions for ease O-Duct maintenance or maintenance positions. What's remarkable about the movement track is that the cowl opening system (COS) and thrust reverse actuated system (TRAS) can move respectively but the moving distance has been limited with slider length and maintenance stop. So we do further research on O-Duct for its maintainability aspects, by studying the simulation of virtual workers for maintenance activities in virtual environment in DELMIA. Then maintainability evaluation can be made to find impact on accessibility analysis, visibility analysis, disassembling and assembling time estimation and so on if we choose O-Duct.

3 Application of DELMIA-virtual maintenance on thrust reverser

3.1 Virtual maintenance introduction and ergonomics application

With the development of virtual reality technology, applications of virtual maintenance technology in the field of maintainability design, maintenance support capability and maintenance training are on the increase. Traditional maintainability analysis usually requires demonstration of maintenance works, which means simulating the maintenance process of real products on physical prototype in actual maintenance environment [5]. Then analysis on visibility, accessibility, mechanics performance for maintenance personnel should be done, while problems occur that work lags behind product design and personal security can not be guaranteed. After virtual maintenance introduced, maintainability design personnel can carry out maintainability analysis and evaluation in parallel; detect problems and defects timely; reduce impact of human factor on aviation maintenance.

Ergonomics is a subject to study the interrelation among human, machine and environment, aiming to reach the optimization results for work efficiency and human-machine security, healthy and comfortable [6]. We combine ergonomics and virtual simulation together to demonstrate maintain works and apply the results to maintainability analysis.

3.2 Virtual maintenance simulation scene construction

3.2.1 The digital prototype model establishment

Virtual prototype is a maintenance object of virtual maintenance operations, which belongs to the three-dimensional solid model. The disassembly and assembly work of virtual prototype is usually involved in maintenance simulation [7]. We install the two different structures of the nacelles (O-Duct and D-Duct) on a certain type of civil aircraft as the digital prototype, adopting the same installation position, height and outline dimension profile(such as outer contour diameter of inlet cowl, fan cowl and thrust reverser cowl). It can be assured that the result differences are mainly caused by the different structures in the same spatial constraints conditions.

Due to many design attributes included during modelling, miscellaneous data may affect the processing speed. So the model is simplified appropriately to remove geometric information of parts which maintenance task does not involve. At the same time, we eliminate some of the complex parts of the task without affecting the modelling, because the aircraft, especially the engine structure is very complex.

3.2.2 Maintenance personnel modelling

Virtual human body model is an important part in maintenance tasks, it is needed to complete all of the ergonomics analysis. Use the Human Builder module in DELMIA to create the full size model of the maintenance personnel, 50th percentile, Taiwan. The establishment of the human body model needs to consider the NASA human element engineering specification. The change of human body posture can be achieved by the manipulation of joint movement, in which complex human body structure includes 68 joints, 6 pairs couple points and 148 degrees of freedom for real joint motion and restriction, in addition, human operator IK(inverse kinematics) function based on forward and reverse kinematics [8].

3.2.3 Maintenance tools and equipment

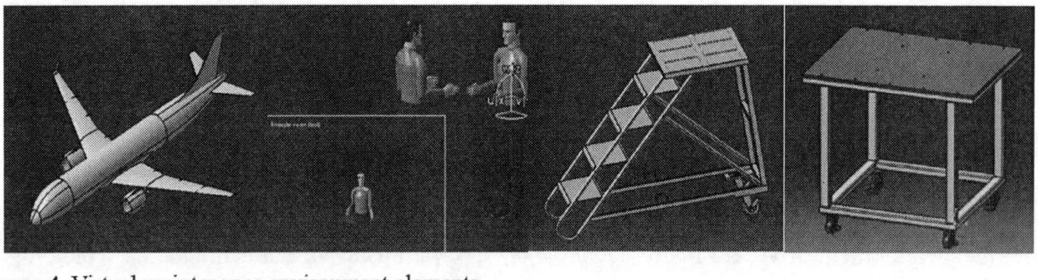

Figure 4. Virtual maintenance environment elements.

The choice of maintenance tools and equipment shall be subject to the specific maintenance tasks. As is shown in the picture 4, take the operation of removal/installation of translating cowl as an example in daily inspection and maintenance, which divided into two cases of O-Duct/D-Duct, and there are some differences between the required equipment. Through the analysis of model maintenance task, ensure the safeguards equipment and tools for implementation of process.
D-Duct: working ladder operating platform hydraulic pump minitype trunk
O-Duct: working ladder operating platform special ground support equipment(GSE)

3.3 Process of virtual maintenance simulation

The virtual simulation process mainly depends on DPM-Assembly process simulation and Human Task Simulation module in DELMIA. The maintenance activities and process simulation are actualized by action of the digital prototype and manikin. It's a dynamic environment as reconstruction of maintenance activities under virtual environment.

3.3.1 Maintenance example on O-Duct

Firstly, finish the simulation for inspection on O-Duct. Load the service material into the P.P.R tree, and all process nodes of simulation are established under ProcessList, the digital prototype model is loaded into ProductList, manikins, support equipment, area and others are placed on the ResourceList. Open the fan cowl doors before open the thrust reverser cowl.

Then move the thrust reverser cowl.

① Make cautions before opening. Make sure that the thrust reverser can't be operated.

② Open the lock at 6 O'clock structure.

③ Then the translating cowl and cascade components move rearward to maintenance position actuated by two synchronized COS.

Next open the core cowl. The core cowl is composed of two parts: one is FCC (Forward Core Cowl), and the other is ACC (Afterward Core Cowl). FCC is D-type and divided into LH(left half) and RH(right half), it has three latches at the bottom of FCC. ACC or called rear sliding cowl is O-type like translating cowl, which can move backward along the track.

① Open the main latch assurance door and release the bottom latches.

② Open FCC. Remove the forward HOR and attached to fan case on one end and IFS on other end. Then attach afterward HOR, one end detaches from TB and the other attaches to IFS.

③ Move the ACC backward along the slider.

The O-Duct translating cowl must be removed if need further repair on core engine or thrust reverser.

① Front and rear fairings have to be removed before removing O-Duct from pylon.

② Release stop bolt at rear of pylon.

③ The special GSE shall be used to hold out the cowl and carry it away from pylon.

④ Inspection of core engine. Here are the simulation scene and PETR chart.

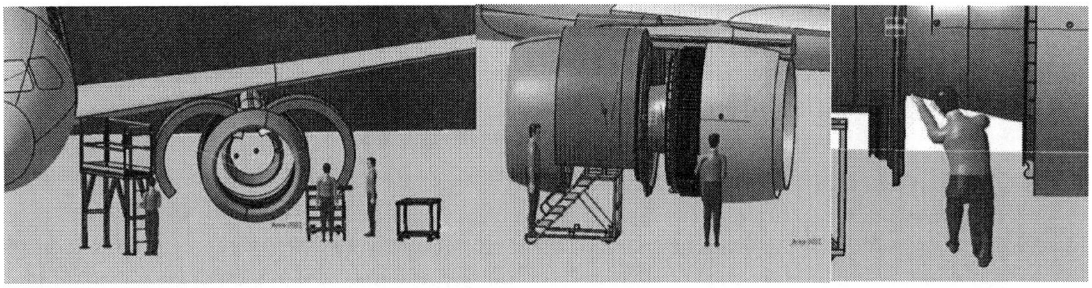

Figure 5. Simulation scene of O-Duct maintenance.

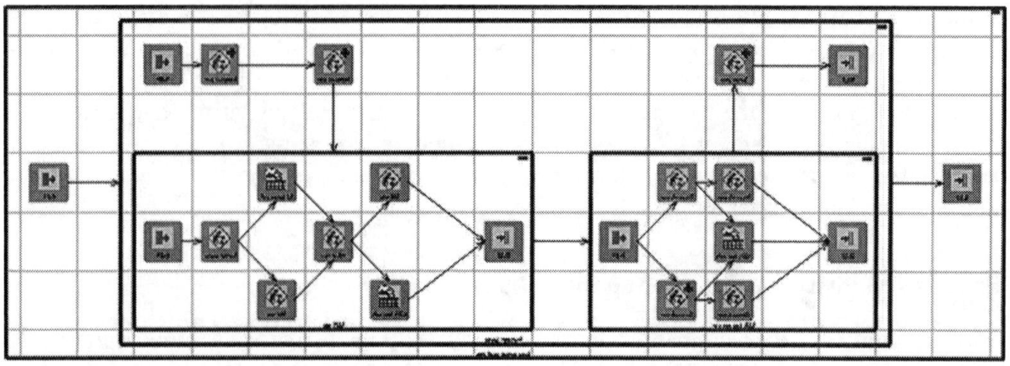

Figure 6. PETR chart of inspection on O-Duct.

The figure 5 shows the simulation scene of O-Duct maintenance, and the figure 6 is PETR chart of inspection on O-Duct.

3.3.2 Maintenance example on D-Duct

Compared with O-Duct in above process, D-Duct has the same operation when open the fan cowl doors, for both O-Duct and D-Duct has D-type form. There are some distinctions in opening the thrust reverser cowl, as D-Duct consists of left and right translating sleeves. The procedures to open the cowl are shown in figure 7.

① Make cautions before opening.(same with O-Duct)

② Unlock the latches along the lower edge of the cowls.

③ Connect the hand pump.

④ Operate the hand pump to open the half cowling to the normal maintenance position. The doors can be opened to 33 degrees with the wing leading edge slats extended. Maximum opening is 45 degrees.

⑤ Inspection of core engine

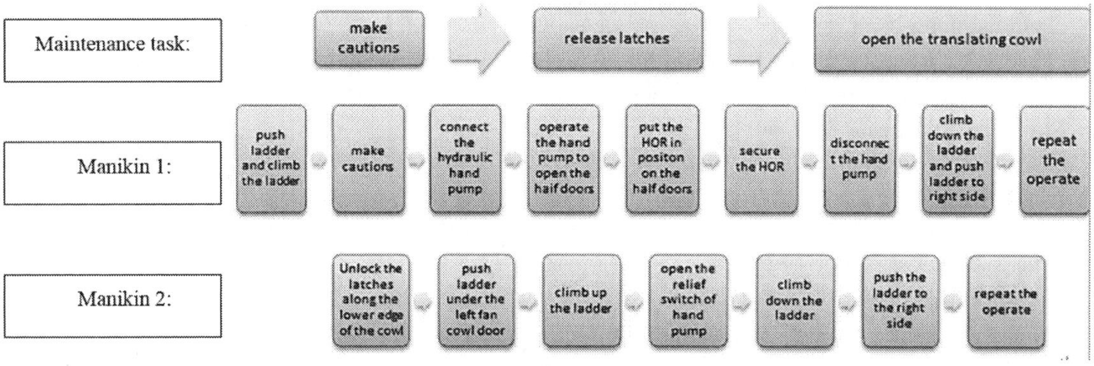

Figure 7. Maintenance program on D-Duct thrust reverser.

4 Maintainability analysis and evaluation

4.1 Accessibility analysis

Maintenance accessibility is the measurement of difficult degree for reaching target components when repairing product [9]. Including entity reaching (for example a region of the body or by means of tools get in touch with the maintenance position), as well as enough operational space.

4.1.1 Qualitative analysis process

Qualitative analysis for accessibility is conducted by observing the virtual maintenance process, and adjusting manikin's working posture to make maintenance object to reachable range. If the maintenance object is out of range, readjust the position of manikin to find the suitable position and new point of view. Maintenance design may be defective when maintain manikin fails to touch the target or difficult to do that after the adjustment. Figure 8 shows the reachable envelope and accessibility analysis for O-Duct and D-Duct.

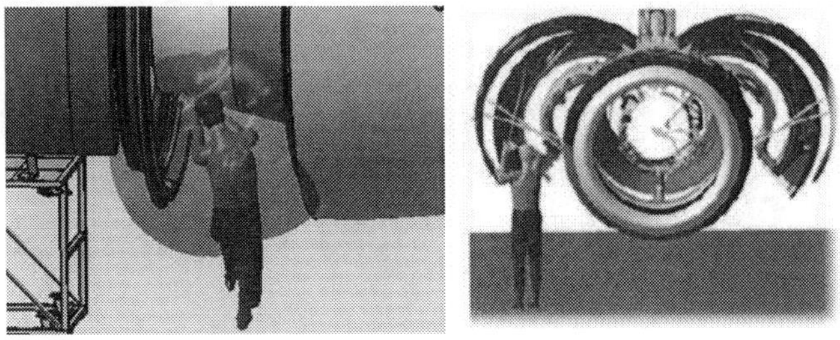

Figure 8. Analysis of reachable envelope and accessibility.

From the analysis of upper envelope diagram, the operated positions are all covered by reachable envelope upper limb, so physical accessibility can be achieved in O-Duct maintenance. However, compared with D-Duct, more operations need to be finished with O-Duct in maintenance tasks. So D-Duct performs better than O-Duct in accessibility theoretically for its inner wall connected with translating cowl by the upper and lower bifurcation. When the translating cowl is opened as the picture shows, fan case and core engine can be checked easily. And engine removal/installation is also simple after that. When it comes to O-Duct, removal of translating cowl, FCC and ACC leads to more complex works. Here we need some quantitative analysis

4.1.2 Quantitative analysis process

Quantitative analysis for accessibility can be measured by reachable coefficient K_π

$$K_\pi = 1 - \frac{n_\pi}{n_0 + n_\pi} \tag{1}$$

n_0 is the basic operation time of maintenance task (refers to most simple action in the process of work). n_π is the basic operation time of additional task (disassembly of interference part). As can be seen in picture ,n_0=17, n_π=5, so K_π =0.77,the value is in a specific interval [0.75, 1], so we regard maintenance process as in good accessibility. K_π =0.81 in the task of inspection of core engine in D-Duct, better than in O-Duct.

4.2 Collision and interference analysis

Collision and interference qualitative analysis can be judged directly during virtual maintenance process. Quantitative evaluation can be assessed by r .Measuring and calculating workspace V enclosed with dashed line in virtual environment. Rectangular the minimum workspace V_{min} related to a certain part of body and tools [10].

$$r = \frac{V}{V_{min}} \tag{2}$$

The maintenance part workspace ratio is good if r > 1.5, and generally not having collision and interference in the process of maintenance. Obviously, the maintenance part has good workspace both in O-Duct and D-Duct.

4.3 Visibility analysis

DELMIA has vision window function in Human Task Simulation module which can judge whether operational position is entirely visible or in best visual area. As is seen in picture 9, visual range can be intuitively observed on different position through the manikin's eyes. Specific operation should be seen during maintain time, besides enough space to accommodate maintenance manikin's hands and arms, a proper gap is available to observe the target.

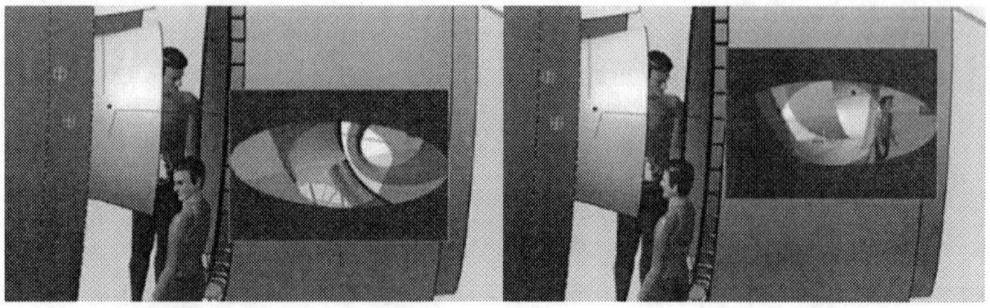

Figure 9. Analysis of visibility.

Comparative study has been made during the simulation to get the conclusion that D-Duct owns better vision as less barrier structure in the task.

4.4 Disassembling and assembling time estimation

Disassembling and assembling time can be estimated by three methods: checking simulation process Gantt graph; studying the time when demonstrating the whole simulation process; or examining "simulation periods time" in maintenance process simulation [11]. The Gantt chart in figure 10 is the most intuitive way and can show all the time information of sub-process and entire process, sequence of activities is also obtained through Gantt chart.

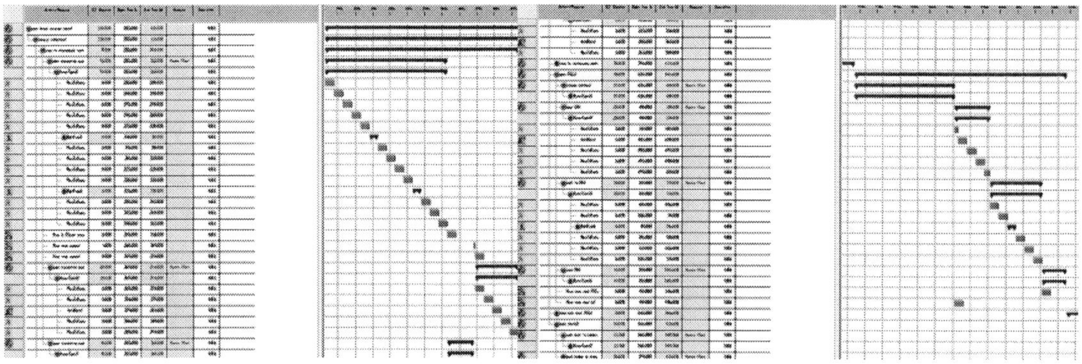

Figure 10. Each motion consume time in Gantt graph.

The maintenance time is not only an important indicator to measure the quality of maintenance, but also an important basis for maintenance hours and maintenance evaluation. Comparing the operation time on two devices, O-Duct costs more time in opening thrust reverser cowl.

4.5 Working posture analysis

DELMIA software can generate report and finish maintain efficiency analysis on main maintenance work posture(raise/place analysis, push/pull analysis, walk, carry analysis, biomechanical and so on). Analysis report is to judge if the maintenance person is in suitable working posture, and whether the maintenance operation would decrease the maintenance person's work efficiency or have some behavior leading to fatigue [12].

5 Concluding remarks

Higher request on engine nacelle design and performance has been put forward since the majority of civil trunk airliners adopt wing-mounted engine structure in modern society. [13]The new type nacelle owing IPS with O-Duct of the translating cowl has different performance compared with traditional D-Duct thrust reverser. Analysis has been done through the maintainability design analysis and tasks simulation in DELMIA. In this paper, information about product's accessibility, rationality of the space distribution and the time to accomplish certain tasks are evaluated by means of virtual maintenance. Since a lot of new technology has been introduced in those O-Duct thrust reversers, they will require highly specialized repairs, and more detailed information has to be obtained by concrete feedback from airline companies.

References

1. Y.F. Zhang, H.X. Chen, S. Fu, et al., CJA. **33**(11), 1993-2001 (2012)
2. H.Z. Wang, Q.G. Zhai, J.F. Yu, et al., Journal of Civil Aviation University of China. **27**(4), 56-59 (2009)
3. W. Wang, Engineering Science and Technology II, **S1** (2011)
4. J. Ke, X.F. Qi, P. Wang, et al., Civil Aircraft Design & Research, **108**, 14-17 (2013)
5. Z.H. Wang, Q.G. Zhai, Mechanical & electrical technology, 36-53 (2009)
6. Z.H. Wang, Q.G. Zhai, Journal of Changsha Aeronautical Vocational and Technical College, **9** (4), 36-41 (2009)
7. Z.G. Guo, Z.X. Liu, H. Li, Aircraft Design, **32**(3), 62-66 (2012, 6)
8. W. Jiang, *Master Theses of NUDT* (2009, 6)
9. Y.Z. Sun and H.W. Deng, Journal of Traffic and Transportation Engineering, **6**(1), 93-112 (2006)
10. Z.X. Yang, Master Theses of JUST
11. H. Liu, J.P. Hao, Journal of System Simulation, **18**(2), 378-394 (2006)
12. A. Amundarain, Mecanique & Industries, **5**, 4 (2004)
13. X.F. Zhao, S.G. Tan, W. Shen, et al., Advences in Aeronautica Science and Engineering, **4**(3), 268-273 (2013)

Study on adult career planning on adult education program and assessment of adaptive design model

I-Chan Kao[a]

General Education Center, Open University of Kaohsiung, Taiwan

Abstract. This study explores adults' on-the-job current career planning and the selection trends of future development as the criteria for a training program design for adult education and the basis of the design model of adaptive assessment. This study regards the contents of the TTQS training process, the items, indicators, and standards of the Workforce Development Agency, and the occupational categories of the "Occupational Competency Standards", as disclosed by the "Integrated Competency and Application Platform", as the framework for the reference model; and probes into the impact of globalization, adult career planning, adult occupational training program design, and adaptive assessment, and adopts a questionnaire survey to understand the design model of the adaptive adult education program, in order to effectively implement adaptive assessment. The purposes are to assist adults to accomplish adaptive career planning, achieve their goals of vocational cultivation, and solve the problems of adult unemployment, in order to lead to successful adult employment and transfer to globalization workplaces to create positive career development.

1 Introduction

1.1 Research motives

Under the influence of global free markets and workplace environments, adult career planning and development are increasingly diverse and complex. Participating in competency development and enhancing learning activities is one of the ways to avoid becoming redundant in the face of constant changes in the market and workplace. Presently, Taiwan is confronted with globally competitive workplace environments, where expatriates, downsizing, layoffs, unpaid leaves, outsourcing and dispatch, reduced salaries and welfare, and other issues, have impact on workplace environments. In addition, the increased speed of technological development and the convenience of life and workplace operations have changed the operational trend of business organizations; for example, the "virtual office" refers to non-traditional offices that engage in work related to traditional offices[1].

Through literature review and questionnaire survey, this study examines the dramatic changes faced by adult learners in the workplace, particularly in the face of continuous changes in the global economic sector and the gradual complete replacement of laborers with a mechanical force for comprehensive automatic production, as caused by "Industry 4.0". In addition, global energy conservation and carbon reduction environmental protection policies encourage countries to

[a] Corresponding author : kic0929@ouk.edu.tw

constantly develop green energy and photovoltaic technology. Fueled by different emerging industry development policies, the industry chain of traditional petrochemical energy production faces its most serious crisis since the first generation of the industrial revolution. Confronted with the aforesaid re-shuffling and integration of the global industrial environment, how adult learners can effectively develop career planning and master the competencies required by contemporary workplaces is an important topic for all current in-service training programs and pre-service education, as well as the motive of this study.

1.2 Research purposes

Based on the above research motives, this study has the following research objectives:

1. To understand the connotations of adult career planning, adult competency training programs, and assessment adaptability.
2. To explore the overview of the cognitive differences of adults with different backgrounds in adult competency training programs and assessment adaptability.
3. To analyze the relationship between adult competency training programs and assessment adaptability.
4. To propose feasible strategies for adult competency training programs and assessment adaptability of adult career planning, as based on the research results.

1.3 Research purposes

This study first employs literature review to collect books, periodicals, magazines, journals, papers, and other important literature regarding adult career planning, adult education programs, and adaptive assessment, and then, analyses such literature in order to obtain the basic concepts of this study. Secondly, this study sorts and summarizes the relevant literature, reviews and recommendations of scholars and experts regarding the studies and practices of adult career planning and adult education programs, as well as their adaptive assessment. Then, this study edits *the Questionnaire of the Overview of Adult Competency Training Programs and Assessment Adaptability*, and uses this questionnaire survey to investigate adult students receiving in-service training at the Open University of Kaohsiung. Through collection and statistical analysis of the data collected from the questionnaire, this study aims to understand adults' perceptions and cognitions of the status of competency training programs and adaptive assessment. This questionnaire survey is conducted in both paper and online questionnaire forms.

2 Connotations of adult career planning, adult competency training programs, and assessment adaptability

2.1 Probe into the orientation of adult career planning and related issues

The impacts of globalization on adult learning and education include the elimination of individualization, the collapse of original communities, delocalization, the rise of post-colonialism, rampancy of cultural hegemony and symbolic violence, and the mechanism of the capital-based market [2-4]. The scope that refers to adult learners is relatively broad, including adult students, mature students, non-traditional students, and other adult learners studying in the education system, as well as adults who engage in self-directed learning, informal learning, and other self-learning. The differences of learning behaviors between adult learners and general traditional learners are primarily resulted from the effects of adult learners' physical, psychological, and social factors, as well as personal characteristics [5-8].

Randall (1987) maintained that when an organization's commitment was low, employees were likely to have low career development and low promotion, which may result in resignation or striving to cause organizational goals to fail. Moreover, along with other changes in the labor force, the female labor force joined the market, which contributed to changes in business management strategies and social policies; and business owners were only willing to provide more stable and reasonable working conditions to core staff. However, it was difficult for non-core staff to get promoted, and such employees may be outsourced or dispatched, thus, long-term staff were replaced by "temporary employees" [10]. When workplace environments change too fast, and people experience different needs, develop cross-professional and cross-organizational career development models, and gradually break away from the traditional career development of "maturity" towards "adaptability "[11], it indicates that employees must equip themselves with various capabilities and determine whether their careers have reached the goals established for this stage, in order to effectively respond to current career trends.

Prior to the 1980s, career orientation was dominated by "linear direction", meaning individuals were promoted in organizations based on the positions planned by the organizations, individuals did not have the sense of "self", and regarded the goals of the company as their goals. However, with the rise of new concepts, such as protean careers and boundary less careers, after the 1980s, changes in the global environment were increasingly dramatic and factors of uncertainty increased, contributing to the "nonlinear" development of career orientation [12-13]. The multiple career concept model, as proposed by Driver et al. (1994), divided career orientation into linear, expert, spiral, and transitory, thus, careers were no longer regarded as linear, and the new concept was that a career was recyclable, reversible, nonlinear, and multi-trajectory [15]. Contemporary adults are in an era of changeable global workplaces, and careers have become unpredictable, fragile, and multi-oriented, leading to more complex career planning [1].

2.2 Discussion of the connotations of adult competency development and related topics

Competency is derived from McClelland (1973), who replaced competence measurement with intelligence, and used it as a predictor of personal success. The connotations of competency encompass knowledge, skills, and feelings, and competence is measure by external behaviors. Additionally, competence can be subdivided into the abilities "acquired" by individuals, and means the acquisition of knowledge, feelings, skills, and "demonstrated" abilities, namely, the successful completion of certain tasks according to the actual behavioral performances [16]. Cheng et al. (2009) defined competency as "an ability to use knowledge, attitude, and skills to perform a task, as well as all behaviors, motives, and knowledge about the success or failure of work [17]. In other words, the scope and context of competency are broader than those of ability, and are primarily related to the high-performance conditions required by a particular role or work, including the basic ability of the lowest standard [18-20].

Competency models refer to the combination of knowledge, skills, abilities, and other characteristics (KSAOs) necessary for high performance at work [21]. In addition, competency can be classified into professional competency, core competency, and management competency. Professional competency referred to the duty abilities of employees engaged in specific works, which they must possess in order to achieve specific work objectives. Secondly, core competency was common competency, also known as "common core competency", and was the competency each member of an organization must possess. Thirdly, management competency meant the duty abilities assumed by management in an organization, which they must possess to implement each stage of management [22-28].

With reference to the data of different countries, Taiwan's Workforce Development Agency (2015) established a reference structure of occupation categories with 16 areas, including (1) natural resources, food, and agriculture; (2) manufacturing; (3) building construction; (4) marketing and sales; (5) logistics and transportation; (6) leisure and sightseeing tourism; (7) information technology; (8) finance and accounting; (9) science, technology, engineering, and mathematics; (10) government public affairs; (11) judiciary, laws, and public security; (13) health care; (14) personal and social services; (15) arts and audio and video communication; (16) business operation and management [29]. In addition, the Occupational Competency Standard (OCS) meant that, without the restriction of specific work tasks, but with the scope of a few occupational competency standard units, the descriptions of work scopes were sorted, work tasks were developed, and the universality and necessity of the connotations of competency, as required by industry, were demonstrated. Moreover, grades were classified based on the maturity of occupational performance, which was the basis for the degree of professional occupational performance [29]. The items, indicators, and standards of occupational competency standards and competency grading are considered as reference for human resources training and career development, as well as a reference standard for developing adult employment, entrepreneurship, occupational change, and other competencies.

2.3 Discussion of the models of adult competence training programs, connotations of assessment adaptability, and related topics

2.3.1 Models of adult competence training programs

The American Society for Training and Development placed emphasis on the ADDIE model for training development, which includes the five stages of analysis, design, development, implementation, and evaluation [30-31], and renders the program design structured and systematic. Taiwan's Workforce Development Agency incorporated ISO10015, European vocational education and training (VET), UK Investors in People (IIP), Australia's active vocational training policy, and the challenges of a globalized knowledge-based economy, as faced by Taiwan, and formulated the "Talent Quality Management System" suitable for Taiwan [32-35].

TTQS stresses the construction of a strategic human resource development system, and improves the system with the training quality cycle of Plan → Design→ Do → Review → Outcome (PDDRO), which includes detailed indicators and specifications, and can be used as an Indicator reference for engaging in the human resources training of an organization. In particular, "Plan" refers to the relevance of training planning and business development objectives, as well as the operational ability of the training system. "Design" refers to the systematic design of the training programs (including the participation of stakeholders, the degree of meeting demands, the standards for selecting programs, and procurement standards and procedures). "Do" refers to the degree of training implementation and the systematization of training records and management. "Review" refers to the regular implementation of analysis, whole-process monitoring, and exception handling of training. "Outcome" refers to the level and completeness of the assessment of training results and continuous training improvements [36]. There are similarities in the framework and processes of ADDIE TTQS, and PDDRO, and ADDIE has become the competency training operating system generally adopted by Taiwan's public institutions and training institutions at present.

In response to the industrial trend of Industry 4.0, adult career planning and competency training should be planned, designed, and executed by strengthening training customization, operational intelligence, educational efficiency, program integration, and other directions [37]. Tsai and Hsu (2004) and Chen (2013) carried out analysis of training programs for adult learning fields, which was divided into planning the following four learning fields: (1) Learning field of business organizations for workplace participation, where adult learners developed adult learning programs for competency orientation through growth learning and adaptability, and by means of training; (2) Learning field of education and training organizations for independent development, where adult learners developed skill training programs for application orientation through growth learning and innovative learning, and by means of education and training; (3) School learning field of participation in recurrent education, where teaching programs were developed through growth learning and transformative learning, and by means of education; (4) Life learning field of residential communities participated by residential communities, where life development learning programs were developed through transformative learning and adaptive learning, and via informal and liberal arts-oriented learning community awareness and social actions.

Dalziel, et al. (2016) proposed the concept map of learning design (see Figure 1), which held that, regarding the formulation of learning maps for adult learners' career planning and competence development, creating specific learning objectives and teaching methods that meet adults' learning experiences can be used as the challenge framework of the learning design concept, where the adult learners' learning feedback, learning test assessment, learning behavior analysis, learning effectiveness evaluation, and other learner responses, were the bases for developing learning instruments and resources. Subsequently, the image-based demonstration of guiding knowledge and technological learning concepts were developed, and the shared core concept of the learning design was regarded as the recycling benchmark for the operation of various teaching activities, such as "design and planning of teaching activities → implementation of teaching activities → reflection feedback and improvement → teaching professional competency development". Additionally, the operational recycling of teaching activities must apply a philosophical foundation of education, teaching theories, and teaching methods, as well as the characteristics and values of learning environments, learning activities, theme discussions, curriculum modules, learning plans, and other teaching structures and levels, as the principle and reference structure of teaching operations, in order to attain the objectives of teaching activity design.

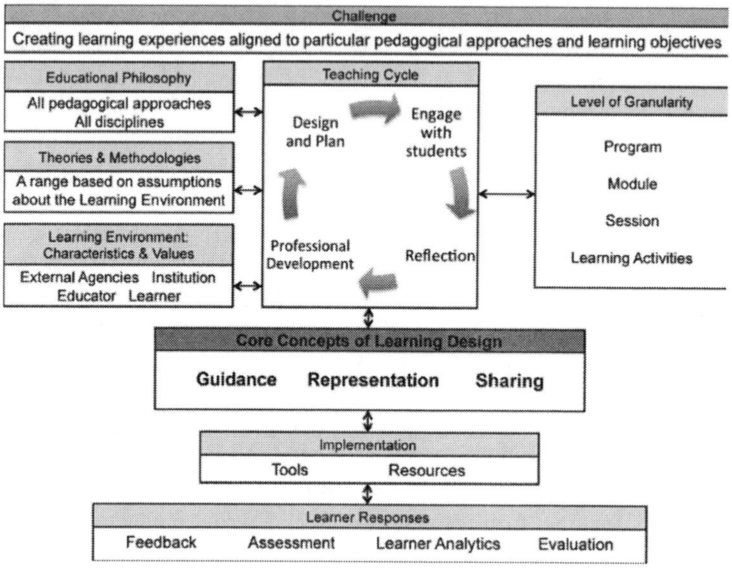

Figure 1. Concept Map of Learning Design. Source: Dalziel, et al., 2016, 10

2.3.2 Assessment adaptability of adult competency training programs

In addition to meeting the characteristics of adult learning, the adaptive assessment of adult competency training programs attaches greater importance to achieving the objectives of learning tasks and occupational performance as the specifications of learning effectiveness. The methods for collecting the results of assessment information regarding adult competency training include rigid information, such as output, time, cost, and quality, as well as soft data, such as work habits, skills, work atmosphere, pioneering initiative, and attitude. The six commonly used methods of information collection are, as follows: (1) Questionnaires: open, check, alternative, multiple choice questions, and rating; (2) Test: pen and paper tests - operation of examination, diagnosis, and preparation; oral tests - work procedures, quality and speed of knowing how to do tasks, and doing them right and skillfully; (3) Interview: course response, learning effects, job application, and behavioral performance improvement; (4) Observation: multiple observations of many people from the assessment center; (5) Focus group: in-depth assessment information about training courses obtained from group discussions; (6) Experimental design: one pre-test group and one post-test group [34].

The learning effectiveness of adult competency training programs include the following three major areas: (1) Cognition, meaning what trainees think they have learned from the training? Why they had to learn the content? How does the training help them and their work? (2) Affectivity, indicating whether the willing attitude of the trainees in training can cause resonance, and whether they can accept the training ways and contents? (3) Skill-based results, suggesting that trainees can systematically and gradually transfer what they have learned to their work to constantly strive for perfection; additionally, after their study contents serve the practical purposes, they must also understand whether their career planning and occupational performance have been improved, in order to demonstrate the final results of progress in personal and organizational performance after training, such as increased sales, increased production capacity, profit growth, reduced costs, improved quality, and decreased staff turnover rate. In addition, the cognition and feelings of high-level executives about the training developments in the organization are indispensable assessment items for the assessment system of performance-oriented training results, and the viewpoints of such high-level executives regarding training performance can be understood by regular questionnaire surveys and visits. Secondly, information about the internal operational performance of enterprises should be collected and analyzed in order to grasp the achievements of the business performances of different departments. Finally, the high-level executives refer to the executives that produced key influences on business objectives, strategies, and allocation of training resources [1].

The training result assessment, as proposed by Kirkpatrick (1998), involved reaction, learning, behavior and results, and the operation of TTQS and ADDIE training models and processes, which considered this assessment method as the main operational reference and basis. The adaptive assessment system of adult competency training programs could assess learning effectiveness in accordance with training procedures, and the content operation is, as follows [1], [39]:

1. Pre-training analysis: This assessment focuses on training needs and objectives, including whether the proposal, summarization, analysis, and identification of training needs are appropriate? Whether the analysis of the standards for achievable results through training is reasonable? Whether the selection of training objectives and participating employees is consistent with the business strategies of organizations?
2. Pre-training planning and development: This assessment concentrates on training program design, including whether the training plan is consistent with the business strategies of the organizations? Are the formulated training programs complete?
3. Implementation of training programs: This assessment focuses on the operation of training programs, including whether the program arrangements meet the training needs? Whether the training hours and time arrangements are appropriate? Whether the lecturers' expertise properly matches the training programs? Whether the training plan complies with the annual plan? In case of any differences, what are the reasons for such differences? What are the changes in the original planning and implementation? In case of any changes, what are the differences? Whether training funds and actual allocation are reasonable? Whether the written and oral tests are consistent with the teaching objectives?
4. Status of the objective achievements of trainees after training: This assessment concentrates on whether the trainees achieve their learning objectives, including the trainees' understanding of the theories and concepts in learning? Whether trainees grasp knowledge and skills in learning? How trainees improve their work attitudes and behaviors through learning?
5. Return to work of trainees after training: This assessment focuses on the application and effectiveness of training and learning, including how trainees' work attitudes and behaviors are improved after training? Whether the training helps trainees to improve their job performance? Whether the training helps trainees to hone their work abilities?

3 Research hypotheses and tools

3.1 Research hypotheses

This study integrates and summarizes the following hypotheses, as based on the discussions of the relevant topics of the abovementioned literature:

H1: Adults with different backgrounds have significant differences in their cognition and feelings regarding program design, program contents, and assessment adaptability.

H2: Adults are significantly related to program design and contents.

H3: Adults have significant predictive power of the assessment adaptability of the program design

H4: Adults have significant predictive power of the assessment adaptability of the program contents.

3.2 Research tools

Through the abovementioned literature exploration and analysis, this study summarizes and edits the *Preliminary Questionnaire about the Overview of Adult Competency Training Programs and Assessment Adaptability*, and employs adult competency training practitioners, as well as scholars and experts of university research in adult education, as the experts of the expert validity of the questionnaires. With regard to adult competency training practitioners, one director from labor unions, chambers of commerce, and community development associations are invited, for a total of 3 practitioners. Secondly, in terms of university professors and scholars in charge of adult education or occupational training are invited, including seven experts, respectively, from National Pingtung University, National Pingtung University of Science and Technology, Lunghwa University of Science and Technology, National Kaohsiung First University, Aletheia University, National Open University, and Open University of Kaohsiung, who acted as the review experts for the expert validity of this study.

This study regards one class of adult learners receiving in-service training in counties and cities, as well as the Open University of Kaohsiung in Taiwan, as the sample unit, and employs a questionnaire survey of random sampling, in order to understand different personal background variables, such as gender, age, educational background, and occupation. The adult competency training program design includes five stratification factors: program needs, program planning, program development, program implementation, and program assessment, for a total of 25 questions. The program contents include three stratification factors: core competency, management competency, and professional competency, for a total of 25 questions. The assessment adaptability of learning effectiveness includes three stratification factors: learning satisfaction, improvement of knowledge-ability, and ensuring the study serves the practical purposes and competence performances, for a total of 20 questions. Through questionnaires, this empirical study regards the listed items as the hypotheses to verify this study, and the results as the operational system to improve future adult competency training programs and adaptive assessments, in order to enhance the benefits of adult competency training, improve occupational performance, and increase competitive advantages in the workplace.

4 Research results

4.1 Descriptive statistical results

The descriptive statistical results of this study are shown in Table 1. Regarding respondents' gender, there are 175 males, accounting for 52.1% of the total respondents, and 161 women, accounting for 47.9%. In terms of the respondents' age, the number of "18-40 years old" is the largest, totaling 223 persons and accounting for 66.4% of the total number of respondents; 107 respondents are "41-60 years old", representing 31.8% of the total number of respondents, and 6 respondents are at least 61 years old, accounting for 1.8%. Regarding the respondents' educational background, the number of respondents graduating from "senior high school and vocational school" is 147, accounting for 43.8% of the total number of respondents; a total of 146 respondents graduated from "college and university", accounting for 43.5%, and a total of 26 respondents from "graduate school" account for 7.7% of the total number of respondents. The least number of respondents graduated from "below (including) junior high school" account for only 5.1% of the total number of respondents. With regard to the occupational distribution of respondents, 188 respondents work in "information technology", accounting for 56.0% of the total number of respondents, and the least number of respondents in "farming, forestry, and fishing animal husbandry total 7 people, accounting for only 2.1% of the total number of respondents.

Table 1. Questionnaire about Program Design Models, Contents, and Competence Performance of Adult Competency Training.

Variable	Group	Number of people	Percentage	Cumulative Percentage
Gender	Male	175	52.1	52.1
	Female	161	47.9	100.0
Age	18—40 years old	223	66.4	66.4
	41—60 years old	107	31.8	98.2
	Above 61 years old	6	1.8	100.0
Educational background	Below (including) junior high school	17	5.1	5.1
	Senior high school and vocational school	147	43.8	48.8
	College and university	146	43.5	92.3
	Graduate school	26	7.7	100.0
Occupation	Farming, forestry, fishing animal husbandry	7	2.1	2.1
	Manufacturing	12	3.6	5.7
	Business and service	92	27.4	33.0
	Information technology	188	56.0	89.0
	Civil servants	27	8.0	97.0
	Unemployed	10	3.0	100.0

N=336

4.2 Confirmatory factor analysis of questionnaires

4.2.1 Validation of convergent validity

Confirmatory factor analysis (CFA) is part of SEM analysis. Thomopson (2004) pointed out that this measurement model could correctly reflect the dimensions of a study, thus, analysis of the measurement model must be carried out prior to analysis of the SEM model. This study conducts SEM model testing according to the two-stage model analysis method proposed by Kline (2005). First, the measurement model is tested to determine whether the results show that the goodness of fit indicator of the measurement model is acceptable, then, assessment of the overall SEM model is carried out [50]. The CFA measurement indicators adopt the standards recommended by Hair, Anderson, Tatham, and Black (2009) and Fornell and Larcker (1981), as follows: (1) Factor loading is greater than .50; (2) Composite reliability is greater than .60; (3) the average extraction number of variables is greater than .50. The CFA analysis results of this study illustrate that the factor loadings of the dimensions range between .71 to .94, the composite reliability is from .91 to .97, and the average extraction number of variables ranges from .68 to .84 (Table 2), indicating that the dimensions of this study have convergent validity.

Table 2. Summary of Verification Information about Convergent Validity.

Model Parameter Estimates						Convergent Validity			
Latent variables	Observation variables	Non-standardized factor load	Standard error S.E.	t-value	P	Standardized factor load	SMC	C.R	AVE
Program needs	Aa01	1				0.88	0.78	0.93	0.72
	Aa02	1.06	0.04	24.90	***	0.92	0.84		
	Aa03	0.98	0.04	22.93	***	0.88	0.78		
	Aa04	0.95	0.05	20.81	***	0.84	0.71		
	Aa05	0.87	0.06	15.56	***	0.71	0.50		
Program planning	Ab01	1				0.82	0.67	0.94	0.77
	Ab02	1.12	0.05	20.84	***	0.90	0.82		
	Ab03	1.10	0.06	19.23	***	0.86	0.74		
	Ab04	1.08	0.05	20.11	***	0.88	0.78		
	Ab05	1.15	0.06	20.83	***	0.90	0.82		
Program development	Ac01	1				0.89	0.80	0.95	0.79
	Ac02	1.02	0.04	25.51	***	0.91	0.82		
	Ac03	0.95	0.04	24.00	***	0.88	0.78		
	Ac04	1.00	0.04	24.21	***	0.89	0.79		
	Ac05	0.92	0.04	22.73	***	0.86	0.74		
Program implementation	Ad01	1				0.94	0.88	0.96	0.84
	Ad02	0.99	0.03	32.37	***	0.93	0.86		
	Ad03	0.94	0.03	30.58	***	0.91	0.83		
	Ad04	0.99	0.04	27.23	***	0.88	0.77		
	Ad05	0.97	0.03	30.35	***	0.91	0.83		
program assessment	Ae01	1				0.87	0.76	0.94	0.76
	Ae02	1.02	0.04	23.73	***	0.90	0.82		
	Ae03	1.05	0.06	19.15	***	0.81	0.65		
	Ae04	0.95	0.04	21.80	***	0.87	0.75		
	Ae05	1.09	0.05	23.80	***	0.91	0.82		
Core competency	Ba01	1				0.84	0.71	0.97	0.77
	Ba02	1.03	0.05	20.86	***	0.87	0.75		
	Ba03	0.96	0.05	20.44	***	0.86	0.73		
	Ba04	1.01	0.05	20.81	***	0.87	0.75		
	Ba05	1.14	0.05	22.74	***	0.91	0.83		
	Ba06	1.09	0.05	21.58	***	0.88	0.78		
	Ba07	1.05	0.05	21.09	***	0.87	0.76		
	Ba08	1.10	0.05	22.32	***	0.90	0.81		
	Ba09	1.05	0.05	21.88	***	0.89	0.79		
Management competency	Bb01	1				0.82	0.67	0.97	0.79
	Bb02	1.16	0.06	20.38	***	0.88	0.78		
	Bb03	1.11	0.05	21.21	***	0.90	0.82		
	Bb04	1.09	0.05	20.19	***	0.88	0.77		
	Bb05	1.16	0.06	20.92	***	0.90	0.80		
	Bb06	1.19	0.05	22.08	***	0.92	0.85		
	Bb07	1.13	0.05	20.95	***	0.90	0.80		

Model Parameter Estimates						**Convergent Validity**			
Latent variables	Observation variables	Non-standardized factor load	Standard error S.E.	t-value	P	Standardized factor load	SMC	C.R	AVE
	Bb08	1.12	0.06	20.43	***	0.88	0.78		
	Bb09	1.13	0.05	21.25	***	0.90	0.82		
Professional competency	Bc01	1				0.83	0.69	0.96	0.79
	Bc02	1.10	0.05	22.07	***	0.91	0.83		
	Bc03	1.07	0.05	21.59	***	0.90	0.81		
	Bc04	1.13	0.05	21.72	***	0.91	0.82		
	Bc05	1.05	0.05	20.67	***	0.88	0.77		
	Bc06	1.07	0.05	21.04	***	0.89	0.79		
Learning satisfaction	Ca01	1				0.81	0.66	0.95	0.79
	Ca02	1.09	0.05	20.03	***	0.89	0.79		
	Ca03	1.14	0.05	21.81	***	0.94	0.89		
	Ca04	1.07	0.06	19.49	***	0.88	0.77		
	Ca05	1.07	0.05	20.60	***	0.91	0.8273		
Improvement of knowledge-ability	Cb01	1				0.82	0.67	0.91	0.68
	Cb02	1.18	0.06	20.26	***	0.90	0.81		
	Cb03	1.27	0.06	21.17	***	0.93	0.86		
	Cb04	0.53	0.05	11.42	***	0.59	0.35		
	Cb05	1.10	0.06	18.40	***	0.84	0.71		
Making study serve the practical purpose	Cc01	1				0.84	0.71	0.95	0.81
	Cc02	1.14	0.05	22.36	***	0.90	0.82		
	Cc03	1.21	0.05	22.69	***	0.91	0.83		
	Cc04	1.17	0.05	22.64	***	0.91	0.87		
	Cc05	1.19	0.05	23.02	***	0.92	0.84		
Competence performance	Cd01	1				0.84	0.71	0.94	0.75
	Cd02	1.09	0.05	21.80	***	0.90	0.81		
	Cd03	1.26	0.06	20.08	***	0.86	0.73		
	Cd04	1.11	0.06	18.48	***	0.82	0.66		
	Cd05	1.17	0.05	23.01	***	0.93	0.86		

***P<.001

4.2.2 Validation of discriminant validity

Table 3. Bootstrap Correlation Coefficient in 95% Confidence Interval.

Parameter			**Estimate**	**Bias-corrected**		**Percentile method**	
				Lower Bound	Upper Bound	Lower Bound	Upper Bound
B0 program contents	<-->	C0 Learning effectiveness	.30	.22	.42	.21	.41
B0 program contents	<-->	A0 Program design	.31	.24	.42	.23	.41
A0 Program design	<-->	C0 Learning effectiveness	.29	.21	.41	.21	.39

There are two estimates for the AMOS confidence interval: the Bias-corrected Percentile Method and Percentile Method [49]. In this study, the bootstrap confidence interval is used to validate the discriminant validity of the dimensions, and the 1000 repetitions are estimated under the 95% confidence interval. If the correlation coefficient of dimensions is less than 1 in the 95% confidence interval, then the dimensions have discriminant validity [47], [49], [58]. The results of this study are presented in Table 3, and the confidence interval between dimensions is less than 1, showing discriminant validity in the three dimensions.

4.2.3 Goodness of fit analysis

In this study, the goodness of fit is based on the recommendations of Wu (2007), Hsu (2010), Bagozzi and Yi (1988), and Hair et al. (1998), verified with p^2, and includes the ratio of o^2 to degree of freedom, goodness of fit (GFI), adjusted goodness of fit (AGFI), root mean square error of approximation (RMSEA), comparative fit index (CFI), incremental fit index (IFI), and non-normed fit index (NNFI). However, when the sample size of SEM exceeds 200, the chi-square value (χ^2 = (n-1) Fmin) is easily too large. In other words, a greater sample size results in a greater chi-square value, thus, the p value is easily refused [48-49]. hence, Bollen and Stine (1992) suggested using Bootstrap for model correction. As the chi-square value becomes smaller after correction, all GFIs must be re-calculated, and the calculation results are shown in Table 4, which suggest that the GFI of the model meets the standards for judgment.

Table 4. Summary of GFI of the Model.

Fitness Indicator	Standard Value		Model GFI Judgment
χ^2	The smaller the better	4652.78	Passed
χ^2/df	<3	2.06	Passed
GFI	>.9	.87	Accepted
AGFI	>.9	.86	Accepted
RMSEA	<.08	.06	Passed
TLI (NNFI)	>.9	.92	Passed
CFI	>.9	.93	Passed

4.3 Analysis of the differences of different personal background variables in adult competency training programs and assessment adaptability

The purpose of independent sample t testing is to infer whether the average value of the original two populations have statistical methods with significant differences after the average value of the two independent groups of samples are calculated [51]. In this study, the differences of different genders in the variables of the adult competency training program models are analyzed by independent samples t-testing. One-way ANOVA is used to test a single variable under different influences, and observe whether the average values of the variables have significant differences. If the test results show that the F value reaches significant level, it means that there is significant difference in at least one set of averages between the groups, and that "multiple-comparison post-hoc" analysis must be used to confirm the differences [45].

4.3.1 Analysis of the differences of gender in adult competency training programs and assessment adaptability

The gender differences of the research subjects in program design, program contents, and assessment adaptability are compared and analyzed. In accordance with the independent samples t-test results, gender difference shows no significant differences regarding the averages of program needs (t = -1.19, p = .85), program planning (t = 1.02, p = .31), program development (t = -1.41, p =.16), program implementation (t=.72, p=.47), program assessment (t=-1.44, p=.15), core competency (t=1.62, p=.11), management competency (t=-1.50, p=.13), professional competency (t=1.18, p=.24), learning satisfaction (t=-1.35, p=.18), improvement of knowledge-ability (t=1.12, p=.26), ensuring study serves the practical purpose (t=-1.45, p=.15), or competence performance (t=.78, p=.44). In other words, different genders show no significant differences in adult competency training programs, program contents, or assessment adaptability.

4.3.2 Analysis of the differences of age in adult competency training programs and assessment adaptability

The age differences of the research subjects in program design, program contents, and assessment adaptability are compared and analyzed. According to the one-way ANOVA verification results, age difference shows no significant differences in the averages of program needs (F =2.40, p=.09), program implementation (F=1.13, *p*=.33), program assessment (F =1.41, *p*=.25), core competency (F =1.99, *p*=.14), management competency (F =-2.28, *p*=.10), professional competency (F =2.53, *p*=.08), learning satisfaction (F =-1.87, *p*=.16), improvement of knowledge-ability (F =1.77, *p*=.17), ensuring study serves the practical purpose (F =-1.83, *p*=.16), or competence performance (F =1.76, *p*=.17). In other words, different ages show no significant differences in adult competency training programs, program contents, or assessment adaptability. However, program development (F=3.19*, *p*=.04) reaches significant difference, and after Scheffe post-hoc comparison, all ages fail to reach significant differences.

4.3.3 Analysis of the differences of educational background in adult competency training programs and assessment adaptability

The educational background differences of the research subjects in program design, program contents, and assessment adaptability are compared and analyzed. According to the one-way ANOVA verification results, educational background difference shows no significant differences in the averages of program development (F =2.62, *p*=.05), program implementation (F =1.09, *p*=.35), or professional competency (F =1.97, *p*=.12). In other words, different educational backgrounds show no significant differences in adult competency training programs, program contents, or assessment adaptability. However, program demand (F =3.03*, *p*=.03), core competency (F =3.24*, *p*=.02), learning satisfaction (F =4.00*, *p*=.01), and improvement of knowledge-ability (F =4.48*, *p*=.00) reach significant differences. After Scheffe post-hoc comparison, the results show the differences of "graduate school" are greater than those of "college and university"; "ensuring study serves the practical purpose (F =5.86*, *p*=.00), and competence performance (F =6.90*, *p*=.00) reach significant differences. After Scheffe post-hoc comparison, the results show the differences of "graduate school" are greater than those of "senior high school and vocational school" and "college and university"; however, the differences of "senior high school and vocational school" and "college and university" fail to reach significant difference. Program planning (F =2.96*, *p*=.03) and management competency (F =-2.94*, *p*=.03) reach significant difference, and after Scheffe post-hoc comparison, no educational backgrounds reach significant differences.

4.3.4 Analysis of the occupation differences in adult competency training programs and assessment adaptability

The occupation differences of the research subjects in program design, program contents, and assessment adaptability are compared and analyzed. According to the one-way ANOVA verification results, the age difference shows no significant differences in the averages of program needs (F=.87, *p*=.50), program planning (F=.42, *p*=.84), program development (F=1.10, *p*=.36), program implementation (F=.72, *p*=.61), program assessment (F=.82, *p*=.53), core competency (F=.91, *p*=.47), management competency (F=.91, p=.43), professional competency (F=.99, *p*=.42), learning satisfaction (F=1.84, *p*=.11), knowledge-ability (F=1.33, *p*=.25), ensuring study serves the practical purpose (F=1.05, *p*=.39), or competence performance (F=1.16, *p*=.33). In other words, different occupations show no significant differences in adult competency training programs, program contents, or assessment adaptability.

4.4 Verification of study hypotheses

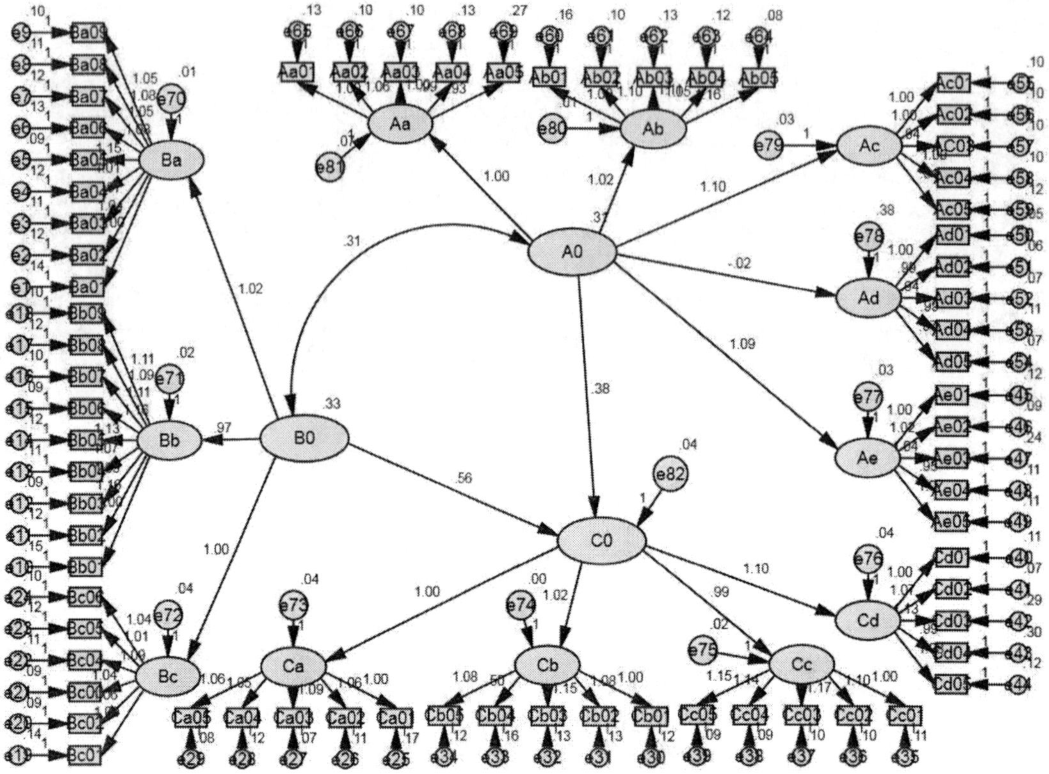

Figure 2. Statistical Model of Adult Career Planning Program Design, Program Contents, and Adaptive Assessment Learning Effectiveness.

In this study, there are no significant differences in the cognition and feelings of adults with different backgrounds regarding program design, program contents, or assessment adaptability. According to the calculation results, as shown in Table 5 and Figure 2, adult career planning program design and program contents have a path relationship with assessment adaptability. Adults are positively significantly related to program design and program contents; adults have significant predictive power of assessment adaptability of program design; adults have significant predictive power of assessment adaptability of program contents. To sum up, the validation results of the hypotheses in this study are detailed, as follows:

- H1: The hypothesis that adults with different backgrounds have significant differences in their cognition and feelings about program design, program contents, and assessment adaptability is partially supported.
- H2: The hypothesis that adults are significantly related to program design and program contents is supported.
- H3: The hypothesis that adults have significant predictive power of the assessment adaptability of program design is supported.
- H4: The hypothesis that adults have significant predictive power of the assessment adaptability of program contents is supported.

Table 5. Summary of Empirical Results of Research Hypotheses.

Hypothesis	Path Relationship			Path Value	C.R.	*p*	Supported Hypothesis
2	Program design	<-->	Program contents	.31	10.22*	***	Supported
3	Learning effectiveness	<--->	Program design	.38	2.96*	***	Supported
4	Learning effectiveness	<--->	Program contents	.56	4.40*	***	Supported

*P<.05 ***P<.001

5 Conclusion and suggestions

Global, diversified, and uncertain workplace environments of the economic industry causes adults, in the face of career planning, to evolve from the "linear trend" to the "nonlinear and changeable trend", in an effort to develop proper career development goals and re-establish their goals at any time, and then, learn and sharpen their skills, in order that they do not become obsolete in the workplace. In addition, as science and technology are constantly developing, adult learners must re-adjust their career planning, as well as the items and connotations of professional and core competency, and learn new knowledge and new technology, especially as human resources may deviate from the trend of Industry 4.0. Under the impacts and challenges of workplace environment, if adults fail to actively identify the functions and positioning of human resources in the global industrial chain, it is difficult for them to persistently develop their career and competency.

Therefore, this study proposes the following adult learner attributes as the foundation for career planning and competency development program design: develop design architectures of program learning concepts for adult learners' career planning and competency development, and formulate a performance-oriented diversified result assessment system and other feasible strategies for performance-oriented training for career planning and career development. These items can provide recommendations and references for adult education institutions regarding adult learner professional career and competency development programs, as well as teaching activity designs, and assist adult learners to construct adaptive program learning maps, which include cross-domain, integrated, contingent, and appropriate career planning and development competency items. Consequently, when confronted with the current paradoxical world context, adults should take a positive attitude, keep up with the times, and continuously hone their skills, in order master the best career development opportunities.

References

1. I.C. Kao, Journal of Agriculture and Leisure Industry Research, **7**(1), 17-38 (2017)
2. S.J. Wang, *Globalization and Adult Education*, 99-122 (2002)
3. I.C. Kao, Empirical Study on the Learning Purposes, Contents, and Methods for Adults in Response to Globalization. Storage Management Solutions, **1**, 199-220 (2016)
4. I.C. Kao, Academy of Education Institute, **1**, 2-10 (2003)
5. I.C. Kao, Journal of Urbanology, **1**(1), 85-113 (2010)
6. F.S. Huang, *Introduction to Adult Education* (Wu Nan Books, Taipei, 2000)
7. Adult Education Institute of Republic of China, *Globalization and Adult Education*(ShiTa Books, Taipei, 2014)
8. P.C. Tsai and W.Y. Wu, *Adult Education* (Li Wen Publishing, Kaohsiung, 2010)
9. D.M. Randall, Academy of Management Review, **12**(3), 462 (1987)
10. T.H. Yeh, Research into the Impact of Workplace Changes on Adult Career Planning(master's thesis), Graduate Institute of Guidance and Counseling of National Chi Nan University (to be published)

11. H.L. Peng, M.H. Chiang and Y.S. Lin(trans), *Adult Career Development — Concepts, Topics and Practices (Original author: Spencer G. Niles)* (Psychological Publishing, Taipei, 2009)
12. P. Herriot and C. Pemberton, Human Resources Management Journal, **7**(1), 45-56 (1997)
13. N. Nicholson, Academy of Management Review, **23**(1), 40-51 (1996)
14. R. Driver, A. Squires, P. Rushworth and V. Wood-Robinson, *Making Sense of Secondary Science* (Routledge, London, 1994)
15. D.K. Duys, J.E. Ward, J.A. Maxwell and L.E. Comerford, The Career Development Quarterly, **56**, 232-241 (2008)
16. C.L. Kang, Employment and Training, **15**(6), 3-9 (1997)
17. S.C. Wu, Psychological Testing, **60**(2), 397-428 (2013)
18. C.J. Chiu, T&D FASHION, **202**, 1-32 (2015)
19. W.C. Liao, Competency Implementation and Packaged Training, *2010 Taiwan Labor Forum*(Taipei, Taiwan, 2010)
20. P.W. Huang, Competency Analysis and Development. Published in Affiliated Vocational Training Center of DerJian Enterprises Ltd., Alltalent Education Development Co., Ltd. (ed.), *Training Manager Level 5 Certification Course Materials* (39-139) (Tainan, Taiwan, 2014)
21. M.A. Campion, A.A. Fink, B.J. Ruggeberg, L. Carr, G.M. Phillips and R.B. Odman, Personnel Psychology, **64**, 225-262 (2011)
22. L.M. Spencer and S.M. Spencer, *Competence at work: Models for superior performance* (New York, USA, 1993)
23. P.J. Guglielmino, Training & Development Journal, **33**(4), 12-14 (1979)
24. R.L. Katz, Harvard Business Review, **33**(1), 33-42 (1955)
25. M.J. Huang and Y.C. Cheng, The Journal of Guidance & Counseling, **30**(1), 61-87 (2008)
26. Y.S. Lan, Bottleneck and breakthrough in the Use of Competency (1), November 18, 2014, from SUNNET: http://www.sun.net.tw/knowledge/knowledge.php?type= 0 (2009)
27. J.J. Li and J.C. Cheng, Construction of Management Competency Model, W.C. Lin(Host), Management Competency Development, 14th Presentation of Special Research Results into Enterprise Human Resource Management Practices, National Central University (2008)
28. G.H. Deng, Establishment of Evaluation Table of Executives' Management Ability and Validity Analysis - A Case Study of a Commercial Bank(master's thesis), Institute of Human Resource Management of National Central University (to be published)
29. Workforce Development Agency, "Occupation" and "Class", December 15, 2015, from the website of iCAP Competency Development Platform: https://icap.wda.gov.tw/Knowledge/knowledge_introduction.aspx (2015)
30. J.C. Kuo, Employment Security, **9** (1), 60-66 (2010)
31. C. Hodell, *Basics of Instructional Systems Developmen*(ATD, USA, 2005)
32. W.T. Lin, C.Y. Kung and L.L. Lin, Analysis of Differences in IIP, ISO 10015 and TTQS, Quality Journal, **45**(4), 52-56 (2009)
33. C.S. Lin, Human Capital Development and National Training Quality Assurance System Employment Security, **5**(2), 44-51 (2006)
34. Workforce Development Agency, TTQS Evaluation Indicators. November 12, 2014, from the website of TTQS: http://ttqs.wda.gov.tw/ (2014)
35. J.C. Liao, Discussion on Improving Education, Training and Human Resource Quality from International Standards ISO 10015 and IIP, Quality Journal, **40**(10), 28-31 (2004)
36. I.C. Kao, Construction of Human Resource Development Strategy of Community Empowerment by Project Management, TOC and TTQS, Storage Management Solutions, **2**, 42-65 (2014)
37. J.C. Chang, "Technical and Vocational Education" in Response to Industry 4.0, May 18, 2017, from the website of Craftsmanship Insights: http://www.tvet3.info/vocational-education-4-0 (2017)
38. J. Dalziel, et al., Journal of Interactive Media in Education, **2016**(1):7, 1-24 (2016)

39. Workforce Development Agency, TTQS Uniform Textbooks for Indicator Programs for Enterprises and Institutions, November 12, 2014, from the website of TTQS, http://ttqs.wda.gov.tw/ (2014)
40. S.C. Cheng, P.C. Lin and Y.C. Cheng, Psychological Testing, **56**(3), 397-430 (2009)
41. D.C. McClelland, American Psychologist, **28**(1), 1-24 (1973)
42. T.L. Chen, T&D Fashion, **161**, 1-21(2013)
43. P.C. Tsai, Y.H. Hsu, Construction of Adult Learning System in the Trend of Globalization. Chinese Community Education Association (Chief Editor), *Globalization and Adult Education*, 331-355 (Normal University Press, Taipei, 2004)
44. M.L. Wu, *Structural equation modeling: Operation and application of AMOS* (Wu Nan Books, Taipei, 2007)
45. M.L. Wu and C.T. Tu, *SPSS & the Application and Analysis of Statistics* (Wu Nan Books, Taipei, 2005)
46. M.C. Hsu, Journal of Sport and Recreation Management, **7**(1), 174-18 (2010)
47. M.C. Hsu, F.C. Fan and C.M. Huang, Journal of Taiwan Society for Sport Management, **11**(2), 85-107 (2011)
48. M.C. Hsu and H.P. Yen, International Journal of LISREL, **6**(1), 24-56 (2013)
49. W.H. Chang, *Techniques of thesis writing* (Tingmao, Taipei, 2011)
50. S.Y. Chen, M.C. Hsu and F.E. Li, Journal of Leisure and Recreation Industry Management, **5**(3), 1-21 (2012)
51. K.Y. Chen and C.H. Wang, *Thesis statistical analysis practice SPSS and operation of AMOS* (Wu Nan Books, Taipei, 2011)
52. C. Fornell and D.F. Lacker, Journal of Marketing Research, **18**, 39-50 (1981)
53. J.F.Jr. Hair, R.E. Anderson, R.L. Tatham and W.C. Black, *Multivariate data analysis (7th ed.)* (Prentice Hall, New Jersey, 2009)
54. R.B. Kline, *Principles and practice of structural equation modeling (2nd ed.)* (Guilford Press, New York, 2005)
55. B. Thomopson, *Exploratory and confirmatory factor analysis: Understanding concepts and applications* (American Psychological Association, Washington DC, 2004)
56. R.P. Bagozzi and Y.Yi, Journal of the Academy of Marketing Science, **16**, 74-94 (1988)
57. J.F.Jr. Hair, R.E. Anderson, R.L. Tatham and W.C. Black, *Multivariate data analysis (5th ed.)* (Prentice Hall, New Jersey, 1998)
58. G. Torkzadeh, X. Koufteros and K. Pflughoeft, Structural Equation Modeling, **10**(2), 263-275 (2003)

Empirical Study on elderly learners' needs and obstacles in network learning by "3C products"

I-Chan Kao[a]

General Education Center, Open University of Kaohsiung, Taiwan

Abstract. This study treated elderly learners at Open University of Kaohsiung who participated in online distance learning as samples. By a literature review and questionnaire survey, it probed into elderly learners' perceived reactions and feedback for network learning using consumer electronics products and analyzed the general situation and relationship between elderly learners' needs and obstacles in network learning with consumer electronics products. By empirical research, this study investigated a total of 472 elderly learners aged at least 55 years old in elderly programs at Open University of Kaohsiung, for-credit continuing education courses, and on-the-job courses of universities in different counties and cities around Taiwan. This study found that elderly learners with a lower educational level encounter more significant learning obstacles in network learning using consumer electronics products. In addition, this study found that when elderly learners' needs to apply consumer electronics products are higher, their network learning obstacles will be lower. Based on the findings, this study suggested that it shall help elderly learners with lower educational level reinforce learning knowledge and competence to apply consumer electronics products to avoid obstacles for network learning.

1 Introduction

1.1 Research motives

3C refers to computers, communication (mostly mobile), and consumer electronics[1]. Many studies have found that the elderly mainly apply 3C products and IT to entertainment, communication, shopping, learning, and health care, and the quality and satisfaction with life of the elderly are improved due to the use of digital information technology[2-8].

Dorin (2007) argued that online learning made it possible for the elderly with illnesses or difficulty in moving to study at home, and benefited elderly learners living in remote areas or too busy to attend offline courses. Furthermore, Dorin emphasized that current online learning environments contain audio and video functions, meaning the elderly could interact with others and realize informal learning and social contact. Githens (2007) pointed out that, when designing digital learning materials for the elderly, teachers should pay special attention to technical issues, usability issues, the curriculum design framework, and the application of new technology. Renaud and van Biljon (2008) interviewed the elderly aged between 60 and 92 to learn the factors influencing their use of digital tools, where

[a] Corresponding author : kic0929@ouk.edu.tw

such factors included social impact, usefulness, supporting facilities, and ease of learning and use, and proposed the Senior Technology Acceptance and Adoption Model (STAM).

However, Lin (2002) asserted that learning obstacles were caused by a lack of confidence, thinking oneself too old, lack of learning motivation, easily becoming nervous or anxious, and psychological factors. Huang (2004) deemed that the learning problems of the elderly included: health problems, response time, slow movement, failing eyesight and hearing, and poor memory. Chen (2006) probed into physical barriers, mental disorders, and learning disabilities. Tsai and Yeh (2009) considered some elderly learners of the Songnian University as their subjects, and discovered that the learning obstacles of elderly learners included: psychological, physiological, institutional, curriculum and teaching, and situational. In addition, Chiang (2006) believed that network learning of the elderly led to learning obstacles in physics, psychology, family, and basic computer abilities; in particular, failing eyesight was the greatest obstacle in learning.

Elderly learners have great demands and many chances to use 3C products in their lives and study; however, their gradually aging sensory functions inevitably cause learning obstacles. Therefore, this study attempts to understand the needs of and obstacles in network learning with "3C products" for elderly learners, and effectively assists them in using 3C products via diversified and suitable methods, overcoming learning obstacles, meeting learning demands, and improving learning outcomes, which are the motivations of this study.

1.2 Research motives

Based on the aforementioned research motivations, the purposes of this study are described, as follows:

1.2.1 This study intends to explore the learning characteristics of elderly learners.

1.2.2 This study aims to learn the needs of and obstacles in network learning with "3C products" for elderly learners.

1.2.3 According to the research results, this study attempts to propose feasible strategies to help elderly learners effectively use "3C products" to overcome their learning obstacles, meet their learning needs, and achieve their learning goals.

1.3 Research method

This study first employs literature review to collect important literature, such as books, journals, magazines, and papers on learning characteristics, as well as the needs of and obstacles in using "3C products" for the network learning of the elderly; summarizes and analyses such literature to form the basic concepts; summarizes the opinions and suggestions of scholars and experts, and develops a "Questionnaire on the Application of "3C Products" by the Elderly to Network Learning". This study investigates the elderly learners of the Open University of Kaohsiung through the questionnaire, and based on the statistics and analysis of the data collected, learns the status quo of the needs of and obstacles in using "3C products" for the network learning of the elderly. Copies of the questionnaire are distributed via paper and electronic forms.

2 Exploration of the influence of the learning characteristics of the needs and obstacles of the network learning of elderly learners

2.1 Learning characteristics of elderly learners

Due to their physical, psychological, and social states, as well as periodic development tasks, elderly learners are affected by physical aging, life phenomena, development stages, social culture, and psychological maturity; hence, their learning is unique. Huang (2004) suggested that there were 10 kinds of learning characteristics of the elderly: learning motivation was triggered by events or transition in life; the learning contents were mostly practical; the learning framework was question-centered; there was low confidence in learning; low learning pressure; lifelong learning ability; close place to study; learning during the daytime; comprehensive and top-down learning methods; the learning goal was to find the significance and value of life. Wei et al. (2008) assumed that: the learning of the elderly was learner-centered, their purpose was to adapt to changes; they dislike too much pressure caused by learning activities; they explored the significance of life; their learning effect was to promote healthy aging.

Tsai (1996), Wu (1997), Chang et al. (2002), Wei et al. (2008), Chen (2013), and Wu et al. (2014) deemed that elderly learners had the characteristics of instant learning motivation, lifelong and continuous learning, unique learning style, learning to solve practical issues, influenced by life experience, self-concept positioning, and self-orientation. In fact, adult and underage learners are greatly different in learning background factors, such as the learning of prerequisite knowledge, experience, physical and mental status, and learning motivation. Thus, teachers must teach students in accordance with their aptitude, allow students to have meaningful learning results, and help them meet their learning needs and realize their learning goals.

Chang (2007) pointed out that, in order to achieve the purpose of the adaptive teaching of the elderly, flexible curriculum teaching principles were important to successfully teach the elderly, including: (1) flexible learning time; (2) flexible learning methods; (3) flexible learning progress; (4) flexible learning guidance plan; and (5) flexible learning assessment. As elderly learners require more help and encouragement from their teachers and peers, suitable teaching methods shall be adopted in line with the characteristics and learning contents of students. Adaptive teaching represents social fairness and justice, meaning the status of each student should be considered, and those lagging behind are not overlooked, as all elderly learners could benefit from learning. Huang and Chang (2010) asserted that adaptive education allowed students to learn according to their unique characteristics and needs, develop their potentials, and realize themselves. As elderly learners have diversified learning characteristics and different experiences, abilities, interests, aptitude, cultures, and styles, different teaching methods shall be adopted. Hence, teachers must teach in accordance with the different characteristics of elderly learners, in order that all elderly learners can meet their learning needs and effectively reach their goals.

Chang and Chen (2013) and Tseng (2004) argued that, in terms of physiology, elderly learners suffered from failing eyesight and hearing, long response time, declining memory, and worse excretory system. In terms of psychology, they had stable intellectual development, strong self-esteem, low confidence in learning, active participation in learning, and the learning motivations of interests and social contact. Moreover, they preferred to engage in practical learning activities and did not stress speed. In terms of society, their learning goals were to complete development tasks and enrich their social and life experiences. Lin (2002) deemed that the personal backgrounds of elderly learners, such as age, education, and residence, would also affect their learning motivations. Tsai and Juan (2011) determined the learning needs of the elderly that should be considered in activity design or planning.

McClusky (1971) held that education intervention could improve the quality of life of the elderly, and classified the learning needs of the elderly as: handling, performance, contribution, influence, and surpassing needs. Hu (2012) believed that, with respect to the design of learning activities for the elderly, in addition to normal scenario analysis, needs assessments should be conducted. Through questionnaires, interviews, observations, testing, seminars, opinions of experts, and review of research reports, materials for needs assessment should be collected, in order to collect various information and observe and assess the learning behaviors of the elderly. Thus, if one considers the learning patterns and needs of the elderly in the design and planning of a curriculum, it would facilitate better understanding of the needs of elderly learners and the appropriate design of adaptive teaching strategies.

As it has become increasingly common to apply digital network systems to learning, more and more elderly learners must access networks to obtain information [30]. Ren (1999), and Wu and Tsai (2006) assumed that teachers, peers, and family members should help elderly learners develop better network usage strategies, offer encouragement and feedback, help them hold positive attitudes regarding the use of the Internet, improve their efficiency, confidence, and times in using the Internet, and facilitate them to obtain better outcomes. Lee (2008) pointed out that, when elderly learners felt they could study according to their own way, plan, and judgment, their efficiency in using the Internet would be improved, which helped the learners to study better.

Lu and Lu (2005) held that the Open University of Kaohsiung, which is an adult education institution that implements remote adult education, can design its teaching strategy in the following ways: (1) It could provide digital network services and diversified teaching environments. Digital network services could offer digital network teaching and discussion, in order to enrich teaching environments and allow learners to have individual choices; (2) It could offer a digital platform, cooperate with off-campus institutions, promote lifelong education, promote the digital learning platform through the cooperation of industry and education, and offer continuing education opportunities to the public; (3) It could store electronic teaching materials that are convenient for learners to read anytime, and learners and teachers can update the contents anytime to improve interactions with each other; (4) A learning process system could be established for digital learning, in order that the learning process of students can be clearly recorded on learning websites to form individual learning profiles; and (5) A digital learning network could integrate administrative and teaching organization systems, and combine administrative and teaching resources, in order that the various departments of a school could unite to improve teaching and administration efficiency.

However, through interviews with teachers of elderly learners, Lai (2016) deemed that elderly learners still preferred the face-to-face learning method. The interviewed teachers pointed out that, one of the important factors for the elderly to study was social contact and exchange with teachers and peers, meaning complete online learning environments lack interpersonal interaction. Elderly learners tend to feel isolated and lack the support of peers. Moreover, as elderly learners were not confident using technologies, they required the help of teachers; hence, it was better for digital courses to become teaching aids, and the elderly could preview or review the online courses prepared by teachers at home. In addition, as elderly learners suffer from apparent aging physiological functions, it was better to make digital courses simple and intuitive, meaning most of the course contents should be images, text should be large, and operational steps should be reduced. In this way, elderly learners could confidently operate and browse digital courses.

2.2 Needs of and obstacles in the network learning of elderly learners

Gagne (1977) used the relevant learning theory to develop a teaching model, and emphasized the establishment of teaching procedures through motivation, understanding, memory, apperception, execution, and feedback. Therefore, the processes of elderly learners participating in flipped learning activities include: First, understanding learning needs, assessing the category of learning needs, collecting and proposing learning goals and activities; second, setting up learning goals with order and levels based on the individual difference principle; third, practical and useful learning activity contents, and combining courses contents with the previous experience of students to maintain learning motivation; finally, evaluating learning outcomes via self-evaluation and peer exchange. In this way, learning needs can be met, as elderly learners are assisted to have peak learning experiences and realize the ideal learning goal of "self-realization". In addition to the awareness of becoming a lifelong learner, participating in various learning groups and cultivating friendly interpersonal interactions are also major factors supporting the elderly to participate in flipped learning [37].

Tsai (1995) proposed the principle of assessing the learning performance of adults, which is applicable to the elderly, deemed that evaluation should be based on teaching goals, diversified evaluation methods should be adopted, and the learning characteristics of individuals should be taken into consideration. Both learning processes and results should be considered in evaluation. The interpretation of evaluation should stress comparison with oneself rather than others, meaning scores and levels should not be overly stressed, and learners should be able to participate in the evaluation to obtain the sense of achievement, as based on the evaluation results. Additionally, if the elderly already have experience in using intelligent products, when coupled with cloud services, the elderly could record their thoughts on learning through an online learning platform (in the form of video or text), and teachers and peers could provide feedback via the platform. Moreover, the platform could provide links to relevant video and audio websites, in order that the elderly could browse all learning resources and exchange and share with their peers online.

This researcher teaches at the Open University of Kaohsiung, and during the first semester of the 2016 school year, conducted informal semi-structural interviews with eight elderly learners over the age of 60, and proposed the launch of a "large face-to-face" (a mixed teaching activity integrating online remote teaching and four face-to-face classes in one semester) course at the university. This method allows students to best accept, adapt to, and absorb the teaching contents and methods of teachers. The opinions of the interviewed students are comprehensively summarized. Some students hoped that teachers could provide attractive oral explanations, ask questions for students to discuss and solve, give plenty of time to students to think and finish homework, and conduct case studies. In addition, they hoped that teachers could demonstrate the operation of computers, where students could be divided into groups to complete homework together, audio and video materials could be added to digital teaching materials, outdoor fieldtrips could be organized, and the presentation of learning achievements and results could be arranged.

The key factors for the success of the network learning of elderly learners include: First, they will achieve better performance if they can adjust their own study progress. Second, if teachers can effectively organize teaching materials in terms of space and time, correlation, and contents, elderly learners will improve their outcomes. Third, through various media, such as combinations of text, information, images, and stories, students will form links to new and previously learned knowledge, which is suitable for individual learning modules. Fourth, it is a must for teachers to help elderly learners understand their own needs and offer them meaningful contents. Finally, teachers shall use as few as possible unified assessment criteria, instead, they can use several assessment methods, such as self-evaluation, to reduce the sense of frustration of the elderly, and offer support and praise, which is the best feedback for elderly learners [38].

3 Research hypotheses and tools

3.1 Research Hypotheses

Based on the above literature review, this study proposes the following hypotheses in terms of the needs of and obstacles in network learning with "3C products"for elderly learners:

H1: There are significant differences in the needs of and obstacles in network learning with "3C products" among elderly learners with different backgrounds.

H2: Elderly learners are highly correlated in terms of the needs of and obstacles in network learning with "3C products".

3.1 Research Tools

On the basis of the above-mentioned literature review, this study summarizes and edits the Questionnaire on the Application of "3C Products" by the Elderly to Network Learning, and invites practitioners teaching elderly learners, as well as researchers and experts on elderly education, to serve as the experts of this study and review the validity of the questionnaire. Regarding the practitioners teaching elderly learners, a total of three teachers are invited from an elderly university, a community university, and an Evergreen school. Regarding the scholars for education of the elderly, seven scholars are invited from National Kaohsiung Normal University, National Pingtung University, National Pingtung University of Science and Technology, Lunghwa University of Science and Technology, Aletheia University, National Open University, and Open University of Kaohsiung.

This study selects samples from elderly learners over 55 years old from the Open University of Kaohsiung attending elderly university programs, promotes education programs, and regular university programs. Copies of the questionnaire are randomly distributed to learn their gender, age, education, major, and profession. In terms of the reasons for using 3C products for network learning, there are seven questions on adaption to the trend of digital times, fast learning, effective learning, convenience in exchange and interaction with teachers and classmates, no need to go to school, preference for computer or mobile products, and compliance with network education rules. Regarding the obstacles in network learning, there are seven questions on being unfamiliar with operational procedures, insufficient knowledge of the functions of a computer or mobile, poor eyesight, poor hearing, lack of actual interaction with teachers and classmates, not enough physical strength to operate for a long time, and difficulty in operating courses. The data collected through the questionnaire are used to verify the hypotheses of this study, while the results are utilized to improve the network learning of elderly learners using 3C products, help them smoothly solve problems, meet their learning needs, and effectively achieve their learning goals.

4 Research results

4.1 Descriptive statistical results

Table 1. Questionnaire on the Application of "3C Product" by the Elderly to Network Learning

Variable	Group	Number of people	Percentage	Cumulative Percentage
Gender	Male	248	52.5	52.5
	Female	224	47.5	100.0
Age	55--59 years old	444	94.1	94.1
	60--64 years old	13	2.8	96.8
	Above 65 years old	15	3.2	100.0
Educational background	Below (including) junior high school	18	3.8	3.8
	Senior high school and vocational school	202	42.8	46.6
	College and university	210	44.5	91.1
	Graduate school	42	8.9	100.0
Department	Department of Law & Political Science	72	15.3	15.3
	Department of Industrial and Business Management	70	14.8	30.1
	Department of Technology Management	182	38.6	68.6
	Department of Culture and Arts	14	3.0	71.6
	Department of Mass Communication	19	4.0	75.6
	Department of Foreign Languages and Literature	25	5.3	80.9
	Unselected	90	19.1	100.0
Occupation	Farming, forestry, fishing animal husbandry	6	1.3	1.3
	Manufacturing	19	4.0	5.3
	Business and service	101	21.4	26.7
	Information technology	19	4.0	30.7
	Civil servants	262	55.5	86.2
	Unemployed	44	9.3	95.6
	Retired	21	4.4	100.0

N=472

The descriptive statistical results of this study are shown in Table 1. In terms of gender, there are 248 males, accounting for 52.5% of the total subjects, and 224 females, accounting for 47.5%. Regarding age, most of the subjects (444) are between "55 and 59 years", accounting for 94.1%; followed by 15 over "65 years", accounting for 3.2%; and 13 between "60 and 64 years", accounting for 2.8%. Regarding education, most of the subjects (210) graduated from "college or university", accounting for 44.5%; followed by 202 from "senior high school and vocational school", accounting for 42.8%;

42 from "graduate school", accounting for 8.9%; and 18 "below (including) junior high school", accounting for 3.8%. Regarding major, most of the subjects (182) are from the "Department of Technology Management", accounting for 38.6%, followed by 14 from the "Department of Culture and Arts", accounting for 3.0%. Regarding occupation, most of the subjects (262) are "civil servants", accounting for 55.5%; followed by 101 from "industrial and commercial sectors", accounting for 21.4%; and 6 from "farming, forestry, and fishing animal husbandry", accounting for 1.3%.

4.2 Confirmatory factor analysis results

4.2.1 Validation of convergence validity

Confirmatory factor analysis (CFA) is a part of SEM analysis. Thomopson (2004) proposed that, as a measurement model could correctly reflect the accuracy of a dimension, it was necessary to analyze the measurement model before SEM analysis. The overall model evaluation of this study is based on the Kline (2005) two-stage correction method. First, the goodness-of-fit of the measurement model is evaluated, and the results show that the goodness-of-fit is acceptable. Then, this study evaluates the overall SEM model [46]. According to Hair, Anderson, Tatham, and Black (2009) and Fornell and Larcker (1981), CFA measurement standards include: (1) Factor load greater than .50; (2) composition reliability greater than .60; (3) average variance extraction greater than .50. The CFA analysis results of this study indicate that the factor load of the dimensions is between .0.69 and .87. Composition reliability is between .91 and .93. Average variance extraction is between .59 and .66 (as shown in Table 2). The results imply that the dimensions of this study have convergence validity.

Table 2. Summary of the data for convergence validity verification

Estimated Model Parameters						Convergence Validity			
Potential variable	Observational variable	Non-standardized factor load	Standard error S.E.	t-value	P	Standardized factor load	SMC	C.R	AVE
Network learning need	A01	1				.85	.72	.93	.66
	A02	1.07	.04	24.42	***	.87	.76		
	A03	1.10	.05	24.16	***	.87	.75		
	A04	1.06	.05	22.67	***	.83	.69		
	A05	0.87	.05	18.38	***	.73	.53		
	A06	0.95	.06	17.17	***	.69	.48		
	A07	0.94	.04	21.96	***	.82	.67		
Network learning obstacle	B01	1				.74	.55	.91	.59
	B02	0.99	.06	16.25	***	.75	.56		
	B03	1.02	.06	16.72	***	.77	.59		
	B04	1.02	.06	16.73	***	.77	.59		
	B05	0.94	.06	16.61	***	.77	.59		
	B06	1.08	.06	18.08	***	.83	.69		
	B07	1.04	.06	16.39	***	.76	.57		

***P<.001

4.2.2 Validation of discriminant validity

AMOS provides two kinds of confidence interval estimation methods, the Bias-corrected Percentile Method and Percentile Method [45]. This study uses the bootstrap confidence interval method to test the discriminant validity of the dimensions. Repeated estimation is conducted 1,000 times under 95% confidence. If the correlation coefficient among the dimensions is smaller than 1, then the dimensions have discriminant validity [43, 45, 54]. The estimation results of this study are shown in Table 3. The confidence interval among the dimensions is smaller than 1, thus, the dimensions have discriminant validity.

Table 3. Bootstrap correlation coefficient 95% confidence interval

Parameter			Estimation	Bias-corrected		Percentile method	
				Lower Bound	Upper Bound	Lower Bound	Upper Bound
Network learning need	<-->	Network learning obstacle	-.06	-.12	-.02	-.12	-.01

4.2.2 Verification of goodness-of-fit

The goodness-of-fit indices of this study are based on Wu (2007), Hsu (2010), Bagozzi and Yi (1988), and Hair et al. (1998). $\chi 2$ Test, $\chi 2$-to-degree of Freedom Ratio, Goodness-of-fit Index (GFI), Adjusted Goodness-of-fit Index (AGFI), Root Mean Square Error of Approximation (RMSEA), Comparative Fit Index (CFI), Incremental Fit Index (IFI), Non-Normed Fit Index (NNFI), and Normed Fit Index (NFI) are adopted as the indicators to verify the overall goodness-of-fit of the model. When the SEM samples exceed 200, it is prone to have too big a chi-square value ($\chi 2=(n-1)Fmin$). In other words, too many samples will increase the chi-square value, resulting in rejection of the p-value [44-45]. Bollen and Stine (1992) proposed Bootstrap for model correction. As the chi-square value becomes smaller after correction, all the goodness-of-fit indices shall be re-calculated. See Table 4 for the results. All the goodness-of-fit values of the model in this study meet the criteria.

Table 4. Model goodness-of-fit indices

Goodness-of-fit Indexes	Standard Value		Model Goodness-of-fit
χ^2	The smaller the better	118.11	Passed
χ^2/df	<3	1.55	Passed
GFI	>.9	.98	Passed
AGFI	>.9	.96	Passed
RMSEA	<.08	.03	Passed
TLI (NNFI)	>.9	.99	Passed
CFI	>.9	.99	Passed
NFI	>.9	.98	Passed

4.3 Differences in the needs and obstacles of network learning with 3C products for elderly learners with different backgrounds

The purpose of t-testing the independent samples is a statistical method that first calculates the average values of the two independent sets of samples, and then, deduces if the average values of the two parent mean values have significant difference [47]. This study utilizes t-testing to analyze the differences in the variables of adults of different genders using 3C products for network learning. One-way ANOVA tests a single variable affected by different levels, and observes whether the mean values of the variables have significant differences [47]. If the test results show that the F value reaches a significant level, it means that the average of at least one group has significant difference; therefore, it is necessary to conduct "Post Hoc" analysis to confirm the difference [41].

4.3.1 Difference analysis of the influence of gender on the variables of needs and obstacles of network learning

Regarding the difference analysis of the influence of gender on the variables of needs and obstacles of network learning, t-tests of independent samples imply that gender does not have significant difference in the averages of learning needs (t= -.66, p=.51) or network learning obstacles (t=.05, p=.96). In other words, there are no significant differences in the variables of the learning needs and obstacles of different groups using 3C products for network learning.

4.3.2 Difference analysis of the influence of age on the variables of needs and obstacles of network learning

Regarding the difference analysis of the influence of age on the variables of needs and obstacles of network learning, one-way ANOVA implies that age has significant differences in learning needs (F= 7.88*, p=.00). Scheffe post hoc comparison shows that the F value of "over 65 years old" is greater than that of "between 55 and 59 years old", and reaches significant difference. The average of network learning obstacle (F=.67, p=.51) shows no significant difference. In other words, there are no significant differences in the variables of the learning needs and obstacles of different ages of elderly using 3C products for network learning.

4.3.3 Difference analysis of the influence of education background on the variables of needs and obstacles of network learning

Regarding the difference analysis of the influence of education background on the variables of the needs and obstacles of network learning, one-way ANOVA implies that different education backgrounds do not lead to significant differences in the averages of learning needs (F= 1.32, p=.29). In other words, there are no significant differences in the variables of the learning needs of different education backgrounds of elderly using 3C products for network learning. However, different education backgrounds lead to significant differences in the averages of learning obstacles (F= 6.35*, p=.00). Scheffe post hoc comparison shows that, the F value of "Senior high school and vocational school" is greater than those of "College and university" and "Graduate school". There is no significant difference between "College and university" and "Graduate school", which implies that elderly learners with lower education backgrounds face more apparent network learning obstacles.

4.3.4 Difference analysis of the influence of major on the variables of needs and obstacles of network learning

Regarding the difference analysis of the influence of major on the variables of needs and obstacles of network learning, one-way ANOVA implies that different majors do not lead to significant differences in the averages of learning needs (F= 2.12, p=.51). In other words, there are no significant differences in the variables of the learning needs of different majors of elderly using 3C products for network learning. However, different majors lead to significant differences in the averages of learning obstacles (F= 3.34*, p=.00). Scheffe post hoc comparison does not show significant difference.

4.3.5 Difference analysis of the influence of occupation on the variables of needs and obstacles of network learning

Regarding the difference analysis of the influence of occupation on the variables of the needs and obstacles of network learning, one-way ANOVA implies that occupation has significant differences in learning needs (F= 2.15*, p=.45). Scheffe post hoc comparison does not show significant difference. The average of network learning obstacles (F=.79, p=.58) shows no significant difference. In other words, there are no significant differences in the variables of the learning needs and obstacles of different occupations of elderly using 3C products for network learning.

4.4 Verification of hypotheses

Verification of the hypothesis shows that there are significant differences in the needs and obstacles of the networking learning of elderly learners using "3C products" and with different backgrounds, meaning there are significant differences when the elderly have different education backgrounds. The learning obstacles of elderly learners graduated from "Senior high school and vocational school" are significantly greater than those from "College and university" and "Graduate school". According to the calculation results of Table 5 and Figure 2, the path value of learning needs and obstacles of elderly learners using "3C products" for network learning is -0.06, which reaches a significant level, indicating that there is negative correlation between learning needs and obstacles. In summary, the verification results of the hypotheses are, as follows: Hypothesis 1: It is valid that there are significant differences in the needs and obstacles of networking learning of elderly learners using "3C products" and with different backgrounds. Hypothesis 2: It is valid that there is significant correlation between the learning needs and obstacles of elderly learners using "3C products" for network learning.

Table 5. Empirical results of hypotheses

Hypothesis	Path Relation			Path Value	C.R.	*p*	Hypothesis Valid
2	Network learning obstacle	<-->	Network learning need	-.06	-2.59*	**	Validity

*P<.05 **P<.01

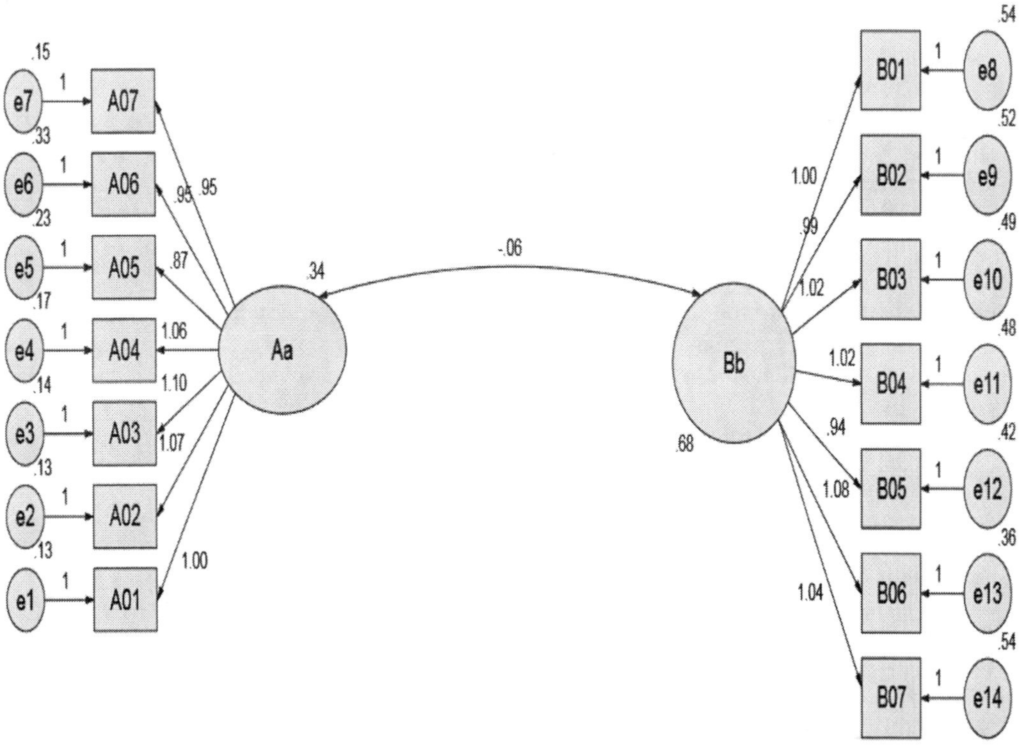

Figure 1. Statistical model of the learning needs and obstacles of elderly learners using "3C products" for network learning.

5 Conclusion and suggestions

According to findings of this study, elderly learners with a lower educational level encounter more significant obstacles in network learning when applying consumer electronics products. Therefore, the reinforcement of individual learning guidance can help elderly learners with a lower educational level to use consumer electronics products and strengthen their familiarity and convenience to apply consumer electronics products in network learning to lower their learning difficulties and enhance learning effectiveness. In addition, it can develop consumer electronics-based learning tools and devices suitable for elderly learners and provide them with the optimal learning aids to enhance the convenience of consumer electronics products for elderly learners in learning. In the process of design and manufacturing of teaching materials for network learning, Open University of Kaohsiung should be concerned with elderly learners' degeneration of their physiological functions and develop online digital learning materials with complete audio and light effects to lower their network learning obstacles. Programs could be based on a blend of face-to-face teaching and network modules in order to upgrade their learning participation and activeness.

The institutions and instructors engaged in the education of the elderly must first understand the basic knowledge of elderly learners, and then diagnose and analyze their learning needs and possible obstacles during network learning, in order to determine suitable course targets, plan learning guidance, re-plan existing courses and teaching materials, ask elderly learners to preview at home, and encourage them to finish preview tasks before the face-to-face classes. Next, according to the physical and metal statuses of elderly learners, digital network teaching materials can be prepared, and learning records can be maintained to help guide students. Third, mind maps covering the key points of a course should be developed, and classroom discussion topics should be prepared. Fourth, elderly learners should be encouraged to study online audio and video materials. Lastly, elderly learners should be encouraged to express their opinions via digital learning tools and actively provide feedback in order to stimulate elderly learners' learning enthusiasm and motive. Hence, when elderly learners enjoy participating in learning activities and their learning needs are satisfied, they will effectively accomplish the learning objectives.

"It is never too late to learn" is more than a slogan, but a practice advancing with the times. It has been said that knowledge is power. Higher education backgrounds bring about more learning opportunities, richer knowledge, and technologies, as well as more effective solutions to problems and obstacles in study and life. Hu (2015) pointed out that education for elderly learners could be conducted via the following steps: establishing an atmosphere beneficial for study→creating a mechanism for mutual planning→ diagnosing learning needs→ developing education goals→ designing a model of learning experiences→ guiding learning activities based on appropriate technologies and teaching materials→ assessing learning outcomes and re-diagnosing learning needs. Based on above, by stressing the learning needs of the elderly, improving the design specifications and operating models of consumer electronics products, and providing suitable network learning systems and aids, it is possible to help elderly learners use consumer electronics products, enjoy gaining knowledge online, solve various problems and obstacles, reinforce learning effectiveness, and successfully pursue keys of intelligence in life.

References

1. Wikipedia, 3C products, April 24, 2017, Retrieved from: https://zh.wikipedia.org/zh-tw/3C(2017)
2. J. Erickson and G.M. Johnson, Canadian Journal on Aging/la Revue canadienne du vieillissement, **30**(2), 197-209 (2011)
3. E. Etchemendy, R.M. Baños, C. Botella, D. Castilla, M. Alcañiz, P. Rasal and L. Farfallini, Computers & Education, **56**(1), 275-279 (2011)
4. H. Russell, Time and Meaning in Later-Life Learning, Australian Journal of Adult Learning, **51**(3), 547-565(2011)
5. F. Wang, B.B. Lockee and J.K. Burton, Journal of Educational Technology Systems, **40**(1), 45-58 (2011)
6. H.P. Yueh, W.J. Lin, M.C. Lee, H.C. Lin and Y.C. Luo, A Study on Using iPad with Reading Operations for the Elders, Instructional Technology & Media, **101**, 65-78(2012)
7. C.K. Huang, Journal of National Kaohsiung Normal University, **32**, 135-157 (2012)

8. H.J. Lai, Journal of Gerontechnology and Service Management, **4**(3), 367-376 (2016)
9. R.P. Githens, Quarterly Review of Distance Education, **8**(4), 329-338 (2007)
10. K. Renaud and J. Van Biljon, *Proceedings of the 2008 annual research conference of the South African Institute of Computer Scientists and Information Technologists on IT research in developing countries: riding the wave of technology*, 210-219 (2008)
11. M. Dorin, Educational Gerontology, **33**(2), 127-143 (2007)
12. C.M. Lin, Obstacles and Difficulties in Learning of Elderly Learners, Adult Education, **65**, 44-50(2002)
13. F.S. Huang, *Elderly Learning* (Wu Nan Books, Taipei, 2004)
14. Y.S. Chen, The Study of the Elders' Learning Experience in the Open University of Kaohsiung(master's thesis), National Kaohsiung Normal University (to be published)
15. C.C. Tsai and H.C. Yeh, *The Study of learning difficulty of the Elderly*. May 12, 2017. Retrieved from: http://ir.meiho.edu.tw/ir/bitstream/987654321/1316/1/%E9%AB%98%E9%BD%A1%E8%80%85%E5%8F%83%E8%88%87%E7%B5%82%E8%BA%AB%E5%AD%B8%E7%BF%92%E9%9A%9C%E7%A4%99%E4%B9%8B%E7%A0%94%E7%A9%B6.pdf(2009)
16. C.L. Chiang, Research of obstacles and coping strategies of older people engaging in web-based learning (master's thesis), National Chung Cheng University (to be published)
17. W.C. Wu, Study on the Factors Influencing the Participation Motivation and Learning Behavior of Elderly Learners of National Open University, No. 22 research report of National Open University. New Taipei: National Open University (1997)
18. M.L. Wu, Successful Adult Education Strategies, December 8, 2016. Retrieved from: http://learning.ace.ncnu.edu.tw(2014)
19. T. L. Chen, Application of "Learning Design" Thinking for Promoting Effective Adult Learning, T&D Fashion, **161**, 1-21(2013)
20. S.T. Chang, T.K. Chu, D.Y. Chang, M.W. Hung, Y.C. Hu, Y.L. Fang and P.Y. Hu, *Education Theory* (National Open University, New Taipei, 2002)
21. H.C. Wei, M.C. Hu, A.T. Li, C.S. Huang, C.C. Tsai and P.I. Lin, *Training Manual for Elderly Learning Planner—Introduction* (Ministry of Education, Taipei, 2008)
22. D.Y. Chang, On the Importance of Lifelong Learning of the Elderly and Mechanism Establishment Principle, Taipei Lifelong Learning, **38**, 17-22 (2007)
23. J.J. Huang and C.Y. Chang, Successful Learning: Concept and Strategy of Adaptive Course and Education, Curriculum & Instruction Quarterly, **13** (3), 1-22 (2010)
24. T.Y. Chang and P.L. Chen, Observation and Evaluation of Learning Behavior of Elderly Learner, T&D Fashion, **163**, 1-19 (2013)
25. Y.J. Tseng, Study on the Learning Pattern and Relevant Factors of Elderly Learners of Universities for Senior Citizens(master's thesis), National Taiwan Normal University (to be published)
26. L.H. Lin, The study of relationship between participating learning activity and life satisfaction of older adult (master's thesis), National Chung Cheng University (to be published)
27. C.C. Tsai and H.Y. Juan, The Study on Learning Needs with Willingness, April 28, 2017. Retrieved from: http://ir.meiho.edu.tw/ir/handle/987654321/1314(2011)
28. H.Y. McClusky, The adult as learner, In S. E. Seashore & R. J. McNeill (Eds.), *Management of the urban crises* (The Free Press, New York, 1971)
29. M.C. Hu, Design of Learning Activities for the Elderly. Published on Department of Adult and Continuing Education Department, National Chung Cheng University (Chief Editor), *Training Materials for the Education Professionals of Elderly Education of the Ministry of Education*, 25-30 (Ministry of Education, Taipei, 2012)
30. M.L. Hilt and J.H. Lipschultz, Educational Gerontology, **30**(1), 57-72 (2004)
31. W.H. Ren, Reference & User Services Quarterly, **38**(3), 283-291 (1999)
32. Y.T. Wu and C.C. Tsai, CybePsychology & behavior, **9**(4), 441-450 (2006)

33. H.Y. Lee, The Study of the Relationships among the Self-Directed learning, Family Support and Internet Self-efficacy of Middle-Aged and Elderly Adults (master's thesis), National Taiwan University of Science and Technology (to be published)
34. L.T. Lu and Y.J. Lu, Study on the Education Strategies Adopted by National Open University from the Perspective of Lifelong Learning. T&D Fashion, **33**, 1-21(2005)
35. R.M. Gagne, *The conditions of learning and theory of instruction* (Holt, Rinchart and Winston, New York, 1977)
36. H.M. Tsai, Adult Learning Assessment. Published on Adult Education Institute of Republic of China (Chief Editor), *Effective Adult Education*, 381-408 (Shita Books, Taipei, 1995)
37. S.P. Hsiao, The Strategies of the Elders' Service-Learning, Ming Hsin Journal, **34**(1), 345-359 (2008)
38. T.C. Chiu, *Elderly Education* (Psychological Publishing, Taipei, 1993)
39. M.C. Hu, Strategy and Method for Adult Education, December 20, 2016. Retrieved from: www.zzcc.tp.edu.tw/1012teacher/teaching materials,htm(2015)
40. M.L.Wu, *Structural equation modeling: Operation and application of AMOS* (Wu Nan Books, Taipei, 2007)
41. M.L. Wu and C.T. Tu, *SPSS & the Application and Analysis of Statistics* (Wu Nan Books, Taipei, 2005)
42. M.C. Hsu, Journal of Sport and Recreation Managemen, **7**(1), 174–186 (2010)
43. M.C. Hsu, F.C. Pan and C.M. Huang, Journal of Taiwan Society for Sport Management, **11**(2), 85–107 (2011)
44. M.C. Hsu and H.P. Yen, International Journal of Lisrel, **6**(1), 24-56 (2013)
45. W.H. Chang, *Techniques of thesis writing* (Tingmao, Taipei, 2011)
46. S.Y. Chen, M.C. Hsu and F.E. Li, Journal of Leisure and Recreation Industry Management, **5**(3), 1-21 (2012)
47. K.Y. Chen and C.H. Wang, *Thesis statistical analysis practice SPSS and operation of AMOS* (Wu Nan Books, Taipei, 2011)
48. C. Fornell and D.F. Lacker, Journal of Marketing Research, **18**, 39-50 (1981)
49. J.F.Jr. Hair, R.E. Anderson, R.L. Tatham and W.C. Black, *Multivariate data analysis (7th ed.)* (Prentice Hall, New Jersey, 2009)
50. R.B. Kline, *Principles and practice of structural equation modeling (2nd ed.)* (Guilford Press, New York, 2005)
51. B. Thomopson, *Exploratory and confirmatory factor analysis: Understanding concepts and applications* (American Psychological Association, Washington DC, 2004)
52. R.P. Bagozzi and Y. Yi, Journal of the Academy of Marketing Science, **16**, 74-94 (1988)
53. J.F.Jr. Hair, R.E. Anderson, R.L. Tatham and W.C. Black, *Multivariate data analysis (5th ed.)* (Prentice Hall, New Jersey, 1998)
54. G. Torkzadeh, X. Koufteros and K. Pflughoeft, Structural Equation Modeling, **10**(2), 263-275 (2003)

Safety analysis for integrated modular avionics based on blueprints

Jiayun Chu, Xiaohong Bao, Tingdi Zhao[a] and Fuchun ren

School of Reliability and Systems Engineering, Beihang University, China

Abstract. The Integrated Modular Avionics System (IMA) has been a core technology for the new generation of aircrafts in recent years. It consists of a set of reusable and interoperable common functional modules. However, the highly coupled relationship of resources makes it difficult to identify and control dangers. As an effective and efficient way, the blueprints are used to describe and manage the IMA system. Owing to the system management functions provided by the blueprints, we can accurately determine the system resources configuration status, which is very crucial for safety analysis. In this paper, we explore the possibilities to conduct safety analysis based on blueprints. A safety analysis method based on blueprints is proposed, which applies mathematical logic to describe the logical relationship between targets and resources provided by the blueprints and uses semi-tensor product of matrix theory to simplify the logical expressions. Based on the mathematical model, we can conduct the fail safety analysis and identify resources failures that may undermine the IMA system safety.

1 Introduction

The avionics system is a comprehensive control information-intensive system consisting of hardware and software such as mission management, display control, detection sensors and weapons [1].From the end of the last century, the integrated modular avionics system is gradually applying to the modern aircraft design process, such as F22, A380. Different from the original discrete or federated avionics system, the IMA system implements a wide range of physical synthesis and functional integration. As shown in Figure 1, each function is not located within special processor or line replacement unit (LRUs) in the IMA system [2]. While the system is running, the resources are dynamically allocated to different targets, and it can be regarded that there are many variable virtual subsystems. Although there are many benefits offered by the resources sharing mechanism, such as improving the mission performance and operational performance, reducing the life cycle cost [3], several problems need to be solved urgently. One of the most important problem is how to conduct safety analysis after sharing resources widely. And the allocation of the applications to hardware in an effective and efficient way is also a critical issue. In order to solve the problems above, a concept called "blueprints" is used to describe the IMA system.

[a] Corresponding author : ztd@buaa.edu.cn

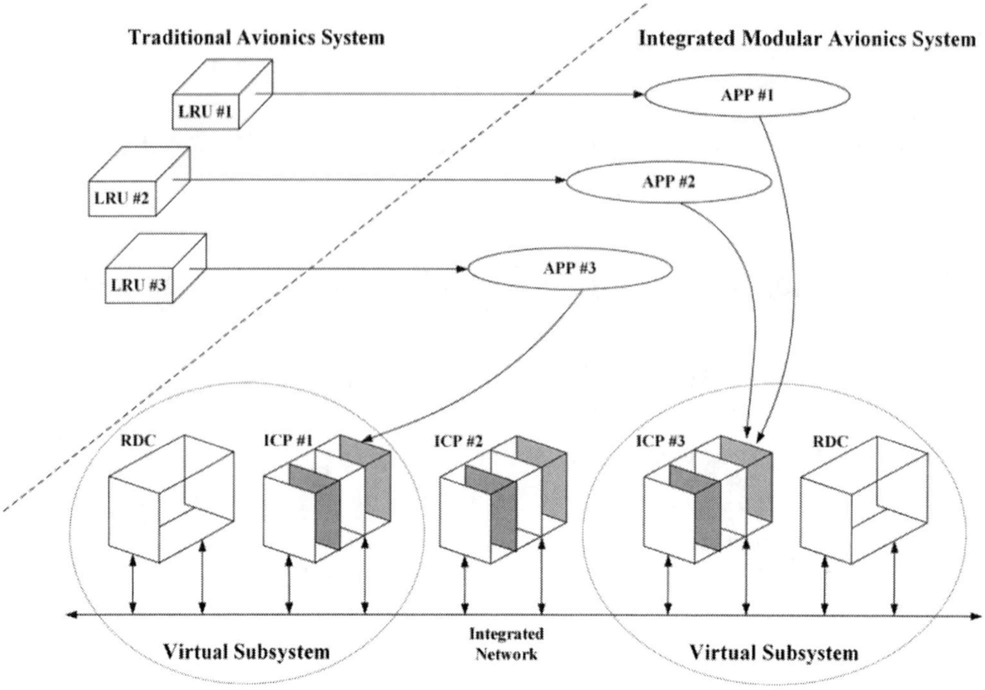

Figure 1. The Basic Schematic Graph of IMA System

In 1997, A.Marchetto [4] proposed an IMA system management methods using blueprints. He described blueprints as a means of centralizing and organizing the system definition information in such a way that changes of system integration decisions can be transferred, in a controlled and automatic or semi-automatic way, to the target system, simply altering the appropriate blueprints. Allied Standards Avionics Architecture Council (ASAAC) incorporated the blueprints into the recommended technology for IMA system management in its guidelines, and fault monitoring and management functions were added to the blueprints. In the same year, Graham Jolliffe and DM Nicholsan [2] explored the possibilities towards a preliminary safety case for IMA blueprints. Li Qian etc. [5] applied the blueprints technology to TV-command-guided system in 2009. Furthermore, there are also many researches on the IMA system blueprints design internationally. Shen Y [6] proposed a method for the design and implementation of IMA system blueprints using AADL in 2008, while Haotian Wang etc. [7] modelled the IMA system blueprints based GSPN and LP in 2013.

Recently, the blueprints technology is widely regarded as a method to centrally organize and manage the system definition information. And the blueprints are implemented by one or a group of management software located in the operating system. Except the system management functions, blueprints can also provide support for the safety analysis due to the definition of system information. In this paper, we apply mathematical logic to describe the logical relationship between targets and resources provided by the blueprints. And semi-tensor product of matrix theory is used to simplify the logical expressions. Based on the mathematical model, we can conduct the fail safety analysis and identify resources failures that may undermine system safety.

The rest of the paper is organized as follows. In Section 2, a brief introduction of the IMA blueprints is given. Section 3 discusses some basic concepts and related properties of matrix algebra at first. And then according to the semi-tensor product theory, the IMA systems model based on blueprints is proposed. In Section 4, we apply the method proposed above to a simple IMA system case. Section 5 is the concluding remarks.

2 IMA blueprints

Blueprints technology is a method to centrally organize and manage the system definition information. And they are implemented by one or a group of management software located in the operating system. The blueprints can implement the resources configuration and reconfiguration automatically or semi-automatically, which means the deterministic management of system resources. The IMA blueprints can be designed by AADL or ADA tools. As shown in Figure 2, the blueprints are usually subdivided into three parts recently, including software blueprints, hardware blueprints and system blueprints. And the system blueprints can be further divided into static configuration blueprints and dynamic runtime blueprints.

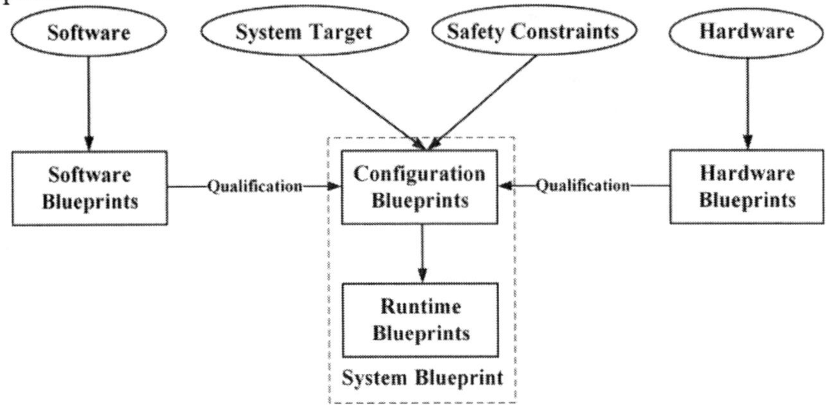

Figure 2. The IMA System Blueprints

The software blueprints describe the function software resources in terms of runtime requirements, processing and memory requirements and communication requirements. And the hardware blueprints describe the physical system [8]. The software blueprints and hardware blueprints together describe the resources set of the IMA system, which can be also called as resources pool [9].

The system blueprints guaranteed the safe operation of the IMA system. System designers make decomposition of system targets and determine the resources needed to complete the targets. The decomposition result can be given in the form of a tree as shown in Figure 3. According to the decomposition result and the resources capacity in the resources pool, the initial resources configuration plan is determined, and designers record it in the configuration blueprints. System designers also record the reconfiguration plans when some resources fail in the configuration blueprints. The monitoring and management of the running process are completed by the running blueprints. For example, while the resources R1.3 fails in the Figure 4, the system configuration changes into the configuration 2.

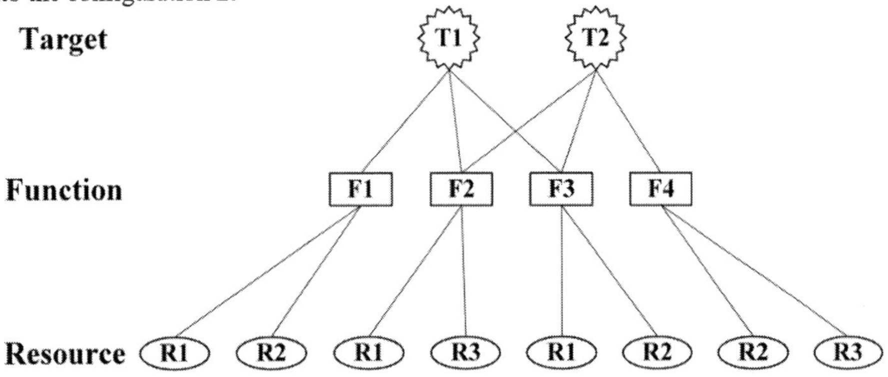

Figure 3. Target Decomposition Diagram

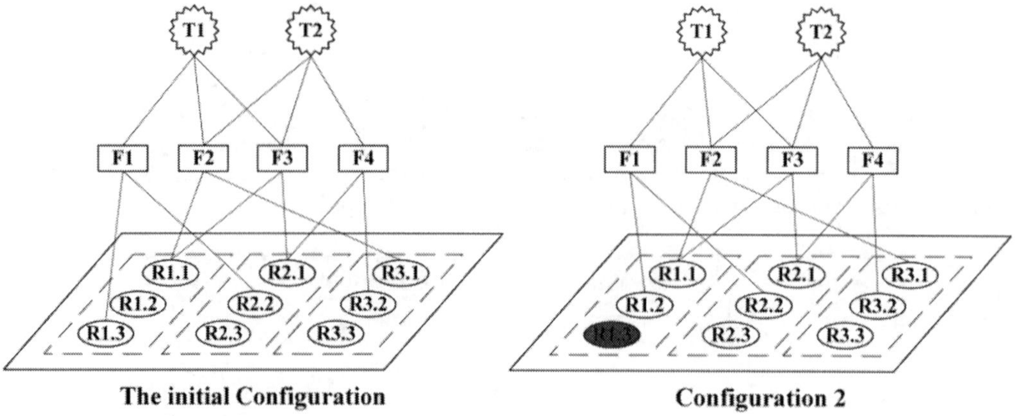

Figure 4. The Configuration Blueprints

3 The IMA system model based on blueprints

3.1 Matrix algebra

This paper is aimed at the logical relationship between targets and resources provided by the blueprints, and tries to simply the logic expressions using semi-tensor product of matric theory. In this section, we present some basic concepts and related properties of matrix algebra first [10-12].

Definition 1: Assume that $A=(a_{ij})\in M_{m\times n}$ and $B=(b_{ij})\in M_{p\times q}$. Then the Kronecker product of ***A*** and ***B*** is represented as $A\otimes B$ and defined by

$$A\otimes B=\begin{pmatrix} a_{11}B & a_{12}B & \cdots & a_{1n}B \\ a_{21}B & a_{22}B & \cdots & a_{2n}B \\ \vdots & \vdots & \ddots & \vdots \\ a_{m1}B & a_{m2}B & \cdots & a_{mn}B \end{pmatrix}\in M_{mp\times nq} \tag{1}$$

Definition 2: Assume that $A=(a_{ij})\in M_{m\times r}$ and $B=(b_{ij})\in M_{n\times r}$. Then the Khatri-Rao product of ***A*** and ***B*** is represented as $A*B$ and defined by

$$A*B=\left(Col_1(A)\otimes Col_1(B)\quad Col_2(A)\otimes Col_2(B)\quad \cdots \quad Col_r(A)\otimes Col_r(B)\right)\in M_{mp\times nq} \tag{2}$$

Definition 3: Assume that $A=(a_{ij})\in M_{m\times r}$ and $B=(b_{ij})\in M_{n\times r}$ .then the left semi-tensor product ***A*** and ***B*** is represented as $A\propto B$ and defined by

$$A\propto B=\left(A\otimes I_{t/n}\right)\left(B\otimes I_{t/p}\right)\in M_{(mt/n)\times(tq/p)} \tag{3}$$

While $t=lcm(n,p)$ is the least common multiple of n and p, and I_k is the k degree identity matrix.

$A\otimes B$, $A*B$ and $A\propto B$ testify the associative law and the distributive law. Furthermore, because the point product is a special case of the left semi-tensor product, this paper makes the following assumption:

$$A\propto B=AB \tag{4}$$

Theorem 1: Assume that $X\in R^t$ is a column vector, $Y\in R^t$ is a row vector, and, A is an arbitrary matrix. Then we have

$$XA = (I_t \otimes A)X, AY = Y(I_t \otimes A) \tag{5}$$

Definition 4: Assume that $W \in M_{mn \times mn}$, and it ranks with double index mark in turn. We define W as transposition matrix if it safeties

$$W_{(I,J),(i,j)} = \begin{cases} 1 & I = i \ and \ J = j \\ 0 & else \end{cases} \tag{6}$$

Theorem 2: Assume that $X \in R^m, Y \in R^n$ are two column vectors, and $A \in R^m, B \in R^n$ are two row vectors. Than we have

$$W_{[m,n]}XY = YX, ABW_{[m,n]} = BA \tag{7}$$

Definition 5: Assume that $D = \{1,0\}$, and $x_i (i = 1,2,\cdots,k) \in D$ is a group of logic variable. A mapping $f : D^n \to D$ is defined as a logic function below

$$y = f(x_1, x_2, \cdots, x_n) = Mx \tag{8}$$

While M is defined as structure matrix, $x = x_1, x_2, \cdots, x_n$. And M is called as logic operator when $n \leq 2$.

Definition 6: Assume that M_c, M_d and M_n respectively represent and($\wedge$),or ($\vee$) and not($\neg$). Then we have

$$M_c = \begin{pmatrix} 1 & 0 & 0 & 0 \\ 0 & 1 & 1 & 1 \end{pmatrix}, M_d = \begin{pmatrix} 1 & 1 & 1 & 0 \\ 0 & 0 & 0 & 1 \end{pmatrix}, M_n = \begin{pmatrix} 0 & 1 \\ 1 & 0 \end{pmatrix} \tag{9}$$

Definition 7: Assume that $R_k^P = diag(\delta_k^1, \delta_k^2, \cdots \delta_k^k)$ with $\delta_k^i = \left(\overbrace{\underbrace{0,0,\cdots,1}_{i}\cdots,0}^{n} \right)$. Then we define R_k^P as reduction matrix if it safeties

$$x^k = R_k^P x \tag{10}$$

3.2 IMA blueprints model and safety analysis

Based on the matric algebra theory shown in Section 3.1, we propose a method to model the IMA blueprints. And with the IMA blueprints model, we can find the combination of dangerous faults which may influence the system safety. Refer to definition 5, we represent the logical nodes (targets, functions and resources) as binary logical variables. According to the logical relationship between the nodes provided by the blueprints, we build the logical expressions, and simply them with the help of semi-tensor product of matrix theory. As long as we identify the structure matrix of the key targets which may influence the system safety, we can discuss the combination of dangerous failures by solving the logical equations. The flow chart of modelling and analysing is shown in Figure 5.

 https://doi.org/10.1051/matecconf/201816901029

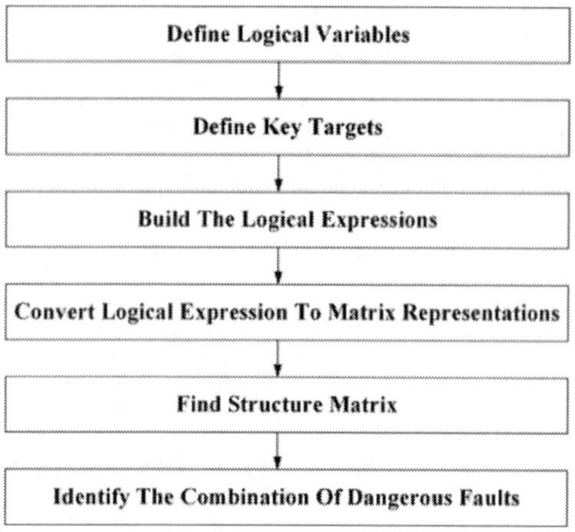

Figure 5. The Flow Chart of Modelling and Analysing

Taking a simply example shown in Figure 6 to present the process of finding structure matrix, we represent the logical nodes as t_1, f_1, f_2, r_1, r_2.

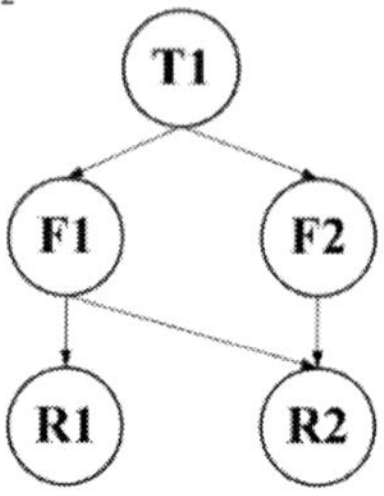

Figure 6. A Simple Example of Logical Graph

Then we can describe the logical relationship as follows.

$$t_1 = f_1 \wedge f_2 = \left(r_1 \wedge r_2\right) \wedge r_2 \tag{11}$$

According to the theory introduced in Section 3.1, we can convert formula 11 to formula 12.

$$t_1 = f_1 \wedge f_2 = M_c\left(M_c r_1 r_2\right) r_2 = M_c^2 r_1 r_2^2 = M_c^2 r_1 R_2^P r_2 = M_c^2\left(I_2 \otimes R_2^P\right) r_1 r_2 \tag{12}$$

While M_c , R_2^P are defined in definition 6 and 7, and I_2 is an identity matrix.

The structure matrix of t_1 is represented as M_{t_1} , and then we have

$$M_{t_1} = M_c^2\left(I_2 \otimes R_2^P\right) = \begin{pmatrix} 1 & 0 & 0 & 0 \\ 0 & 1 & 1 & 1 \end{pmatrix} \tag{13}$$

4 A simple IMA system case

In this section, we take the radio data processing function as an example. As shown in Figure 7, the AAP1 represents voice data processing function while the APP2 represents general data processing function. And the APP1 has a higher priority than the APP2. And we consider the APP1 as the key target.

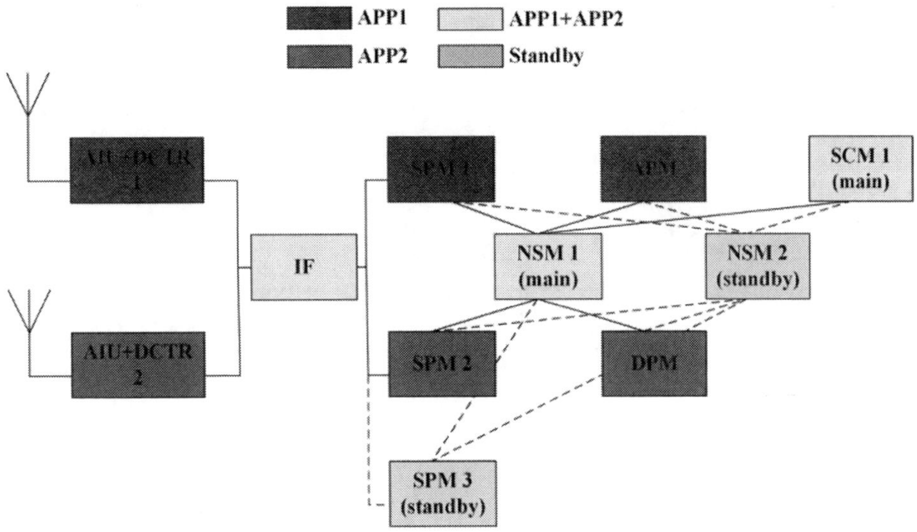

Figure 7. A Simple IMA System Case

According to the order of functions priority, the pre-set reconfiguration strategies which are record in the blueprints are as follows:

1) The APP1 has a higher priority than the APP2;

2) While the AIU+DCTR1 fails, stop the APP2 and reallocate the AIU+DCRT2 to APP1;

3) While the SPM1 fails and the SPM2 doesn't fail, use the SPM3 to replace the SPM1; While the SPM1 doesn't fail and the SPM2 fails, use the SPM3 to replace the SPM2; while the SPM1 and the SPM2 fail together, there isn't any change;

4) While the NSM1 fails, use the NSM2 to replace the NSM1;

Similar to Figure 4, we describe the initial configuration plan as follows.

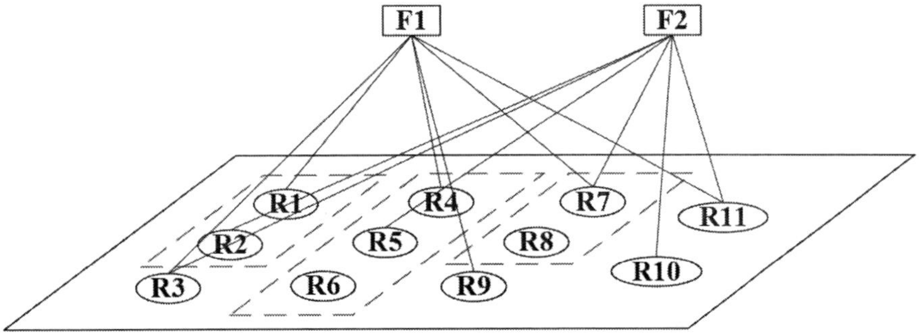

Figure 8. Initial Configuration Plan

While the meanings of each node in Figure 8 are shown in Table 1.

Table 1. The Meanings of Nodes

Number	Item	Number	Item
f_1	APP 1	r_6	SPM 3
f_2	APP 2	r_7	NSM 1
r_1	AIU+DCTR 1	r_8	NSM 2
r_2	AIU+DCTR 2	r_9	APM
r_3	IF	r_{10}	DPM
r_4	SPM 1	r_{11}	SCM 1
r_5	SPM 2		

Because we only consider the APP1 as the key target, we can build the following logical expression according to the initial configuration plan.

$$f_1 = r_1 \wedge r_3 \wedge r_4 \wedge r_7 \wedge r_9 \wedge r_{11} \tag{14}$$

As we consider the pre-set reconfiguration strategies, we can modify formula 14 to formula 15.

$$f_1 = (r_1 \vee (\neg r_1 \wedge r_2)) \wedge r_3 \wedge (r_4 \vee (\neg r_4 \wedge r_5 \wedge r_6)) \wedge (r_7 \vee (\neg r_7 \wedge r_8)) \wedge r_9 \wedge r_{11} \tag{15}$$

And then we covert the logical expression to matrix representation.

$$f_1 = M_c^5 (M_d r_1 M_c M_n r_1 r_2) r_3 (M_d r_4 M_c^2 M_n r_4 r_5 r_6)(M_d r_7 M_c M_n r_7 r_8) r_9 r_{11} \tag{16}$$

The simplification process is as follows.

$$M_d r_1 M_c M_n r_1 r_2 = M_d (I_2 \otimes M_c M_n) R_2^P r_1 r_2 = \begin{pmatrix} 1 & 1 & 1 & 0 \\ 0 & 0 & 0 & 1 \end{pmatrix} r_1 r_2 = M_1 r_1 r_2 \tag{17}$$

$$M_d r_4 M_c^2 M_n r_4 r_5 r_6 = \begin{pmatrix} 1 & 1 & 1 & 1 & 1 & 0 & 0 & 0 \\ 0 & 0 & 0 & 0 & 0 & 1 & 1 & 1 \end{pmatrix} r_4 r_5 r_6 = M_2 r_4 r_5 r_6 \tag{18}$$

$$M_c r_7 M_d M_n r_7 r_8 = M_c r_7 r_8 = \begin{pmatrix} 1 & 1 & 1 & 0 \\ 0 & 0 & 0 & 1 \end{pmatrix} r_7 r_8 = M_3 r_7 r_8 \tag{19}$$

The final simplification result is shown as formula 20.

$$f_1 = M_c^5 M_1 r_1 r_2 r_3 M_2 r_4 r_5 r_6 M_3 r_7 r_8 r_9 r_{11} = M_c^5 M_1 (I_8 \otimes M_2)(I_{64} \otimes M_3) r_1 r_2 r_3 r_4 r_5 r_6 r_7 r_8 r_9 r_{11} \tag{20}$$

The structure matrix of f_1 is represented as M_{f_1}, and then we have

$$M_{f_1} = M_c^5 M_1 (I_8 \otimes M_2)(I_{64} \otimes M_3) \tag{21}$$

If we only consider the possible safety problems caused by SPM resources failures, we make $r_1 = r_2 = r_3 = r_7 = r_8 = r_9 = r_{11} = (1 \quad 0)^T$, and solve the equation $f_1 = (0 \quad 1)^T$, that is

$$\begin{pmatrix} 0 \\ 1 \end{pmatrix} = \begin{pmatrix} 1 & 1 & 1 & 1 & 1 & 0 & 0 & 0 \\ 0 & 0 & 0 & 0 & 0 & 1 & 1 & 1 \end{pmatrix} r_4 r_5 r_6 \tag{22}$$

According to the equation solution, we can find that only when $r_4 = (0 \quad 1)^T$, and at least one of $r_5 = (0 \quad 1)^T$ and $r_6 = (0 \quad 1)^T$ holds, then the f_1 fails. The analysis result is in line with expectations.

5 Conclusions

In this paper, we propose a method of safety analysis for the IMA system based on blueprints, which applies mathematical logic to describe the logical relationship between targets and resources provided by the blueprints and uses semi-tensor product of matrix theory to simplify the logical expressions. Based on the mathematical model, we can conduct the fail safety analysis and identify resources failures that may undermine system safety. Adopting the method proposed above, the value of the blueprints is further tapped. Owing to the system management functions provided by the blueprints, we can accurately determine the system resources configuration status, which is very crucial for safety analysis.

References

1. I. Moir, A. Seabridge and M. Jukes, *Civil Avionics Systems* (West Sussex, UK, 2013)
2. G. Jolliffe and DM. Nicholson, Constituents of Modern System-safety Thinking, 163-181 (2005)
3. C. Yin, Y. Renliang and Z. Li, *Integrated Technology of Avionics Modular Integrated System* (Beijing, China, 2013)
4. A. Marchetto, Dasc. Aiaa/ieee. IEEE, **18**, 3.2-24-31 (1997)
5. L. Qian, F. Jinfu, P. Bo and Z. Jiaqiang, Computer Engineering, **35**, 225-227 (2009)
6. Y. Shen, Computer Engineering, **34**, 66-71 (2008)
7. W. Haotian, H. Feng and X. Huagang, Aerosp. Sci. Technol., **26**, 111-119 (2013)
8. D. Suo, J. An and J. Zhu, Digital Avionics Systems Conference. IEEE, 1C4-1-1C4-12 (2011)
9. Y. Wonkeun and Y. Baeck-jun, Computer Standards & Interfaces, **36**(6), 889-898 (2014)
10. C. Daizhan, X. Yuanqing, M. Hongbin, Y. liping, *Matrix Algebra, Control and Game* (Beijing, China, 2016)
11. C Daizhan, Q Hongsheng and X Ancheng, Journal of Systems Science & Complexity, **20**, 304-322 (2007)
12. C Daizhan and Q Hongsheng, IEEE Trans. Autom. Control, **55**, 2251-2258 (2010)

Human reliability research of numerical control machining based on optimized Holistic Decision Tree

Lu Chen[1,2], Run-Jiang Zhao[1], Jian Jiao[1,a] and Yan-Bo Zhang[1]

[1]*School of Reliability and Systems Engineering, Beihang University, China*
[2]*China Astronautics Standards Institute, China*

Abstract. Human reliability has been a focus in system reliability and safety researches since 1960s, and now human factor has become an increasingly important reason which affects system reliability. Numerical control machining system characterized by man-machine interaction can be applied to automatic parts machining with high efficiency. However, professional and technical staffs are still needed for clamping, positioning, inspection and others. So there is no doubt that the quality of the parts might be influenced by personnel behaviour. What's more, on-site staff is responsible for handling potential fault directly. The study on human error plays an important role to ensure the quality and reduce industrial accident. Holistic Decision Tree (HDT) is a dynamic Human Reliability Analysis (HRA) method, emphasizing the wholeness of man-machine interaction. This paper tries to use an optimized HDT method to investigate human reliability in numerical control machining system. In order to obtain a rational order of the importance of each IF, this paper incorporates entropy weight method to the conventional expert method to determine the relative weight value of each human behaviour influence factor (IF), and finally calculates the human error probability (HEP).

1 Introduction

Numerical control machining system is a typical complicated man-machine interaction system, which needs relevant professional and technical staff for clamping, positioning, inspection and others, so there is no doubt that the quality of the parts produced by the system might be influenced by personnel behaviour. What's more, on-site staff is responsible for handling potential faults directly, and human error also plays an important role to ensure the parts quality and reduce industrial accident. However, the existing studies on the reliability of the numerical control machining system mostly focus on equipment failures and the improvement of the man-machine interface [1-3].

Human reliability, a term used to express personnel performance, can be described as the probability of personnel completing a task successfully within the specified minimum time limit (if there have time requirements), and at any operation stage of the system [4]. The research on human reliability analysis (HRA) began in the 1950s [5]. In the late 1950s and early 1960s, researchers started to discuss the influences of human error on the system reliability, and there are more than twenty types of HRA methods up to now. The first generation HRA methods such as Technique for

[a] Corresponding author : jiaojian@buaa.edu.cn

Human Error Rate Prediction (THERP) [6], Human Error Assessment and Reduction Technique (HEART) [7] and Success Likelihood Index Method (SLIM) [8] are widely used in risk analysis of industrial system. However, there are some insufficiencies in the consideration of human cognitive error and situational factors. Nevertheless, the effect of environment or work scenario on human behaviour is generally investigated in the second generation HRA methods such as A Technique for Human Error Analysis (ATHEANA) [9], Cognitive Reliability and Error Analysis Method (CREAM) [10] and Holistic Decision Tree (HDT) [11]. ATHEANA and CREAM focuses mainly on retrospective analysis and predictive analysis. The analytical processes are complicated and a large amount of data resource input is required, which limits their applicability and practicality in industrial production field.

As a dynamic HRA method, Holistic Decision Tree (HDT) method was proposed based on simulator's analysis results of the "Paks" nuclear power station by A.J. Spurgin [12] in the 1990s, emphasizing the wholeness of man-machine interaction. In the form of decision tree, various behaviour influence factors (IFs) are considered to judge their affects the human error probability in this method. In recent years, researchers tried to apply the HDT method to other man-machine interaction industries and fields, such as air traffic control, aircraft maintenance, automatic drive [13-15] etc. Lots of qualitative and quantitative conclusions got from HDT analysis have been adopted to improve man-machine system design, development and to reduce human errors.

This paper tries to use an optimized HDT method to investigate behaviour Ifs, such as mental states of the operators, personnel skills, man-machine interface, working environment, operation specification, etc., to analyze and assess human reliability in Numerical control machining system. In order to obtain a rational order of the importance of each IF, this paper incorporates the objective entropy weight method to the conventional expert method to determine the relative weight value of each IF, and finally calculates the human error probability (HEP), which can provide reasonable and reliable basis for making preventive and control measures for the potential accident or production damage. This optimized HDT method is more objective and efficient, and is available for improving operational efficiency and operational safety of numerical control machining industry.

2 Principle of the HDT method

2.1 Basic principle of the HDT method

The basic principle of HDT is that accident scenarios and staffs' situation are used to determine HEP. The HDT model can be expressed as the following equations [5, 16], which can link staffs' scene with HEP.

Human error probability = Function of the staffs' situation

Staffs' situation = Function of the influence factors

Human error probability = Function of the influence factors (man-machine interface, leadership quality, etc.)

A typical decision tree is constituted by some main elements such as IFs, branch-like structure, weight value of each IF and quality description value etc., as shown in Figure 1. The titles of the X axis are IFs, along with the corresponding weight value of each scenario. The notes on the tree's branches are quality description (QD) levels, along with the corresponding quality values (QV). QV is the quantified expression of QD levels, which describes the increasing degree of HEP accelerated by the corresponding QD. Just like building an Event Tree, the structure of the HDT is divided into several paths according to the titles of the X axis. The branch value is corresponding to different IF's QD level, and each path is corresponding to a different combination of quality descriptions.

2.2 Principle of the entropy weight method

The concept of entropy was produced in thermodynamics, which is used to describe the phenomenon of irreversible process. In 1948, C.E. Shannon introduced the concept of entropy into information theory as a measure parameter. According to the basic principle of information theory, information is used to measure the order degree of a system while entropy measures disorder. The increasing of information means the decreasing of entropy; and information is inversely proportional to entropy [17, 18].

As an objective way to determine the weighted value of indicators, entropy weight method can use multiple indicators to evaluate multiple objects, and has the following three characteristics [19]: (1) Entropy weight method, based on the "differentia-driven" theory, highlights the local differences and obtain the optimal weight value through the actual data of all samples, which avoids the influence of human error, and has high confidence level; (2) The weight process has transparency and reproducibility; (3) Normalization method is used to deal with the dimensionless processing of data, which has a good robustness and can keep monotonicity, scaling of independence and total constancy.

3 The optimized Holistic Decision Tree

3.1 Procedure of HDT analysis

The general procedure of HDT contains 9 steps:

(1) List all of the possible influence factors (IFs).

(2) Distinguish between IFs for overall situation and IFs for related scenarios. The value of human error probability is influenced by the scenario (local) and global factors, where the latter have influence on every scenario.

(3) Arrange the order of the IFs according to their importance, and select the most important one.

(4) Scrupulously select and define the quality description levels of these factors, and draw the tree structure.

(5) Determine the weighted value of IFs. Expert judgment method is recommended to get the weighted value, and different participants' judgment can be used as the measurement of uncertainty.

(6) Select the minimum and maximum calibration values of human error probability.

(7) Rank each IFs of each scenario with quality description (QD).

(8) Determine the corresponding quality values (QV) of QD.

(9) Calculate of human error probability (HEP).

3.2 Incorporate entropy weight into the expert method

Entropy weight method [20, 21] is an objective method based on entropy calculation of each IF to measure the amount of information, and then to determine the weight value. Weighting each evaluation index with entropy weight can avoid human interference, which makes the evaluation results more objective. Entropy weight method ensures that the established indicators can reflect most of the original information, so that the result is more in line with the actual situation.

Expert method integrates experts' industry experience of all relevant fields to give each IF a weight value. Because it is the subjective judgment of decision makers, the weight value may be more or less affected by the lack of experience or knowledge of decision makers. Therefore, this paper tries to incorporate entropy weight method to the conventional expert method to determine the relative weight value of each IF to improve its accuracy.

Firstly, we can get a $n \times m$ matrix of the original data from the determined n different samples and m evaluation indexes, as follows:

$$X = \begin{bmatrix} x_{11} & x_{12} & \cdots & x_{1n} \\ x_{21} & x_{22} & \cdots & x_{2n} \\ \vdots & \vdots & \vdots & \vdots \\ x_{n1} & x_{n2} & \cdots & x_{nm} \end{bmatrix}_{n \times m} \quad (1)$$

where x_{ij} means the j_{th} evaluation index of the i_{th} sample, $i=1,2,...,n; j=1,2,...,m$.

Due to the different dimensions of the various indexes, normalization is needed for each index. Take the ratio between column vector and the sum of all elements in the matrix X as the normalized value, and its computation formula is as follows:

$$z_{ij} = \frac{x_{ij}}{\sum_{j=1}^{m} x_{ij}} \tag{2}$$

where z_{ij} are the elements of the normalized matrix.

The computation formula to determine the entropy value of evaluation index is as follows:

$$H(x_j) = -k\sum_{i=1}^{n} z_{ij}\, lnz_{ij} \tag{3}$$

where k is the adjustment coefficient; z_{ij} means the j_{th} index's standardized value of the i_{th} evaluation unit, $i=1,2,...,n; j=1,2,...,m$.

Then, transform the entropy value of evaluation index into weighted value $w1_j$.

$$w1_j = \frac{1-H(x_j)}{m-\sum_{j=1}^{m} H(x_j)} \tag{4}$$

where $0 \le w1_j \le 1, \sum_{j=1}^{m} w1_j = 1$.

Finally, incorporate the weighted value $w1_j$ determined by entropy weight method to the weighted value $w2_j$ determined by experts to obtain the final weighted value w_j.

$$w_j = \alpha \cdot w1_j + (1-\alpha) \cdot w2_j, 0 < \alpha < 1 \tag{5}$$

3.3 Calculation of the human error probability (HEP)

After many times practice in recent years, Spurgin and his partners adopted the following logarithmic interpolation formula to calculate HEP:

$$S_i = \sum_{i=1}^{n} (QV_i) I_i \tag{6}$$

$$\ln(HEP_i) = \ln(HEP_l) + \ln(HEP_h / HEP_l)\left(\frac{S_i - S_l}{S_h - S_l}\right) \tag{7}$$

Thereinto,

S_i : The i_{th} path weighted value. S_i is the weighted value of each IFs of the corresponding path's human error probability. We considered that factors of the path are work multiplied when calculating the weighted value S_i. Assuming the weighted value of each IF after normalization as I_i (i =1~n, n is the number of IFs), I_i corresponds to the quality description value QV_i ;

S_l : Minimum path weighted value;

S_h : Maximum path weighted value;

HEP_i : The i_{th} path's human error probability;

HEP_l : Minimum calibration value of human error probability;

HEP_h : Maximum calibration value of human error probability.

4 Case study

4.1 Introduction of the numerical control machining industry

Originating from the needs of aviation industry, numerical control technology is a kind of machining method using digital information control parts and cutters' displacement. The control system gives commands and instructions to make cutting tools do all kinds of cutting motion conforming to the requirements. Numerical control machine will express the shape and size of the workpiece and other technical requirements and process requirements in the form of numbers and letters. It's an effective way to address parts machining problems about changing varieties, small batch, complex shape and high precision and to realize automatic processing efficiently.

However, professional and technical staffs are still needed for clamping, positioning, inspection and other duties. So there is no doubt that the quality of the parts might be influenced by personnel behaviour. What's more, on-site staff is responsible for handling potential fault directly. The study on human error plays an important role to ensure the production quality and reduce industrial accident. This paper will take the numerical control operators as case study to investigate human reliability by using the optimized HDT proposed in Section 3.

4.2 Select the behaviour IFs of staffs

The main duties of numerical control personnel are listed as follow: checking machines before starting in accordance with the operating rules; clamping, positioning and a series of preparations before processing; determining the processing status; operating numerical control machine tool parts; loading & un-loading parts and tooling. IFs that affect the operating personnel are selected as follow: mental state of the operators, personnel skills of the operators, quality of the man-machine interface, working environment of the workshop, operation specification of the workshop.

These IFs are analyzed to classify operators' behaviour factors accurately and the QD levels of various Ifs are determined.

(1) Mental state of the operators

Mental state of the operators (non-capacity factors) refers to work attitude and professionalism of the operators, which means whether the operating personnel's subjective initiative is at play. In this case, mental state of the operators includes two levels: good and poor.

(2) Personnel skill of the operators

Personnel skill of the operators (capacity factors) refers to individual psychological quality, physiological quality, professional skill and the operator training. Only the strong personal psychological quality can keep up in an emergency, and make the right decision. Only a good physiological quality and healthy body can competent to shift work for a long time. Training can improve the operation personnel's understanding of numerical control machining and the function of machine tools, and enhance operators' processing experience. Operators training can also improve the understanding of safe operate of machine tools. In this case, personnel skill of the operators is classified into three levels: skilled, enough and lack.

(3) Quality of the man-machine interface

Large numerical control workshop has different requirements on parts, and needs different numerical control machine tools. The numerical control machine tools manufactured by different manufacturers vary widely, so the degree of friendly man-machine interface has a great influence on operator's behaviour. A good man-machine interface can offer distinct content display, sensitive alarm display and fluent information transfer, while a bad interface of some relatively old machine may not supply enough information, and can lead to some wrong operation. In this case, quality of the man-machine interface is divided into three levels: helpful, available and adverse.

(4) Working environment of the workshop

Working environment of the workshop refers to environmental noise, high and low temperature, lighting, space and other effects on operator. Some measures such as noise reduction, the installation of air conditioning system, supplying bright and spacious workshop environment can effectively improve operator's work ability. In this case, working environment of the workshop is divided into two levels: comfortable and uncomfortable.

(5) Operate specification of the workshop

The machining of a large numerical control machine is generally a complicated process where a tiny mistake may produce great loss. Thus, operators have to follow a series of operation specification, including boot checking specification, equipment cleaning specification and processing steps inspection, etc. In this case, operate specification of the workshop is classified into two levels: helpful and available.

4.3 Rank the importance of each IF

After the IFs are selected, the next step is to determine the relative weighted value of them. Table 1 incorporate entropy weight method to the conventional expert method to determine the relative weight value of each IF, where in accordance with the distribution of 4:6. Relevant data come from a manufacturing plant of aviation products.

Table 1. Weight values of IFs.

IFs	Mental state of the operators	Personnel skill of the operators	Quality of the man-machine interface	Working environment of the workshop	Operate specification of the workshop
Entropy weight method	0.44	0.37	0.09	0.08	0.02
Expert Method	0.45	0.35	0.10	0.05	0.05
Comprehensive weights	0.446	0.358	0.096	0.062	0.038

4.4 Select quality factors & identify the importance of IFs

The IFs can be described into three QD levels in practical analysis generally. If it is divided into two levels, the results are slightly rough; if it is divided into four levels, the middle two levels may be difficult to distinguish in actual operation.

In this case, there are five IFs. If the QD of each IF is divided into three levels, the structured HDT will have 3^5=243 branches in theory; if the QD of each IF is divided into two levels, the structured decision tree will have 2^5=32 branches. A huge decision tree will obstruct the analysis while a small one cannot supply enough information, so an appropriate size should be determined carefully. According to the actual situation, different divisions on the QD of the five IFs are chosen as follows:

(1) Mental state of the operators: good and poor.
(2) Personnel skill of the operators: skilled, enough and lack.
(3) Quality of the man-machine interface: helpful, available and adverse.
(4) Working environment of the workshop: comfortable and uncomfortable.
(5) Operate specification of the workshop: helpful and available.

4.5 Select quality description value

After the determination of the QD of each IF, HDT method focuses on how the different qualities affect reliability. Define a unified value "1" for the best QD level (such as good, skilled and helpful) of each IF. Lower QD level will increase the human error probability, which can be transformed as quality factor values greater than "1". This paper use the order quantity "1~9" to compare and measure the importance of quality factor with reference to the analytic hierarchy process (AHP). "3" is selected as intermediate gradient factor to measure the quality description level from good to middle and from middle to poor. Therefore, in this case, the QVs of the IFs (shown in Table 2) are expressed as: "1" for good, "3" for middle and "9" for poor.

Table 2. Quality description (QD) levels and corresponding quality values (QV) of the IFs.

QD levels of IFs	Mental state of the operators	Personnel skill of the operators	Quality of man-machine interface	Working environment of the workshop	Operate specification of the workshop
"1" for good	good	skilled	helpful	comfortable	helpful
"3" for middle	poor	enough	available	uncomfortable	available
"9" for poor.	——	lack	adverse	——	——

4.6 Construct the Holistic Decision Tree (HDT)

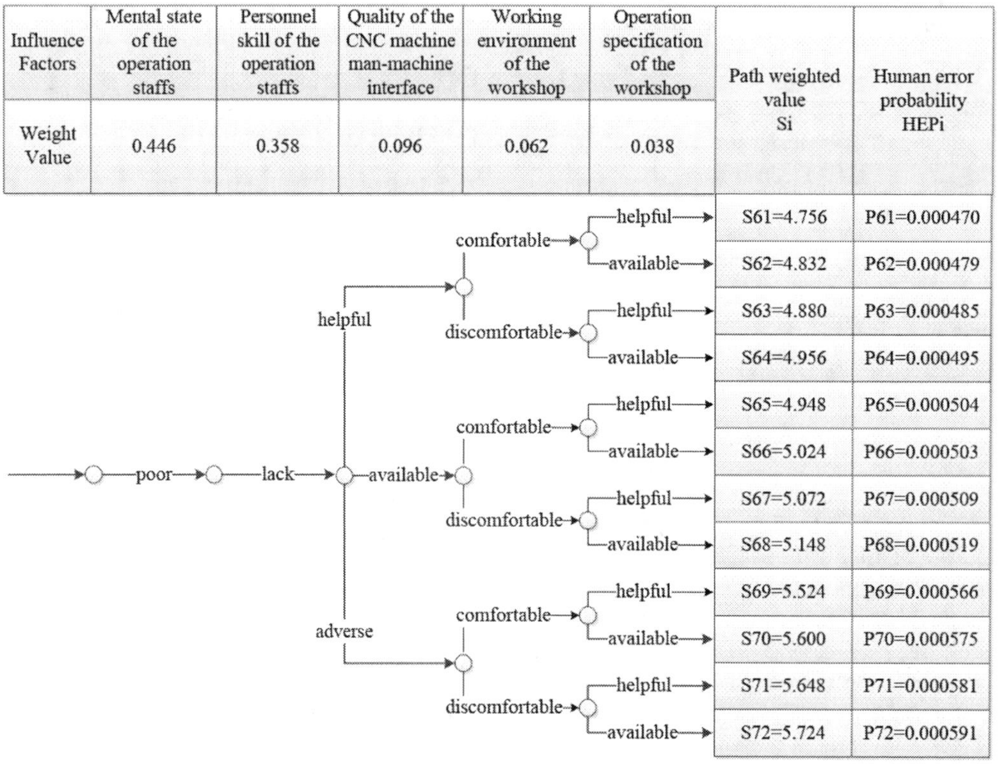

Figure 1. Last part of the HDT.

A typical HDT is constructed using some main elements such as the list of the key IFs, the relative weight of each IF, the list of the QD factors and corresponding QVs, and the calibration values. Assume that IFs are relatively independent, so they can arbitrarily arranged in the decision tree. After the decision tree's structure is built up, human error probability values can be calculated according to the relative weight, quality value, and the calibration value. It can be concluded from the typical decision tree that the lower position in the tree, the higher HEP value is. Thus, the HEP values at upper and lower boundary respectively are corresponding to the best and the worst IF's quality description respectively, and those two values are called the calculated calibration values.

The HDT is constructed according to the division of quality with a total of 72 branches. Figure 1 shows the last part of the HDT with the highest HEP value.

4.7 Select the calibration values of human error probability

Maximum calibration value of HEP assumes that all of the IFs are on the worst level, and staffs often make mistakes. If fail to get any help in the accident, the failure rate is 1. Minimum calibration value of HEP assumes that all of the IFs are on the best level. The minimum calibration values can change, and its uncertainty can be included in the analysis. In most scenarios, the 5% quantile of minimum calibration value is set as 10^{-4}, while 95% quantile is set as 10^{-3}; In some slight severe cases, the 5% quantile of minimum calibration value is set as 10^{-3}, and 95% quantile is 10^{-2}; In some serious cases, the 5% and 95% quantile of minimum calibration value are 10^{-2} and 0.1 respectively. So, in this case, we set the 5% quantile of minimum calibration value as 10^{-4}, and 95% quantile as 10^{-3}.

4.8 Calculate the human error probability (HEP)

After determining the weight value, the quality description value and the calibration value of the IFs, the HEP of each path can be figured out.

First, the i_{th} path weighted value S_i is calculated through formula (6). S_i is the weighted value of each IF for HEP in the corresponding path. Then, the HEP value HEP_i can be figured out through formula (7). As shown in the last two columns of Figure 1, the HEP rises along with the worse of quality description level.

4.9 Discuss the analysis results

Resourcing historical statistics of human error of numerical control machining workshop's accident investigation reports, the average QVs of each IF were got. In which, mental state of the operators is usually good, recommended value is 1; personnel skill of the operators is usually enough, recommended value is 3; quality of man-machine interface is usually available, recommended value is 3; working environment of the workshop is usually comfortable, recommended value is 1; operate specification of the workshop is usually helpful, recommended value is 1. Querying the constructed HDT and combining with the above results we have gotten, the quality description value of the five IFs and the final HEP value under the usual scene's HDT path is figured out as 0.000114, as shown in Table 3.

Table 3. HEP value under usual scene's path.

Influence factors (IFs)	Weighted value of IFs(*Ij*)	Quality description (average level)	Quality value (*QVi*)	*(QVi)Ij*
Mental state of the operators	0.446	Fine (good)	1	0.446
Personnel skill of the operators	0.358	Enough (middle)	3	1.074
Quality of man-machine interface	0.096	Available (middle)	3	0.288
Working environment of the workshop	0.062	Comfortable (good)	1	0.062
Operate specification of the workshop	0.038	Helpful (good)	1	0.038
Path weighted value (***S****i*)	—	—	—	1.908
Human error probability (***HEP***)	—	—	—	**0.000114**

Through the HDT analysis, it can be concluded that personnel skill and mental state of the operators are the two most critical IFs in numerical control machining industry. Especially, the lack of operative skill can accelerate the increasing degree of HEP by 56.3% while the poor mental state of the operators accelerates the increasing degree of HEP by 23.4%. Based on the analysis results, it is recommended that more professional training is required to enrich operators' skills and overtime

 https://doi.org/10.1051/matecconf/201816901030

working should be decreased to ensure a good mental state of the operators to reduce the human error probability.

5 Conclusions

Operators' behaviour influence factors (IFs) and their relationship in the numerical control machining industry is analyzed based on the optimized HDT proposed in this paper. Human error probability (HEP) values under different scenario are figured out and the two most critical IFs in numerical control machining industry is found out, which are personnel skill and mental state of the operators. Suggestion and measure for improvement are given specifically for purpose of the reduction of HEP. This optimized HDT method can provide reasonable and reliable basis based on analysis and calculation results for making the prevention and control measures of product damage and accident, and then improve the efficiency and safety of the industrial production process.

References

1. X.C. He, International Journal of Advanced Manufacturing Technology, **85**(1-4), 115-131 (2016)
2. T.D. Huang, et al., Journal of Shanghai Jiaotong University, **21**(5), 635-640 (2016)
3. X. Wu, Research on the Reliability Analysis and Growth of the CNC Machining Center, Journal of Guangxi University for Nationalities (2016)
4. B.S. Dhillon, *Human Reliability: With Human Factors* (Oxford, UK, 1986)
5. X.H. He and X.R. Huang, *Human reliability analysis in engineering system: theory, methods and applications* (Beijing, China, 2007)
6. A.D. Swain and H.E. Guttman, Handbook of human reliability analysis with emphasis on nuclear power plant applications, Technical Report (1983)
7. J.C. Williams, *Proceedings of the IEEE 4th conference on human factor in power plants*, June 6-9, (1985)
8. D.E. Embrey, P.Humphreys and E.A. Rosa, et al., SLIM-MAUD: an approach to assessing human error probabilities using structured expert judgment, Volume I (1984)
9. S.E. Cooper, et al., A technique for human error analysis (ATHEANA), Office of Scientific & Technical Information Technical Reports, NUREG/CR-6350 (1996)
10. E. Hollnagel, *Cognitive Reliability and Error Analysis Method CREAM* (1998)
11. A.J. Spurgin, Frontiers Science Series, **2**, 1343-1350 (2000)
12. A.J. Spurgin and M.V. Frank, *Developments in HRA technology from nuclear to aerospace Probabilistic Safety Assessment and Management* (London, UK, 2004)
13. H.F. Li, W.Q. Li and X.P. Wu, Chinese Journal of Ergonomics, **2**, 34-39 (2010)
14. X.Q. Zhang and H. Wang, China Safety Science Journal, **22**(1), 100-105 (2012)
15. L.Q. Lin, K.M. Huang and S.J. Liao, Communications Standardization, **42**(19), 75-80 (2014)
16. F. Sun, S. Zhong and Z.Y. Wu, Chinese Journal of Nuclear Science and Engineering, **28**(3), 268-272 (2008)
17. D.L. Mon, C.H. Cheng and J.C. Lin, Fuzzy Sets and Systems, **62**, 127-134 (1994)
18. Y. Fu, X.P. Wu and Q. Ye, ACTA Electronica Sinica, **38**(7), 1489-1494 (2010)
19. L.J. Sun, X.J. Xing and D.Q. Zhou, Statistics and Decision, **21**, 153-154 (2010)
20. W. Li, S.W. Gong and X.B. Li, China Safety Science Journal, **18**(3), 166-170 (2008)
21. S. Zhang, M. Zhang and G.T. Chi, Chinese Journal of Management, **1**, 34-42 (2010)

Optimum distance for voice navigation in walking support system for visually impaired

Akinari Ohfuji[1], Yoshihiro Taniura[2], Dai Hanawa[1,3] and Kimio Oguchi[1, 2,a]

[1]*Information Networking Laboratory, Faculty of Science and Technology, Seikei University, 180-8633 Tokyo, Japan*
[2]*Information Networking Laboratory, Graduate School of Science and Technology, Seikei University, 180-8633 Tokyo, Japan*
[3]*Graduate School of Design and Architecture, Nagoya City University, 464-0083 Nagoya, Japan*

Abstract. Voice navigation with information and communication technologies allows the visually impaired (VI) to grasp the current situation promptly. While instruction prior to the action e.g. turning a corner, seems to be effective, the optimum lead distance remains unclear. This paper verifies experimentally the optimum lead distance to support VI while walking. In the experiment, each subject (normal vision) walked while wearing an eye mask to simulate VI, holding a white cane, and received voice guidance. After each trial, the subject scored the lead distance in terms of acceptability. 28 young healthy males/females participated. Results showed the highest score was for 3 m while the lowest was for 5 m. Distance range with score over 3 (average) was 2 to 4 m. No clear significant difference was observed in terms of subject height. As walking speed was around 0.55 m/s, 1 m was deemed insufficient. 5 m or more seemed to be too long as subjects failed to predict accurately how far they had moved. This experiment verified that VI voice navigation turn instructions are effective if given more than 1m and less than 4 m before the turning point.

1 Introduction

Due to the spread of diabetic disease, the number of the visually impaired (VI) is increasing. As more than 80 % of the information obtained by healthy people is generally derived from visual information [1], any vision impairment complicates daily life. The VI emphasizes the remaining senses of touching, hearing, smell and taste [1]. There are two main types; totally blind and partially blind. More than 285 million (M) people worldwide suffer from vision loss that interferes with daily activities. Among them, 39 M are totally blind. In Japan, about 0.3 M people are visually impaired and the number is increasing year by year [2].

Most VI use a white cane in daily life while walking outside the house, or inside of shopping malls/office buildings, however, its extremely limited range limits its effectiveness. Use of ICTs (Information and Communication Technologies) is an interesting and attractive solution [3]. As more of the VI are now using ICT devices such as tablet, smartphones [4], guidance systems based on such devices seemed effective.

[a] Corresponding author : oguchi@st.seikei.ac.jp

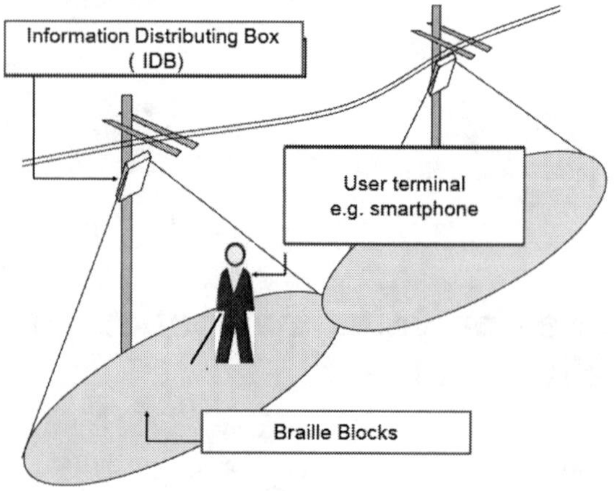

Figure 1. Proposed walking support system image on the street.

Many other researches are being performed. Some are as follows as: Voice guidance using information gathered from RFID (Radio Frequency Identification) tags embedded in braille blocks [5, 6]. Cloud base outdoor assistive navigation system [7]. A smart and intelligent cane developed that detects an object or obstacle using ultrasonic sensors and gives audio instructions for guidance [8]. Haptic feedback device and voice command used for moving freely realized by data captured through the Kinect device [9]. Make a warning through both voice and vibration when determine the range of obstacles by an ultrasonic device [10]. They all are systems based on voice navigation.

Voice navigation allows the VI to grasp the situation promptly. While instruction prior to the action e.g. turning a corner, seems to be effective, the optimum lead distance remains unclear. Therefore, this research verifies experimentally the optimum lead distance to support the walking VI.

The structure of this paper is as follows; Section 2 clarifies the final proposed VI walking support system with voice navigation. Section 3 describes the first experiment; it clarifies whether guidance that uses only distance is enough for voice navigation. In addition, it shows the importance of optimizing the lead distance for voice navigation, the subject of the second experiment. Section 4 describes outlined functional blocks for the future system image. Conclusion and future work are then given.

2 Proposed system image

The final target of this research, a VI walking support system, is shown in Fig. 1. VI walks with a mobile terminal device as smartphone or voice navigating terminal that can communicate with any information distribution box (IDB) located on street poles. Radio frequency (RF) signal is likely to be used for such communication. Brail blocks [5, 6] on the road might be another guidance technology them.

3 Experiment

3.1 Lead distance information only (First experiment)

3.1.1 Experimental methodology

The first experiment verifies whether just the lead distance is enough to assist the VI when walking.

The experiment proceeded as follows;

1-1. Each subject was told a distance to move (e.g. 3 m) randomly prior to walking,

1-2. The subject walked straight ahead wearing an eye mask for the indicated distance,

1-3. Each subject walked 15 times in total (5 different distances x 3 times each), and

1-4. The distance actually walked was then compared the instructed distances.

The subjects were 11 young healthy people as listed in Table 1. The experiment was performed in a corridor of the university.

Table 1. Attributes of 11 subjects: the first experiment.

Subject	Gender	Age (year)	Height (m)
1a	male	24	1.70
1b	male	21	1.73
1c	female	19	1.53
1d	male	22	1.75
1e	male	22	1.63
1f	male	19	1.78
1g	male	22	1.75
1h	female	20	1.60
1i	male	21	1.72
1j	female	21	1.55
1k	male	19	1.72

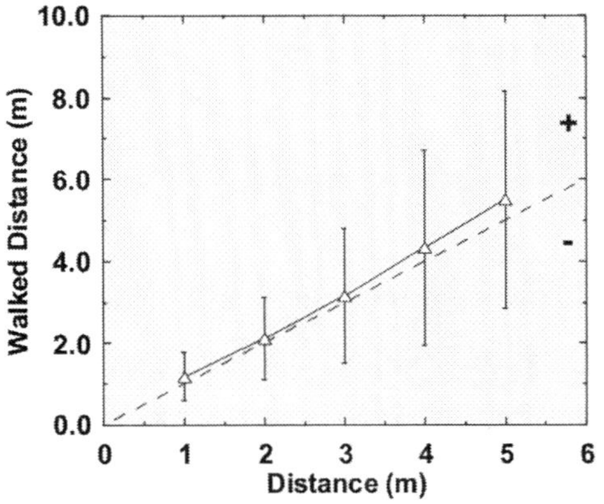

Figure 2. Results when only distance information was provided. Measured average values and deviations are

3.1.2 Experimental results

Figure 2 plots the average walked distance versus the instructed distances of 1, 2, 3, 4, and 5 m, plus their standard deviations. Here, "+"means a subject walked longer than the instructed distance, while "–" indicates a shorter distance.

As seen in the figure, all subjects walked longer regardless of the distance and their deviations increased with distance. Distance perspective with eye mask, or accuracy of walked distance, depends strongly on the subject. For example, some subjects walked more than twice the required distance for 3, 4, and 5 m. These results imply that a simple statement of distance is not enough for VI navigation assistance.

3.2 The second experiment

3.2.1 Experimental methodology

As the first experiment confirmed that just distance information is not enough, the second experiment used two voiced instructions.

The procedures are as follows;

2-1. Each subject walked while holding a white cane (1.27 m length, folding type) while wearing an eye mask to simulate VI,
2-2. The subject received voice guidance such as "left turn is in N m" at N m before the corner, and then the instruction "turn left here" at the corner, as seen in Figure 3. N is 1, 2, 3, 4 and 5 m,
2-3. After each trial, the subject evaluated the acceptability of the lead distance using scores of 1 to 5 (very bad, bad, average, good, and very good, respectively).
2-4. Each subject performed 3 trials for each N.

Subjects participating in the second experiment were 28 young healthy people whose attributes are listed in Table 2.

This experiment was performed in the same university corridor.

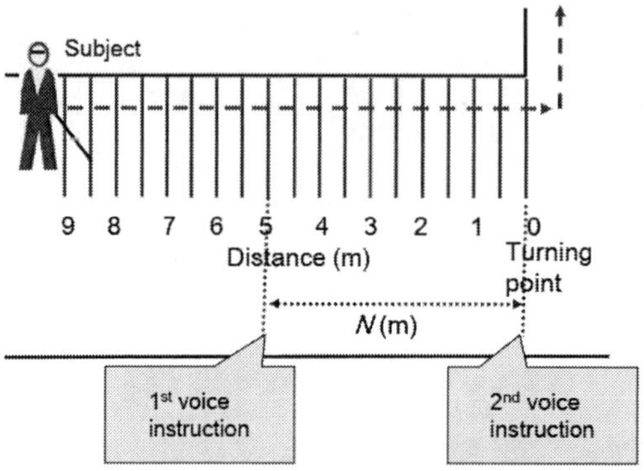

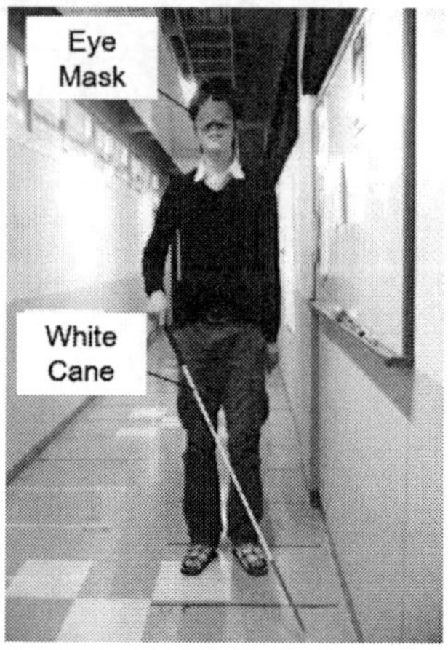

Figure 3. Experimental procedures (schematic view), and a photo of a subject walking with a white cane.

Table 2. Attributes of 28 subjects: second experiment.

Subject	Gender	Age (year)	Height (m)	Foot size (cm)
2a	male	20	1.65	25.0
2b	male	20	1.71	26.5
2c	female	20	1.70	23.5
2d	female	19	1.53	24.5
2e	male	20	1.67	26.5
2f	male	21	1.77	28.0
2g	male	21	1.70	26.0
2h	male	21	1.72	26.0
2i	male	22	1.75	28.0
2j	male	19	1.78	27.5
2k	female	20	1.60	24.0
2l	male	20	1.75	26.5
2m	male	22	1.66	26.0
2n	female	21	1.60	23.5
2o	female	21	1.55	23.5
2p	male	19	1.64	27.0
2q	male	22	1.63	25.0
2r	male	22	1.63	26.0
2s	male	19	1.74	27.0
2t	male	19	1.70	27.0
2u	male	19	1.61	27.5
2v	male	19	1.69	26.0
2w	male	19	1.75	26.0
2x	male	19	1.82	27.0

2y	male	19	1.66	27.0
2z	male	19	1.79	27.5
2aa	male	20	1.76	27.0
2ab	male	21	1.77	27.0

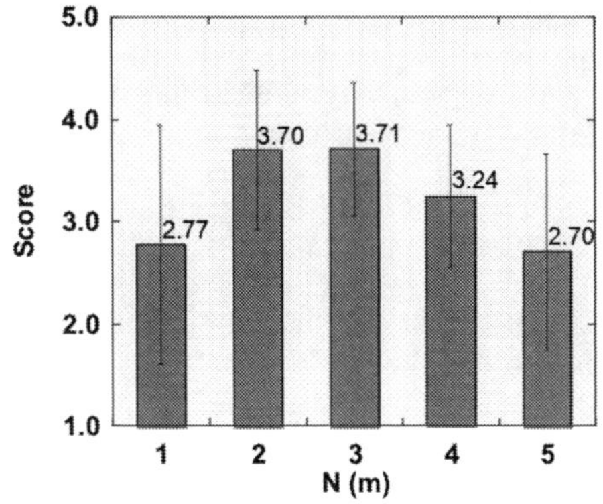

Figure 4. Measured results of the second experiment. Average score and its deviation (error bar) are depicted for the N values of 1, 2, 3, 4, and 5m.

3.2.2 Results

Figure 4 shows the results of average score for each N and its deviation. Highest score (3.71) was for N=3 m while the lowest (2.70) was for 5 m. Distance range with score over 3 (average) was 2 to 4 m.

The significant difference in the results was identified and analyzed. The sign test found a significant difference between N=3 and 4, and between N=3 and 5. These results indicate that preferred lead distance is 1 < N < 4 m. As walking speed was around 0.55 m/s, N=1 (around 2 sec) was deemed insufficient. N=5 or more seemed to be too long as subjects failed to predict accurately how far they had moved.

Dependency on height was also analyzed using two groups, one with height of $1.60 \leqq h < 1.70$ m, and the other with $1.70 \leqq h < 1.80$ m (gender neutral). Here, all data were separated into the two groups and analyzed.

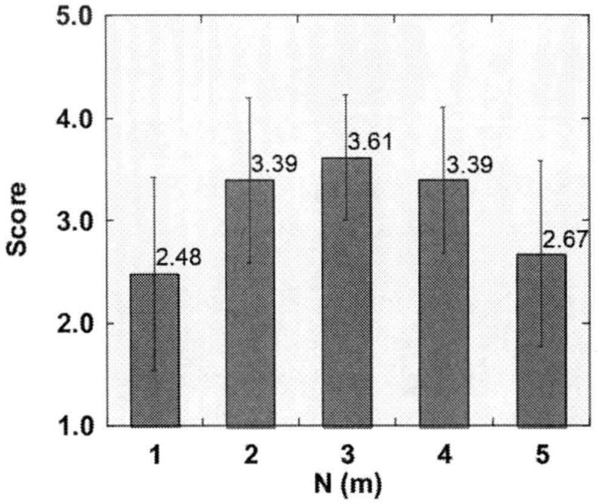

Figure 5. Measured results for the shorter group: average score for each N and its deviation.

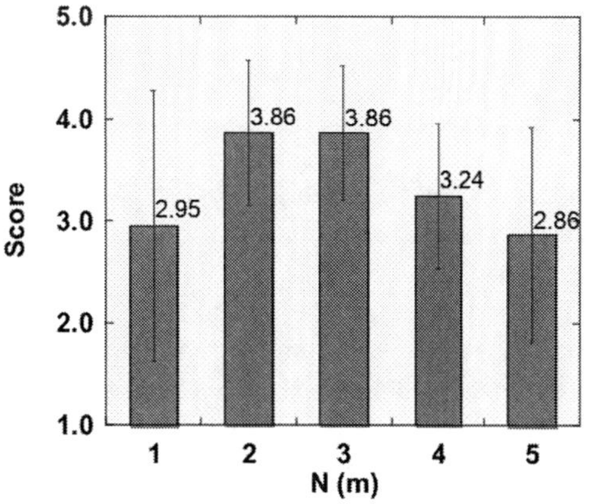

Figure 6. Measured results for the taller group: average score for each N and its deviation.

Figure 5 shows the results for the shorter group with average score for each N and its deviation. Highest score (3.61) was for N=3 m while the lowest (2.48) was for 1 m. Distance range with score over 3 (average) was the same, 2 to 4 m. No significant difference was observed. These results indicate that preferred lead distance is 1 < N < 5 m.

Figure 6 shows the results for the taller group with average score for each N and its deviation. Highest score (3.86) was for N=2 and 3 m while the lowest (2.86) was for 5 m. Distance range with score over 3 (average) was the same, 2 to 4 m. These results indicate that preferred lead distance is 1 < N < 5 m. With the sign test, a significant difference was observed between N=2 and 1, and between N=3 and 1, between 3 and 4, and between 3 and 5. These results indicate that N=1, 4, and 5 have significant differences. Therefore, the preferred lead distance is more than 1 m and less than 4 m.

This analysis of subject height dependency shows that it is small, if it exists. Therefore, the preferred lead distance is more than 1 m and less than 4 m for all body heights.

4 Outlined functional blocks

Figure 7 shows outlined functional blocks for the VI supporting system (see Figure 1) where the administration manages information distributing boxes (IDBs) via network. VI requests e.g. location information to the nearest IDB, and then the information is provided through a wearable device. Home of the VI provides the path information to be planned and knows the current location via network. The wearable device may use the GPS (Global Positioning System) information.

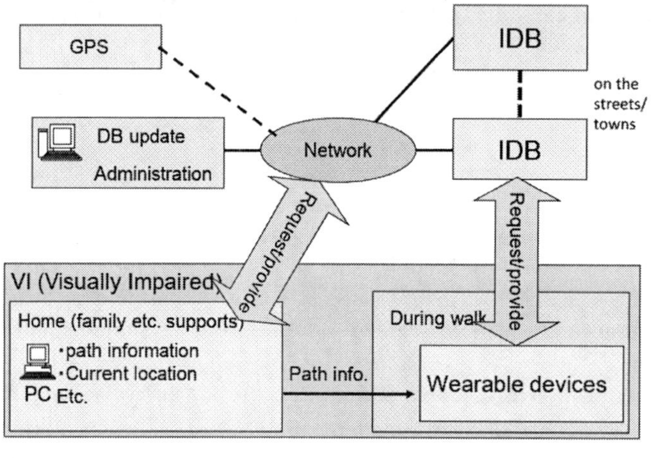

Figure 7. Outlined functional blocks for VI supporting system.

5 Conclusion

In order to realize a more effective walking support system with voice navigation for the visually impaired, we experimentally examined the optimum lead distance for voiced movement (distance) instruction.

The first experiment clarified that simply providing just distance information once is not enough to realize effective voice navigation. The second experiment examined the effectiveness of two voice instructions. A total of 28 subjects simulated the movements of the visually impaired and scored the preferred distance. These experiments verified that VI voice navigation is effective if the distance instruction is given more than 1m and less than 4 m before the turning point. No clear significant difference was observed in terms of subject height.

Future work includes an experiment with actual VIs to clarify how different they are and for other case e.g. at the step, and development of a more comprehensive guidance system.

References

1. M. Tsuda, When the visually impaired walk around the city (Toshi Publication, 1999)

2. Government of Japan, Status of persons with disabilities, http://www8.cao.go.jp/shougai/whitepaper/h25hakusho/zenbun/furoku_08.html
3. Z. Farid, R. Nordin and M. Ismail, Journal of Computer Networks and Communications, **2013**, 1-12 (2013)
4. T. Watanabe, T. Yamaguchi and K. Minatani, A Survey on the Use of Mobile Phones, Smartphones, Tablets, Personal Computers by Blind and Visually Impaired Persons 2013, The Telecommunications Advancement Foundation, The study Investigation Grant Achievement Report (2014)
5. NPO Kotoba no Michiannai, Tokyo Metropolitan Government Bureau of Construction, Experiment of a guide system for the visually impaired in Tokyo, https://www.hido.or.jp/14gyousei_backnumber/2011data/1111/1111kotonavi-metro.pdf
6. Japanese Industrial Standard, JIS T9251:2014, Guidelines for older persons and persons with disabilities -- Shapes, dimensions and patterns of raised parts of tactile walking surface indicators for persons who are blind or with seeing impairment (2014)
7. A. N. Lapyko, L-P. Tung and B-S. P. Lin, *2014 7th IFIP Wireless and Mobile Networking Conference (WMNC)*, 1-8 (2014)
8. A. Noorithaya, M.K. Kumar and A. Sreedevi, *International Conference on Circuits, Communication, Control and Computing*, 177-181 (2014)
9. F. Ahmad, Tanveerulhaq, I. Ishaq, D. Ali and M. F. Riaz, *2016 International Conference on Bio-engineering for Smart Technologies (BioSMART)*, 1-4 (2016)
10. N. Mahmud, R.K. Saha, R.B. Zafar, M.B.H. Bhuian and S.S. Sarwar, *2014 International Conference on Informatics, Electronics & Vision (ICIEV)*, 1-5 (2014)

Research on coalescence and probabilistic failure physics modeling of coplanar semiellipse cracks

Sufen Li, Yufeng Sun[a], Weiwei Hu and Zhaoxia Liu

Reliability and System Engineering School, Beihang University, China

Abstract. Multiple fatigue cracks, causing transient fatigue fracture and resulting in great loss of life and property, are very commen in engineering structures. In order to predict the fatigue life of engineering alloys more accurately, the interaction and coalescence of adjacent cracks was investigated and the four stages of cracks growth were discussed in this paper. The SIF(stress intensity factor) value was found to growth rapidly in the ligament failure, and simulation method was used to find out the coalescence point. Then, the Paris' law was chosen to discribe the crack growth process. Taking random factors like structure size, material properties, environment stress parameter into account, the probability physical of failure model of alloys can predict the failure life more accurate. Accordingly, the uncertain parameters of the Paris' law were updated via Bayesian update method. The posterior distributions of uncertain parameters were obtained by the MCMC-Gibbs sampling. Finally, the probabilistic failure life models before and after coalescence were established by Monte Carlo simulation method. The probabilistic modeling method, which can greatly reduces the historical data or experimental data needed for the reliability evaluation of small sample products, can also be applied to other aspect of physic of failure , effectively saving time and cost.

1 Introducition

Fatigue failures, which occures at stress below the tensile strength or yield strength of alloys is hard to be detected. Since most engineering components are or similarly are subjected to alternating stress. Multiple fatigue cracks causing transient fatigue fracture and resulting in great loss of life and property, are very commen in engineering structures. Crack mainly initiats in localized areas of stress concentratin or material defects[1,2]. Then these initial cracks grow together by interacting with each other, meanwhile shortening the fatigue life with an accelerated crack growth rate[3]. Interaction and coalescence processes of adjacent cracks is the key issue in the study of coplanar fatigue cracks.

Various theories have been propounded to assess the growth of adjacent surface cracks growth. The British Standard Institution[4] proposed a prediction method assuming cracks to be indepentend. However this idealize simple model mainly considering the geometric condition of cracks, turned out to be too conservative[5]. Iida[6] assumed that when the cracks touch, a semi-elliptical crack immediately drawn through the outer tips of the two adjucent cracks and through the deepest point of the deeper crack. The study neglected the interaction in calculation between adjucent cracks(the interaction factors γA to F =1) and proposed a simple realistic prediction method. However the

[a] Corresponding author : syf@buaa.edu.cn

SIF(stress intensity factor) approximation and the material constents used may result in unsafe assessments. Leek & Howard[7, 8] developed methods to calculates the interaction factors γA to F, which quantifies the interactions between the stress intensity factors of adjucent cracks. But the coalescence point have not figured out.

The Physics of failure(PoF) model can characterize the relation of the time of failure, the degradation of the performance parameters, the structure,the material and so on. The PoF model can describe the failure cause from the failure mechanism level and predict the failure time. [9-11]. After twenty-first Century, researchers gradually realized that the deterministic model could not describe the inevitability of material, stress and other conditions. Thus the PoF model tend to develope into the Probabilistic Physics of failure(PPoF), which describes the probability function relation between the external environment and the time of failure. [12-14]

In this paper, 1) the coalescence of two adjucent coplanlar semiellipse cracks is researched via simulation method; 2) probabilistic failure physics(PPoF) models for cracks before and after coalescence are established; 3) life distribution of probability density curves are obtaind by Monte Carlo simulation.

2 Methods

2.1 Crack interaction and coalescence process

According to Bayley[1], fatigue crack coalescence means two adjacent crack increase in size and finally grow together to form a single lager crack. The adjacent and coplanar surface cracks under LEFM conditions can interact. And Leek[7]et al. estimated the interaction factors for two coplanar surface cracks.The interaction and coalescence between adjucent cracks can critically increase the crack growth rate(Fig. 1). To assess the fatigue life of engineering components safely, crack interaction and coalescence must be taken into consideration. As illustrated in Fig. 2, the interaction and coalescence process can be described in four stages:

1) Independent stage: the spacing of crack tips are large enough that the two adjacent cracks can be regarded as independent.
2) Interaction stage: the two crack get closer and start to influences the stress field of each other.
3) Recharacterization stage: the neiboring cracks are close enough to form a concave crack. The front of the concave crack growth rapidly due to the high SIF value, which is called the failure of ligament. The stage ends when the concave crack is recharacterized as an envelope crack.
4) Post-coalescence stage: the envelope crack tends to resume its semi-cicular shape. Afterwards, it continues to popagate as it did before the interaction.

It's not easy to define the recharacterization stage, for it's hard to be observed. Moukawsher et al.[16] demonstrated that it is the yeild failure of the ligament that caused the failure of small fatigue cracks. And Swift[17] applied a yeild criterion to a linear alastic solution of the crack tip stress field. In the studies of Jeong[18], it was assumed that the yielded volumes was resulted from the stress field at each crack tip. Owing to the constraint of the surrounding material preventing complete plastic deformation, the stress concentrated at the center of the ligament. Once the average ligament stress reach a critical value that is equivalent to the ultimate tensile strength, the ligament failure occurs.

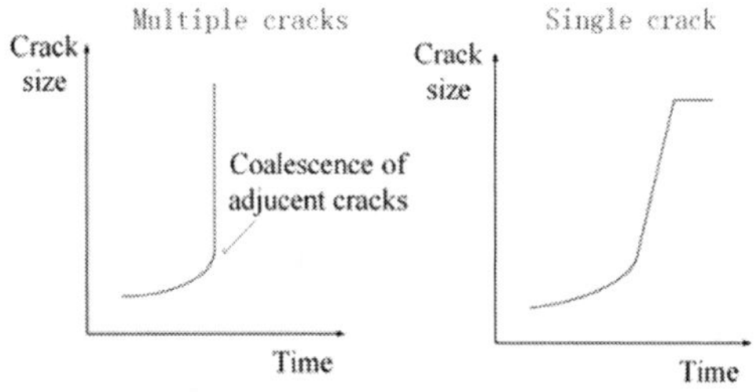

Figure 1. The crack growth rate of multiple cracks and single cracks.

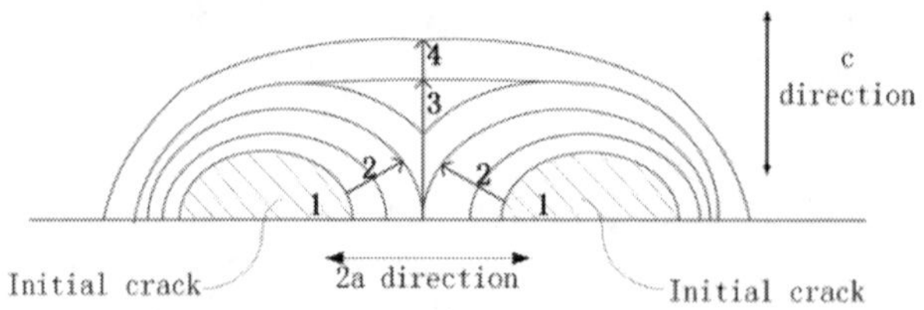

Figure 2. The interaction and coalescence process of adjucent cracks.

2.2 The line spring model of multi coplanar cracks

To study the interaction and coalescence of multiple co-plane cracks, the numerical simulation method is used to calculate the stress intensity factor. Compared with the crack size, the inelastic region of the crack tip is very small, so linear elastic fracture mechanics (LEFM) can be applied to solve the fatigue problem. Rice proposed that the issue of three dimensional surface crack can be simplified to the plate and shell structure with line spring model.The line spring model is a simple and effective method to solve the problem of co-plane multiple cracks[3, 14, 15].

In the line spring model, there is a infinite plate with n coplanar surface cracks under force N^{∞} and bending moment M^{∞}(Fig. 3(a)) ai and ci refers to the depth and half length of the crack i; bi and di means the coordinates of the crack tips on the front surface of crack i, where i=1,2,3,......n. As shown in Fig. 3(b) the model is simplified to a plate containing n semi - elliptical cracks, the discontinuities of cracks are treated with line springs..When $x \in [b_i,\ d_i]$, the line springs can be discribed with a plane strain slab. The SIF of the surface crack front is equal to the SIF of the corresponding position of the edge crack on the slab.

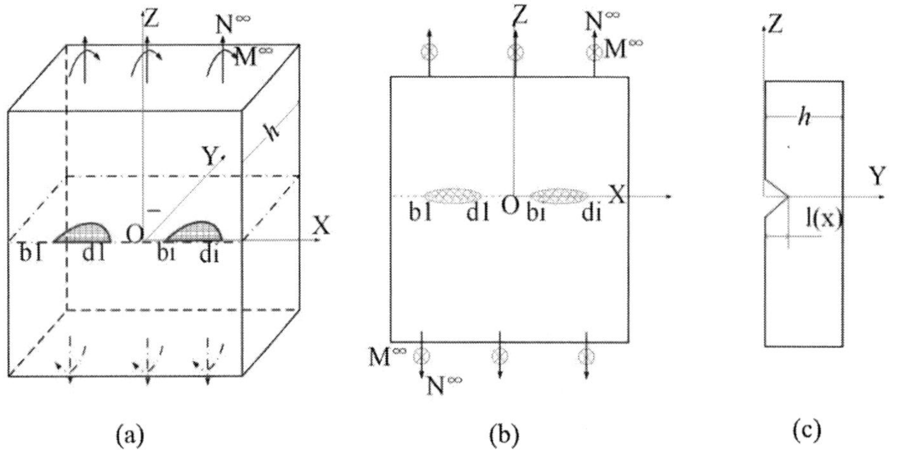

Figure 3. The line spring model

Accroding to Sun[3], the dimensionless integral characteristic equations of the plate with *n* cracks are:

$$\begin{cases} \sum_{j=1}^{n}\int_{-1}^{1}\frac{1}{t_j-x_i}\left(\frac{c_j}{c_1}\right)\mu_{j1}(\bar{t})d\bar{t} = \frac{2\pi}{E}\left[\sigma_{iM}(\bar{x_i}) - \sigma_M{}^{\infty}\right] \\ \sum_{j=1}^{n}\int_{-1}^{1}\left[\frac{1}{t_j-x_i} + k(t_j,x)\right]\left(\frac{c_j}{c_1}\right)\mu_{j2}(\bar{t})d\bar{t} = \frac{4\pi}{E}\left(\frac{c_1}{h}\right)\left[\sigma_{iB}(\bar{x_i}) - \sigma_B{}^{\infty}\right] \end{cases} \tag{1}$$

where $t_j = \frac{c_i}{c_1}\bar{t} + \frac{b_{i+d_i}}{2c_1}$, $x_j = \frac{c_i}{c_1}\bar{x} + \frac{b_{i+d_i}}{2c_1}$; $\bar{t},\bar{x}[-1,1]$; E refers to the modulus of elasticity; $k(t_j,x)$ is known kernel function; ci refers to half length of the crack; h is the slab thickness; $\sigma_M{}^{\infty} = N^{\infty}/h$, $\sigma_B{}^{\infty} = 6M^{\infty}/h^2$; $\mu_{j1}(\bar{t})$, $\mu_{j2}(\bar{t})$ are unknown dislocation density function; $\sigma_{iM}(\bar{x_i})$, $\sigma_{iB}(\bar{x_i})$ are unknown stesses; i=1,2,3,...,n.

The line spring constitutive relation is determined by the relationship between the generalized force suffered by the plane strain edge slab and the additional generalized displacements caused by cracks. Taking crack i for example:

$$\begin{Bmatrix} \sigma_{iM}(\bar{x}_i) \\ \sigma_{iB}(\bar{x}_i) \end{Bmatrix} = \frac{E}{1-v^2} \begin{bmatrix} \gamma_{MM}^{(i)} & \gamma_{MB}^{(i)} \\ \gamma_{BM}^{(i)} & \gamma_{BB}^{(i)} \end{bmatrix} \begin{Bmatrix} \frac{c_i}{h}\int_{-1}^{1}\mu_{i1}(\bar{t})H(\bar{x}-\bar{t})d\bar{t} \\ \frac{1}{6}\frac{c_i}{c_1}\int_{-1}^{1}\mu_{i2}(\bar{t})H(\bar{x}-\bar{t})d\bar{t} \end{Bmatrix} \tag{2}$$

where the matrix $[\gamma_{\lambda\mu}^{(i)}] = [\alpha_{\lambda\mu}^{(i)}]^{-1}$, $H(\bar{x}-\bar{t})$ are unit step functions; $\alpha_{\lambda\mu}^{(i)} = \int_0^{\xi} g_\lambda^{(i)}(\xi) g_\mu^{(i)}(\xi) d\xi (\lambda, \mu = M, B)$; $g_M^{(i)}(\xi)$ and $g_B^{(i)}(\xi)$ are known function.

Under the general force $N(x)$ and $M(x)$, the stress intensity factor of the edge crack slab can be described as：

$$K_I = \sqrt{h}[\sigma_M(\bar{x})g_M(\xi) + \sigma_B(\bar{x})g_B(\xi)] \tag{3}$$

where $\xi = l(x)/h$, and $l(x)$ means the distance between the tip of the edge crack and the axis of the ellipse(Fig. 3(c)). Substitut equation (2) into equation (3), the basic integral equations for the surface crack line spring model with residual stress are acquired and it can be solved by the Gauss-Chebyshev methods[2]. Afterwards, the stress intensity factor near the crack tip can be solved by using equation (1) and equation (3).

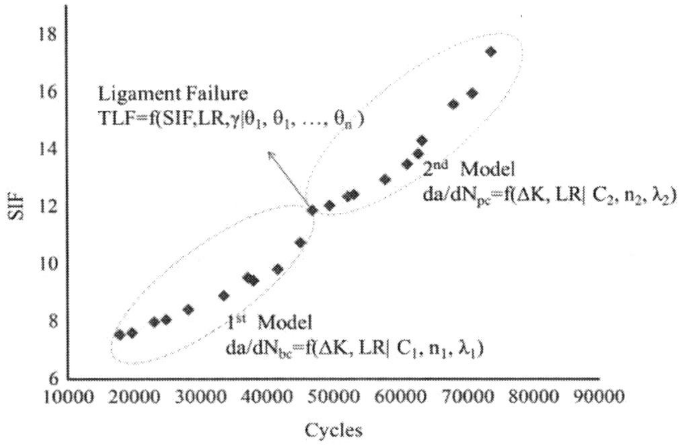

Figure 4. The simulation results of SIF of coplanar cracks.

Simulations have been done to observe the change of the SIF of coplanar cracks under the stress 290 Mpa. As can be seen from Fig. 4, the SIF value increases with the fatigue cycling number. And there is a sudden increase of the SIF near where fligament failure is expect to occur. Cracks growth with different crack propagation speed before and after the fligament failure. Thus, there are two failure physics model to describe the fatigue life.

2.3 Physic of failure model

2.3.1 The Paris law

The opening mode caused by tensile stress, the most common cracks in engineering, will be studied in this paper. Based on the experiment, the Paris's law was promoted by Paris and Erdogan in 1963[22]. Taking the dislocation dynamics theory, the thermal activation energy theory and the rate process into account, sevral investigator revealed the physical essence of the Paris equation[23, 24]:

$$\frac{da}{dN} = \frac{A_0}{f^\lambda}(\Delta K)^\lambda exp(-\frac{\lambda U}{kT} + \frac{\alpha\lambda ln\Delta K}{kT}) \tag{4}$$

da/dN means the crack growth rate and f is the lading frequency; ΔK is the stress intensity factor. And

$$A_0 = b\gamma(m)(\frac{A}{4bG\epsilon^{1/2}})^{\lambda} \quad (5)$$

where b is the Burgers vector; γ(m) is a dimensionless quantity about m; G, τ_a means shear modulus and shear stress respectively; ε is the placement under mean stress; $\lambda = (m+1)\ /\ (m+2)$;

Denote as:

$$\mathrm{C} = \frac{A_0}{f^{\lambda}} exp\left(-\frac{\lambda U}{kT}\right) \quad (6)$$

$$\mathrm{n} = \lambda + \frac{\alpha\lambda}{kT} \quad (7)$$

Thus, the famous Paris equation is abtained:

$$\frac{da}{dN} = C(\Delta K)^n \quad (8)$$

To take the loading ratio (lr) into consideration, a parameter m is introduced in this paper:

$$\frac{da}{dN} = C(\Delta K)^n (lr)^m \quad (9)$$

2.3.2 Model parameters

In the Paris equation, there are two uncertain parameter: *C, n* and *m*, whose prior data are collected by researching literatures.

The prior information matrix are as following:

$$\theta_{C,n} = \begin{bmatrix} C & n \\ 2.06E-11 & 4.59 \\ 6.11E-14 & 4.42 \\ 5.13E-10 & 3.61 \\ 3.00E-10 & 2 \\ 4.53E-10 & 2.09 \\ 1.13E-10 & 2 \end{bmatrix}$$

After fitting these data, the prior distribution of the parameter C and n are obtained: C_B~Normal(2.33E-10,2.41E-10), n_B~Normal(3.10, 1.19). Besides, the uncertain parameter *m* is assumed as uniform distribution: m~ Uniform(-4,4).

2.4 Probabilistic failure physics

The PoF model can establish the relationship between failure time or the degradation of performance parameters with structure, material, work environmental conditions and function, describing the failure cause from the failure mechanism level and predicting the product failure time. However, owing to the uncertainties of structure size, material properties, environment stress parameter, the product life is supposed to follow a certain distribution. Based on the PoF model and taking these random factors into account, the probability failure of physical model can describ the failure life more detail.

Fig. 5, shows the realization way for PPoF. Firstly, analysis the failure mechanism of product according to its working environment and stress conditions, and choose a appropriate PoF model. Secondly, analysis and research on the uncertainties of the failure mechanism, and get the prior distribution of the uncertain parameters by fitting the investigation data. Furthermore, collect evidence data by simulation experments and obtain the likelihood function of crack growth rate. Then, update the parameters according to the Bayeslaw to get the posterior distribution; due to the difficulty of high dimensional integral calculation process, the MCMC-Gibbs sampling method is used to aquire the posterior distribution of the parameters. Finally, the posterior distribution of the failure life is acquired through Monte Carlo sampling.

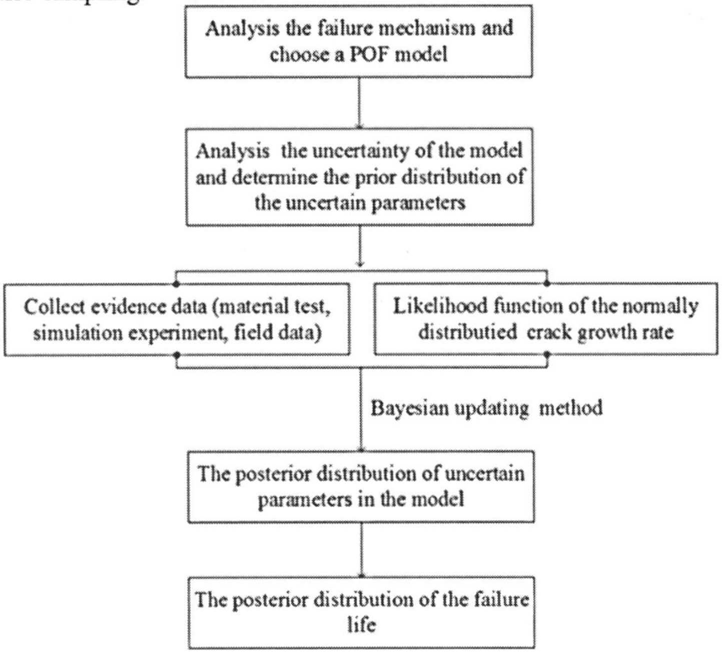

Figure 5. The flow-process diagram of realization way for PPoF.

3 Results

The evidence data (Ds) required in the Bayes updating comes from the simulation experiments. Assuming the likehood function of the crack growth rate that the data obey normal distribution:

$$L(C,l,m,\sigma|Ds) = \prod_{i=1}^{T} \frac{1}{\sigma\sqrt{2\pi}} e^{\frac{\frac{da}{dNi}-(C\Delta K^{n}(lr)^{m})}{2\sigma^{2}}} \tag{10}$$

According to the Bayes equation, the posterior distribution of the parameters are as following：

$$\pi(C,m,n,\sigma|Ds) = \frac{L(C,m,n|Ds)\pi_0(C,m,n)}{\iiint L((C,m,n|Ds))\pi_0(C,m,n)dCdmdn} \tag{11}$$

Obviously, it is difficult to solve the above equation analytical with integral method. However, with the usage of WinBUGS, it can be solved by using Markoff Montecalo method. Combination the prior information and test data with crack propagation, the Bayesian parameter updating method can obtain the probabilistic failure physics model. To study the change of the parameters, model parameters before and after the failure of the ligament are updated respectively.

3.1 The model before coalescence

Fig. 6 presents the posteir distribution of uncertain parameters before coalescence, which are required in the Bayesian parameter updating. The mean value of uncertain parameters in Table 1. are used to establish the failure life model. And the mean value of C, m, n are 2.376E-10, 0.1997, 2.09, respectively. Thus the mean model for cracks before coalescence is formula (12).

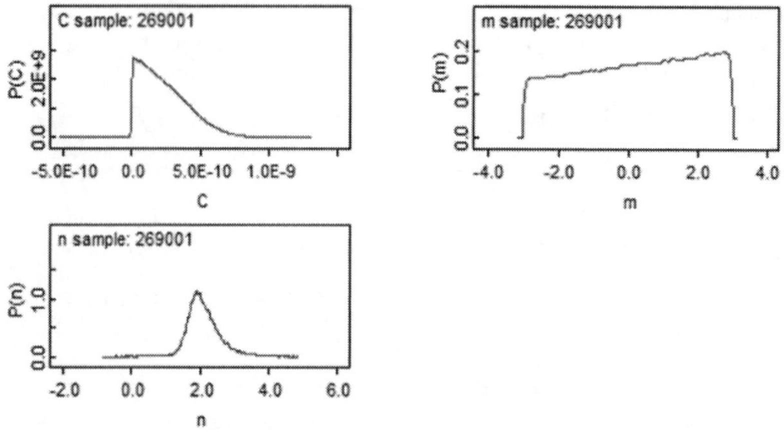

Figure 6. The posteir distribution of uncertain parameters before coalescence

Table 1. The updating results of uncertain parameters before coalescence

	Mean	sd	MC_error	Val2.5pc	median	Val97.5pc
C	2.376E-10	2.027E-10	2.986E-12	8.788E-10	2.047E-10	6.474E-10
m	0.1997	1.722	0.009009	-2.813	0.292	2.876
n	2.09	0.46	0.01052	1.393	2.019	3.216

$$\frac{da}{dN} = 2.376E - 10\Delta K^{0.197}(lr)^{2.09} \tag{12}$$

3.2 The model after coalescence

The Bayesian parameter updating results are presented in Fig. 7. As can be seen, the parameter distributions have changed. The mean value of uncertain parameters in Table 2. are used to establish the second failure life model. And the mean value of C, m, n are 3.33E-10, -0.04588, 1.828, respectively.

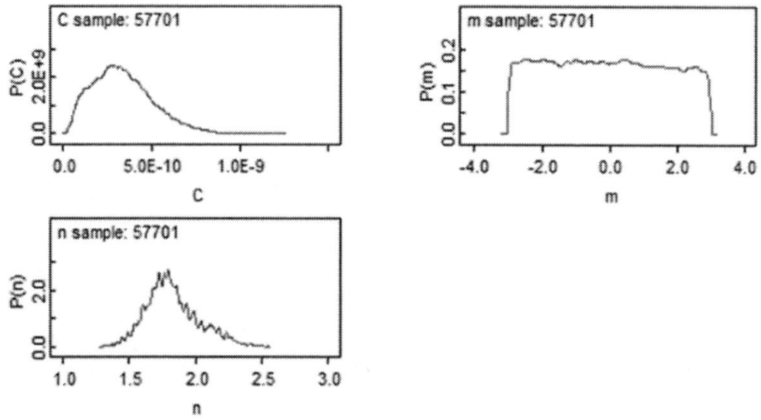

Figure 7. The posteir distribution of uncertain parameters after coalescence

Table 2. The updating results of uncertain parameters after coalescence

	Mean	sd	MC_error	Val2.5pc	median	Val97.5pc
C	3.33E-10	1.954E-10	7.136E-12	7.381E-10	3.16E-10	7.071E-10
m	-0.04588	1.716	0.03485	-2.85	-0.06547	2.839
n	1.828	0.1909	0.009572	1.495	1.801	2.251

So the mean model for cracks after coalescence can be written as:

$$\frac{da}{dN} = 3.33E - 10\Delta K^{-0.04588}(lr)^{1.828} \quad (13)$$

3.3 The probability failure model

According to the PoF model, Monte Carlo simulation of the 10000 sets of parameters has been done to get the probability density distribution histogram of failure life. And the life distribution of probability density curve can be acquired by connecting each midpoint(Fig. 8). The mean failure life are 6223.1 and 7531.82 cycles for cracks considering or ignoring the interactions, respectively. It's apparently that the failure life considering the cracks interactions is shorter than that of ignoring the interactions. For safer failure life assessment of engineering structural parts, the interactions between adjucent cracks are not negligible.

Fig. 9 shows the failure probability curve for cracks considering and ignoring the interactions between adjucent cracks. At first, the two curve have nearly the same failure probabilites before point A where the interaction begins. Then the crack propagation accelerates slowly untile point B. And in the final stage, the cracks propagation further speed up.

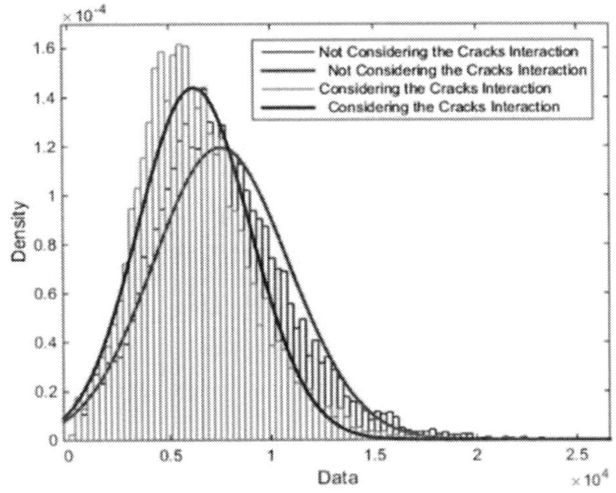

Figure 8. The probability density distributions of cracks

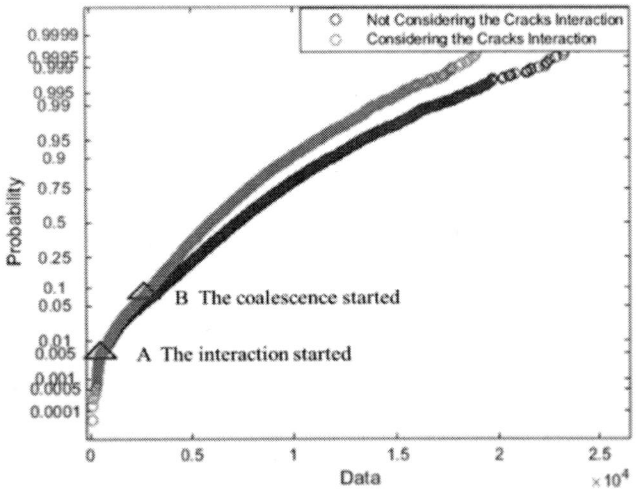

Figure 9. The failure probability curve

4 Conclusion

By investigating the process of interatction and coalescence of adjacent cracks, we found out that the stress intensity factor will increase suddenly in the ligament failure. The time when the coalescence happen was detected through the simulation of the SIF. The posterior distributions of the uncertain parameters were obtained via Bayesian update method. And the MCMC-Gibbs sampling process was realized via WINBUGS.

The PPoF models before and after coalescence were established by Monte Carlo simulation method. Starting from the root cause of failure and taking uncertainty like materials propertie, loading rate and so on into account, the PPOF model can describe the failure life more detail. The probabilistic modeling method can also be applied to other aspect of physic of failure which can greatly reduces the historical data or experimental data needed for the reliability evaluation of small sample products, effectively saving time and cost. The research of coplanar semiellipse cracks propagation in this paper promoted a new thought of studying the interaction and failure propagation in a complex engineering systerm. What's more, the effective life predictting method for multiple fatigue cracks will be a guarantee for personnel and products safety.

References

1. C.J. Bayley and R. Bell, Int. J. Fatigue, **21**, 33-37 (1999)
2. D.L. Su, *Mechanical properties of engineering materials* (Beijing, China, 2015)
3. J.B. Duan, Chin. Quart. Mech., **32**, 6-11 (2011)
4. B.S. Institution, *Guidance on methods for assessing the acceptability of flaws in fusion welded structures* (London, British, 1991)
5. T.H. Leek and I.C. Howard, Int. J. Pres. Ves. Pip., **68**, 181-201 (1996)
6. K. Iida, *I.C.F. International Symposium on Fracture Mechanics* (Beijing, China, 1983)
7. T. Howes, Russ J Genet, **39**, 382-385 (1990)
8. T.H. Leek and I.C. Howard, Int. J. Pres. Ves. Pip., **60**, 307-321 (1994)
9. J.R. Black, Phys. Electromig., **1**, 142-149 (1974)
10. N.C. Wilmington, I.C. Chen, S. Holland and C. Hu, 23rd Annual Proceedings of Reliability Physics[C], 24-31 (1985)
11. J.F. Archard, J. Appli. Physi, **24**, 981-988 (1953)
12. Z. Matik and V. Sruk, The Physics-of-Failure approach in reliability engineering[C], 745-750(2008)
13. M Chookah M. Nuhi and M. Modarres, Relia. Eng. Sys. Saft., **96**, 1601-1610 (2011)
14. T. Nagy and T. Turanyi, Relia. Eng. Sys. Saf., **107**, 29-34 (2012)
15. E.A. Debartolo and B.M. Hillberry, Int. J. Fatigue, **20**, 727-735 (1998)
16. E.J. Moukawsher, M.B. Einimann and A.F. Randt, J. Aircraft, **33**, 1003-1013 (2012)
17. T. Swift, Int. J. Fatigue, **16**, 75-94 (1994)
18. D.Y. Jeong and J.C. Brewer, Eng. Fract. Mech., **51**, 233-238 (1995)
19. J. R. Rice, N. Levy, J. Appl. Mech., **39**, 185-194(1972)
20. F. Delale and F. Erdogan, Int. J. Eng. Sci., **19**, 1331-1340 (1981)
21. J.H. Yuan, G.J. Tang and X.C. Song, Chin. Quart. Mecha., **21**, 8-14 (2000)
22. P. Paris and F. Erdogan, J. Basi. Eng. Trans. ASME, **85**, 528-533(1963)
23. M.L. Duan, C.T. Liu and H.J. Zhao, Acta mech. Sinica, **35**, 5 (2003)
24. K. Krausz and A.S. Krausz, Int. J. Fracture, **36**, 23-28 (1988)

Parametric study on soil-pile interaction failure using numerically research

Baydaa Hussain Maula[1,a], Ahmed Raad Al-Adhadh[2]

[1]*aInstitute of Technology, Middle Technical University, Baghdad, Iraq*
[2]*College of Engineering, Al Muthanna University, Sammawa, 00964, Iraq*

Abstract. The numerical approach became a great analytical tool due to the close matching between recorded data, experimental and computed one numerically by FEM package program, the effect of the flow and mechanical conditions that are responsible for pore water pressure dissipated and generated which concludes; changing soil permeability, degree of saturation for the liquefy layer, and permeability of soil strata contrast will be examined through a parametric study with the 2-D FE model. This paper will portray a few discoveries of an examination concerning the pile dynamically failure in light of one of past studies; three FE models were carried out. Basic comments will be made on the current understanding of pile failure. This mechanism, based on confirmed utilizing numerical dynamic shaking table tests on flatter ground. The principal discoveries identified with a parametric effect which takes the responsibilities of weakening the soil and produce pressure build up, variation of lateral pressure and displacement.

1 Introduction

The pile is essentially affected by liquefaction throughout earthquakes. The doubtful related to the effects of various soil characteristics of soil – pile interaction leads to the primary goal of the current study to numerically investigate the functional part of soil permeability and its.
Relation with liquefaction in order to respect the main finding of case studies simulated by finite element as a kind of soil improvement against liquefaction phenomena.

Nemours researcher thinks about these phenomena have been done, including shake-table experiments [1-6] dynamic centrifuge tests [7-9]. Through these examinations, the shake-table analysis is broadly used to distinguish and evaluate the dynamic response and failure mechanisms in the soil amid seismic tremors. The outcomes are regularly utilized as adjustment benchmarks for clear strategies or to give a database to the approval of numerical methodologies. Permeability- Seismic behaviour conducted that issues in geotechnical quake practice are limited issues. Parametric study plays a vital part in the generation of internal pore water pressure in the soil strata [10-13] and consequently impact of dynamic reaction enough using two-dimensional (2D) numerical models. In 2D FE modelling, the utilization of fitting parametric conditions is imperative in numerical. In the following sections, the utilized numerical framework is exhibited and the carried out results of 2D FE model for simulating the dynamic soil-pile response in an 11.0 m soil field are depicted. The domino

[a] Corresponding author : dr.maula@itbaghdad.edu.iq

effect of the conducted parametric study is outlined, emphasizing the worth of various permeability conditions in the FE model.

The point and targets of this investigation are introduced initially, followed by the procedures and parameters adopted during the study. Next, the numerical techniques utilized as a part of this research examination illustration of one of pile foundations with the baseline parameter set is exhibited to show the investigation approach totally. Results from all of the pile foundations with the baseline set of input parameters are then summarized. The impact of different approximations or presumptions that contrast from the baseline set of parameters is investigated and an affectability study is performed to recognize the input parameters that all impact the analysis results. The motivation behind the numerical investigations of the pile foundation is to survey the potential impact of received configuration parameters, to distinguish the impact of alternative affected conditions that are difficult to adopt through experiments of either shaking table or centrifuge test and to identify the input parameters that mostly impact the consequences of the analyses. As a final point, insights and conclusions are drawn based on the reported results. This paper gives a rule data and sets the outline constraints in which the analyses were performed.

2 Factors prone to liquefaction

Sladen [14] offers a description "Liquefaction is a phenomenon wherein a mass of soil loses a large percentage of its shear resistance, when subjected to monotonic, cyclic, or shocking loading, and flows in a manner resembling a liquid until the shear stresses acting on the mass are as low as the reduced shear resistance". Since liquefaction occurs higher when soil grain pattern to rearrange when sheared, if its movement trapped due to some limitations and constraints (flow and mechanical properties) that builds the soil resistance. Again the vital limitations are depth as overburden pressure, "characteristics of the soil grains like distribution of shapes, sizes, composition, etc." [15].

Permeability relation with liquefaction can be explained as the ability of dissipated or generated pore water pressure. Well-drained soil is a key variable for pore water pressure to dissipate. If pore water flows within a soil of poor drainage, water flow upward due shaking will gather on the soil deposit. Clay soil is an exceptionally dense type of soil; its particles are firmly packed together, which can keep the rapid drain of excess pore water pressures (EPWP) "that may have generated in the adjacent saturated sand deposit. Sufficient drainage above or below a saturated deposit may inhibit the accumulation of excess pore water pressure and hence liquefaction" [16]. Soil drainage depends on soil categorization structure and classification system (coarse-grained and fine-grained) each soil has its classifier of a specific level of capacity to permit water from going through it. Soil drainage might figure out which types of soil that will liquefy.

A big number of researches have recently investigated the permeability consequences on the behaviour of soil-pile foundation by experiment (centrifuge and shake-table model). All documented, evidence, reported, collected data and proof gathered information demonstrated that the permeability of the liquefiable soil layer has an impact effect on controlling liquefaction onset. This look is expected to give an incredible support to concentrate on parameters that can be enhanced to diminish liquefaction; by improving strength, density and drainage attributes of the soil.

3 Simulation of Soil-pile interaction modelling

3.1 Introduction to OpenSees and user-interface for nonlinear 2D analysis

OpenSees "is an object-oriented framework for finite element analysis" [17]. OpenSees' got a great resonance among research and scholarly papers. A key feature of OpenSees is the library of materials, elements and analysis commands makes the software product powerful tool for the numerical simulation of nonlinear structural and Geotechnical systems. The software enables the user to create their own property fits in with the input material data and their numerical model analysis type (linear and nonlinear). Regarding its powerful simulations that enable the method of analyses, for example, static pushover, and dynamic time-series analyses can be accomplished. According to paper investigation and concern, the OpenSees can provide an excellent environment for conducting research topic in terms of the change some properties that are already selected for the diagnosis liquefaction and offers a considerable effort, time and money needed for the testing process and collect site data.

All FE investigations were led to utilizing the open-source computational platform OpenSees [18]. The simulation done in platform OpenSees took into consideration (1) exploring of the 2D soil-pile contact; (2) choosing of the boundary conditions see Figure 1; (3) contribution of quake excitation through the option of analysis (4) choice of soil materials from an accessible menu of multi-yield soil models; and (5) advantages of utilization to direct liquefaction parameters.

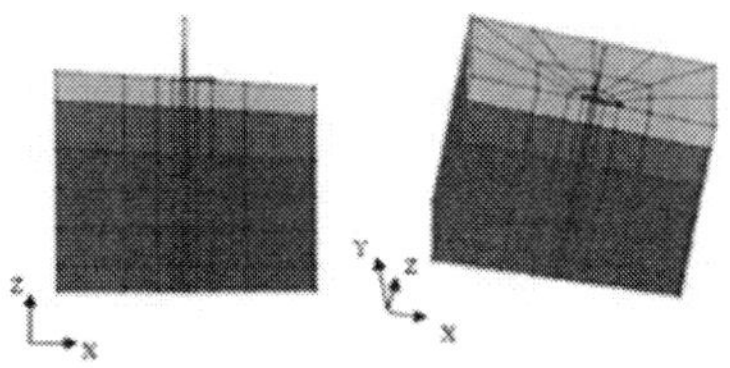

Figure 1. FEM of the SP system

3.2 Finite element model

3.2.1 FE formulation

In OpenSees, the soaked soil framework is demonstrated as a two-phase material in view of the Biot (1962) theory for permeable media. Known as u-p formulation, was executed. In the u-p definition, displacement of the soil skeleton u, and pore pressure p, are the essential questions. The u-p formulation as defined by Chan (1988) comprises of: i) equation of motion for the solid-fluid mixture, and ii) equation of mass conservation for the fluid phase, incorporating equation of motion for the fluid phase and Darcy's law [19].These two representing conditions can be communicated in the finite element matrix form as follows (Chan 1988):

$$\mathbf{M\ddot{U}} + \int_{\Omega} \mathbf{B}^{\mathrm{T}}\boldsymbol{\sigma}'\,\mathbf{d\Omega} + \mathbf{Qp} - \mathbf{f}^{s} = \mathbf{0} \tag{1a}$$

$$\mathbf{M\ddot{U}} + \int_{\Omega} \mathbf{B}^{\mathrm{T}}\boldsymbol{\sigma}'\,\mathbf{d\Omega} + \mathbf{Qp} - \mathbf{f}^{s} = \mathbf{0} \tag{1b}$$

Where M is the mass matrix, U the displacement vector, B the strain-displacement matrix, σ' the effective stress tensor (determined by the soil constitutive model described below), Q the discrete gradient operator coupling the solid and fluid phases, p the pore pressure vector, S the compressibility matrix, and H the permeability matrix. The vectors f^s and f^p represent the effects of body forces and prescribed boundary conditions for the solid-fluid mixture and the fluid phase respectively [19, 20].

3.2.2 Investigation featured for the example simulation

A baseline set of input parameters and some of them are systematically varied to demonstrate their influence on soil- pile response. The standard parameters are given in Table 1. The soil profile selected for study belongs to the Muthanna province south of Iraq. The 11.0 m "According to the Unified Classification System, the subsoil profile from (N.G.S) to depth of (23-24) m .The underground water level was encountered at a depth of (1.25-3) m below natural ground surface (N.G.S) during the time of boring. The properties of the soil are shown in Tables 1and 2 [21].

Table 1. Input parameters for benchmark case under scaled motion of 0.2g [21]

B.H No.	Depth (m)	Modulus of Elasticity E_s (kN/m^2)	Direct shear		Unit weight (gm/cm^3)	
			C (kN/m^2)	(degree)	Wet	Dry
1	1.0-1.5	11420.3	9.35		1.974	1.715
	5.5-6.0	22915	26.5		2.01	1.81
	8.0-8.5	13286.7	11.64		1.94	1.78
	11.0-11.5	51937	0		1.83	1.69
	11.5-12.0	10538	17		1.83	1.69
	16.5-17.0	22579.43	0		2.05	1.84
2	1.5-2.0	14325	14		1.95	1.72
	3.5-4.0	39468	1.71		1.88	1.67
	6.0-6.5	11689	11.9		1.97	1.79
	9.0-9.5	10576	0		1.96	1.77
	12.0-12.5	17344.3	0		2.15	1.84
	23.0-23.5	37897.47	0		1.83	1.62
3	4.0-4.5	91235.8	8.9		1.932	1.726
	7.0-7.5	21689.2	17.13		1.959	1.787
	9.0-9.5	15029.8	12.35		1.917	1.798
	9.5-10.0	59541	35		1.517	1.32
	11.5-12.0	45076	20		1.533	1.34
4	5.0-5.5	49851.2	3.16		1.905	1.763
	7.5-8.0	23164.9	17		1.968	1.733
	10.0-10.5	13275.8	11.78		1.876	1.618
	13.0-13.5	83307.4	8.18		1.526	1.313
	13.5-14.0	16352.23	0		1.13	0.95
	18.5-19.0	27091.33	0		1.24	1.06

The tested layers modeled with a single pile of 2.0 m diameter subjected to harmonic motion of 1Hz sinusoidal wave scaled with a peak acceleration of 0.2 g with 20 s time, the reality of the situation will become obvious eventually considered as a sample of reenactment test; will be considered as an example of simulation test. The investigation will be done and consider for the case of full saturation (S=100%) and a case of non-saturation.

Table 2. Overburden and pre-consolidation pressure with depth [21]

Depth (m)	Initial void ration (e_o)	Compression Index (C_c)	Swelling Index (C_r)	σ'_{v0} (kN/m^2)	$\Delta\sigma'$ (kN/m^2)	σ'_P (kN/m^2)	$\sigma'_{v0} + \Delta\sigma'$ (kN/m^2)
0.0-2.5	0.752	0.225	0.048	21.25	128.052	51	149.302
2.5-4.0	0.768	0.190	0.045	34.00	82.754	77	116.754
4.0-6.5	0.779	0.189	0.053	55.25	58.767	85	114.017
6.5-10.5	0.903	0.265	0.069	89.25	34.658	88	123.908
10.5-15.5	0.796	0.277	0.054	111.75	21.119	106	132.869
15.5-20.5	0.768	0.212	0.053	136.75	14.294	110	151.045
20.5-24.5	0.760	0.245	0.055	168.75	10.989	133	179.739

3.2.3 Significance impact of degree of saturation

In this parameter, mechanical soil properties (impact of soil permeability) on liquefaction generation with a uniform soil deposit are investigated; the 11.0 m layers modeled with a single pile of 2.0 m diameter subjected to harmonic motion of 1Hz sinusoidal wave with peak acceleration of 0.2 g with 20 s time will be tested. The investigation will be done and considered for the case of full saturation

(S=100%) and case of non-saturation. The saturation analysis introduces the permeability of sand is taken into account as k= 10^{-2} m\s while the other case represents an unsaturated condition (S=0%). Subsequently, the response of the sand layer is controlled exclusively by the mechanical properties of the materials [22]. From Figure 2 it shows that no critical amplification happens while the motion engenders to the surface. Figure 3 analyzes the time history of excess pore water pressure ratio, (r_u) for varied depths for both (saturated) drained and (partially) conditions. As it can be seen, for the partially condition, the (r_u) is higher at the deepest depths (e.g., 6.0 and 11.0 m) while the lower (r_u) is seen at the surface. This attitude changes (in saturated condition) at the upper parts concerning case at 4.0 m depth close to the surface, where $r_u \approx 0.6$ for partially saturated condition contrasted with $r_u \approx 1.2$ for the completely saturated condition. This unexpected result indicated that the water infusion results in soil loosening\ weakening and reduction in its liquefaction resistance, and this outcome recommended that the more extreme seismic tremor - incited harms at/close ground surface of uniform liquefied deposits are emerging from soil loosening during shaking. Figure 4 shows the surface lateral displacement for the fully and (partially) saturated condition (for the case with k = 10^{-2} m\s). The surface maximum lateral displacement is 0.42 m and 0.38 m in partially and (fully) saturated conditions, respectively. Along these lines, the partially condition results in larger displacements that reach out to deepest depths. Another arrangement of analyses (in the saturated condition) was conducted with 10 to 1000 times diminish in permeability of the liquefied sand subjected to the same motion event. Responses are appeared for these two extreme cases, in terms of EPWP increase rate, dissipation rate and timing, and displacements. Figure 4, compares the time histories of (r_u) for soil permeability of k = 10^{-2} m\s and. The permeability reduction results in a changed liquefaction propensity.

It also causes a delay in EPWP dissipation. The consequences of further and additional analyses with a with a scope of assortment in permeability values (from k= 10^{-2} m\s to k= 10^{-7} m\s) in terms of (r_u) time histories for the ground surface (i.e., 4.0 m depth) in conjunction with surface lateral displacement versus, soil permeability are shown in Figures 5 and 6. For permeability's of an order of magnitude typical of most sandy soils, flow effects speed up liquefaction activating at upper parts, that makes liquefaction to generally occur first close to the top and work its way down.

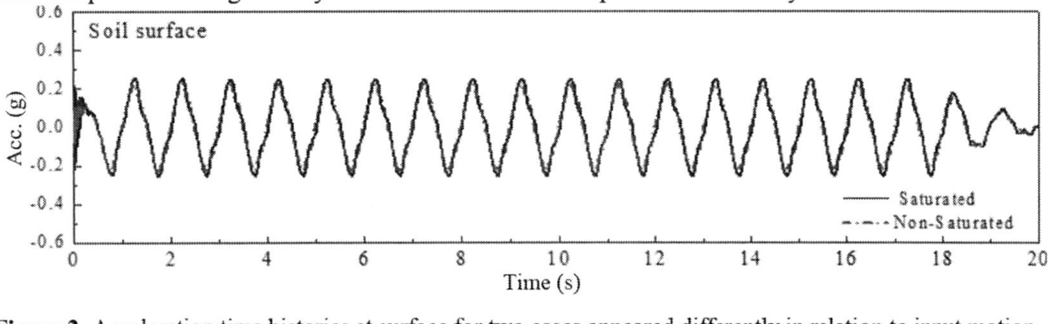

Figure 2. Acceleration time histories at surface for two cases appeared differently in relation to input motion

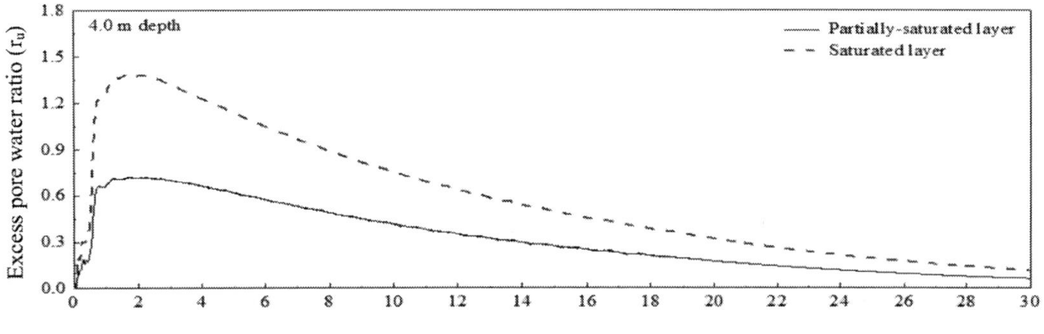

https://doi.org/10.1051/matecconf/201816901033

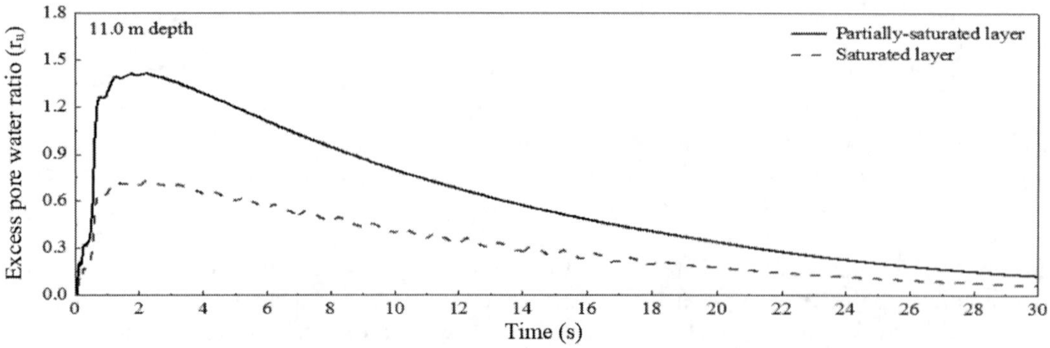

Figure 3. Time history of EPWP ratio, r_u for Flow-on and Flow-off condition investigations

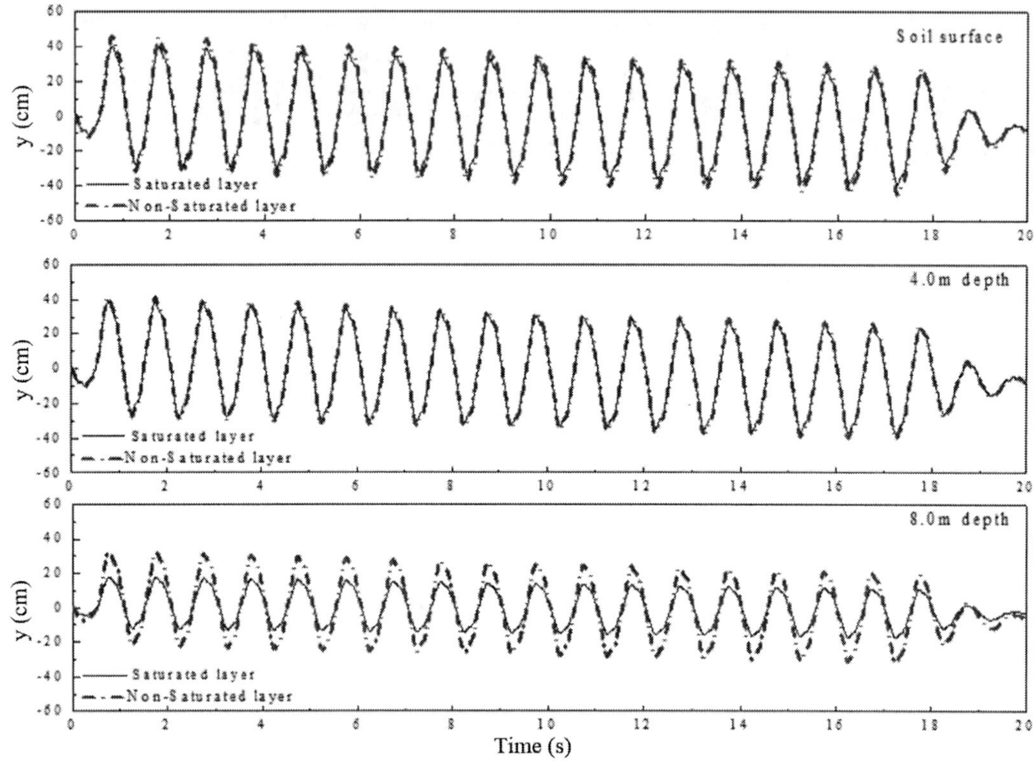

Figure 4. Time history of Lateral displacements for Flow-on and Flow-off condition analyses

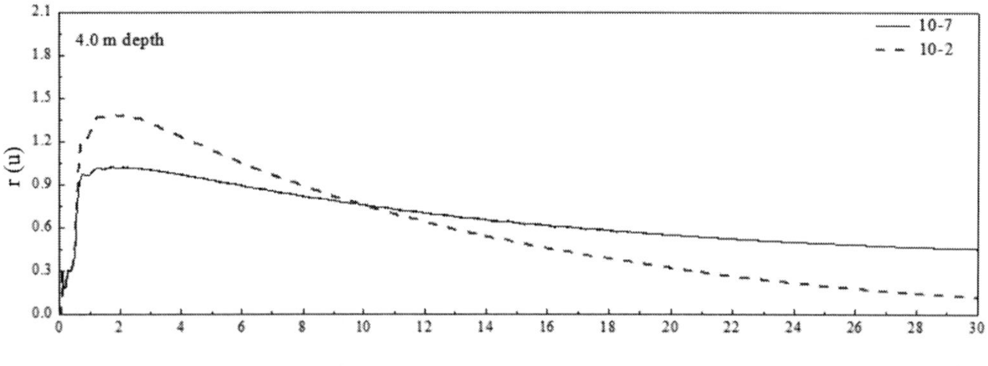

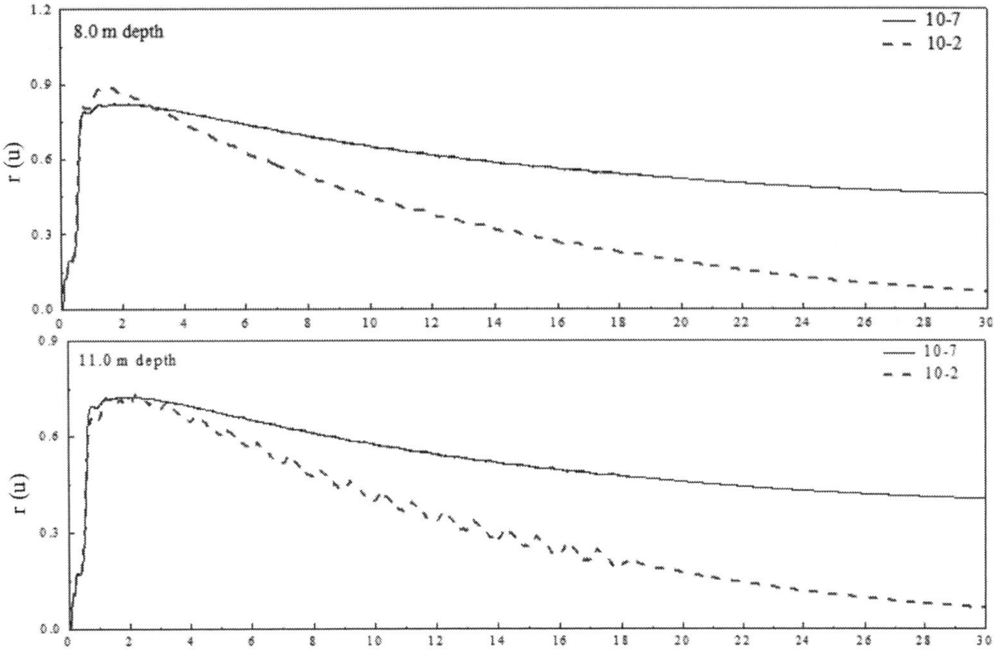

Figure 5. Pattern of excess pore water pressure generation,(a) k= 10^{-2} and (b) k= 10^{-7}

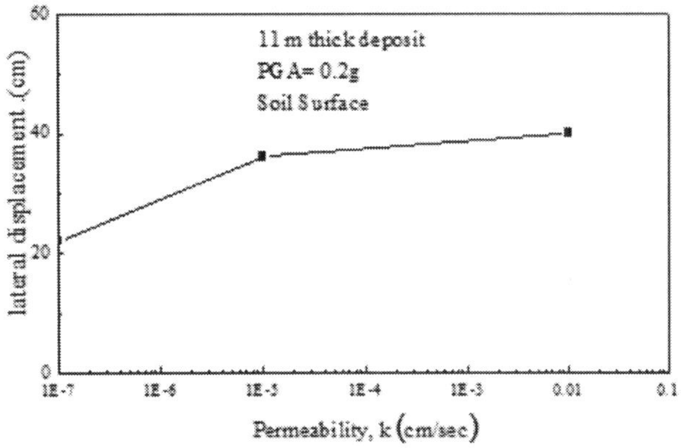

Figure 6. Effects of liquefied soil permeability reduction at ground surface displacement

3.2.4 Significance of the permeability of liquefiable soil layer

In this section, the aftereffects of soil permeability on liquefaction onset with a sand deposit are inspected and tended to. Consider the same 11.0 m deposit layer of liquefiable soil without obstruction layer (uniform deposit modeled with a single pile of 2.0 m dia) subjected to harmonic motion of 1Hz sinusoidal wave with peak acceleration of 0.2g with 20s time. The issue was broken down for two conditions; 1) water table at the surface (Saturated layer) and 2) water table at 8m under the ground level (Non- saturated layer).

The main investigation shows a more reasonable condition where pore water will flow during shaking (partially drained condition) attributable to hydraulic gradient. The permeability of sand is considered as k= 10^{-2} m\s. The second case represents an unsaturated condition. Therefore, the response of the sand layer is controlled solely by the mechanical properties of the materials. Figure 7 shows the expected time histories for surface acceleration compared with that of base motion. It indicates that no significant amplification happens while the motion propagates to the surface.

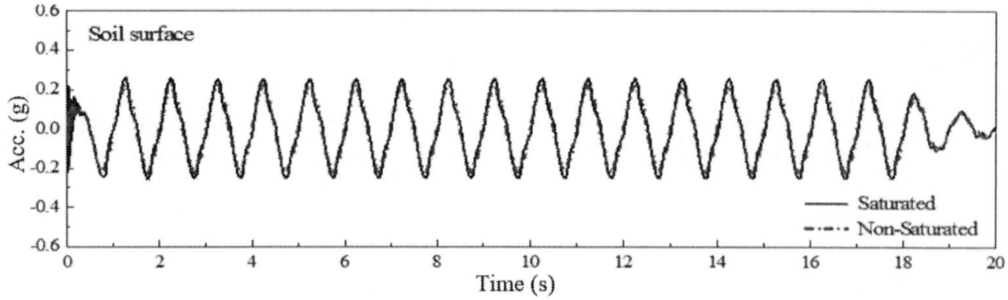

Figure 7. Acceleration time histories at surface for two cases compare to input motion

3.2.5 Impact of decrease in liquefiable layer permeability

The impact of permeability contrast was investigated through diminishing liquefiable layer permeability ratios, reaching that of the non-liquefiable soil layer, going from 10^{-1} to10^{-3} times for the (liquefied) with perfect crust- layer, Figure 8. The effect of permeability on seismic conduct of uniform liquefiable layers is specified in this section, the impact of diminishing in permeability of the liquefiable layer while the permeability contrast is constant ($k_{(sand)} \backslash k_{(crust-layer)}$) =1000 is exhibited.

The 10 m soil profile, subjected to the harmonic base motion was analyzed, with the same soil properties as in the benchmark case (recorded in Table 1-2), aside from the permeability; $\mathcal{K}$ was reduced by a specific value (e.g., 100 times). Figure 8 shows the surface lateral displacement in respect to the benchmark case versus permeability for the three analyzed cases. It indicates that the permeability reduction results in an increasing in the deformations rate due to a lower flow rate, as in the case of uniform layer, through the vicinity of crust- layer aggravates this impact.

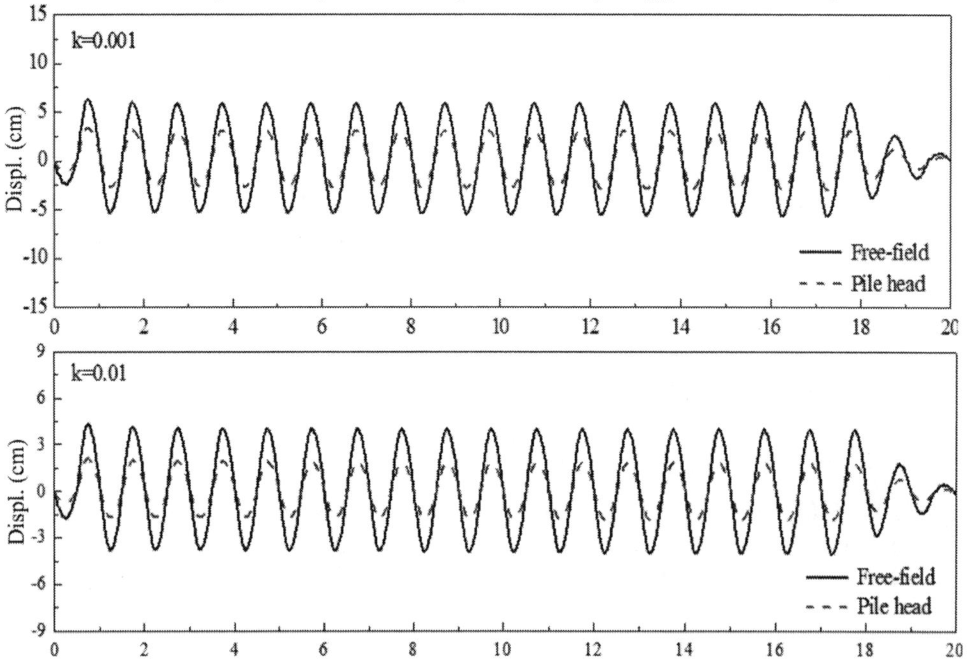

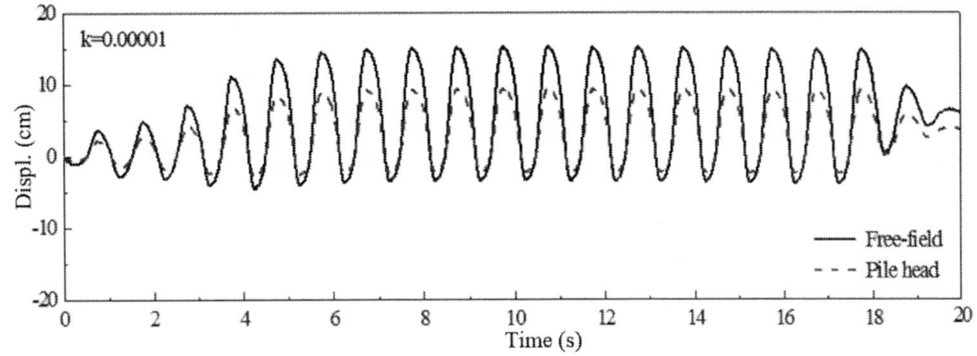

Figure 8. Increase of lateral displacement with reduction in permeability of liquefiable soil

3.2.6 *Impact of Permeability Contrast within the sand layer ($k_{(sand)} \setminus k_{(crust-layer)}$)*

This area presents consequences of investigations that investigate the impacts of (hydraulic conductivity) permeability contrast between the liquefiable sand layer and the non-liquefied layer with regards to the characteristic response of the 3°degree infinite slope. Permeability of the non-liquefied layer was increased from 1000 times less than that of the underlying sand layer (i.e. $k_{(crust-layer)} \setminus k_{(sand)}$ =0.001) to crust layer at 4 m of the 10 m soil profile was reduced from 1m to 0.1 m in the analyses. Every single other parameter, i.e., ground slope; and the base motion were kept the same.

The impacts of permeability of the non-liquefied layer (relative to that of the liquefiable soil, permeability ratio), as far as excess pore water pressure generation and surface lateral displacements, are displayed in the accompanying areas.

3.2.6.1 *Impact of permeability Contrast on Excess pore water pressure*

Figure 9 demonstrates pattern changes in generation of excess pore water pressure, in terms (r_u) time history, within the soil profile arising from a decrease in the permeability ratio; ($k_{(sand)} \setminus k_{(crust-layer)}$). As it is seen, EPWP is generated in the deep parts with nearly the same pattern, in spite of permeability differentiation. However, the dissipation rate is affected by the permeability ratio, which is more pronounced at shallower depths. This prompts high excess pore water pressures lasting for a longer time after shaking ceases, beneath, when the permeability contrast comparatively great ($k_{(sand)} \setminus k_{(crust-layer)}$) *>100.*These examinations demonstrate that permeability reduction substantially affects the scattering rate of the dissipation for EPWP, remarkably, for the upper parts close to the sub-layer, and the same effect was observed earlier for a uniform layer. This impact; notwithstanding, is exacerbated by the vicinity of low-permeability crust-layer.

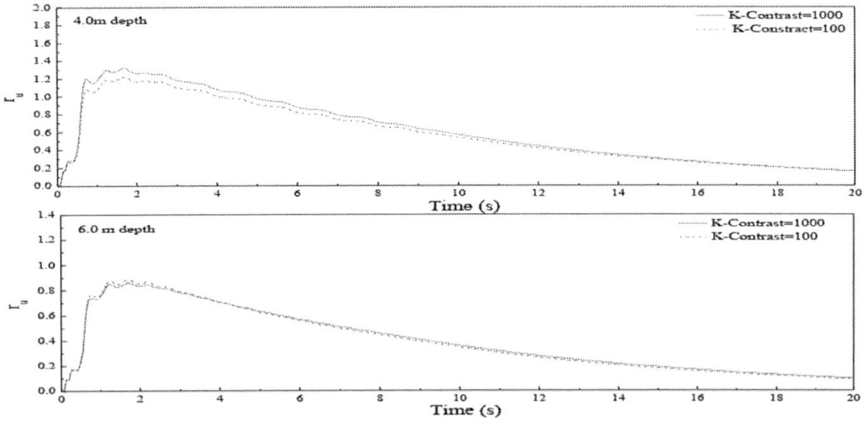

Figure 9. Time histories of r_u at different depths of the soil layer for different permeability contrast

3.2.5.2 Effect of Thickness of low permeability sub layer (barrier)

The effects of low permeability sub-layer thickness, in terms of surface displacements, are shown in Figure 10 for different permeability contrast. As the sub layer thickness increases, from 0.1m to1m, it results in about 33% increase in surface displacements. The rate of displacement increase decreases the barrier thickness when the barrier thickness is greater than 0.4 m for the soil profile. From the figure, the permeability ratio has nearly identical effect on displacements regardless of the barrier thickness. Figure 10 shows the surface lateral displacement vs. permeability contrast, times the barrier layer thickness. This means that the effect of permeability contrast is more significant that of displacement.

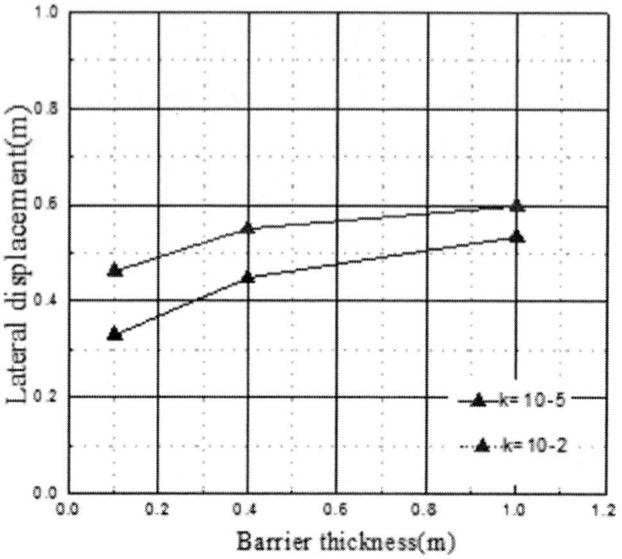

Figure 10. Change of surface displacements with barrier layer

4 Conclusion

A FE analysis was conducted to simulate the soil-pile interaction on the dynamic response of a level ground. The analyses were made on the basis of OpenSees, as a user interface. The goal of this paper was to examine the impact of the permeability -parametric cases on the soil behaviour. Displayed the consequences of a parametric study on the impacts of permeability contrast ratio; on seismic behaviour of a typical liquefiable infinite slope with pile foundation. Reliable with the outcomes, the work through this paper results brings about the consequent conclusions:

Some vital conclusions from this study are outlined briefly below.

1. There is an inverse relationship between EPWP and examined depth of soil profile, from the outcomes it simple perceive that EPWP disseminate at deeper depth while it demonstrates a higher value at shallower depth due to the less path of flow.
2. The relation between lateral displacement and permeability is a Positive relationship.
3. Partially saturated layer can be considered one of soil improvements against shaking, therefore, it is necessary to lower water table level.
4. Similar test must be performed to investigate the available options of openSees software especially the boundary condition (shear, rigid) box and periodic boundary. This must be examined through calibration with experiments.
5. It is recommended to find the effect of permeability under multiple earthquake dimension influence, on the generation of p-y curves; pore water pressure; vertical-longitudinal and lateral displacement; pile pin effect; and shear stress.

6. It is a highly recommended to find empirically equations based on numerical investigation that deal with the effects of geometric and seismic parameters on liquefaction-induced lateral spreading.

Acknowledgement

The author gratefully acknowledges the support for this research by the National Center for Construction Laboratories in Baghdad-Iraq, Architectural and Scientific Bureau in the Muthanna University-Iraq.

References

1. L. Tang, P-Y Model of Dynamic Pile-soil Interaction in Liquefying Ground, PhD Thesis, Dissertation, Harbin Institute of Technology, Harbin, China (2010)
2. B.H. Maula, X.Z. Ling, T. Liang and P.J. Xu, Research Journal of Applied Sciences, Engineering and Technology, **3**(4), 344-355 (2011)
3. B.H. Maula, Dynamic Response and Simplified Analysis Method for Bridge Pile Foundations in Liquefiable Sloping Ground, Ph.D. Thesis, Dissertation, Harbin Institute of Technology, Harbin, China (2013)
4. L. Tang, B.H. Maula, X. Ling and L. Su, Earthquake Engineering and Engineering Vibration, **13**(1), 171-180 (2014)
5. B.H. Maula, T. Liang and A.M Gazal, Research Journal of Applied Sciences, Engineering and Technology, **7**(24), 5163-5171 (2014)
6. K.L. Wang and M.L. Lin, Engineering Geology, **122**(1-2), 106-114 (2011)
7. G. Chen, H. Zhuang, X. Du, L. Li and S. Cheng, Journal of Earthquake Engineering and Engineering Vibration- Chinese Edition, **27**(2), 171 (2007)
8. M. Maugeri, G. Musumeci, D. Novità and C.A. Taylor, Soil Dynamics and Earthquake Engineering, **20**(5-8), 435-444 (2000)
9. K. Fukutake, A. Ohtsuki, M. Sato and Y. Shamoto, Earthquake Engineering & Structural Dynamics, **19**(7), 977-992. (1990)
10. B.H. Maula, Journal of Emerging Trends in Engineering and Applied Sciences (JETEAS), **7**(1), 45-55 (2016)
11. L.L. González Lagos, Centrifuge modeling of permeability and pinning reinforcement effects on pile response to lateral spreading (Doctoral dissertation, Ph. D. Thesis, Rensselaer Polytechnic Institute, Troy, New York) (2005)
12. L. González, T. Abdoun, A.M. ASCE and R. Dobry, Journal of Geotechnical and Geoenvironmental Engineering, **135**(1), 62-73 (2009)
13. J.M. Ramirez, Influence of soil permeability on liquefaction-induced lateral pile response, Thesis submitted in partial satisfaction of the requirements for the degree of Master of Science, University of California, San Diego (2010)
14. J.A. Sladen, R.D. D'hollander and J. Krahn, Canadian Geotechnical Journal, **22**(4), 564-578 (1985)
15. A.F. Rauch, EPOLLS: an empirical method for predicating surface displacements due to liquefaction-induced lateral spreading in earthquakes (Doctoral dissertation, Virginia Tech) (1997)
16. A. Biswas, Study on liquefaction of soil, A project report submitted in partial fulfillment of the requirements for the degree of Bachelor of Technology in Civil Engineering, Department of Civil Engineering National Institute of Technology Rourkela (2010)
17. OpenSees Sensitivity Reliability Optimization, http://jaguar.ucsd.edu/OpenSees.htm
18. A.Elgamal, J. Lu and D. Forcellini, Journal of geotechnical and geoenvironmental engineering, **135**(11), 1672-1682 (2009)
19. J. Lu, J. Peng, A. Elgamal, Z. Yang and K.H. Law, Earthquake Engineering and Engineering Vibration, **3**(1), 23-37 (2004)

20. J. Peng, J. Lu, K.H. Law and A. Elgamal, Simulation of Earthquake Liquefaction Response on Parallel Computers, In Structures 2004: Building on the Past, Securing the Future, 1-10 (2004)
21. H.Y. Aziz and J.L. Ma, Advanced Materials Research Trans Tech Publications, **243-249**, 3676-3684 (2011)
22. M. Seid-Karbasi, Effects of void redistribution on liquefaction-induced ground deformations in earthquakes: a numerical investigation (Doctoral dissertation, University of British Columbia) (2009)

Apply data mining to analyze the rainfall of landslide

Chou-Yuan Lee[1], Zne-Jung Lee[2,a], Bin-Yu Peng[3,4], Chen-chen Lin[2], and Hsiang Huang[4]

[1]*Department of Information Technology, Fuzhou University of International Studies and Trade, Fujian, China*
[2]*Department of Information Management, Huafan University, Taiwan*
[3]*Department of Information Management, University of Kang Ning, Taiwan*
[4]*Department of Mechatronic Engineering, Huafan University, Taiwan*

Abstract. Taiwan is listed as extremely dangerous country which suffers from many disasters. The disasters from the landslide result in the loss of agricultural productions, life and property and so on. Many researchers concern about the disasters of landslide, but there are few discussions for the threshold of rainfall for landslide. In this paper, data mining is applied to establish rules and the threshold of rainfall for landslide in Huafan University, Taiwan. These used variables include rainfall, insolation, insolation rate, averaged humidity, averaged temperature, wind speed, and the tilt of inclinometer. The inclinometer is an important instrument for measuring tilt, elevation or depression of an object with respect to gravity. There are 26 inclinometers in Talun mountain area of Huafan University. In this research, the used data were collected from January 2008 to July 2014. In the proposed approach, the regression analysis is used to predict rainfall first. Then, decision tree is used to obtain decision rules and set the threshold of rainfall for landslide. The output of this approach can provide more information for understanding the change of rainfall. The threshold of rainfall could also provide useful information to maintain the security for Huafan University.

1 Introduction

The United Kingdom risk management consultant company, Verisk Maplecroft, announced that United States, Japan, China, and Taiwan were suffered extreme dangerous natural disasters in 2011 [1]. These above countries are still suffering severe damages due to the impact of climate in 2015. The flood is one of the severe damages that concentrate in Asia, especially in India, Bangladesh, China, and Taiwan. These floods cause 75% risk of death in the world, and about annual 2.2 million people are affected by the effect of mountain landslide. Additionally, there are about annual 78 million people expose in the dangerous of tropical cyclone [2]. Taiwan is listed as extremely dangerous country which suffers from many natural disasters. The rainfall of tropical cyclone is one of the most important issues in Taiwan, because it direct links to voluminous rainfalls [3].

Recently, there are many researches focus on rainfall. These researches include statistical approaches, ensemble, Bayesian model, and data mining [3-12]. Above techniques provide useful information for the rainfall. Unfortunately, there are no information of inclinometer and few discussions for the threshold of rainfall for landslide. The inclinometer is an important instrument for measuring tilt, elevation or depression of an object with respect to gravity. It provides the direct

[a] Corresponding author : johnlee@cc.hfu.edu.tw

response to the settlement and displacement of various stratigraphic depth distributions via the slopes of the tilt tube [13]. There are 26 inclinometers at Huafan University where is located on Talun mountain area of Shihding District in New Taipei City, Taiwan. Talun mountain area belongs to slope land. Because data mining techniques provide useful information for the rainfall of landslide [14], it is an important issue to analyze the rainfall of landslide and inclinometer data for Huafan University. In this paper, data mining is applied to analyze the rainfall of landslide for Huafan university.

The rest of this paper is organized as follows. Because regression analysis and decision tree play important roles in the proposed approach, section 2 provides the brief literature overview of multiple regression and decision tree. Section 3 describes the proposed approach. Section 4 outlines the simulation results. It also provides detailed comparisons. Finally, conclusions are drawn in the last section.

2 The brief description of multiple regression analysis and decision tree

In this paper, multiple regression analysis is used to provide the predict model of the rainfall of landslide. Furthermore, decision tree is used to find the threshold of rainfall. In this section, multiple regression analysis and decision tree are brief described first.

For multiple regression analysis, the data consist of m observations on a dependent variable Y and n independent variables, $x_1, x_2, \cdots, x_n$, The relationship between Y and $x_1, x_2, \cdots, x_n$ is expressed as Eq. (1).

$$Y = \alpha_0 + \alpha_1 x_1 + \alpha_2 x_2 + \cdots + \alpha_n x_n + \varepsilon \quad (1)$$

Where $\alpha_0, \alpha_1, \cdots, \alpha_n$ are constants, α_0 is referred to as the intercept. $\alpha_1, \alpha_2, \cdots, \alpha_n$ represent the corresponding n regression coefficients, and ε is the random error. Using matrix in Eq. (1), the model equation can be expressed as [15]

$$Y = XB + E \quad (2)$$

The normal equation can be expressed as

$$(X'X)\tilde{B} = X'Y \quad (3)$$

Where $\tilde{B}$ is a vector of least squares estimates of B. The solution to matrix equation is written as [16]

$$\tilde{B} = (X'X)^{-1}X'Y \quad (4)$$

Decision tree is based on the greedy algorithm that utilizes a divide-and-conquer strategy to recursively construct decision rules [17]. It consists of the root node, internal nodes, branches, and leaves. Each decision tree represents a rule which categorizes data according to these attributes [18]. A node specifies an attribute (feature) in the dataset. A branch connects either two nodes or one node and a leaf. Each node has a number of branches which are labeled as the possible value of attribute in the parent node. Leaves are labeled as the decision value of classification. The tree-like structure is composed of a root node, a set of internal nodes, and a set of leaf nodes. When applied to the set of train patterns, *Info(S)* measures the average amount of needed information to identify the class of the pattern *S*.

$$Info(S) = -\sum_{j=1}^{k}\left\{[freq(C_j, \tfrac{S}{|S|})]\log_2[freq(C_j, \tfrac{S}{|S|})]\right\} \quad (5)$$

Where $|S|$ is the number of cases in the training set. C_j is a class for $j = 1,2,\cdots,k$ where k is the number of classes and $freq\left(C_j, \frac{S}{|S|}\right)$ is the number of cases used in C_j. To consider the expected information value $Info_x(S)$ for attribute X to the partition S, it can be stated as:

$$Info_X(S) = -\sum_{j=1}^{n}\left\{[(\frac{|S_j|}{|S|})Info(S_j)]\right\} \quad (6)$$

where n is the number of output for the attribute X, S_j is a subset of S corresponding to the jth output and $|S_j|$ is the number of cases of the subset S_j. The information gain, *Gain(X)*, according to attribute X is shown as:

$$Gain(X) = Info(S) - Info_X(S) \quad (7)$$

For constructing the decision tree, the root node will branch to child node first. The recursive processes will be continued to branch the child node of training datasets until each leaf node in the dataset contains only one class of patterns, or until there is no improvement [18].

3 The proposed approach

The rainfall of Huafan University can be divided into: (a) From December to next April, this dry season rains more days than other periods. However, the cumulative rainfall leads to less rainfall intensity. (b) Plum rain season is from May to June, and the typhoon season is from July to September. From September to November, it is also typhoon season and northeast monsoon. Due to the rainfall levels in typhoon season increased significantly, the intensity of rainfall rises to the top from July to September in one year. Occasional typhoons bring rainfall and the northeast monsoon rainfall from October to November but accumulated rainfall and average rainfall intensity become less. In this paper, the multiple regression analysis and decision tree are applied to analyze the rainfall of landslide in Huafan University.

In this study, these used variables include rainfall, insolation, insolation rate, averaged humidity, averaged temperature, wind speed, and the tilt of inclinometer. The dependent (target) variable is rainfall and others are independent variables. The proposed approach is based on multiple regression analysis and decision tree (DT). The pseudo code of the proposed approach is listed as follow.

Procedure:

```
    Pre-process data
    Begin
       Apply multiple regression analysis;
       Apply decision tree;
       Output the multiple regression results, threshold of rainfall, and decision rules;
    End
```

In pre-process data, the missing values in the inclinometer data are imputed by 0. It means that there are no tilts for these used inclinometers. In the proposed approach, the multiple regression analysis is first used to predict rainfall. After multiple regression analysis, decision tree uses Eq. (7) to generate decision rules. These obtained threshold of rainfall and decision rules are provided for decision-making. Finally, the proposed approach outputs the multiple regression results, threshold of rainfall, and decision rules for decision-making.

4 Results

There are 79 collected datasets and 33 used variables in this study. These collected datasets are from Jan. 2008 to Oct. 2014 in Huafan University. These variables include rainfall, insolation, insolation rate, averaged humidity, averaged temperature, wind speed, and others are inclinometers (*C5~C32*). First, multiple regression analysis is applied to predict the rainfall of landslide in Huafan University. After the process of multiple regression analysis, the predict model of multiple regression analysis is shown in Eq. (8).

$$
\begin{aligned}
Rainfall = {} & (-240.2114) + (-1.88) \times Insolation + (-4.1553) \times InsolationRate \\
& +0.3586 \times \ AverageHumidity + 28.9926 \ \times \ AverageTemperature \\
& + 94.1142 \ \times WindSpeed + (-6.9486) \times C6 + (-5.2393) \times C7 \\
& +1.7772 \times C8 + 9.0128 \times C9 + 5.1209 \times C10 + (-1.2573) \times C11 \\
& +(-0.9317) \times C12 + 3.1387 \times C13 + 5.768 \times C14 + 9.2376 \times C15 \\
& +12.3311 \times C16 + 7.2773 \times C17 + 0.7927 \times C18 + 9.1294 \times C19 \\
& +16.2721 \times C20 + (-27.4859) \times C21 + (-4.3699) \times C22 \\
& +(-5.3602) \times C23 + (-3.1363) \times C24 + (-0.6239) \times C25 \\
& +(-9.1850) \times C26 + (-0.1621) \times C27 + (-3.8095) \times C28 \\
& +(-8.4103) \times C29 + (-6.1479\&) \times C30 + 4.5759 \times C31 \\
& +(-5.2223) \times C32
\end{aligned} \tag{8}
$$

The residual standard error of multiple regression analysis is 132.1029. Thereafter, decision tree is used to generate decision rules. There are total 7 decision rules generated in this study, and these generated decision rules are shown in Table 1.

Table 1. The decision rule of rainfall.

No.	Rule	Rainfall
1	C7>=7.025	616mm
2	C7<7.025, Insolation Rate<12.75	465mm
3	C7<7.025, Insolation Rate>=12.75, Temperature>=19.21, and C32>=2.28	432mm
4	C7<7.025, Insolation Rate>=12.75, Temperature>=19.21, C32<2.28 and C18<-1.325	351mm
5	C7<7.025, Insolation Rate>=12.75, Temperature>=19.21, C32<2.28, and C18>=-1.325	210mm
6	C7<7.025, Insolation Rate>=12.75, Temperature<19.21, and Insolation Rate<19.87	257mm
7	C7<7.025, Insolation Rate>=12.75, Temperature<19.21, and Insolation Rate>=19.87	121mm

In Table 1, these rainfalls for decision rules of No. 3, 4, 5, 6, and 7 are not so large enough, so the decision rules of No. 1 and 2 are set as the threshold. For decision rule of No. 2 (C7<7.025, Insolation Rate<12.75), its rainfall is 465 mm. Finally, the decision rule of No. 1 is set as the important threshold. These threshold values must take more attention to beware of disaster emergency and dangerous. However, it is safe when the rainfall is below 465mm. It needs to prepare all necessary plans for disaster emergency, when the rainfall is between 466 mm and 616 mm. Moreover, it is dangerous as the rainfall is greater than 617 mm.

5 Conclusions

In this paper, multiple regression analysis is applied to provide the model for the predict rainfall of landslide in Huafan University. Decision tree is used to obtain decision rules and set the threshold of rainfall for landslide. There are seven decision rules for the rainfall of landslide in Huafan University. From decision rules, the threshold of rainfall is set as 465 mm and 616 mm. It could provide useful information to maintain the security for Huafan University.

Acknowledge

The authors would like to thank the Ministry of Science and Technology (MOST) of Taiwan for financially supporting this research under Contract No. MOST 106-2221-E-211-002, MOST 106-2632-M-211-001, MOST 106-2622-E-211-002-CC3, and MOST 105-2632-M-211-001.

References

1. Maplecroft, Natural Hazards Risk Atlas 2011, https://maplecroft.com/about/news/natural_hazards_2011.html (2011)
2. M. Dilley, S Chen Robert, D. Uwe, A. L. Lerner-Lam and A. Margaret, Natural Disaster Hotspots: A Global Risk Analysis, World Bank, Washington DC, http://earth.columbia.edu/news/2005/story03-29-05.html (2005)
3. C.C. Wei, Soft computing techniques in ensemble precipitation nowcast, Applied Soft Computing, **13** (2013)
4. S.T. Wang, C.L. Yen, G.T. Chen and S.L. Shieh, The characteristics of typhoon precipitation and the prediction methods in Taiwan area (III) (in Chinese), Hazards Mitigation Program Technical Report 74-51, National Science Council, Taiwan (1986)
5. T. Hall, H.E. Brooks and C.A. Doswell, Weather and Forecasting, **14**(3), 338-345 (1999)
6. M. Nasseri, K. Asghari and M.J. Abedini, Expert Systems with Applications, **35**(3), 1415-1421 (2008)
7. G.A. Fallah-Ghalhary, M. Mousavi-Baygi and M.H. Nokhandan, Research Journal of Environmental Sciences, **3**(4), 400-413 (2009)
8. G. Srivastava, S.N. Panda, P. Mondal and J. Liu, Journal of Hydrology, **395**(3), 190-198 (2010)
9. A. Talei, L.H.C. Chua and T.S.W. Wong, Journal of Hydrology, **391**(3-4), 248-262 (2010)
10. S.A. Asklany, K. Elhelow, I.K. Youssef and M.A. El-Wahab, Atmospheric Research, **101**(1), 228-236 (2011)
11. C.C. Wei, Weather and Forecasting, **27**(2), 438-450 (2012)
12. C.C. Wei, Journal of Hydrometeorology, **13**(2), 722-734 (2012)
13. P.E. Mikkelsen, Advances in inclinometer data analysis, In Symposium on Field Measurements in Geomechanics (2003)
14. M. Kannan, S. Prabhakaran and P. Ramachandran, Rainfall forecasting using data mining technique (2010)
15. H.W. Altland, Regression Analysis: Statistical Modeling of a Response Variable, Technometrics, **41**(4) (1999)
16. Chatterjee, Samprit and Ali S. Hadi. Regression analysis by example, John Wiley & Sons (2015)
17. J.R. Quinlan, Induction of Decision Trees, Machine Learning (1986)
18. J.R. Quinlan, C4.5: Programs For Machine Learning, Morgan Kaufmann Publishers Inc. San Francisco, CA (1993)

Simulation and verification for seed-filling performance of cell wheel precision seed metering device based on discrete element method

Qinghui Lai, Ziwu Hu[a], Jinlong Xing and Wenpeng Ma

Faculty of Modern Agriculture Engineering, Kunming University of Science & Technology, Kunming, China

Abstract. The cell wheel seed metering device was improved and a stirring seed-filling device was added to improve the seed-filling performance of cell wheel pseudo-ginseng precision seed metering devices. Using pseudo-ginseng seeds in Wenshan Prefecture, Yunnan Province as the objects for seed metering, the software application EDEM was adopted based on the discrete element method for the simulation calculation and analysis of the seed-filling performance of the seed metering device under 4 rotational speeds of the cell wheel and 6 rotational speeds of the stir wheel. The simulation results indicate that the filling ratio increases as the rotational speed of the stir wheel increases under a constant rotational speed of the cell wheel. Test verification of the simulation analysis results was conducted on the test bed of the seed metering device. The results indicate that increasing the rotational speed of the stir wheel can obtain a filling ratio of over 90%. The test results display a similar variation trend to that of the simulation analysis with an error of average filling ratio less than 5%. Therefore, it is feasible to analyze the seed-filling performance of the stirring and seed-filling device of the seed metering device with the discrete element method.

1 Introduction

Pseudo-ginseng is a rare traditional Chinese medicinal herb. Its seeds have irregular shapes with high internal friction among themselves, and mis-seeding happens frequently during the seeding of pseudo-ginseng, therefore improving seeding yield has always been a research focus of the pseudo-ginseng precision seed metering device. The cell wheel seed metering device has been widely adopted for the seeding of various crops[1-5]. However, the cell wheel seed metering device has strict requirements for the shape and size of seeds. Its filling ratio is significantly affected by operating speed and its operating efficiency is very low. Many researchers have conducted studies on the cell wheel seed metering device to improve its seed-filling performance. P. P. Garcia[6] found by studies that the seed metering precision is under significant influence of the hole shape and size of the cell wheel seed metering device. W. Hu[7] optimized the hole parameters of the seed metering device according to the size of seeds. L. X. Wen et al.[8] designed an elastic follow-up seed protection device for the hole wheel seed metering device to reduce the damage of rice seeds and the wear on the outside surface of the cell wheel. Y. J. Qun et al.[9] conducted a discrete element simulation analysis on the performance of the hole wheel soybean seed metering device with their self-developed 3D software application

[a] Corresponding author : m15288451629@163.com

CAE and verified its feasibility. T. C. Zhou et al.[10] designed a variable volume hole wheel seed metering device composed of the hole wheel and adjustment ring (tongue) and conducted tests on its performance. By collecting the sample video files of the corn seed flow at the orifice of the cell wheel seed metering device, Q. Ling[11] adopted image processing for the statistics of the performance indexes of the seed metering device. W. Z. Man et al.[12] acquired the seed-filling process of the hole rice seed metering wheel by high-speed photography and analyzed the flow regularity, status and seed-filling trend of rice seeds in the seed-filling area. L. Tao et al.[13] studied the influence of 3 types of hole structures on the seed group disturbance and seed metering performance of the cell wheel rape seed metering device.

The analysis on the performance of the seed metering device by conventional tests costs a long period of time and lots of manpower and material resources. With the development of computer technology, the computer-aided analysis software application EDEM has been widely adopted in the simulation analysis of particulate matters. S. Maitraye et al.[14] conducted a numerical simulation analysis on the prilling technique of fluidized bed by coupling CFD-DEM-PBM. L. Bart et al.[15] conducted a simulation analysis on the separation process of the straw combine harvester by the particle model of flexible straw discrete elements. Z. Xuan et al.[16] studied the flow process of soybean in the chute by the discrete element method. I. Keppler et al.[17] conducted a simulation analysis on the flow of wheat in the mixed gas grain dryer by the discrete element method. L. Y. Qin et al.[18] simulated the air-suction seed metering device of the soybean no-tillage planter by the discrete element method and optimized the working parameters by the response surface methodology.

Starting with the seed-filling performance of the seed metering device, this paper improved the structure of the cell wheel seed metering device to deal with its low seed metering efficiency of pseudo-ginseng seeds. A contact model among the pseudo-ginseng seed particles in the seed case was established by the discrete element theory to analyze the normal force and tangential force from the impact contact among the particles so as to determine the principal factor that affects the seed group disturbance. The influence of the rotational speed of the stir wheel and that of the cell wheel on the filling ratio was investigated with the discrete element numerical simulation software application EDEM to analyze the fluctuation of the average normal force and the average speed of the seed group with time, as well as the variation trend of the filling area with the rotational speed of the stir wheel. The simulation analysis results were verified by tests to determine the feasibility of the discrete element simulation analysis method. The simulation analysis on the seed-filling performance of the cell wheel seed metering device by discrete element numerical simulation can save test costs, reduce the test cycle and provide references for the improvement of pseudo-ginseng precision planters.

2 Discrete element theory

2.1 Establishment of contact model

The discrete element method simulates the propagation of movement in a particulate set. When the seed metering device is in operation, the seeds in the seed case are disturbed by the rotation of the cell wheel, and the movement of seeds will certainly cause impact among seeds, thus generating contact force[19, 20]. In the discrete element analysis method, the contact model can be divided into soft particle contact and hard particle contact according to different contact forms. The soft particle contact model allows a certain period of impact, and it calculates the contact force according to the impact overlap and tangential displacement of particles by taking into account the impact of multiple particles. Such contact force is closely correlated with seed group disturbance. Considering there is no surface adhesion on pseudo-ginseng seeds, the Hertz-Mindlin soft particle contact model was adopted[21-23]. The contact of this model was simplified as the damped oscillation of the spring oscillator with couplers and slides[17,24,25] as shown Figure 1. Its differential equation is:

$$m \cdot \frac{d^2x}{dt^2} + c \cdot \frac{dx}{dt} + kx = 0 \tag{1}$$

Where: m stands for the mass of model particles; x stands for the displacement deviated from the equilibrium position; c stands for the contact damping coefficient; and k stands for the elastic coefficient of the spring.

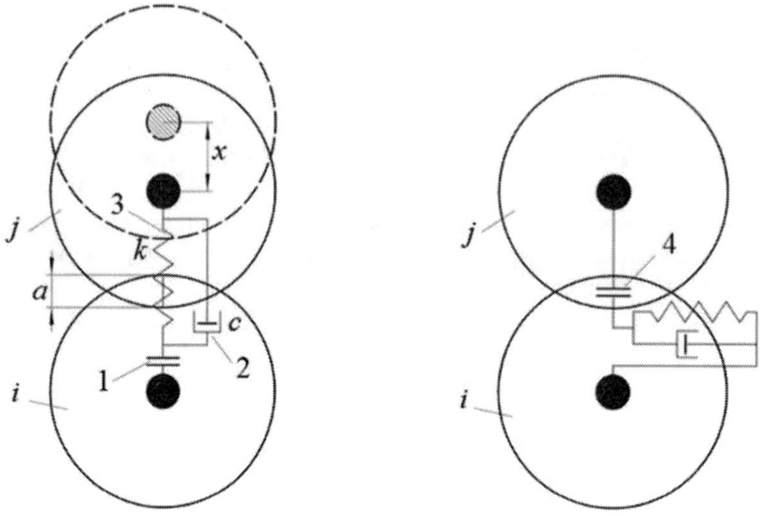

a. Normal contact model b. Tangential contact model

Figure 1. Simplified soft spherical model of contact force among particles.
1. Coupler 2. Damper 3. Spring 4. Slider
Note: i stands for particle i; j stands for particle j

On the normal direction, the coupler determines the pairing of contacted particles and no force is introduced; on the tangential direction, the slider can realizes the slide of two particles by the normal force and friction force[26].

2.2 Analysis of contact force

In the soft spherical particle impact model, the contact force can be divided into the normal force and tangential force, and the normal force is the combination of the elastic force and damping force in the model. As for the 3D spherical particles of pseudo-ginseng seeds, according to the Hertz elastic contact theory, the normal force F_n and the tangential force F_t can be expressed as follows:

$$F_n = (-k_n a^{3/2} - c_n v_{ij} \cdot n)n \tag{2}$$

$$F_t = -k_t \delta - c_t v_{ct} \tag{3}$$

Where: k_n stands for the modulus of normal elasticity; k_t stands for the modulus of tangential elasticity; c_n stands for the normal damping coefficient; c_t stands for the tangential damping coefficient; a stands for the normal overlap; v_{ij} stands for the relative speed of particle i to particle j; v_{ct} stands for the sliding speed of the contact point; δ stands for the tangential displacement; n stands for the normal unit vector of the contact surface between particle i and particle j.

If the following formula is tenable:

$$|F_t| \geq \mu |F_n| \tag{4}$$

Relative sliding occurs between particle i and particle j, then:

$$F_t = -\mu_s |F_n| \cdot n_t \tag{5}$$

Where: μ_s stands for the static friction factor which is determined by the physical properties of pseudo-ginseng seeds; n_t stands for the tangential unit vector. According to formula (5), the internal friction force among seeds is the tangential force in the above contact model. The tangential force is determined by the normal force and is also a principal factor of seed group disturbance.

The internal friction force among pseudo-ginseng seeds is the principal factor that hinders the relative movement or the trend of relative movement among seeds. The internal friction force among pseudo-ginseng seeds is greater and the relative movement is weaker, which are not conducive to the seed-filling of the seed metering device. Therefore, the stirring and seed-filling device added to the cell wheel seed metering device can overcome the friction force among seeds, facilitate the relative movement among seeds and improve the seed-filling efficiency.

3 Discrete element simulation analysis on seed-filling process

3.1 Structure and working principle of seed metering devices

The structure of the seed metering device is as shown in Figure 2. It mainly consists of a seed cleaner, a cell wheel, a seed case, a stir wheel, a hairbrush, a seed protecting plate, a transmission shaft and a shell. There are cell wheel holes with the diameter of 8.0 mm and the depth of 5.5 mm[27]. The simple-structured seed metering device (176 mm long, 42 mm wide and 240 mm high) is light in mass and applicable to the seeding of pseudo-ginseng which requires narrow row space. During operation, the cell wheel of the seed metering device rotates clockwise while the stir wheel rotates anticlockwise. The seed group in the seed case moves with the stir wheel, and the seeds rotate with the cell wheel when they fall into the holes of the cell wheel in the seed-filling area. The excessive seeds will be removed by the hairbrush when reaching it. Then the seeds pass the seed protecting area to the seed charging area. At that time, the seeds detach from the holes by their own gravity, and those cannot fall themselves will be forced out of the holes by the seed cleaner.

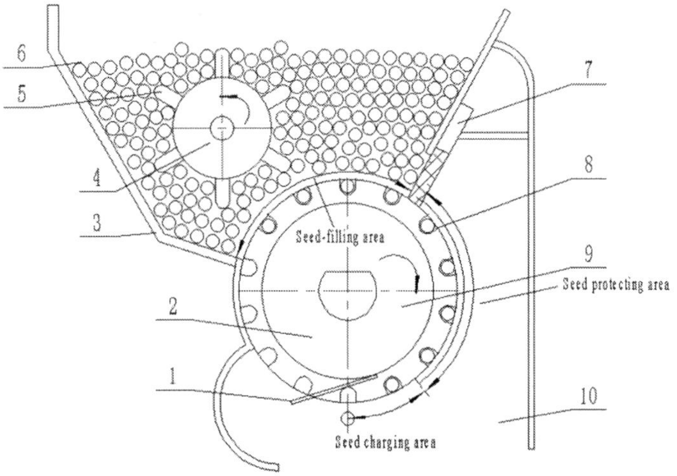

Figure 2. Structure and principle of the seed metering device.
1. Seed cleaner 2. Cell wheel 3. Seed case 4. Stir wheel shaft 5. Stir wheel 6.seeds
7. Hairbrush 8. Seed protecting plate 9. Seed metering shaft 10. Shell

3.2 Discrete element model of pseudo-ginseng seeds

Considering the triaxial size and external physical characteristics of pseudo-ginseng seeds, spheres with different diameters were used for filling and superposition by the discrete element analysis software application EDEM to make them more close to the geometrical size of real seeds[28]. The pseudo-ginseng seed model was established by filling and superposition of 5 spheres in the discrete element software application EDEM as shown in Figure 3.

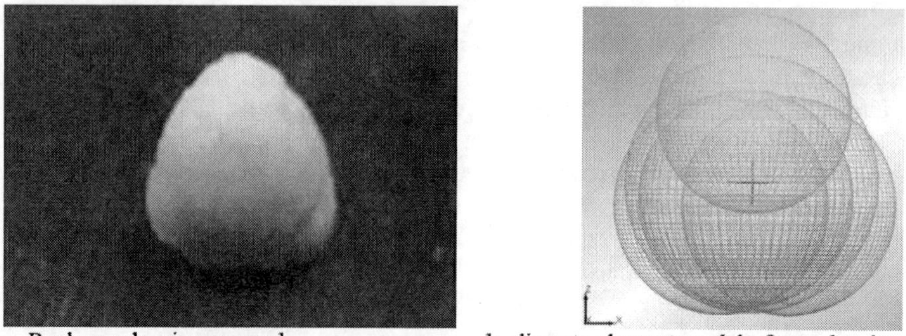

a. Real pseudo-ginseng seed b. discrete element model of pseudo-ginseng seeds

Figure 3. Model of pseudo-ginseng seeds.

The Hertz-Mindlin (no slip) contact model was adopted between seed particles as well as between seeds and the seed metering device, and the particle properties and contact parameters were set up according to Table 1[29]. The generation type of particles was set as dynamic, the mass of the generated particle was 150g and the particle size complies with the normal distribution. The step length of simulation time, simulation time and mesh size were set as 1e-5s, 20s and 2mm respectively.

Table 1. Parameters needed for simulation.

Parameter	**Value**
Particle density/kg·m^{-3}	1100
Poisson's ratio of particles	0,4
Shear modulus of particles/MPa	1,3×10^{7}
Recovery coefficient among particles	0,48
Static friction coefficient among particles	0,32
Rolling friction coefficient among particles	0,085

With the filling ratio as the test index, 6 rotational speeds of the stir wheel (0, 50, 100, 150, 200 and 250rpm) were selected to conduct 3 repeated simulation tests on 4 rotational speeds of the cell wheel (60, 80, 100 and 120rpm). Each group of the test tested 200 holes continuously, and it shall be deemed as miss-filling when there are no seeds in holes in the seed protecting area. After obtaining the statistics of the number of miss-filled holes, the filling ratio A and the average filling ratio is were calculated according to formula 6.

$$A = (1 - \frac{N_m}{N}) \times 100\% \tag{6}$$

Where: Nm stands for the number of miss-filled holes, and N stands for the number of continuously tested holes, N=200.

3.3 Simulation test results and analysis

3.3.1 Influence of rotational speed of stir wheel on filling ratio

Figure 4 shows the variation trend of the filling ratio with the rotational speed of the stir wheel under 4 rotational speeds of the cell wheel. It can be seen from the figure that the filling ratio under each rotational speed of the cell wheel is on the rise as the rotational speed of the stir wheel increases. The maximum filling ratio is 99,50% when the rotational speed of the cell wheel reaches 60rpm, and the filling ratio can still reach 91,33% when the rotational speed of the cell wheel grows to 120rpm. The filling ratio decreases as the rotational speed of the cell wheel increases when the rotational speed of the stir wheel is constant.

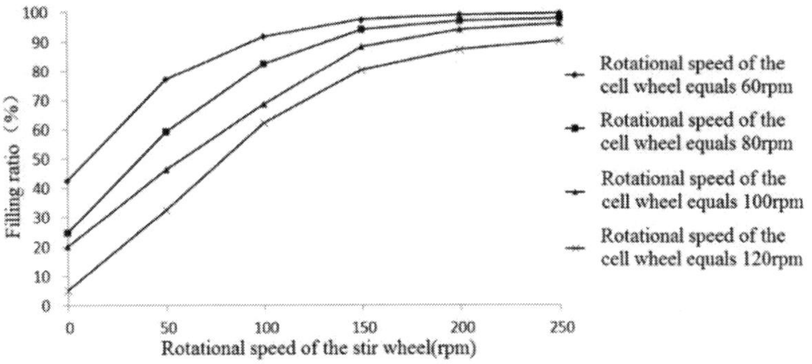

Figure 4. Variation trend of the filling ratio with the rotational speed of the stir wheel under different rotational speeds of the cell wheel.

3.3.2 Influence of rotational speed of stir wheel on seed group disturbance

Figure 5 and Figure 6 shows the fluctuation of the average normal force of contact among pseudo-ginseng seed particles and the average speed of particles within the range of 10s to 11s under different rotational speeds of the cell wheel when the rotational speed of the cell wheel is 60rpm. It can be known from Figure 5 that the average normal force fluctuates irregularly with time under each rotational speed of the stir wheel, and it fluctuates more wildly as the rotational speed of the stir wheel increases. When the rotational speed of the stir wheel is 0, the average normal force among pseudo-ginseng seed particles fluctuate the least with time with an average value of 0,0028N. When the rotational speed of the stir wheel exceeds 50rpm, the average normal force increases significantly with little difference in the average value (around 0,02N) when the rotational speed of the stir wheel lies between 50rpm and 250rpm. According to Figure 6, the average speed of pseudo-ginseng seed particles increases as the rotational speed of the stir wheel grows. It can be known from Figure 4 to Figure 6 that a higher rotational speed of the stir wheel causes higher average speed of seeds, seed group disturbance and filling ratio, indicating that increasing the rotational speed of the stir wheel can help seeds get into the holes to complete the seed-filling process.

3.3.3 Influence of rotational speed of stir wheel on filling area

The filling area is an area formed by the overlap of movement tracks of seeds in the seed-filling area of the seed metering device. A larger filling area indicates lower filling efficiency and higher filling efficiency otherwise. Figure 7 shows the filling area under different rotational speeds of the stir wheel by tracking the movement tracks of seed groups when the rotational speed of the cell wheel is 60rpm. According to Figure 7, the filling area decreases as the rotational speed of the stir wheel increases. When the rotational speed of the stir wheel is 0, the filling area occupies the entire seed-filling area of the seed metering device and seeds are able to enter the holes anywhere in the seed-filling area, thus the seed-filling efficiency is the lowest at that time. When the rotational speed of the stir wheel is 250rpm, the filling area is mainly concentrated at the initial position of the seed-filling area and the holes can finish seed-filling at the initial position, thus the seed-filling efficiency is the highest.

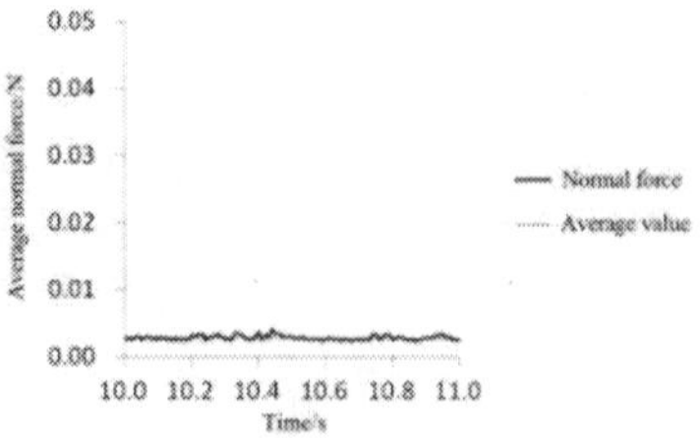

a. Rotational speed of the stir wheel equals 0

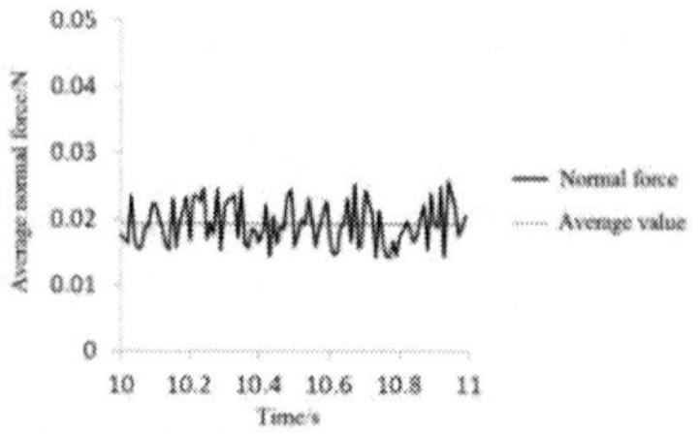

b. Rotational speed of the stir wheel equals 50rpm

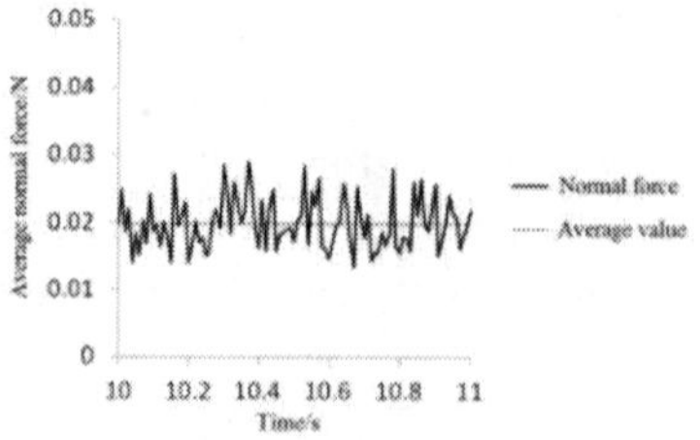

c. Rotational speed of the stir wheel equals 100rpm

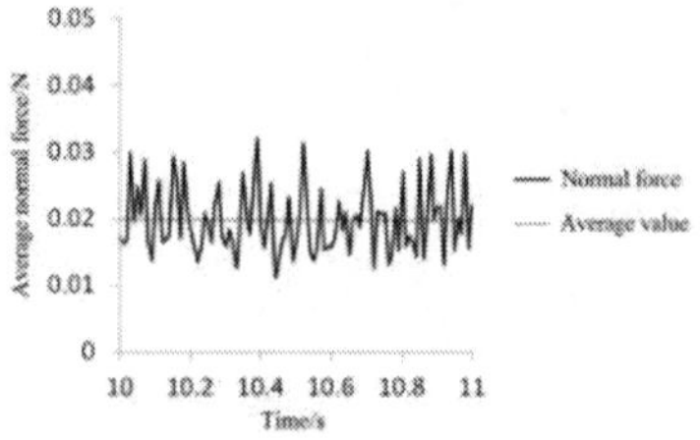

d. Rotational speed of the stir wheel equals 150rpm

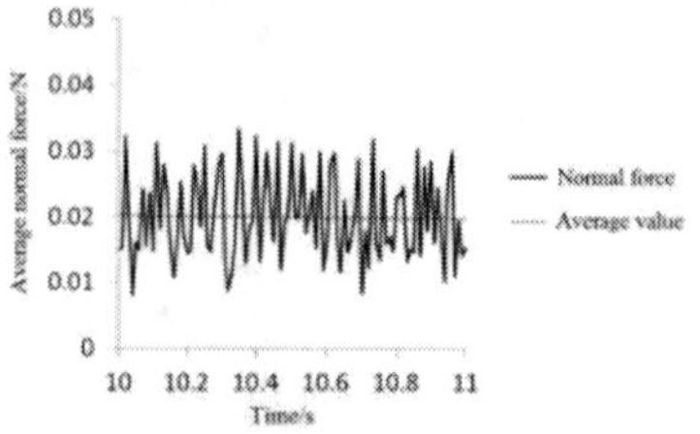

e. Rotational speed of the stir wheel equals 200rpm

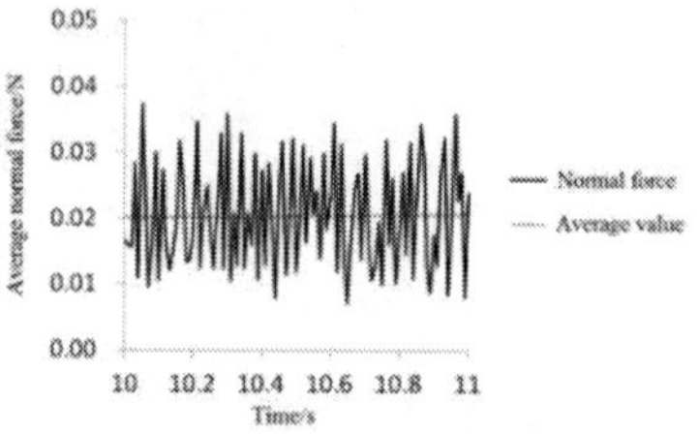

f. Rotational speed of the stir wheel equals 250rpm

Figure 5. Fluctuation of the average normal force with time under different rotational speeds of the stir wheel.

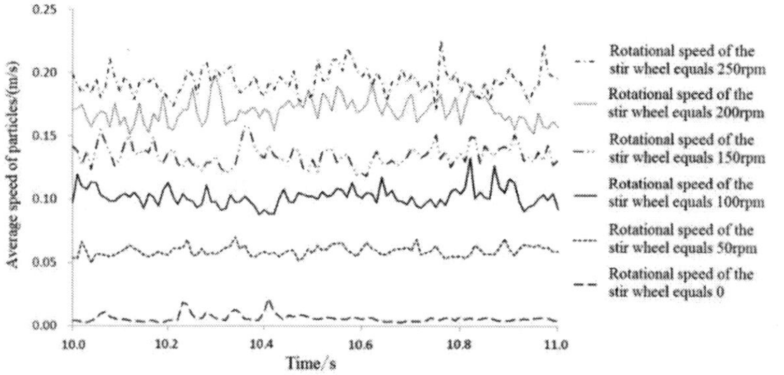

Figure 6. Fluctuation of the average speed of seeds with time under different rotational speeds of the stir wheel.

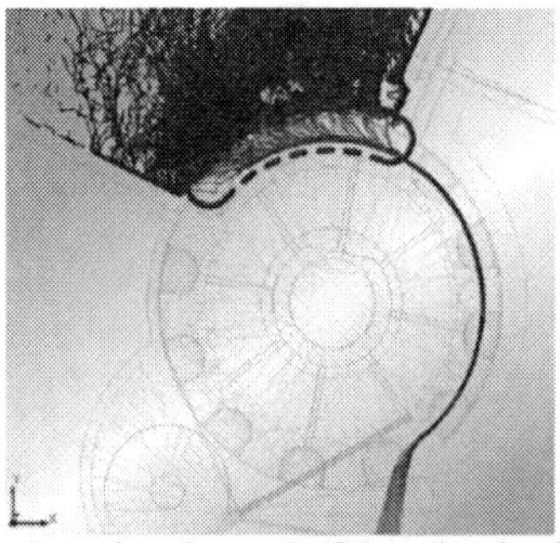

a. Rotational speed of the stir wheel equals 0

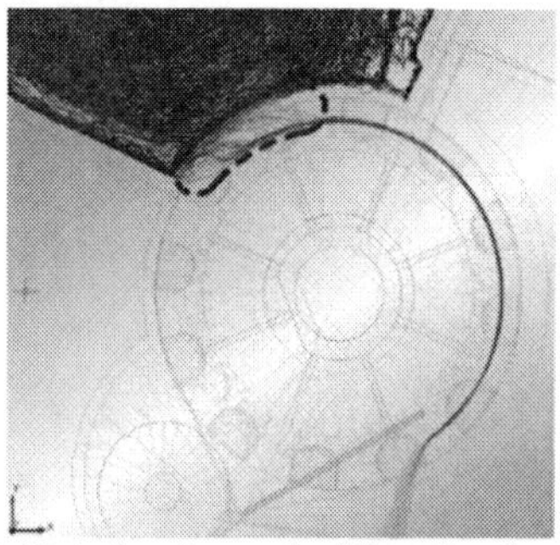

b. Rotational speed of the stir wheel equals 50rpm

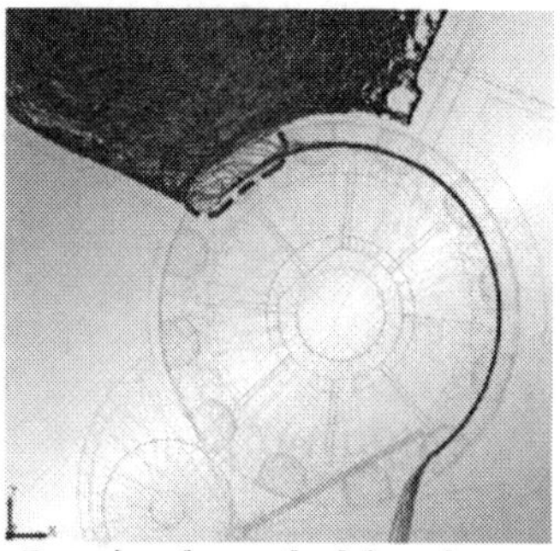

c. Rotational speed of the stir wheel equals 100rpm

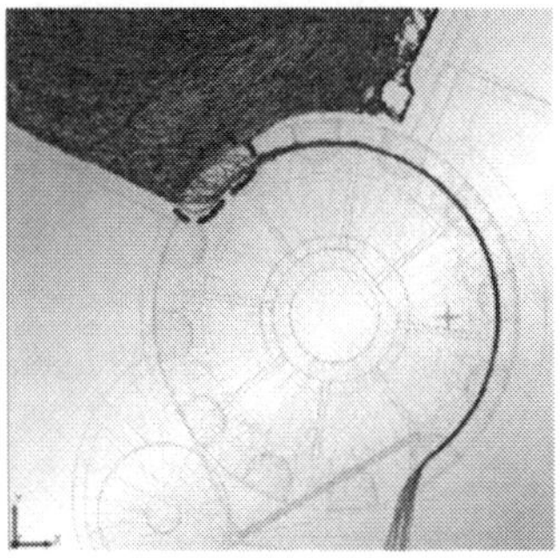

d. Rotational speed of the stir wheel equals 150rpm

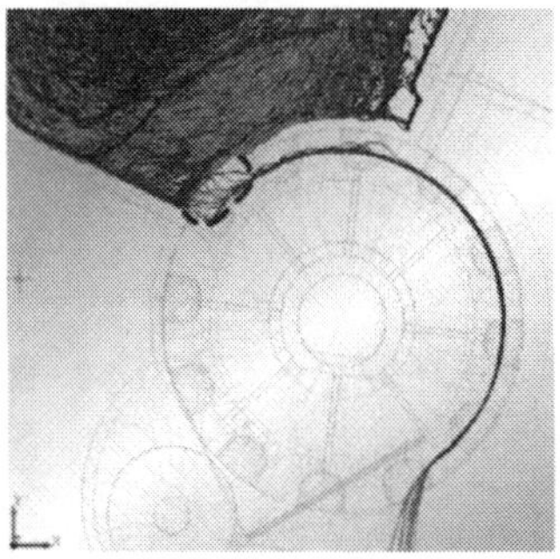

e. Rotational speed of the stir wheel equals 200rpm

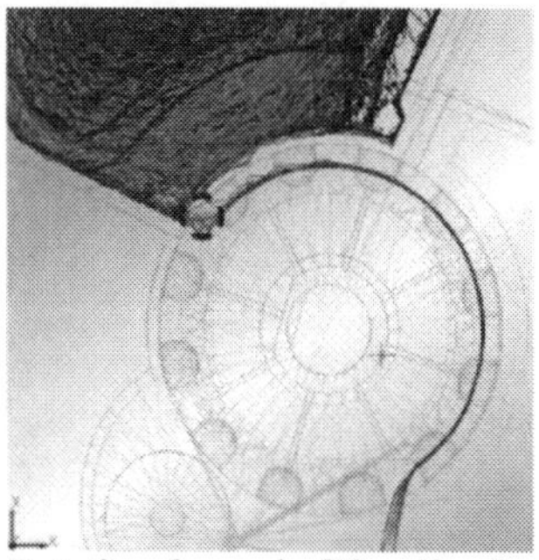

f. Rotational speed of the stir wheel equals 250rpm

Figure 7. Variation of the filling area under different rotational speeds of the stir wheel.

4 Test verification

4.1 Test materials and method

4.1.1 Test materials

A test device of the seed metering device was produced according to the simulation model, and it was tested and verified on the test bed of the JPS-12 computer vision seed metering device. The pseudo-ginseng seeds picked from the pseudo-ginseng cultivation base in Wenshan Prefecture, Yunnan Province were selected as the test materials. The test facilities mainly include a DMK 23G618 high-speed camera, a 57BYGH250C stepping motor, a TB6600 driver, a E30S4-1000-3-N-24 speed sensor, a ME204E electronic balance (with a weighing range of 0-220g and the precision of 0.0001g) and a computer. As shown in Figure 8, the cell wheel is connected to the main shaft of the test bed through chain transmission and is driven by the electric motor to rotate. The output shaft of the stepping motor is connected to the left end of the stir wheel shaft and adjusts the rotational speed of the stir wheel by the driver. The speed sensor is connected to the right end of the stir wheel shaft to acquire the rotational speed of the stir wheel and to facilitate adjustment. The high-speed camera is mounted under the seed cleaning area of the seed metering device to record the images of seed metering. The rotational speeds of the cell wheel and the stir wheel as well as the recorded images can be displayed on the computer interface in real time.

a. Test bed
1. Rack 2. Stepping motor 3. Seed metering device
4. Speed sensor 5. Chain transmission 6. High-speed camera

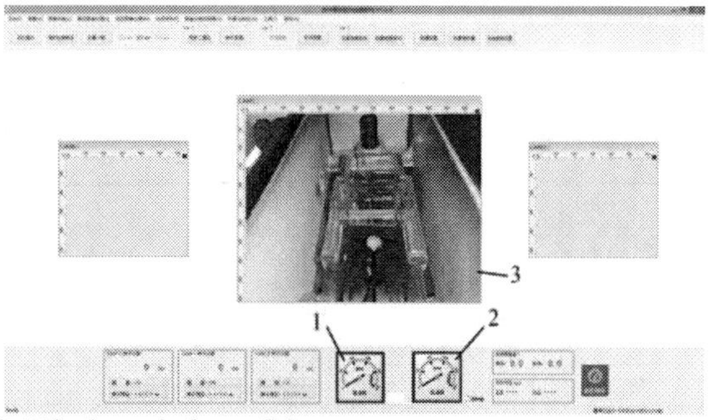

b. Display window of the computer
1. Display interface of the rotational speed of the stir wheel
2. Display interface of the rotational speed of the cell wheel
3. Display interface of the high-speed camera

Figure 8. Test setup.

4.1.2 Test method

The 4 rotational speeds of the cell wheel and 6 rotational speeds of the stir wheel same as the simulation test were selected, and the filling ratio was adopted as the test index. Three repeated variance analyses were conducted on the test results by the two-factor equal repeated tests. Before each group of the test, weigh 150g pseudo-ginseng seeds by the electronic balance and put them into the seed case and then proceed with the tests according to Table 2. The images of the seed cleaning area were recorded with the high-speed camera. Circumstances that there are no seeds in the holes of the seed cleaning area should be deemed as miss-filling, and the filling ratio is calculated according to formula 6. The filling ratio measured in each group of the test was recorded in Table 2.

Table 2. Design and results of the two-factor equal repeated tests.

Test No.	Rotational speed of the stir wheel (X_1)/rpm	Rotational speed of the cell wheel (X_2)/rpm	Filling Ratio (A)/%		
			1	2	3
1	0	60	43,5	47,5	42,5
2	50	60	81,5	77,5	77,0
3	100	60	92,0	92,0	84,0
4	150	60	92,0	92,5	93,5
5	200	60	96,5	94,5	94,5
6	250	60	99,5	98,0	97,0
7	0	80	28,0	26,0	24,5
8	50	80	54,5	60,0	55,5
9	100	80	81,0	81,0	79,0
10	150	80	92,0	89,5	89,0
11	200	80	95,0	91,5	93,5
12	250	80	98,0	97,0	98,0
13	0	100	15,0	18,0	20,0
14	50	100	46,5	47,0	49,0
15	100	100	71,5	71,5	72,0
16	150	100	83,0	86,0	84,5
17	200	100	89,5	90,0	92,5
18	250	100	96,0	96,5	95,0
19	0	120	7,0	9,0	6,5
20	50	120	30,5	34,0	36,0
21	100	120	58,5	55,5	56,0
22	150	120	79,0	74,5	72,5
23	200	120	89,5	87,0	81,5
24	250	120	91,5	89,5	93,0

4.2 Test materials and method

A variance analysis was conducted for the test results by the software application Design-Expert 8.0.6 to test its significance. According to Table 3, the P value of the rotational speed of the stir wheel (X_1), P value of the rotational speed of the cell wheel (X_2) and that of the interaction item (X_1X_2) of the rotational speeds of the stir wheel and cell wheel are all less than 0.0001, suggesting that the rotational speeds of the stir wheel and cell wheel as well as the interaction item of the rotational speeds of the stir wheel and cell wheel affect the filling ratio significantly.

Table 3. Variance analysis table of filling ratio *A*.

Source	**Sum of Squares**	**df**	**Mean Square**	**F Value**	**P Value Prob>F**
Model	52956,38	23	2302,45	418,63	< 0,0001**
X_1	45008,08	5	9001,62	1636,66	< 0,0001**
X_2	5822,79	3	1940,93	352,90	< 0,0001**
X_1X	2125,50	15	141,70	25,76	< 0,0001**
Pure Error	264,00	48	5,50		
Cor Total	53220,38	71			

Note: "*" stands for significant (P<0,05) and "**" stands for extremely significant (p<0,01), similarly hereinafter.

Figure 9 shows the variation trend of the filling ratio with the rotational speed of the stir wheel under 4 rotational speeds of the cell wheel. It can be seen from the figure that the filling ratio grows as the rotational speed of the stir wheel increases, and it has a consistent variation trend with the test results of the simulation. The average error of the filling ratio of the simulation test and that of the verification test is 3,43% (<5%), indicating that it is feasible to analyze the seed-filling performance of the pseudo-ginseng precision seed metering device by the discrete element method[30].

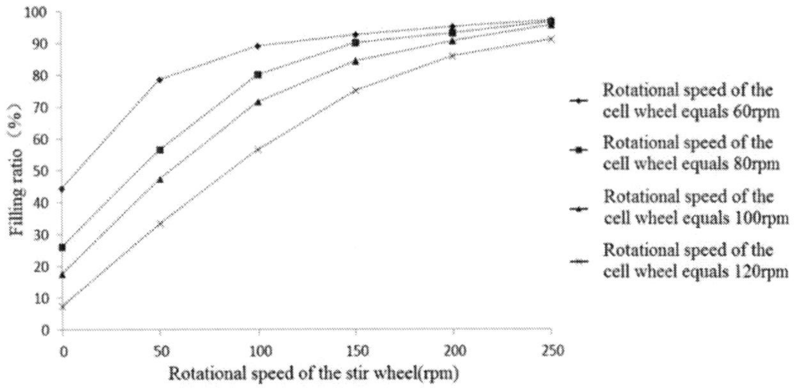

Figure 9. Variation trend of the filling ratio with the rotational speed of the stir wheel under different rotational speeds of the cell wheel.

5 Conclusions

(1) This paper conducted a theoretical analysis on the contact model of pseudo-ginseng seeds and proposed the stirring and seed metering method for the cell wheel seed metering device to overcome the friction force among seeds, improve the seed group disturbance, enhance the relative movement of seeds and improve the seed-filling efficiency. (2) The simulation test indicates that a higher rotational speed of the stir wheel causes higher seed group disturbance, average speed of seeds and filling ratio, thus easier for seeds to enter the holes. The filling area reduces as the rotational speed of the stir wheel increases. A larger filling area leads to lower filling efficiency while a smaller filling area leads to higher filling efficiency. When the filling area is minimal, it is indicated that the hole is filled in the

initial position of the filling area, and the filling efficiency is the highest. When the filling area is maximum, it is indicated that the hole must pass through the whole filling area to have the seed fall into the hole, at this time, the filling efficiency is the lowest. (3) According to the variance analysis on the verification test results, the filling ratio is significantly affected by the rotational speeds of the cell wheel and the stir wheel as well as the interaction item of the rotational speeds of the cell wheel and the stir wheel. When the rotational speed of the cell wheel reaches the maximum, improving rotational speed of the stir wheel can cause a filling ratio of over 90%. The test results are almost identical to the variation trend of the simulation analysis results with an average error of the filling ratio less than 5% (3,43%), indicating that it is feasible to analyze the seed-filling performance of the stirring and seed-filling device by the discrete element method. (4) The adding of a stirring and seed-filling device can improve the filling ratio of the cell wheel seed metering device so as to improve the seeding yield. When the seed metering filling ratio is guaranteed, adding a stirring and seed-filling device can improve the seeding metering efficiency of the seed metering device so as to increase the operating speed of planters.

References

1. Y.W. Sheng, W.C. You and J.C. Qian, Transactions of the CSAM, **40**, 72 (2009)
2. Z.Y. Wen, Transactions of the CSAM, **36**, 51 (2005)
3. W.M. Liang, G.C.Yun, G.X. Yan, et al., Transactions of the CSAE, **26**, 119 (2010)
4. Y.W. Man, W.M. Lin and H.T. Feng. Journal of Agricultural Mechanization Research, **34** (2015)
5. W.Y. Yao, L.J. Dong, W.D. Wei, et al., Transactions of the CSAM, **43**, 54 (2012)
6. P.P. Garcia, Participatory development of a corn seeder, Philippines: Philippines University (1993)
7. W. Hui, Philippines University (1991)
8. L.X. Wen, W.Z. Man, J.E. Chen, et al., Transactions of the CSAM, **39**, 60 (2008)
9. Y.J. Qun, W. Gang, X. Nan and F. Hong. Transactions of the CSAM, **42**, 83 (2011)
10. T.C. Zhou, L.H. Feng, W.M. Liang, et al., Transactions of the CSAM, **26**, 114 (2010)
11. Q. Ling, Agricultural Technology & Equipment, **23** (2014)
12. W.Z. Man, L.X. Wen, H.S. Xing, et al., Transactions of the CSAM, **40**, 56 (2009)
13. L. Tao, H.R. Yin, L. Jing, et al., Journal of South China Agricultural University, **37**, 126 (2016)
14. S. Maitraye, B. Dana, S.Ravendra and R. Rohit, Processes, **89** (2014)
15. L. Bart, A. Thomas, T. Engelbert, et al., Computers and Electronics in Agriculture, **24** (2014)
16. Z. Xuan and L.V. Quoc, Mech Mater, **115** (2000)
17. I. Keppler, L. Kocsis, I. Oldal, et al., Advanced Powder Technology, **824** (2012)
18. L.Y. Qin, Z.M. Quan, L. Fei, et al., Transactions of the CSAM, **47**, 65 (2016)
19. L.Q. Xi, Z.P. Ling, L.Y. Tao, et al., Transactions of the CSAM, **45**, 109 (2014)
20. L.H. Chang, L.Y. Ming, T. Zhong, et al., Transactions of the CSAE, **27**, 117 (2011)
21. L.P. Van, E. Tijskens, E. Dintwa, et al., Powder Technology, **190**, 348 (2009)
22. V. Aleshin and K. Van Den Abeele, International Journal of Non–Linear Mechanics, **15** (2013)
23. Y. Shimizu and P. Cundall, J Eng Mech–Asce, **127**, 864 (2001)
24. A.O. Raji and J.F. Favier, Journal of Food Engineering, **359** (2004)
25. L. Yang, H.Y. Long, J.F. Guo, et al., Acta Phys. Sin., **64**, 114501 (2015)
26. Y.J. Qun, S.Y. Fang, N.X. Ttang, et al., Transactions of the CSAE, **24**, 105 (2008)
27. H.Z. Wu, M.W. Peng, X.J. Long and L.Q. Hui, Journal of south China agricultural university, **37**, 109 (2016)
28. Z. Zhan, W.Y. Fang, Y.J. Jun and T. Zhong, Computers and Electronics in Agriculture, **25** (2015)
29. L.Q. Hui, G.X. Jun and Z.Z. Hong, Simulation and Experiment of Seed–filling Performance of Pneumatic Cylinder Seed–metering Device for Panax notoginseng, **47**, 27 (2016)
30. S.L. Rong, W.J. Min, S. Wei, et al., Transactions of the CSAE, **30**, 40 (2014)

Application of passive damping fixture to the improvement of CNC drilling

Shao-Yi Hsia[a], Chen-Chung Chang and Di-Hung Huang
Department of Mechanical & Automation Engineering, Kao Yuan University, Kaohsiung 82151, Taiwan

Abstract. High precision and high efficiency are the essential competition conditions for multi-axis machine. However, machine vibration problems are always plagued by equipment manufacturers. This study focuses on solving or eliminating the multi-angle drilling vibration of processing machines. By the design and experiment combining CNC multi-axis machine tools with vibration tables, the surface roughness in the multi-angle drilling holes is discussed for the reference of the industry or the academia. The results indicated that adding proper vibration factors could effectively improve the bore surface roughness and smoothness. Roughness *Ra* with the vibration interval time 0.5 sec, could improve the surface roughness 48.95% of the processing hole, comparing to it without vibration. Besides, the resonance phenomenon disappears at the feed to generate good effect. It is expected to reduce various vibration problems faced by processing with machine tools in order to enhance the international competitiveness of processing tool manufacturers.

1 Introduction

Cutting has played a critical role in the machinery-manufacturing environment. However, various countries have actively improved the processing methods with the constant development and advance of high-tech industry. Moreover, Rahman and Kumar [1] experimented and analyzed the micro-milling of pure copper, improved the experimental result through the analyses of cutting force and chip types, and observed the wear of micro-mills. Nouari*et al.* [2] discussed drills with three different materials and distinct coating materials to analyze the tool life, the quality of hole, and the surface roughness of hole wall of the three drills in the aluminum alloy (A2024) drilling. The results revealed that B-type drill presented better performance than the other two types, aiming at above three processing properties. Lee *et al.* [3] utilized a micro-mill (with the diameter 254μm), similar to drilling, for processing micro-holes. It was found that the tool life reduced with increasing burr height. The characteristics of burr formation were therefore discussed and the processing parameters were controlled for the minimum burr formation and further improving the tool life.

Masahiko and Masao [4] investigated a combination of countermeasures to prevent the chipping of the edge of a cutting tool. The results indicated that the chipping of the cutting edge can be effectively prevented and a good surface finish obtained by both continuous and intermittent cutting modes when cutting hardened steels. Altintas and Weck [5] collected and organized various studies to describe flutter principles, various processing modeling and simulation, and suppression tactics.

[a] Corresponding author : syhsia@cc.kyu.edu.tw

There has not been a usable damping method for tapping, broaching, and gear rolling. Marui *et al.* [6] mentioned that the damping of a tool system came from micro slide, the damping of the cutting tool itself, and the friction with the air during vibration. They therefore inserted thin sheets between cutting tool and tool slide and inside the cutting tool in order to enhance the friction force and further promote the damping of the cutting tool system.

From above literatures, a lot of domestic and international experts and scholars have studied and concerned about the improvement of CNC processing industry. However, little research focuses on CNC vibration, and most of them pay attention to the spindle, but rare on the vibration of CNC processing trays. Consequently, this study stresses on the vibration of processing trays in order to provide reference for domestic and international experts and scholars and the industry with more complete processing parameters.

2 Basic theory

2.1 Vibration

The use of machines is inevitable in human life, and the operation of each machine would generate vibration. The major factors in machine damage are mainly resulted from resonance or close-resonance of machines. Such situations might occur in the high-speed operation of machines or the starting and stopping of a machine. Such resonance stress would gradually accumulate to appear cracks as well as increase and expand such cracks to result in machine damage. In this case, the subject of mechanical vibration becomes critical to evaluate the life of machine element in advance and to ensure the life limit of a machine. To understand the factors in machine damage, it needs to understand the reason of machine vibration and to know the design of vibration control.

Moreover, the factors in mechanical system failure or misalignment are the over vibration and resonance of elements and systems. The natural frequency and external stimulation frequency of mechanical structure vibration would appear resonance to result in excessive deformation and system failure. As vibration would cause violent effect on mechanical structure, vibration test has been listed as a standard design step in the design and development of most engineering systems.

2.2 Damper

The principles of damper commonly used in academia and the industry are described as below.

2.2.1 Active damper

An active damper is equipped a sensor to monitor the processing situation. When there is vibration, signals are sent to the actuator which gives an opposite force to the cutting tool and vibration to offset the vibration. As sensor signals need to be immediately handled and transmit orders back to the actuator, the speed to deal with signals and the reliability of vibration judgment are considered critical. It is the difficulty in the active damper design. Besides, the actuator force would affect the damping effect; weak force would not appear effect, while strong force might become the vibration source. It therefore requires experiment confirmation or detailed simulation.

2.2.2 Semi-active damper

Similar to active dampers, a semi-active damper monitors the processing situation with a sensor, but the rigidity and damping characteristic of the cutting tool would change to offset vibration when the vibration occurs. To achieve such an objective, electrorheological and magnetorheological fluid or piezoelectic materials are utilized for changing the properties to change the vibration stability diagram so as to avoid the vibration are and prevent vibration from occurrence.

2.2.3 Passive damper

In comparison with active dampers, a passive damper reveals the advantages of simple structure, low costs, and easy installation. Nevertheless, it could not randomly change to match the processing environment that the effect is worse than an active one. Although some dampers present tunable mechanism to adjust the natural frequency or damping of dampers, it might not be properly adjusted because of unknown field processing parameters.

3 Research methodology and processing step

By summing up the above discussion and theories, the relationship between cutting and vibration is the primary part to control the machining precision that the relative technology R&D and problem overcoming are worth further discussion. The research process is referred to Fig.1 and described as below.

(1) To search data related to CNC comprehensive processing machine, vibration table, and cutting, according to the speed and feed and cutting tool materials, to establish a database.
(2) To cultivate the vibration factor and the use technique of CNC comprehensive processing machine.
(3) Precede the use proficiency of NX modeling and processing modules.
(4) Match and communicate software and hardware with vibration table, CNC comprehensive processing machine, and NX software.
(5) Collect cutting data with cutting speed and feed matching with cutting tool materials.
(6) Calculate suitable cutting values for seriously worn cutting tool materials, feed speed, and vibration frequency by matching with cutting data.
(7) Compare the analysis result and the entity, and revise the cutting and vibration table parameters.
(8) Select cutting numerical control factors for the optimization and influence analyses.
(9) Compare the cutting tool life before and after the optimization and the surface precision of workpiece.

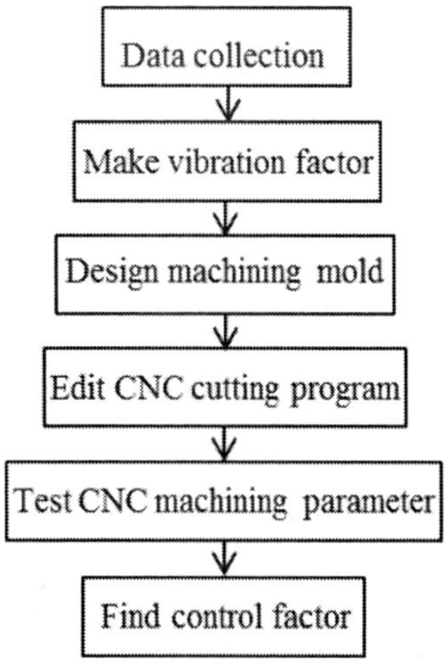

Figure 1. Experimental flow of CNC drilling matching with passive damping fixture.

The relevant equipment and the parameters are described as following.

(1) CNC multi-axis processing machine

CNC multi-axis processing machine developed by Tongtai Machine & Tool Co., Ltd. is used as the machine table in this study.

(2) Vibration table

The vibration table is produced in this study; the designed passive damper entity is shown in Fig.2. The interior mechanism is composed of base, magnetic metal table, electromagnet, four support columns, and damping spring. The motion principle reveals to directly drive the mechanism, through electromagnet power-on, to form the up-down vibration of the spring.

Figure 2. Passive damper entity.

4. Result and discussion

In the beginning of this study, the borehole workpiece without vibration source is produced; the finished appearance and the surface roughness are shown in Fig.3. The mean *Ra* , *Rmax*, *Ry*, and *Rz* after four measurements are 3.379, 39.247, 18.517, and 13.846μm. Since there is no vibration frequency, the processed bore surface roughness is not favorable, the value measured by the surface measuring instrument is rather high, and the surface appears irregular processing lines and unsmooth conditions.

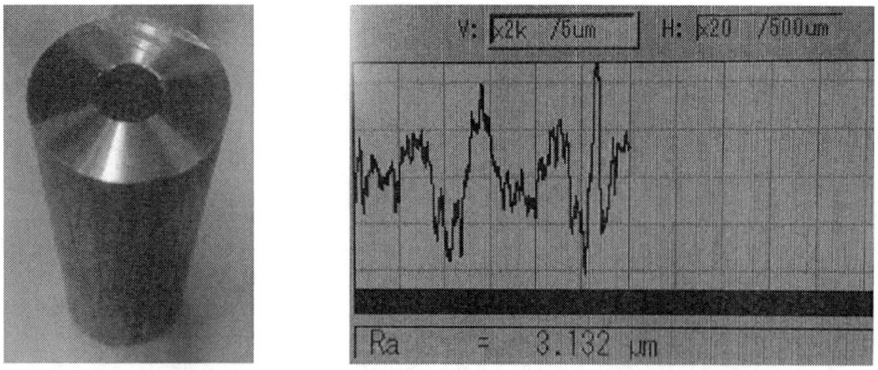

Figure 3. Analysis result of non-vibration factor finished product and surface roughness.

A PLC programmable control device is further utilized for setting various vibration frequency intervals 0.2 sec, 0.5 sec, 1 sec, and 1.5 sec. When the bottom electromagnet of the vibration platform generates magnetic force with 0.2 sec transmission frequency and further generates vibration force and an aluminum alloy rod is processed under the same processing parameters, the acquired finished processing appearance and bore surface roughness are shown in Fig.4. The mean *Ra* , *Rmax*, *Ry*, and

*Rz*appear1.992, 25.798, 12.220, and 9.583μm. Since a 0.2 sec micro vibration factor is opened for this finished product, it is observed that the bore lines are even, the bore smoothness is improved comparing to the finished product without vibration, and the measured *Ra* data are reduced a lot than those without vibration. Apparently, the bore roughness is actually improved. However, it is worth noticing that the borehole feed processing would slightly generate resonance phenomenon to result in hole cracks as the vibration frequency is opened before the borehole processing. Such a phenomenon disappears after drilling for about 1 sec. The feed for the borehole processing therefore presents obvious crack marks, Fig.4.

Processing aluminum alloy rod material by opening 0.5 sec transmission frequency through PLC to acquire the finished processing appearance and the bore surface roughness, Fig.5, the mean *Ra*, *Rmax*, *Ry*, and *Rz*show 1.725, 28.885, 11.443, and 8.254μm. In comparison with non-vibration and 0.2 sec finished products, it is obviously that the 0.5 sec drilling bore lines are evener than 0.2 sec ones, the bore smoothness and *Ra* data are better than those of non-vibration and 0.2 sec ones, the cracks generated by the resonance phenomenon in the drilling feed disappear, while *Rmax* data of 0.5 sec ones are higher than those of 0.2 sec ones. It requires further research.

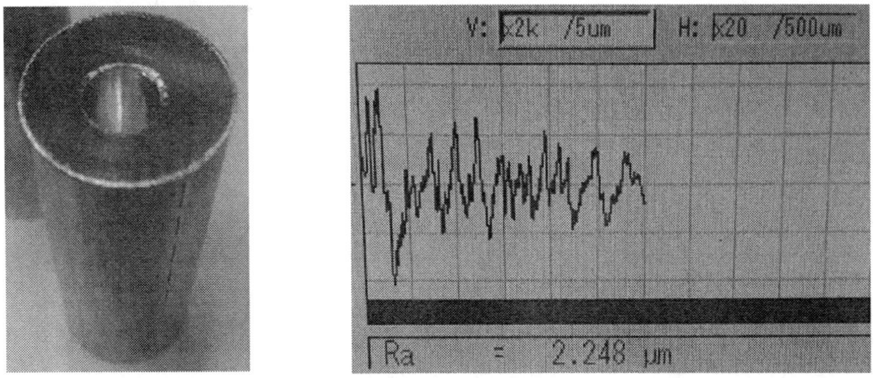

Figure 4. Analysis of finished product with 0.2 sec vibration frequency interval and surface roughness.

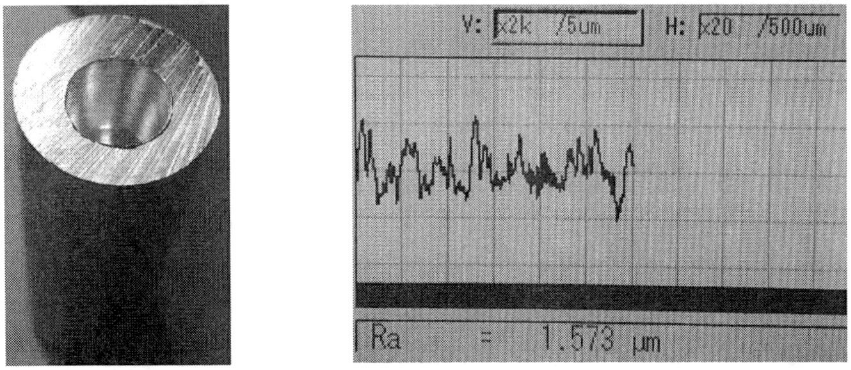

Figure 5. Analysis of finished product with 0.5 sec vibration frequency interval and surface roughness.

1.0 sec transmission frequency is further opened for processing the aluminum alloy rod material. The finished appearance and the bore surface roughness are shown in Fig.6. The mean *Ra* , *Rmax*, *Ry*, and *Rz* are 2.170, 37.377, 13.390, and 10.859μm. The results of the experiment finished product are not as good as expected. The bore drilling lines are obviously improved, and *Ra* data are lower than those without vibration; however, obvious uneven lines appear and the resonance phenomenon is more obvious at the feed. The crack marks at the initial drilling feed are therefore larger and the difference is obvious. Such a phenomenon gradually disappears after processing for 1 sec, when the successive bore lines become evener.

Finally, the finished appearance and the bore surface roughness with 1.5 sec transmission frequency through PLC are shown in Fig.7. The mean *Ra*, *Rmax*, *Ry*, and *Rz*appear 2.830, 37.972,

16.770, and 11.025μm. With the experiences in 1.0 sec experiment finished product and surface roughness, the 1.5 sec finished product is the same as expected. The lines after drilling are similar to those without vibration, the smoothness is about the same as it without vibration, and Ra data are even worse than those without vibration. Moreover, the resonance phenomenon generated in the drilling feed is so larger that the feed port of the finished product generates larger crack marks. The successive bore processing therefore could not match with the vibration frequency and further appear opposite effect.

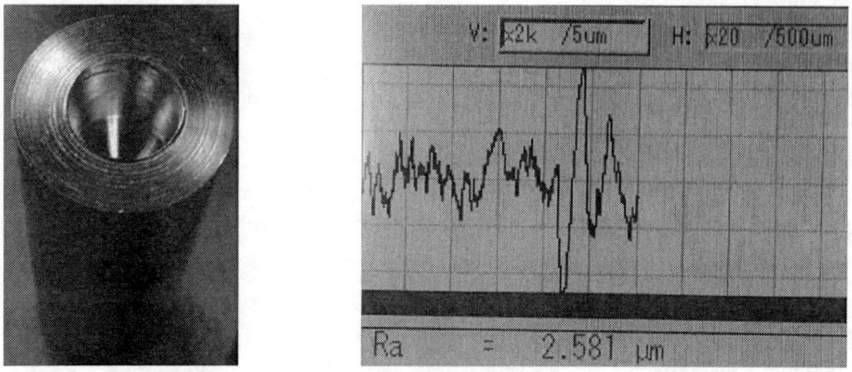

Figure 6. Analysis of finished product with 1.0 sec vibration frequency interval and surface roughness.

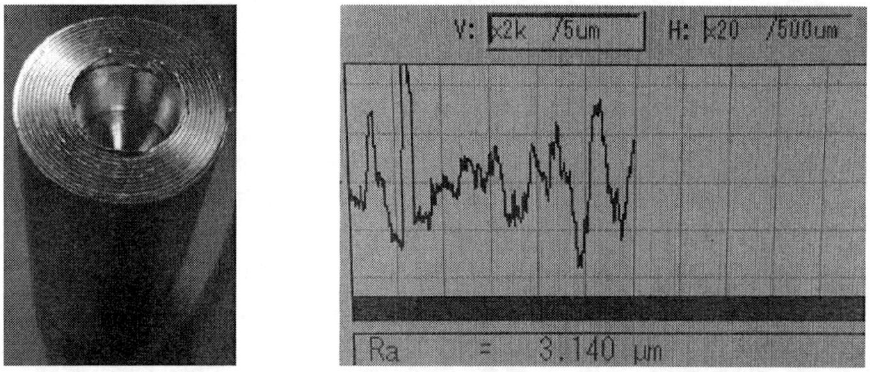

Figure 7. Analysis of finished product with 1.5 sec vibration frequency interval and surface roughness.

Summing up the above results, the mean borehole surface roughness of the processing object, under different vibration frequency intervals after four measurements, is shown in Table 1, where the bottom line represents the lowest in the evaluation standard. According to the table, the following results are concluded.

(1) Adding proper and continuous vibration frequency intervals, under the same processing conditions, could effectively improve the bore surface roughness and enhance the bore smoothness.
(2) Faster vibration frequency intervals do not necessarily present the best improvement. It is essential to match with certain processing conditions and parameter settings to find out the optimal vibration frequency interval.
(3) With slow vibration frequency intervals, the bore surface roughness, after certain area, would not be improved, but might cause opposite effect to worsen the bore roughness.
(4) When the vibration frequency exceeds certain area, the accuracy of the processed workpiece measured with the surface roughness measuring instrument would be largely reduced, possibly because the bore is no longer a proper circle, but appears ellipse or polygon.
(5) The maximum roughness (*Rmax*) is not necessary the best under good average roughness (*Ra*), possibly because of the vibration factor causing the obvious difference between the highest and the lowest vibration.

Figure 8 shows the surface roughness mean values versus vibration interval time. It indicates that roughness *Ra* with the vibration interval time 0.5 sec could improve the surface roughness 48.95% of the processing hole, comparing to it without vibration. At the same times, the best vibration interval timefor the smaller surface roughness valuesof *Ra*, *Ry* and *Rz* is 0.5sec, while *Rmax* is 0.2 sec.

Table 1. Surface roughness mean values.

Roughness	Non-vibration	Vibration interval time (sec)			
		0.2	0.5	1.0	1.5
Ra(μm)	3.379	1.992	1.725	2.170	2.830
Rmax(μm)	39.247	25.798	28.885	37.377	37.972
Ry(μm)	18.517	12.220	11.443	13.390	16.770
Rz(μm)	13.846	9.583	8.254	10.859	11.025

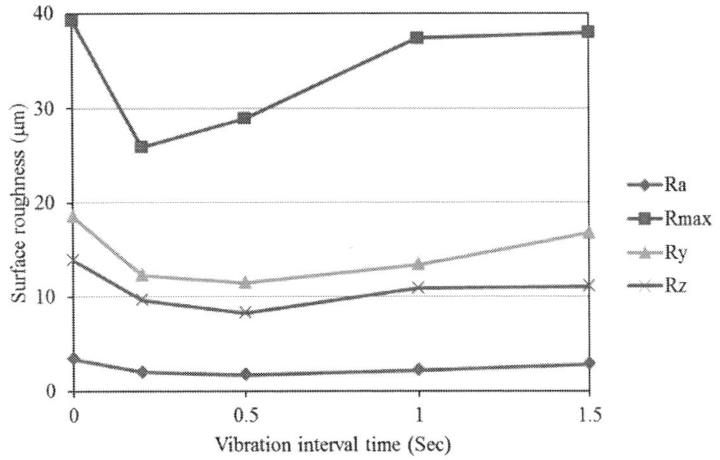

Figure 8. Schematic diagram ofsurface roughnessmean values versus vibration interval time.

5 Conclusion

Aiming at the homemade vibration platform, this study discusses the relationship between the vibration frequency interval and the processing borehole and tests the bore surface roughness of the processing borehole under different frequency. With statistical deduction, the following conclusions are summarized.

(1) Adding proper vibration factors, under the same processing conditions, could effectively improve the bore surface roughness and smoothness. *Ra* with the vibration interval time 0.5 sec, could improve the surface roughness 48.95% of the processing hole, comparing to it without vibration. Besides, the resonance phenomenon disappears at the feed to generate good effect.

(2) It is not necessary that the fast vibration frequency interval would generate better effect. However, adding slower vibration frequency intervals would affect the processing quality and stability, and even worse result than the processing without vibration.

(3) Under small mean roughness (*Ra*), the maximum roughness (*Rmax*) is not necessarily lower. In the experiment, the best vibration intervaltimeof *Ra*, *Ry* and *Rz* is 0.5sec, while *Rmax* is 0.2 sec.

References

1. M. Rahman, A.S. Kumar and J.R.S Prakash, Journal of Materials Processing Technology, **116**, 39 (2001)

2. M. Nouari, G. List, F. Girot and D. Géhin, International Journal of Machine Tools & Manufacture, **45**, 1436 (2005)
3. K. Lee and D.A. Dorn, Precision Engineering, **29**, 246 (2005)
4. J. Masahiko and M. Masao, Journal of Materials Processing Technology, **113**, 342 (2001)
5. Y. Altintas and M. Weck, Ann CIR, **53**, 619 (2004)
6. E. Marui, S. Ema, M. Hashimoto ans Y. Wakasawa, International Journal of Machine Tools & Manufacture, **38**, 1209 (1998)

Research on two dimensional Wiener stochastic degradation model based on the wear model

Hui Zhang[a], Jun Yao and Yan-Lin Zhao

School of Reliability and System Engineering, Beijing University of Aeronautics and Astronautics, Beijing, China

Abstract. One dimensional Wiener degradation process is often used to describe the degradation of product performance. However one dimensional Wiener degradation process doesn't sufficiently consider the relevance of multiple degradation factors and wear process, which lead to inaccurate results. To overcome these problems, a new two dimensional Wiener stochastic degradation model is proposed, which applys to the products on wear process and stochastic degradation process. Combining wear model and two dimensional Wiener stochastic degradation model, a new reliability analytical form is obtained by constructing the Fokker-Planck equation. Then using the relation among wear volume, degradation characteristic lifetime and drift parameter, parameters of two dimensional Wiener degradation model on the basic of wear model can be estimated. Compared with the existing approaches, the proposed method can effectively improve accuracy. Finally, a case study is illustrated the application and advantages of the proposed method.

1 Introduction

Products subjected to the action of the external environment, the material properties of the product will gradually change. And because of the randomness of the external environment and the material properties, the degradation of the product is also random. So stochastic process[1] is usually used to describe the degradation process of product performance. Wiener process[2-5] is the most often used method to describe the degradation of product performance. Tang[6] et al estimated failure time distribution by a Wiener degradation model based on intermediate data. Wang[7] used degradation data of Wiener processes with random effects to study reliability level of the bridge. Peng[8] et al proposed Bayesian method to assess the reliability of the products with Wiener process degradation. Nicolai[9] et al proposed an application to coatings on steel structures by comparing models for measurable deterioration with Wiener processes. Balka[10] et al reviewed and implemented the cure models based on first hitting times of Wiener processes. Si[11] et al estimated remaining useful life based on a nonlinear diffusion degradation process with monitoring degraded signal. Most of the studies on degradation of product performance are based on one dimensional Wiener process degradation, without taking into account the multiple degradation and the physical mechanism of degradation.

Wear is one of the common failure types of components. When the wear occurs, the features of the product can be reduced or failed, reliability and safety will be lost if products are continued to put into

[a] Corresponding author : zhanghui1133018@163.com

use. Many scholars have put forward relevant theories and models about the wear[12,13], which predict the lifetime of the products by calculating the wear volume and wear rate. Hu[14] et al proposed finite element modeling (FEM) approaches with Lagrangian increment method for 3D metal turning of hardened steel H13 by common ceramic tool and ultrafine-grained tool respectively have been investigated by simulation of DEFORM-3D software and turning test. Lin[15] combined the finite element numerical simulation analysis method with the Archard wear calculation model and temperature effect, predicted the most serious wear parts, and proposed a formula for calculating the total wear rate. Gao[16] et al estimated die service life of foging spiral gear based on modified Archard method. Kim[17] et al estimated die service life against plastic deformation and wear during hot forging processes with finite element modeling. Most researches on the wear only consider physical mechanism of wear, predict wear rate and the wear volume by finite element simulation, without considering the impact of the external environment randomness on products degradation.

In this paper, on the basis of the wear model, considering the influence of the randomness of the external environment on the products degradation, combined with the theory of random degradation, a model of two dimensional Wiener stochastic degradation model is proposed based on the wear model. Using 1000h reliability test data of the steam turbine pump as an example, the rationality and accuracy of the method proposed in this paper is verified by comparing with calculation results of other methods.

2 The performance degradation model based on two dimensional Wiener process

2.1 Binary linear Wiener process model

Assuming that product performance degradation process in $\{X(t)=(X_1(t),X_2(t))^T,t\geq 0\}$ [18], namely, the degradations of products are binary. At the time t, it follows:

$$X(t) \sim N(t\mu, t\Sigma) \tag{1}$$

where

$$\mu=(\mu_1\ \mu_2)^T, \Sigma=\begin{pmatrix} \sigma_1^2 & \rho\sigma_1\sigma_2 \\ \rho\sigma_1\sigma_2 & \sigma_2^2 \end{pmatrix} \tag{2}$$

ρ is the correlation coefficient. Assume $\rho \neq 0$, which means the two marginal degradation process $\{X_1(t),t\geq 0\}$ and $\{X_2(t),t\geq 0\}$ are related.

2.2 Reliability Assessment

Initial degradation $X_1(t)=X_2(t)=0$, When one of $\{X_1(t),t\geq 0\}$ and $\{X_2(t),t\geq 0\}$ exceeds its failure threshold, product fails. The lifetime of the product can be represented as:

$$T=\inf\{t: X_1(t)>l_1 \text{ or } X_2(t)>l_2\} \tag{3}$$

Then the reliability function $R(t)$ is derived:

$$R(t)=P\{X_1(s)<l_1, X_2(s)<l_2, 0\leq s\leq t\} \tag{4}$$

When Domine and Pieper[19] studied the first hitting time of binary Wiener process, the Fokker-Planck equation was constructed (Kolmogrov forward equation) based on the initial performance of $X_1(0)=x_{0,1}$ and $X_2(0)=x_{0,2}$, the analytic form of the reliability fuction $R(t)$ was obtained by solving the equation:

$$\begin{aligned}R(t)=\int_0^{\alpha}\int_0^{\infty}\sum_{n=1}^{\infty}&\frac{2r}{\alpha\sigma_2^2(1-\rho^2)t}\sin(\frac{n\pi}{\alpha}\varphi_0)\cdot\\&\exp\left[\frac{\sigma_2\mu_1-\sigma_1\mu_2\rho}{\sigma_1\sigma_2^2(1-\rho^2)}r\cos\varphi-\frac{\mu_2}{\sigma_2^2\sqrt{1-\rho^2}}r\sin\varphi\right]\cdot\\&\exp\left[\frac{(\mu_1\rho\sigma_1\sigma_2-\mu_2\sigma_1^2)x_{0,2}+(\mu_2\rho\sigma_1\sigma_2-\mu_1\sigma_2^2)x_{0,1}}{(1-\rho^2)\sigma_1^2\sigma_2^2}\right]\cdot\\&\exp\left[-\frac{\sigma_1^2\mu_1^2-2\mu_1\mu_2\sigma_1\sigma_2\rho+\sigma_2^2\mu_1^2}{2(1-\rho^2)\sigma_1^2\sigma_2^2}t-\frac{r^2+r_0^2}{2(1-\rho^2)\sigma_2^2t}\right]\cdot\\&\sin\left(\frac{n\pi}{\alpha}\right)I_{\frac{n\pi}{\alpha}}\left(\frac{rr_0}{(1-\rho^2)\sigma_2^2t}\right)drd\varphi\end{aligned}\tag{5}$$

Where $\alpha=\arctan\left(-\frac{\sqrt{1-\rho^2}}{\rho}\right)+\pi$; r_0, φ_0 are the solution of the equation below:

$$\begin{cases}r_0\cos\varphi_0=\frac{\sigma_2}{\sigma_1}(l_1-x_{0,1})-\rho(l_2-x_{0,2})\\ r_0\sin\varphi_0=\sqrt{1-\rho^2}(l_2-x_{0,2})\end{cases}\tag{6}$$

$I_v(z)$ is the modified Bessel function.

$$I_v(z)=\sum_{k=0}^{\infty}\frac{\left(\frac{z}{2}\right)^{2k+v}}{k!\Gamma(k+v+1)}\tag{7}$$

When $x_{0,1}=x_{0,2}=0$, the reliability function $R(t)$, lifetime function $F(t)=1-R(t)$, and density function $f(t)=dF(t)/dt=-dR(t)/dt$ will be obtained. If $x_{0,1}$ and $x_{0,2}$ in formula (5) are regarded as the performance degradation when product run to a certain time, then formula (5) is actually the reliability of the continue running time of product, and $F(t)=1-R(t)$ is the distribution of the remaining lifetime.

2.3 The performance degradation data model

Assuming a total of N products of a degradation test, at the time of $t_{i1}, t_{i2}, \cdots t_{im_i}$, measure the degradation X and Y and get measured data:

$$\begin{matrix}X_{11}, & X_{12}, & \cdots & X_{1m_1}\\ X_{21}, & X_{22}, & \cdots & X_{2m_2}\\ \vdots & \vdots & \ddots & \vdots\\ X_{n1}, & X_{n2}, & \cdots & X_{nm_n}\end{matrix} \quad \text{and} \quad \begin{matrix}Y_{11}, & Y_{12}, & \cdots & Y_{1m_1}\\ Y_{21}, & Y_{22}, & \cdots & Y_{2m_2}\\ \vdots & \vdots & \ddots & \vdots\\ Y_{n1}, & Y_{n2}, & \cdots & Y_{nm_n}\end{matrix}\tag{8}$$

Note $\Delta X_{ij} = X_{ij} - X_{i,j-1}, \Delta Y_{ij} = Y_{ij} - Y_{i,j-1}, i = 1,2,\cdots,n, j = 1,2,\cdots m_i$ as the degradation increment of sample i during moment $t_{i,j-1}$ to $t_{i,j}$, thus the following data is derived:

$$\begin{matrix} \Delta X_{11}, & \Delta X_{12}, & \cdots & \Delta X_{1m_1} \\ \Delta X_{21}, & \Delta X_{22}, & \cdots & \Delta X_{2m_2} \\ \vdots & \vdots & \ddots & \vdots \\ \Delta X_{n1}, & \Delta X_{n2}, & \cdots & \Delta X_{nm_n} \end{matrix} \text{ and } \begin{matrix} \Delta Y_{11}, & \Delta Y_{12}, & \cdots & \Delta Y_{1m_1} \\ \Delta Y_{21}, & \Delta Y_{22}, & \cdots & \Delta Y_{2m_2} \\ \vdots & \vdots & \ddots & \vdots \\ \Delta Y_{n1}, & \Delta Y_{n2}, & \cdots & \Delta Y_{nm_n} \end{matrix} \tag{9}$$

3 Method for estimating parameter μ and σ^2

3.1 Maximum likelihood estimation

ΔX_{ij} , ΔY_{ij} are degradation increment of sample i from $t_{i,j-1}$ to $t_{i,j}$, note $\Delta t_{ij} = t_{ij} - t_{i,j-1}$, $(\Delta X_{ij}, \Delta Y_{ij})^T$ are independent and obey two-dimensional normal distribution $N(\mu\Delta t_{ij}, \Delta t_{ij}\Sigma)$.

Using the maximum likelihood estimation(MLE) method to get the estimated values of $\mu_1, \sigma_1^2, \mu_2, \sigma_2^2$:

$$\hat{\mu}_1 = \frac{\sum_{i=1}^{n} X_{im_i}}{\sum_{i=1}^{n} t_{im_i}}, \quad \hat{\sigma}_1^2 = \frac{1}{\sum_{i=1}^{n} m_i}\left[\sum_{i=1}^{n}\sum_{j=1}^{m_i}\frac{(\Delta X_{ij})^2}{\Delta t_{ij}} - \frac{\left(\sum_{i=1}^{n} X_{im_i}\right)^2}{\sum_{i=1}^{n} t_{im_i}}\right] \tag{10}$$

$$\hat{\mu}_2 = \frac{\sum_{i=1}^{n} Y_{im_i}}{\sum_{i=1}^{n} t_{im_i}}, \quad \hat{\sigma}_2^2 = \frac{1}{\sum_{i=1}^{n} m_i}\left[\sum_{i=1}^{n}\sum_{j=1}^{m_i}\frac{(\Delta Y_{ij})^2}{\Delta t_{ij}} - \frac{\left(\sum_{i=1}^{n} Y_{im_i}\right)^2}{\sum_{i=1}^{n} t_{im_i}}\right] \tag{11}$$

Estimation of the correlation coefficient:

$$\hat{\rho} = \frac{1}{M\hat{\sigma}_1\hat{\sigma}_2}\left[\sum_{i=1}^{n}\sum_{j=1}^{m_i}\frac{(\Delta X_{ij} - \hat{\mu}_1\Delta t_{ij})(\Delta Y_{ij} - \hat{\mu}_2\Delta t_{ij})}{\Delta t_{ij}}\right] \tag{12}$$

In this paper, in order to measure the two degradations at the same time, the equal interval measurement is adopted, which is so called stable measurement, i.e. $\Delta t_{ij} = \Delta t$, so $(\Delta X_{ij}, \Delta Y_{ij})^T$ are independent and identically distributed bivariate normal samples,

$$(\Delta X_{ij}, \Delta Y_{ij})^T \sim N(\mu\Delta t, \Sigma\Delta t) \tag{13}$$

the estimations of $\mu_1, \sigma_1^2, \mu_2, \sigma_2^2$ by the properties of the multivariate normal distribution is derived as follows:

$$\hat{\mu}_1=\frac{\sum_{i=1}^{n}X_{im_i}}{\sum_{i=1}^{n}t_{im_i}},\hat{\sigma}_1^2=\frac{1}{\sum_{i=1}^{n}m_i}\left[\frac{1}{\Delta t}\sum_{i=1}^{n}\sum_{j=1}^{m_i}\left(\Delta X_{ij}\right)^2-\frac{\left(\sum_{i=1}^{n}X_{im_i}\right)^2}{\sum_{i=1}^{n}t_{im_i}}\right] \tag{14}$$

$$\hat{\mu}_2=\frac{\sum_{i=1}^{n}Y_{im_i}}{\sum_{i=1}^{n}t_{im_i}},\hat{\sigma}_2^2=\frac{1}{\sum_{i=1}^{n}m_i}\left[\frac{1}{\Delta t}\sum_{i=1}^{n}\sum_{j=1}^{m_i}\left(\Delta Y_{ij}\right)^2-\frac{\left(\sum_{i=1}^{n}Y_{im_i}\right)^2}{\sum_{i=1}^{n}t_{im_i}}\right] \tag{15}$$

Estimation of the correlation coefficient:

$$\hat{\rho}=\frac{1}{\Delta t\hat{\sigma}_1\hat{\sigma}_2\sum_{i=1}^{n}m_i}\sum_{i=1}^{n}\sum_{j=1}^{m_i}\left(\Delta X_{ij}-\hat{\mu}_1\Delta t\right)\left(\Delta Y_{ij}-\hat{\mu}_2\Delta t\right) \tag{16}$$

3.2 Parameter estimation of two dimensional Wiener stochastic degradation model based on the wear model

3.2.1 Archard Wear Model

The Archard wear model is a classic wear calculation model by Professor J.F.Archard[20]. The model is verified by experiment and it has a good practicability. The model is as follows:

$$V=K\frac{F_nS}{H}\propto a+b\gamma \tag{17}$$

Where V and S are the wear volume and relative slip distance, respectively, F_n is the normal force, K is the wear coefficient, H is the Brinell hardness. Whereas a and b are unknown parameters, γ is a variable.

Archard[21]wear model is a linear model, it can be seen that wear volume is a linear function of time t, considering the randomness in the process of wear, the Archard wear model is combine with the Markov model to form a new wear model:

$$\Delta\propto\alpha_0+\alpha_1t+\varepsilon \tag{18}$$

Where Δ is the wear volume, α_0 and α_1 are determined coefficient, ε is a random diffusion, and $\varepsilon\sim N\left(0,\sigma^2\right)$, t is the characteristic lifetime.

We use T to represent the lifetime of the product, so its expectation is $\bar{T}=E\left(T\right)=\frac{l_1-\alpha_0-\varepsilon}{\alpha_1}$. By measuring the degradation of products at $T_1,T_2,\cdots T_n$, $X_1,X_2,\cdots X_n$ are obtained. So the point estimations of α_0,α_1 are:

$$\hat{\alpha}_0=\bar{X}-\frac{\bar{X}\sum\left(T_i-\bar{T}\right)\left(X_i-\bar{X}\right)}{\sum\left(T_i-\bar{T}\right)^2},\hat{\alpha}_1=\frac{\bar{X}\sum\left(T_i-\bar{T}\right)\left(X_i-\bar{X}\right)}{\sum\left(T_i-\bar{T}\right)^2} \tag{19}$$

3.2.2 Method of parameter estimation of two dimensional Wiener degradation model based on the wear model

In the Wiener process, the distribution of product lifetime T is inverse Gaussian distribution, and the characteristic function of T can be calculated because of the stability of Wiener process and the characteristic of independent increment, which is $E\left[\exp(-\theta T)\right]=\exp\left\{-l\left\{\sqrt{\mu_2+2\theta}-\mu\right\}\right\}$, and its expectation is: $E\left[T\right]=\, {}^{l}\!/_{\mu}$ .It can be seen that the expectation of lifetime T is only related to the drift parameter. When one of the two degradations exceeds its failure threshold, product fails. In this case, the working time is the lifetime of the product. By combining the two expectations, it follows:

$$E\left(T_1\right)=\frac{l_1}{\mu_1}=\bar{T}_1\approx\frac{l_1-\alpha_{11}-\varepsilon}{\alpha_{12}} \quad E\left(T_2\right)=\frac{l_2}{\mu_2}=\bar{T}_2\approx\frac{l_2-\alpha_{21}-\varepsilon}{\alpha_{22}} \tag{20}$$

Estimations of μ_1,μ_2

$$\hat{\mu}_2=\frac{l_2\alpha_{22}}{l_2-\alpha_{21}-\varepsilon},\ \hat{\mu}_2=\frac{l_2\alpha_{22}}{l_2-\alpha_{21}-\varepsilon} \tag{21}$$

Bring them into the formula (21), the reliability function $R\left(t\right)$ and distribution function $F\left(t\right)$ of the model proposed in this paper are obtained:

$$\begin{aligned}
R(t)=&\int_0^{\alpha}\int_0^{\infty}\sum_{n=1}^{\infty}\frac{2r}{\alpha\sigma_2^2(1-\rho^2)t}\sin(\frac{n\pi}{\alpha}\varphi_0)\\
&\cdot\exp\left[\frac{\sigma_2 l_1\beta_{12}\left(l_2-\beta_{21}-\varepsilon\right)-\sigma_1 l_2\beta_{22}\left(l_1-\beta_{11}-\varepsilon\right)\rho}{\sigma_1\sigma_2^2\left(1-\rho^2\right)\left(l_1-\beta_{11}-\varepsilon\right)\left(l_2-\beta_{21}-\varepsilon\right)}r\cos\varphi-\frac{l_2\beta_{22}}{\sigma_2^2\sqrt{1-\rho^2}\left(l_2-\beta_{21}-\varepsilon\right)}r\sin\varphi\right]\\
&\cdot\exp\left[\frac{(l_1\beta_{12}\left(l_2-\beta_{21}-\varepsilon\right)\rho\sigma_1\sigma_2-l_2\beta_{22}\left(l_1-\beta_{11}-\varepsilon\right)\sigma_1^2)x_{0,2}+(\beta_{22}\left(l_1-\beta_{11}-\varepsilon\right)\rho\sigma_1\sigma_2-{}_1\beta_{12}\left(l_2-\beta_{21}-\varepsilon\right)\sigma_2^2)x_{0,1}}{(1-\rho^2)\sigma_1^2\sigma_2^2\left(l_1-\beta_{11}-\varepsilon\right)\left(l_2-\beta_{21}-\varepsilon\right)}\right]\\
&\cdot\exp\left[-\frac{\sigma_1^2 l_2{}^2\beta_{22}{}^2\left(l_1-\beta_{11}-\varepsilon\right)^2-2\left(l_1-\beta_{11}-\varepsilon\right)\left(l_2-\beta_{21}-\varepsilon\right)\sigma_1\sigma_2\rho+\sigma_2^2 l_1{}^2\beta_{12}{}^2\left(l_2-\beta_{21}-\varepsilon\right)^2}{2(1-\rho^2)\sigma_1^2\sigma_2^2\left(l_1-\beta_{11}-\varepsilon\right)^2\left(l_2-\beta_{21}-\varepsilon\right)^2}t-\frac{r^2+r_0^2}{2(1-\rho^2)\sigma_2^2 t}\right]\\
&\cdot\sin\left(\frac{n\pi}{\alpha}\right)I_{\frac{n\pi}{\alpha}}\left(\frac{rr_0}{(1-\rho^2)\sigma_2^2 t}\right)drd\varphi
\end{aligned} \tag{22}$$

$$F\left(t\right)=1-R\left(t\right) \tag{23}$$

4 The example of reliability assessment of steam turbine pump

Steam turbine is the power device of ship, and steam turbine pump is the key part of the steam turbine as the output device of lubricating oil. Once a fault occurs, it will affect the whole ship power system. Now a 1000h reliability test data of the steam turbine pump is used as an example to verify the rationality and accuracy of the method proposed in this paper. The 1000h reliability test of steam turbine pump can be divided into two stages. Firstly, individual diagnostic test for 60 hours. Then in the remaining 940 hours, according to the test pattern for continuous operation test, it is a total of 47 cycles with 20h for each cycle. Each cycle is interspersed with automatic variable conditions, parameter fluctuations, exhaust safety valve and other test items.

Steam turbine pump belongs to a complex mechanical system with continuous, variable condition, and reciprocating motion. Wear, fatigue and fracture of the parts are the main reasons for the lifetime and failure of the steam turbine pump. Therefore, both domestic and foreign reseachers have proposed concepts of life evaluation of key parts. In order to master the wear of the key parts of the test process completely and accurately, reliability test program requires precise measurements of the thickness and size of the key parts before and after test. In this paper the retarder rotor support bearing clearance and the support bearing temperature are selected as key parameters to assess the steam turbine pump lifetime, because these data can reflect the working state of the steam turbine pump. Reliability and lifetime assessment of key parts is an important part of describing the reliability level and service life of the whole machine. At the same time, it also provides reference for the determination of maintenance cycle and maintenance level.

In accordance with the requirements of the test program, the steam turbine pump must be dismantled for inspection when total running time reaches 100, 200, 300, 400, 500, 600, 700, 800, 900 and 1000 hours. Table1 is the data of retarder rotor support bearing clearance and the support bearing temperature. According to the design requirements of the bearing, when the support bearing temperature is higher than 80℃, it is considered that the bearing fails, namely, the failure threshold of the support bearing temperature is 80℃. When the bearing clearance is greater than 0.3mm, it is considered failure, that is to say, the failure threshold of the bearing clearance is 0.3mm. And the reliability index of the steam turbine pump is that the reliability is 0.7 when the working time reach 10000h.

Table 1. Data of retarder rotor support bearing clearance and the support bearing temperature.

Time(h)	100	200	300	400	500	600	700	800
Support bearing temperature (℃)	62.81	63.64	61.67	63.45	61.41	59.89	63.08	63.77
Reducer rotor support bearing clearance (mm)	0.10	0.11	0.11	0.11	0.11	0.11	0.12	0.12

4.1 Reliability assessment of steam turbine pump

We use three methods to assess the reliability of steam turbine pump, respectively, one dimensional Wiener degradation model based on the maximum likelihood method, two dimensional Wiener degradation model based on the maximum likelihood method and the method we proposed where μ_1 and σ_1^2 are parameters to support bearing temperature, μ_2 and σ_2^2 are parameters to reducer rotor support bearing clearance and ρ is the correlation coefficient between the two.

4.1.1 One dimensional Wiener degradation model based on MLE

The parameters of the support bearing temperature and the reducer rotor support bearing clearance are estimated by MLE, results are shown in Table 2

Table 2. Parameters estimator of one dimensional Wiener degradation model based on MLE.

Parameters	μ_1	σ_1^2	μ_2	σ_2^2
Estimator	0.0070	0.0091	2.714×10^{-5}	9.928×10^{-8}

Bringing estimated results of parameter into one dimensional Wiener degradation model, the reliability of support bearing temperature is 0.8370 in 10000h, and the reliability of reducer rotor support bearing clearance is 0.8032 in 10000h. Because when one of the two parameters reaches its failure threshold, the steam turbine pump fails. So choose the result which reach its failure threshold first as reliability assessment result of steam turbine pump. In this example, the assessment of reducer rotor support bearing clearance is taken as result of steam turbine pump, namely, the reliability assessment of steam turbine pump is 0.8032 when the working time reach 10000h. The reliability curve is shown in Figure 1.

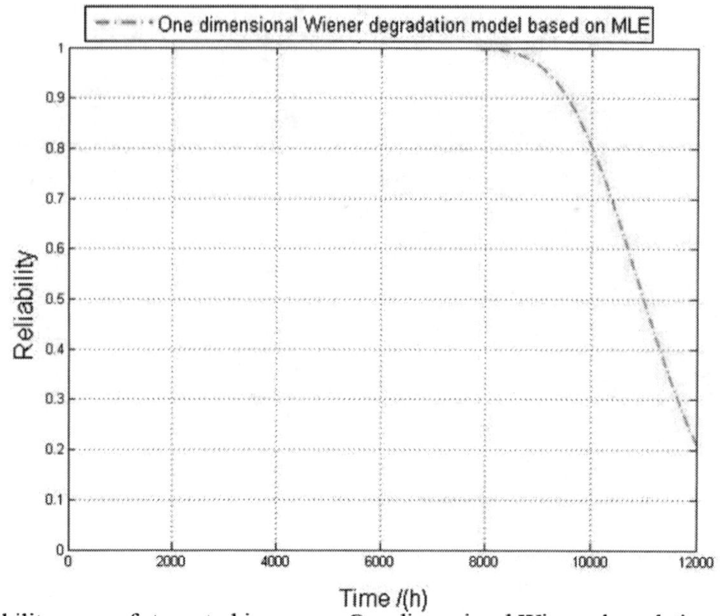

Figure 1. The reliability curve of steam turbine pump- One dimensional Wiener degradation model based on MLE.

4.1.2 Two dimensional Wiener degradation model based on MLE

Table 3. Parameters estimator of two dimensional Wiener degradation model based on MLE.

Parameter	μ_1	σ_1^2	μ_2	σ_2^2	ρ
Estimator	0.0797	0.6043	1.500×10^{-4}	2.000×10^{-6}	0.7997

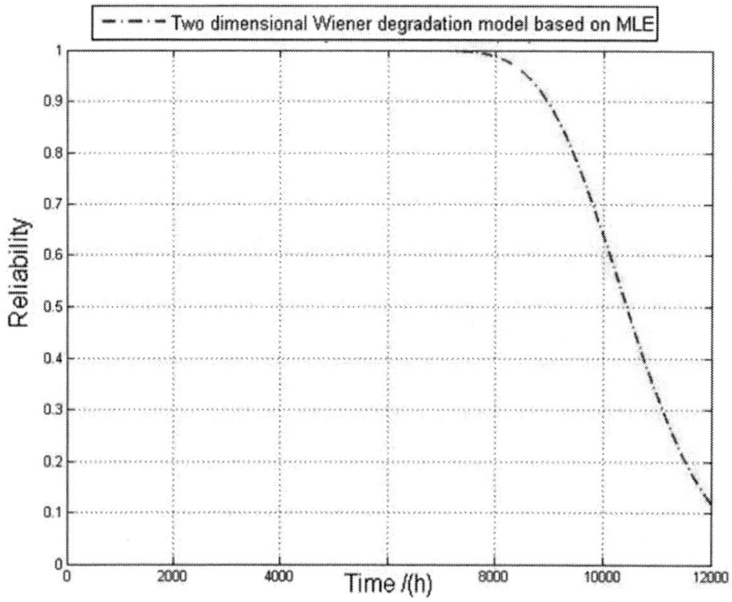

Figure 2. The reliability curve of steam turbine pump- Two dimensional Wiener degradation model based on MLE.

Stable measurement is used to measure two parameters. Considering the correlation between these two parameters, the results of μ_1 σ_1^2 μ_2 σ_2^2 and ρ are derived by bringing the data of two parameters into formula (10)- (12). Results are shown in Table 3

Bringing the parameter estimator into formula (5), the reliability of steam turbine pump is obtained, it is 0.6376 at 10000h. The reliability curve is shown in Figure 2.

4.1.3 Two dimensional Wiener degradation model based on wear model

Bringing the data of two parameters into formula (12) (21), the results of μ_1 σ_1^2 μ_2 σ_2^2 and ρ are obtained. Results are shown in Table 4

Table 4. Parameters estimator of two dimensional Wiener degradation model based on wear model.

Parameter	μ_1	σ_1^2	μ_2	σ_2^2	ρ
Estimator	0.0823	0.6043	1.610×10^{-4}	2.000×10^{-6}	0.8956

Bringing the parameter estimator into formula (22), the reliability of steam turbine pump at 10000h is obtained, which is 0.6920. The reliability curve is shown in Figure 3.

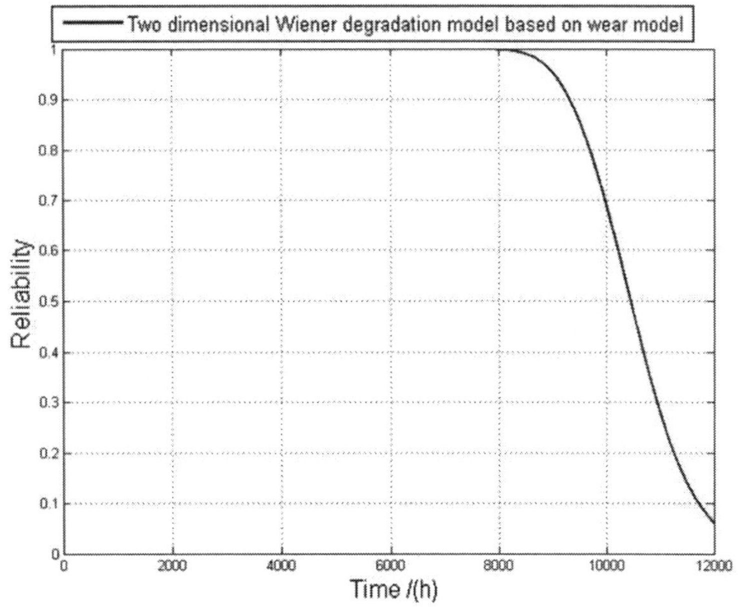

Figure 3. The reliability curve of steam turbine pump- Two dimensional Wiener degradation model based on wear model.

4.2 Comparison and analysis

Taking parameters estimator of three methods into their reliability functions respectively, the reliabilities at 10000h, are obtained, then calculate the relative error based on index. As shown in the Table 5, the reliability curves of the three methods are shown in the Figure 4.

Table 5. Relative errors of the three method.

Method	R	Relative Error
One dimensional Wiener degradation model based on MLE	0.8032	14.74%
Two dimensional Wiener degradation model based on MLE	0.6376	8.91%
Two dimensional Wiener degradation model based on wear model	0.6920	1.14%

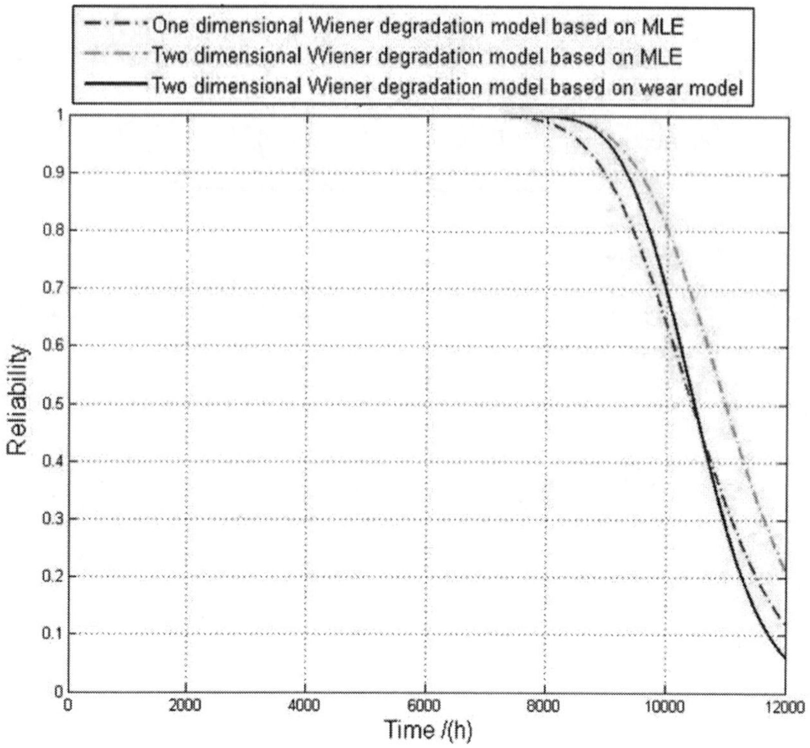

Figure 4. The reliability curves of the three methods.

Comparing and analyzing the reliability calculation results and reliability curves of the three methods, it shows: the relative error of one dimensional Wiener degradation model based on MLE is the maximum which is 14.74%, followed by two dimensional Wiener degradation model based on MLE which is 8.91%. The minimum is the method proposed in this paper, which is 1.14%. It can be seen that the method proposed in this paper is the closest to the real index. From the Fig.4, the reliability falling speed becomes faster over time. As time goes by, support bearing temperature showed a fluctuating upward trend, which will increase the wear of metal components. Then reliability will decrease. Therefore, the method proposed in this paper is more in line with the actual situation.

5 Conclusion

(1) In practical engineering, more than one parameter affects the life of the product. When assess the reliability of the product, only considering one parameter to estimate the whole product is not comprehensive. The evaluation results of the multivariate degradation process are more in line with the actual situation than one dimensional degradation process.

(2) In this paper, from the point of view of two performance parameters, combined with the theory of random degradation, a model of two dimensional Wiener stochastic degradation model is proposed based on the wear model for the products which have a wear process and uncertain stochastic degradation processes. The two dimensional Wiener degradation model is optimized by the method proposed in this paper in a certain extent.

(3) From the result of steam turbine pump reliability assessment, compared to the results of one dimensional Wiener degradation model based on MLE and two dimensional Wiener degradation model based on MLE, the method proposed in this paper has the minimum relative error which is 1.14% on the reliability 0.7 at 1000h. Thus the method proposed in this paper is more in line with the actual situation and more reliable.

References

1. J. Liu, J. Xie, H. Zuo, et al., Acta Aeronaut, Astronaut, Sin, **36**(2), 564-574 (2015)
2. S. Mishra and O.A. Vanli, J. Nondestruct. Eval., **35**(1), 1-13 (2016)
3. W. Kahle and A. Lehmann, *The Wiener Process as a Degradation Model: Modeling and Parameter Estimation* (Birkhäuser Boston, 2010)
4. F. Guérin, M. Barreau, A. Demri, et al., IFAC, **43**(3), 74-79 (2010)
5. P. Sander and R. Badoux, *Bayesian methods in reliability* (Kluwer Academic Publishers, 1991)
6. J. Tang, T.S. Su, NavRes Log, **55**(3), 265 (2008)
7. X. Wang, J. Multivariate Anal, **101**(2), 340 (2010)
8. B.H. Peng, J.L. Zhou and Z.Q. Pan, Syst. Eng. Theory Pract, **30**(3), 543 (2010)
9. R.P. Nicolai, R. Dekker and J.M. Noorwijk, Reliab Eng Syst Safe, **92**(12), 1635-1640 (2007)
10. J. Balka, A.F. Desmond and P.D. McNicholas, Lifetime Data Anal, **15**(12), 147-176 (2009)
11. X.S. Si, W. Wang, C.H. Hu, et al., Ieee T Reliab, **61**(1), 50-67 (2012)
12. F. Antonio Dorini and R. Sampaio, J. Appl Mech, **79**(5), 051008-051014 (2012)
13. Silva CRÁD, Pintaude G. Tribol Int, **41**(6), 473-481(2008)
14. H.J. Hu and W.J. Huang, Int J Adv Manuf Tech, **69**(1-4), 31-39 (2013)
15. G.Y. Lin, D. Feng, X.Y. Zheng, et al., J. Cent South Univ, **40**(5), 1245-1251 (2009)
16. Z.S. Gao, X.Z. Deng, F.X. Chen, et al., China Mech. Eng., **25**(2), 26-229 (2014)
17. D.H. Kim, H.C. Lee, B.M. Kim, et al., J. Mater Process Tech, **166**(3), 372-380 (2005)
18. G. Jin, *Reliability Technology based on Degradation* (Beijing, China, 2014)
19. M. Domine and V. Pieper, Probab Eng Inform Sc, **7**(4), 545-555 (1993)
20. J.F. Archard, J. Appl Phys, **24**(8), 981-988 (1953)
21. X. Shen, L. Cao and R. Li, MACE, 325-329 (2010)

MATEC Web of Conferences **169**, 01038 (2018) https://doi.org/10.1051/matecconf/201816901038
IMETI 2017

Steering test of mid-size bus with flexible-body dynamics model

Jia-Shiun Chen[a] and Hsiu-Ying Hwang

Department of Vehicle Engineering, National Taipei University of Technology, Taiwan

Abstract. The object of this paper is to analyze the vehicle structural vibration responses during the step steering simulation test. With computer simulations, confirming the influence from the difference between configurations of components on mid-size electrical bus and tradition one could be possible. The difference of the vibration responses between flexible multi-body model and rigid one would be discussed. Since the wheel base of mid-size bus is longer than passenger car, the deformation of vehicle body more obviously influences the output of vibration response. The operating condition in practical driving can be expressed by flexible model, and the performance prediction of vehicle can be closer to the real vehicle behavior. The difference between the flexible body simulation and rigid body simulation is 1.17%, and the flexible body model is more practical.

1 Introduction

The design requirements of vehicle handling and comfort are considerably important in the field of vehicle design. The major reason is the safety of driver and passengers, and followed by the driving quality of the vehicle. In order to reduce development time and cost, the prediction of the dynamic characteristics of the designed vehicle is required, so vehicle mathematical model which supports the scientific theory to evaluate the traffic safety and quality of vehicles should be established. With the advance of modern computer, computer-aided engineering (CAE) is applied to the vehicle mathematical model.

In 2014, the worldwide sale of global electric and hybrid buses is accounted for 5.6% of total buses sales, and it is expected to reach 16.9% in 2020 AD. In electric bus, the primary power source is replaced by an electric motor, and the motorizing of powertrain system makes body structure changes along with the increased load on the vehicle battery. The battery weight would change the weight distribution of the vehicle, and cause the impact on handling and comfort during traveling. With CAE's technology, the vehicle dynamic simulation with the overall bus structure could more precisely predict the vehicle behavior during the different maneuver, and improve its handling and comfort.

This research investigated the electric bus stability with different steering inputs, and predicted the dynamic characteristics of the electric bus. The rigid body dynamic simulation model vehicle was established, and then the bus structure was replaced with the flexible body from structure analysis. Both simulation models would be analyzed. This study could present that the application of flexible

[a] Corresponding author : chenjs@mail.ntut.edu.tw

multi-body dynamics simulation based on actual body deformation could affect the behavior of vehicle handling simulation, and it could provide more precisely prediction.

2 Bus model

The bus CAD model was established with Hyperworks simulation software [1-3]. The model file was read into MotionView along with the sets of suspension systems, tires, steering group and connected components to complete the rigid motion model of electric bus body model, as shown in Figure 1. Later, the rigid bodies in the model were substituted by the flexible body of vehicle structure to build the flexible body dynamic model. Detailed modelling process was shown in Figure 2.

Figure 1. Vehicle model diagram.

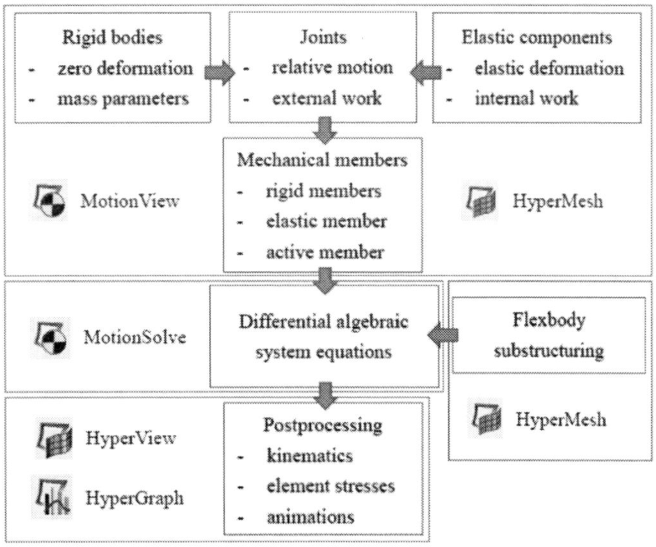

Figure 2. Flowchart of flexible multibody dynamic analysis.

The vehicle model was divided into two major parts: the body structure and chassis structure. The body structure comprised bus body, battery, seat, beams. The chassis structure comprised steering systems, suspension systems, tires, transmission. The body structure was fixed on the top of the chassis structure. The flexible body would replace the rigid body, and it was marked within the red frame. The vehicle structure diagram was shown in Figure 3.

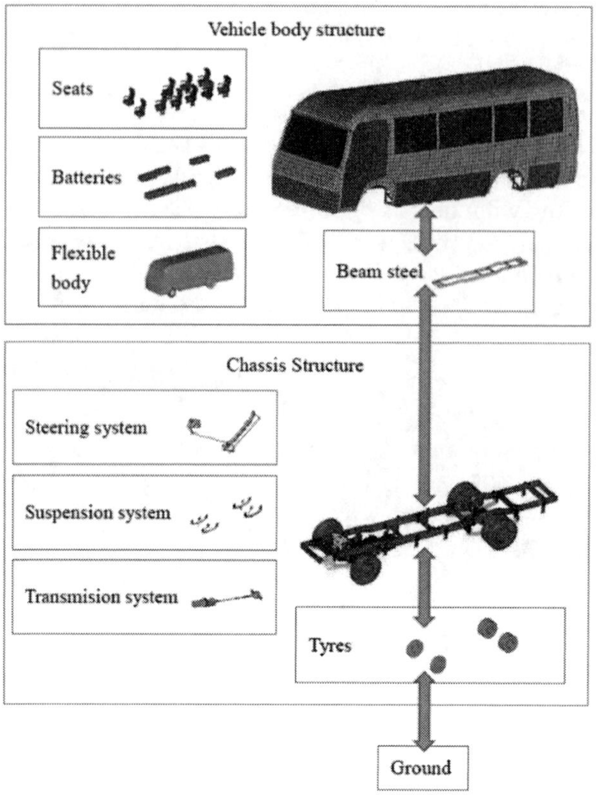

Figure 3. Vehicle structure diagram.

2.1 Vehicle flexible body structure

In this study, flexible body structures would replace the rigid body in vehicle dynamic model. The flexible bus body and flexible beams were shown in Figure 4 and Figure 5, respectively. The major selected modes of flexible body were torsional modes, and longitudinal and lateral bending modes. The first three mode shapes were shown in Figure 6. Since large vibration transmitted to the vehicle body is generally below 50 Hz [4]. The selected modal frequencies were up to 70 Hz, as shown in Table 1.

Figure 4. Bus flexible body.

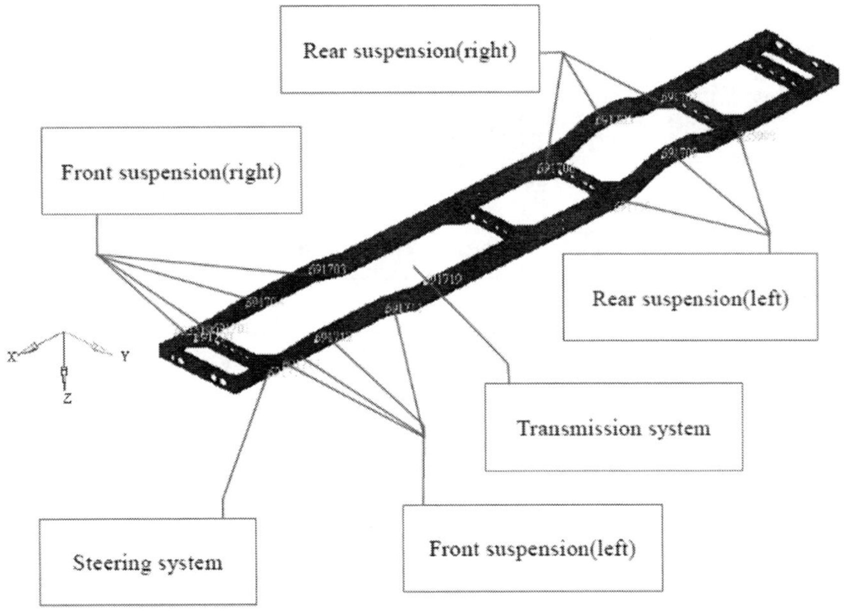

Figure 5. Beam flexible body and binding locations.

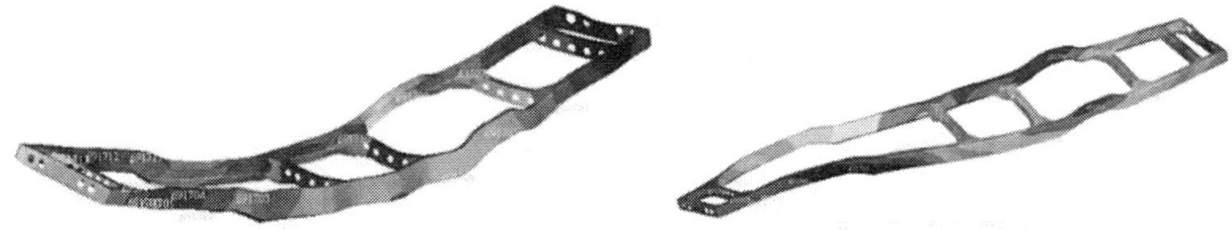

First bend mode (23.84 Hz) First torsion mode (27.84 Hz)

Second Bending mode (35.86 Hz)

Figure 6. First three mode shapes of beam.

2.2 Steering system

The vehicle steering system included a steering gearbox to turn the wheel spindles. All parts were considered as a rigid body. The system was shown in Figure 7.

2.3 Suspension system

The vehicle suspension system of front and rear wheels were semi-elliptical leaf springs. Each suspension was equipped with four plates and a damper. A 10mm thickness of steel plate was on the top, and the other three were 8mm thick plates. The length of front suspension plates was 1247mm, and the center of plates was connected to the fixed contact of the front axle. The length of rear suspension plates was 1366mm, and the center fixed contact was connected to the differential. The gap between each piece was 5mm. Four damper tubes were installed in the vertical direction. The stability bar was attached to the bottom two front damper tubes, as shown in Figure 8.

2.4 Driveline system

The driveline system included a 75 kW permanent magnet synchronous motor, clutch, six-speed manual gearbox, drive shaft, and differential [5], as shown in Figure 9.

Table 1. Flexible body modes.

Flexible body modal frequency		
Order	Mode Frequency(hz)	Mode Type
7	18.258284	Longitudinal torsion(back fixed)
8	22.830112	Vertical bending(back fixed)
9	23.845562	1st lateral bending
10	27.841775	1st longitudinal torsion
11	30.379962	Lateral bending and longitudinal torsion
12	30.925637	Lateral bending and longitudinal torsion (2 fixed)
13	31.372244	Lateral bending and longitudinal torsion (middle fixed)
14	33.339106	Lateral bending(back fixed)
15	33.643737	Longitudinal bending(back fixed)
16	35.645866	Longitudinal bending and vertical bending(back fixed)
17	35.865369	1st Vertical bending
18	37.269394	Vertical bending and lateral bending
19	37.590479	2nd lateral bending
20	39.564707	Lateral bending(2 fixed)
21	40.209258	3rd lateral bending
22	41.208339	Vertical and torsion(front right wheel)
23	41.733302	Vertical bending(front end)
24	42.706628	2nd longitudinal torsion
25	43.900234	Longitudinal torsion (steering system)

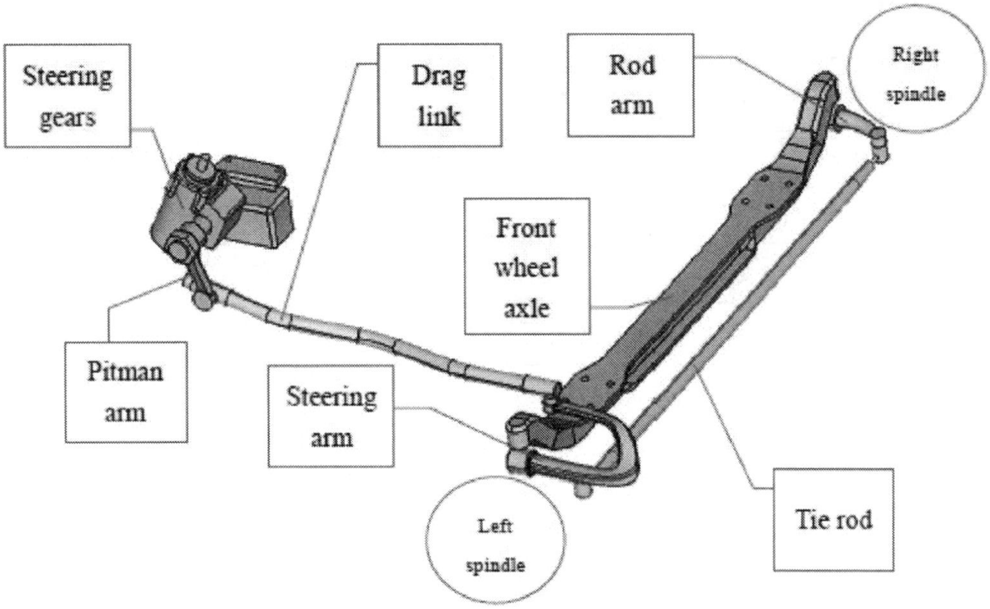

Figure 7. Steering system configuration diagram.

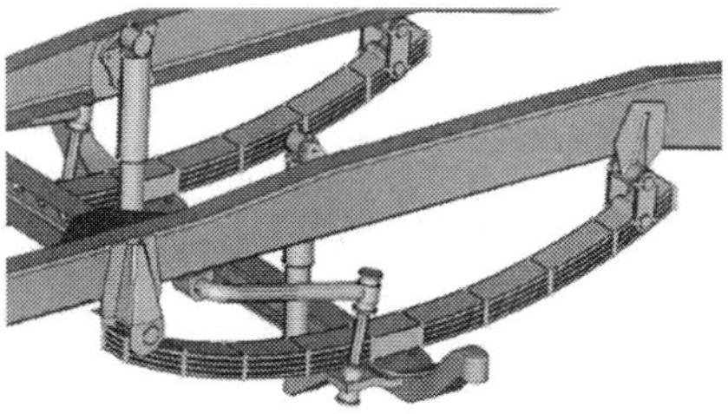

Figure 8. Suspension system configuration diagram.

Figure 9. Suspension system configuration diagram.

3 Simulation and analysis

3.1 Simulation conditions

This research referenced the ISO7401 regulation of the road vehicles transient open-loop response test method which was suitable for passenger cars and light trucks. The vehicle speed must reach 100 kilometers per hour, and the speed difference should not exceed 2 kilometers per hour. The yaw rate should be in steady state with the difference less than 0.5 degrees per second. The steering signal should increase from 10% to 90% within 0.15 seconds, and the throttle position should not change during the steering test. Tests were divided into the left turn and right turn, and the steering wheel angle was determined by the steady-state tests of constant radius circle [5].

The total simulation time was 40 seconds, and steering wheel angle rotated to 25.91 degrees from 10 and to 10.15 seconds, as shown in Figure 10. The left lateral acceleration would reach 400 meters per second after turning the steering. Simulation parameters are shown in Table 2.

Table 2. Steering turn parameters.

Parameters of Simulation	Value
Velocity v_x	100km/hr
Steady state lateral acceleration a_Y	4m/s^2
Correction steering wheel angle δ_{cor}	0.42°
Steady state steering wheel angle $\delta_{h@ss}$	25.919°(L), 27.023°(R)

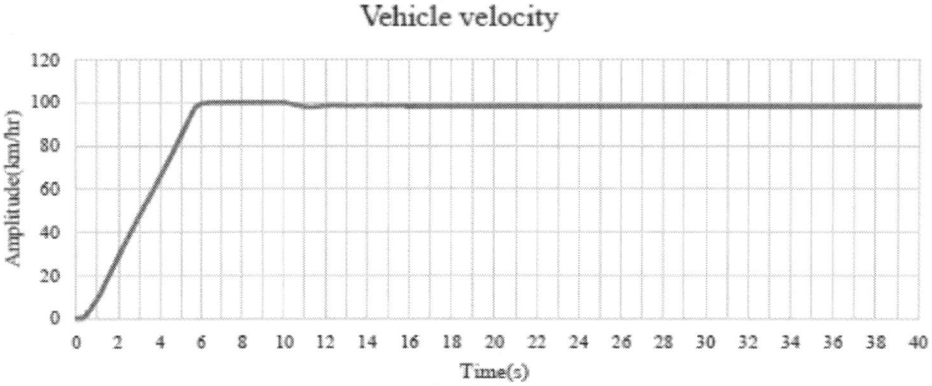

Bus simulation speed

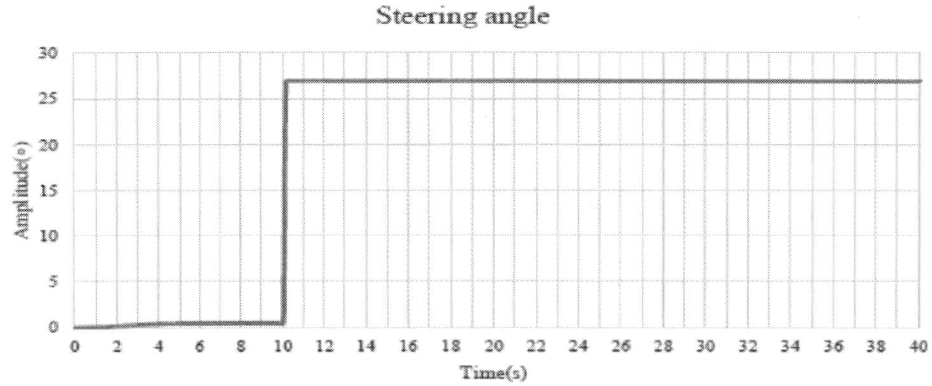

Steering simulation angle

Figure 10. Vehicle speed and steering angle.

3.2 Simulation results and analysis

In this research, the steering step input simulations were analyzed. In addition to the comparison between left and right turns, the comparison between the rigid body and flexible body simulations was also included.

During the right turn, the lateral acceleration of rigid and flexible body simulation was shown in Figure 11. The flexible body simulation data reached the steady state later than the rigid body simulation, which represented the flexible body had a longer response time. The deformation of vehicle body structure caused vibration during the steering turn, so the time to reach the steady-state was delayed. Yaw rate response was shown in Figure 12. The flexible body simulation also showed a longer response time. Over time response of roll angle was shown in Figure 13. Since the flexible of bus structure, the flexible body simulation had longer response time and higher peak value.

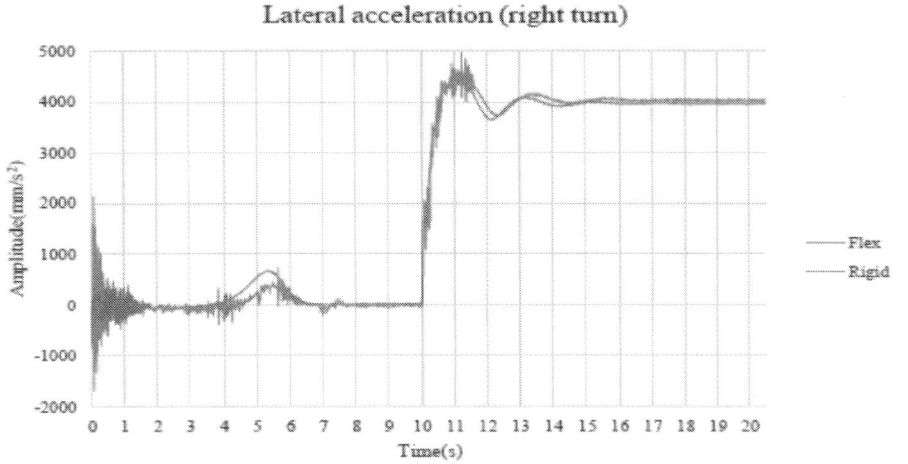

Figure 11. Lateral acceleration during right turn.

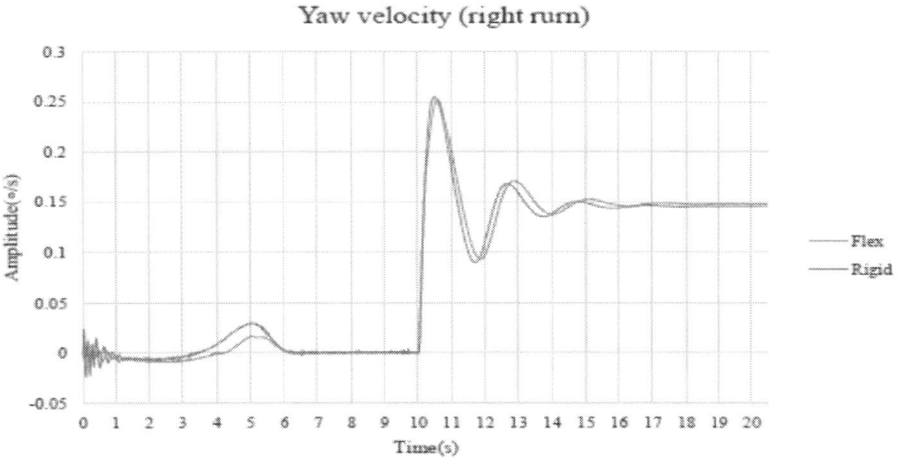

Figure 12. Yaw velocity during right turn.

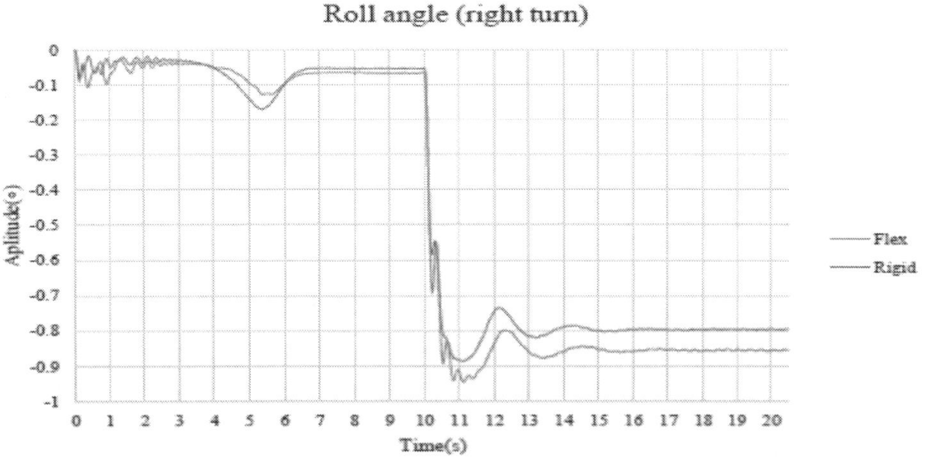

Figure 13. Roll angle during right turn.

For left turn simulation, the lateral acceleration, yaw velocity and roll angle response were presented in Figures 14, 15 and 16, respectively. Since the flexible body dynamic simulation allowed deformations of vehicle body, the response time to reach steady-state of flexible body simulation was longer than that of the rigid body dynamic simulation. The peak values of flexible body simulation were also higher than that of the rigid body simulation.

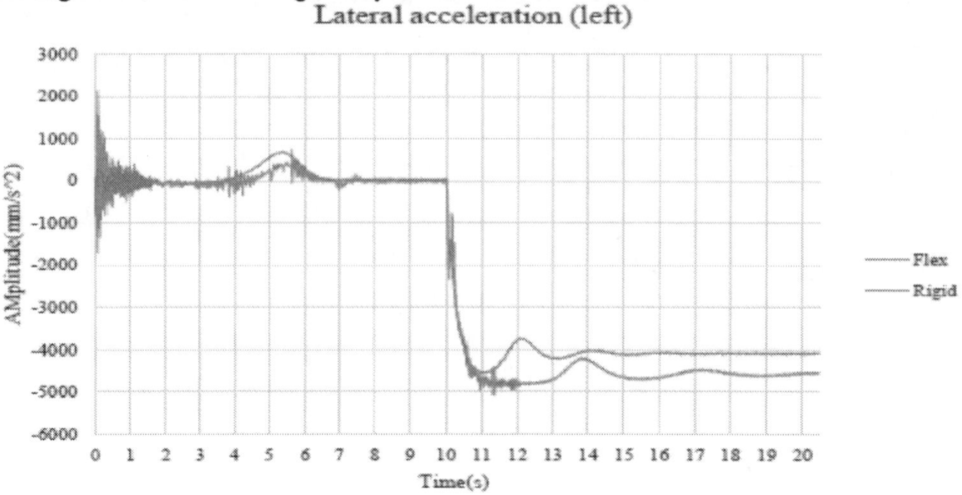

Figure 14. Lateral acceleration during left turn.

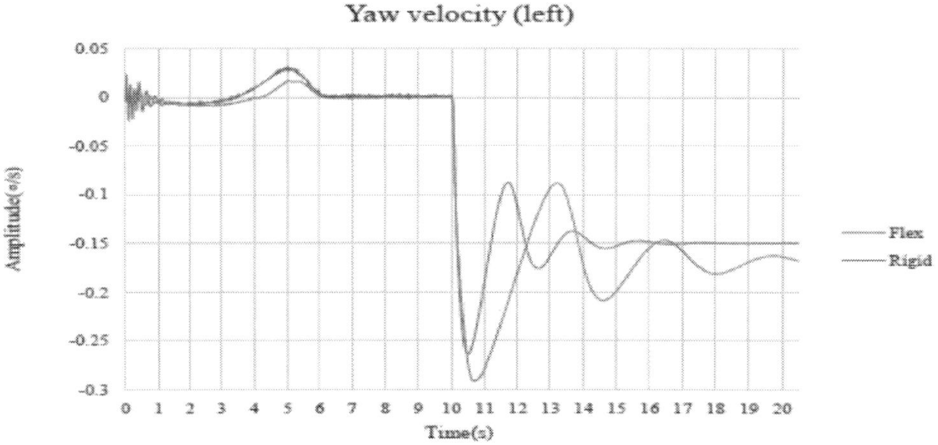

Figure 15. Yaw velocity during left turn.

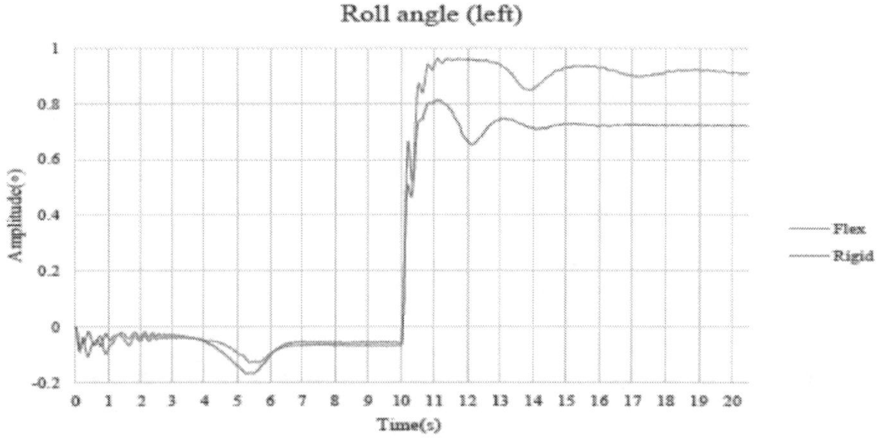

Figure 16. Roll angle during left turn.

The response time of flexible body simulation was longer, because the bus body was deformable which extended the vibration time. With the batteries and seats, the vehicle center of gravity shifted to the left, which caused the higher values of lateral acceleration, yaw velocity and roll angle during the left turn simulation. Table 3 presented the summary difference table of step steering input response between the flexible body and rigid body simulations.

Table 3. Summary table of step steering input response.

Percentage of data difference of Flexible body from rigid body	
Parameter	Percentage of data difference
Steady-state yaw velocity response gain	+7.14%
Lateral acceleration response time	+32.34%
Yaw velocity response time	+8.47%
Roll angle response time	+5.51%
Lateral acceleration peak response time	+15.51%
Yaw velocity peak response time	+33.06%
Roll angle peak response time	+13.43%
Overshoot value of lateral acceleration	-12.13%
Overshoot value of yaw velocity	-6.02%
Overshoot value of roll angle	-35.32%

4 Conclusions

This research presented dynamics simulations of electric bus, and replaced the rigid bodies with flexible bodies in multi-body dynamic model. According to the steering test regulations, dynamics simulations were performed and the results were presented. The conclusions were presented as below.

1. The elastic deformation of the body caused the vehicle vibration response times lengthen. Comparing with the rigid body model which did not have body flexibility, the flexible body model which represented the elastic deformation during steering turning simulation was closer to the actual performance of real vehicle.

2. The flexible vehicle parts could exhibit more detail motion and vibration, and this vibration would pass through vehicle structure.

3. The geometry of vehicle was not symmetric, which caused the higher normal load of right rear tire during the left turn simulation. This higher tire load would act on the vehicle body and caused higher body deformation. This was the reason that left turn had higher yaw velocity and roll angle. This behavior would not be predicted with the rigid body model. So, the flexible body simulation could provide more realistic data.

4. From the tire load data, the vehicle body's longitudinal torque was different between the left turn and the right turn. This would affect the steering performance. With the longer vehicle body, such as bus and truck, the flexible body model provided more accurate results of the simulation.

References

1. J.M.P. Dias and M.S. Pereira, Multibody System Dynamics, **1**(3), 303-322 (1997)
2. J.S. Chen and H.Y. Hwang, Advances in Mechanical Engineering, **5** (2013)

3. J.S. Chen, Advances in Mechanical Engineering, **7**(3) (2015)
4. J.A.C. Ambrósio and J.P.C. Gonçalves, Multibody System Dynamics, **6**(2), 163-182 (2001)
5. ISO 7401 2011, Road vehicles—Lateral transient response test methods—Open-loop test methods
6. Z. Lozia and W. Pieniazek, SAE Technical Paper, No. 2002-01-1184 (2002)
7. S. Chen, D. Wang, A. Zuo, Z. Chen, Y. Sun and J. Zan, *Study on handling and stability of a passenger car based on rigid-flexible coupling model*, 2010 International Conference on Computer, Mechatronics, Control and Electronic Engineering, Changchun, China, 615-620 (2010)
8. B. Göttlicher and K. Schweizerhof, Computers & Structures, **83**(25-26), 2035-2051 (2005)
9. S. Azadi, M. Vaziri and M. Hoseini, Vehicle System Dynamics, **48**(5), 587-617 (2010)

An investigation on the cutting force of milling Inconel 718

Jhy-Cherng Tsai[1,a], Chung-Yu Kuo[1], Zing-Ping Liu[1] and Kelvin Hsi-Hung Hsiao[2]

[1]*Department of Mechanical Engineering, National Chung-Hsing University, Taiwan*
[2]*Intelligent Machinery Center, Industrial Technology Research Institute, Taiwan*

Abstract. Inconel alloy has been widely used in industry but is difficult to machine due to its rapid change in cutting force during machining. This paper investigated the cutting force for milling Inconel 718 as conventional force model is unable to handle the above situation. Theorectical force model is first reviewed and two-phase experiments of slot milling based on dry cutting are designed to measure the cutting force and the specific cutting force. Experiments in phase I are designed based on Taguchi method with spindle speed, feedrate per cutting edge and depth of cut as experimental parameters. The results showed that the first two parameters play more important roles in the cutting force. A phase II exhaust experiments is designed with spindle speed set from 400 to 800 rpm while the feedrate per cutting edge is set from 0.04 to 0.08 mm/tooth. The results are concluded as the following. (i) There exists a strong size effect in milling Inconel 718 as the cutting force changed with the chip thickness. Specific cutting force is larger at small thickness of cut and become smaller when the thickness increases. (ii) A 2nd order non-linear cutting force model, which takes spindle speed N and feedrate f_z into account, for milling Inconel 718 is derived from the measured data and represented as $F_t(N, f_z) = (13910 - 3.1N - 109900f_z - 0.0028N^2 + 23.9Nf_z + 434500f_z^2) \times a_p \times h$. The derived force model compensates the inaccuracy of conventional force model.

1 Introduction

Nicle-based alloy has been widely emplyed in the aerospace and marine industries because of its good physical and chemical properties, including high-temperature strength, thermal stability, corrotion resistance and high endurance strengthy. It is commonly used in jet engine and rotor due to these excellent properties. Such Nicle-based alloys are thus called as super alloys and Inconel is the commonly used one among these alloys. Inconel is a typical anti-correotion material and contains 50% to 55% Nickle(Ni) with 17% to 21% Chromium(Cr) and a few other elements such as Cobalt(Co) and Molybdenum(Mo) for different applications. It maintains high strength at high temperature with good anti-creep and endurance properties. It is therefore difficult to machine due to these properties.

While aerospace components require good precision, machining is the most commonly and widely employed process in fabricating these components. The machinability of Inconel, on the other hand, is poor due to these good performance[1, 2]. Such difficult-to-machine material is hard to machine due to strain hardening and often results in high tool wear rate and built-up-edge thus downgrades surface quality. It is an crucial issue how to solve these problems for aerospace and machining industries. This paper is aimmed to investigate the cutting force of machining Inconel super alloy as cutting force

[a] Corresponding author : jctsai@mail.nchu.edu.tw

reflects the material properties as well as cutting conditions that are highly related to product quality and tool management. As machining of Inconel is often a know-how of a company, very few literatures investigated the detail of cutting force. We start with analysis of cutting force of a milling proccess and machining parameters that plays important role in straining hardening effect. Experiments are then designed, based on the Taguchi method, and conducted to collect cutting forces. A model of the cutting force, as a function of machining parameters, is then derived.

2 Analysis of milling force

Milling process is a conventional material removal process. Figure 1(a) shows the milling process that the workpiece moves across the rotating spindle and material is removed by the cutter attached to the spindle. The tangetail cutting forces $\mathbf{F_t}$, or the effective cutting force in material removal as illustrated in Figure 1(b), of a cutting edge in the process is the specific cutting force, k_s, multiplied by the material removal. The material removed is the depth of cut a_p (in axial dirction) multiplies the instant thickness of cut h (in radial direction). As a result, $\mathbf{F_t}$ can be derived as shown in eqution 1 where $\mathbf{n_t}$ represents the unit vector in tangentail direction. The instant thickness of cut h can be further represented as a function of the roating displacement φ and the feed per cutting edge f_z as shown in equation 2 [3, 4].

$$\mathbf{F_t}=k_s \times a_p \times h\ \mathbf{n_t} \quad (1)$$

$$h = f_z \times \sin\varphi \quad (2)$$

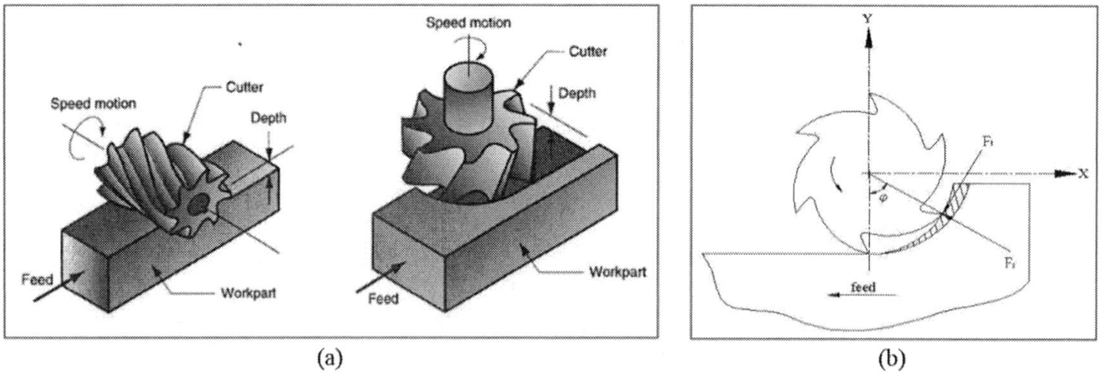

Figure 1. Illustration of (a) milling process and (b) tangential cutting forces in an up milling process [3, 4].

The radial cutting force F_r is usually presented as a ratio k_r to the tangentail cutting force F_t. Tlusty & MacNeil [5] showed that, based on their experimental data, k_r is around 0.3 for most metals. For a milling process with multiple cutting edges engaged, the tangetial and radial cutting forces are the vector sum of tangetial and radial cutting forces of each individual cutting edge. However, Inconel alloy is difficult to maching mainly because of its high strain hardening and high strength at high temperature [6]. The above two equations do not well match experimental data thus a modified model is required.

Rahman et al[1] noticed that tool life decreases when the cutting speed increases in his turning Inconel 718 experiment. He proposed this is caused by the friction heat between the chip and the tool as heat geneartion increases as cutting speed increases. Li[7], in milling Inconel 718, also noticed that flank wear is the major factor reducing tool life and tool wears faster in up-milling. Nalbant et al[8], when turning Inconel 718, found that cutting force increases as cutting speed increases though speed is not an explicit factor affecting cutting force in the theorectical model shown in equation 1. Liao et al[9], in their investigation on the slot milling and side milling, found that high cutting temperature and difficulty in chip disposal are two main problems encountered in high-speed end milling of Inconel 718. Thakur et al [10] found that tool life, in turning Inconel 718, was affected by the cutting speed, feed rate and tool type while depth of cut, in addition to the first two factors, also affects material hardening. A similar result is also reported in D'Addona et al [11] that tool wear increased rapidly at high cutting speed was observed though surface finish does not change much.

Some other researchers, such as Choudhury & El-Baradie [2], Sharman et al [12] and Devillez et al [13] noticed that tool life is highly reduced in machining Inconel alloy if the cutting parameters are not well selected. They used cutting tools with various coating films in machining Inconel tests and found that tools with coating films perfomes better then regular tools when the depth of cut is higher then 1.0mm. Among many cutting parameters, depth of cut and feed rate, rather than cutting speed, are two major factors affecting tool wear. Furthermore, tools coated with TiAlN performs the best among many other coating films [12].

As most literatures emphasized investigating tool wear and tool life, cutting force is the more fundamental index needed to be investigated as it reflects not only tool wear but also machining and material conditions. In particular, a suitable cutting force model based on equations 1 and 2 is required to serve this purpose.

3 Phase I experimental design and results

As the purpose is to find a modified cutting force model considering different cutting parameters, we started with the root cause analysis of cutting force for Inconel 718. In this investigation, we conducted two phases of experiments. In the first phase, we wanted to find important parameters that affecting cutting force as well as their ranges. A list of experiments are then designed to further investigate the effect of major parameters.

In the first phase, we began with slot cutting experimental design by Taiguchi method [14] in order to reduce the number of experiments. While cutting force is strongly related to the material removal rate, the depth of cut a_p, feed rate per cutting edge f_z and spindle speed N are three factors used in designing the experiments. Three levels of each parameter are designed based on machining parameters provided by the cutting tool maker. A list of experiments are designed using the Taguchi $L_9(3^3)$ orthogonal array as shown in Table 1 where the corresponding feed rate and cutting speed are calculated from the cutting parameters with a ϕ16mm end mill.

Table 1. Experimental design by the Taguchi $L_9(3^3)$ orthogonal array in phase I.

No.	spindle speed N (rpm)	feedrate per cutting edge f_z (mm/tooth)	depth of cut a_p (mm)	corresponding feed rate (mm/min)
1	400	0.02	0.1	16
2	400	0.04	0.3	32
3	400	0.06	0.5	48
4	500	0.04	0.1	40
5	500	0.06	0.3	60
6	500	0.02	0.5	20
7	600	0.06	0.1	72
8	600	0.02	0.3	24
9	600	0.04	0.5	48

Experiments are conducted in a CNC machining center with rated power 21KW and maximum spindle speed 14,000rpm. The workpiece material used in the experiment is Inconel 718 and the cutting tool is a ϕ16mm flat end mill with Kennametal Mill4 inserts coated with TiAlN. Cutting force is measured by a Kistler 9124B dynamometer with measuring capacities ±20 KN in radial direction and ±30 KN in axial direction. The natural frequency of the dynamometer is 1 kHz and the sensitivities are 0.5 mV/N and 0.33 mV/N in radial and axial directions. Figure 2 shows the machine tool and the dynamometer used in the experiment. Figure 3 shows experiment setup and detailed dynamometer setup on the spindle.

Figure 2. CNC Machining center (left) and dynamometer (right) used in the experiment.

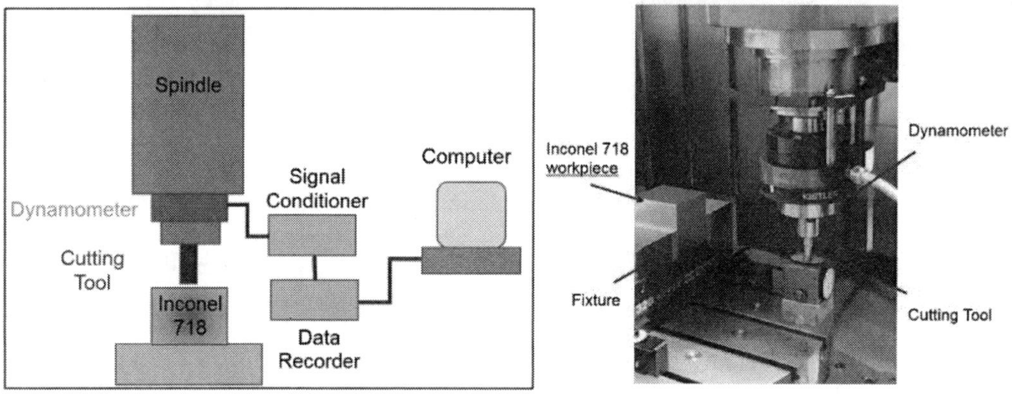

Figure 3. Illustration of experiment setup (left) and setup of dynamometer on the spindle (right).

The measured cutting force signal, as shown in the left in figure 4, is filtered first. The filtered signal is shown at the right of figure 4 which showed the tangential and radial forces in the milling process. Data from the phase I experiment are collected and calculated by ANOVA (analysis of variance) [14]. Table 2 shows the response of each parameter. The result indicated both spindle speed N and feedrate per cutting edge f_z have higher impact on the cutting force. A second phase experiments is then designed in order to investigate the effectiveness of both parameters.

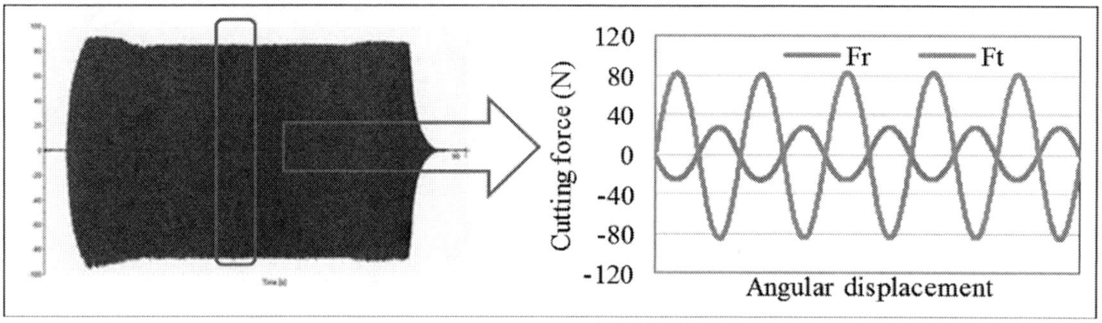

Figure 4. Measured cutting forces (left) and enlarged signal after filtering (right).

Table 2. Response of parameters in phase I experiment by ANOVA.

Parameter / Level	spindle speed N	feedrate per cutting edge f_z	depth of cut a_p
1	-32.28	-32.23	-32.26
2	-32.53	-32.15	-32.32
3	-32.01	-32.44	-32.24
Max-Min	0.52	0.29	0.08

4 phase II experiment and results

In phase II experimental design, we emphosized on the detailed data of cutting forces because the instant thickness of feedrate per cutting edge h keeps change in the milling process. As the machining parameters are reduced to two, we expanded the experiment number to do exhaust experiments with spindle speed set to 400 to 800 rpm, with interval 100 rpm, while the feedrate per cutting edge is set as 0.04 to 0.08 mm/tooth, with interval 0.01 mm/tooth. The parameters of the 25 experiments are shown in table 3. Each parameter set is experimented three times to confirm the repeatability. The evaluation index is also changed from cutting forces to the specific cutting force k_s in order to nomalizing the effect of parameters. k_s is defined in equation 1 can be rewritten to equation 3.

$$k_s = \mathrm{F_t} /(a_p \times f_z \times \sin\varphi) \tag{3}$$

Table 3. Experimental design at constant a_p in phase II.

No.	spindle speed N (rpm)	feedrate per cutting edge f_z (mm/tooth)	No.	spindle speed N (rpm)	feedrate per cutting edge f_z (mm/tooth)
1	400	0.04	14	600	0.07
2	400	0.05	15	600	0.08
3	400	0.06	16	700	0.04
4	400	0.07	17	700	0.05
5	400	0.08	18	700	0.06
6	500	0.04	19	700	0.07
7	500	0.05	20	700	0.08
8	500	0.06	21	800	0.04
9	500	0.07	22	800	0.05
10	500	0.08	23	800	0.06
11	600	0.04	24	800	0.07
12	600	0.05	25	800	0.08
13	600	0.06			

The specific cutting force k_s changed with the instant thickness of cut h which is further a function of rotating displacement. As data collected at the beginning (0 degree) and the end (180 degree) of the cut are relatively small due to small thickness, signal variations at the two places thus become large and are not suitable to use. We calculated k_s starting from 15 degree upto 168 degree . The result of k_s with repect to angular displacement φ at different feedrate is shown in figure 5. This figure shows that k_s is larger at small thickness of cut, the condition near both ends in the curve, and become smaller when the thickness of cut increases (at the center of curve). This size effect is critical for machining super alloys similar to Inconel.

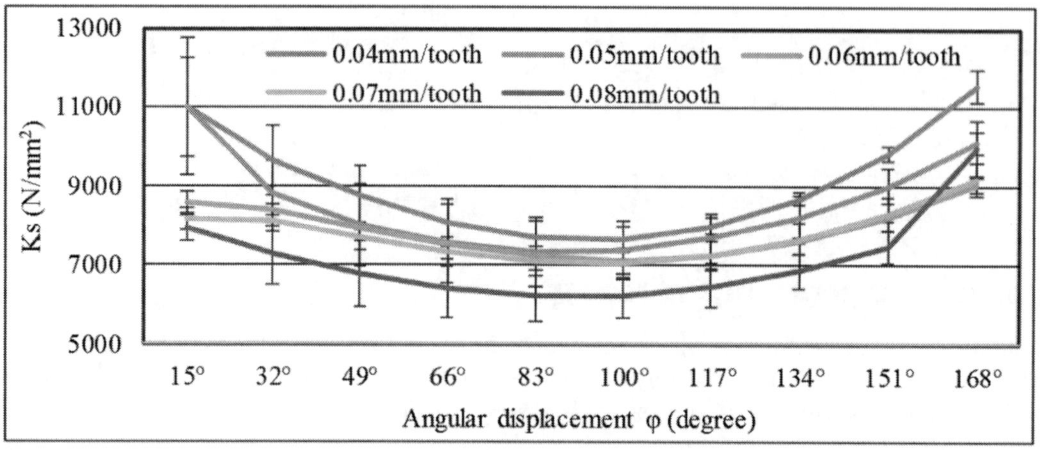

(a)

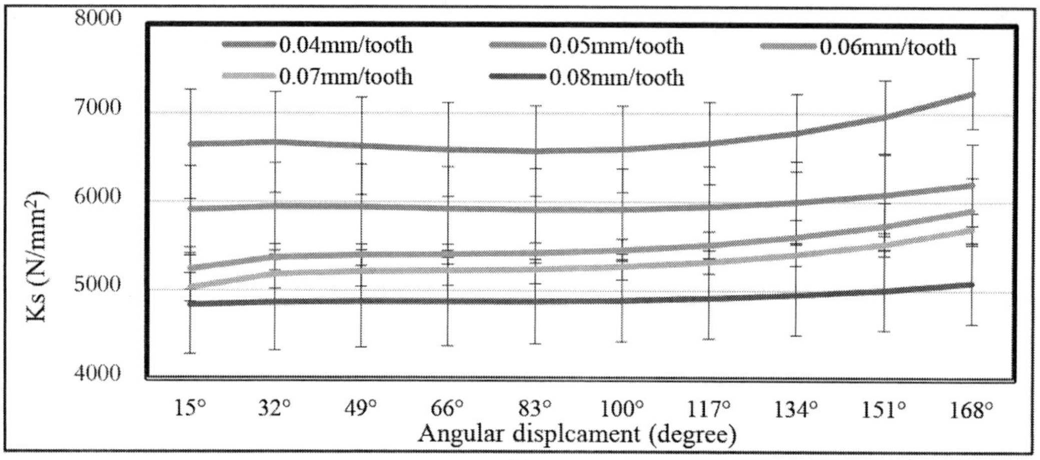

(b)

Figure 5. Specific cutting force k_s changed with spindle angular displacement φ at different feedrate under (a) 400rpm and (b) at 800rpm respectively.

The specific cutting force k_s changed not only with the instant thickness of cut but also with the spindle speed N and the feedrate per cutting egde f_z. Based on the experimental results, k_s is represented as a function of N and f_z, which is plotted in figure 5 where the surface is fitted with measured data shown as blue dots close to the fitted surface. The mean square error (MSE) is 387.9 corresponding to 4.8% error for specific cutting force at 8000 N/mm^2. The fitted function is represented as in equation 4 that corrects equation 1 and takes cutting speed and feed rate into account. Cutting force $\mathbf{F}_t$, based on equation 1, is modelled as equation 6. The equation shows that cutting force is strongly dependant upon machining parameters, in particular the feed per cutting edge f_z and the rotating speed N. f_z represents the size effect that tells the specific cutting force is enlarged at small feed though the material removal is small.

$$k_s(N, fz) = 13910 - 3.1N - 109900f_z - 0.0028N^2 + 23.9Nf_z + 434500f_z^2 \quad (4)$$

$$\mathbf{F}_t(N, fz) = k_s(N, fz) \times a_p \times h \ \mathbf{n}_t \quad (5)$$

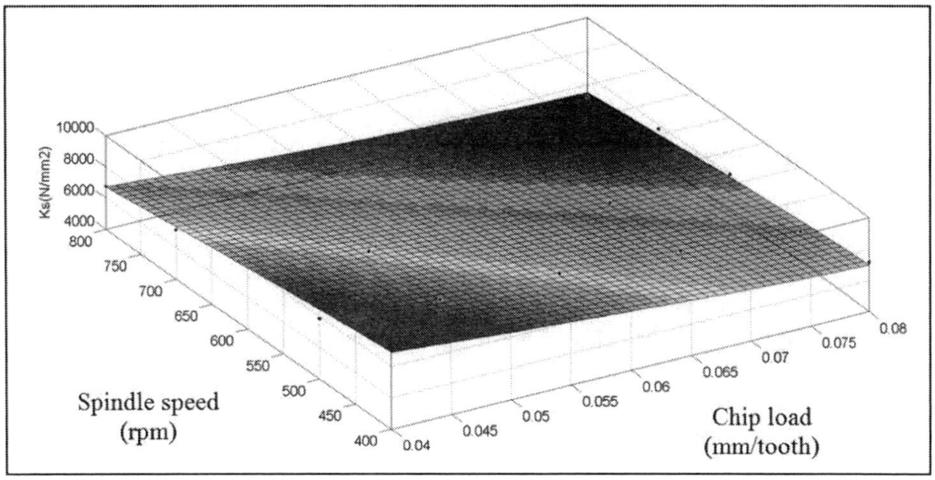

Figure 6. Specific cutting force k_s with respect to spindle speed N and feedrate per cutting edge f_z based on experimental data by curve fitting.

5 Conclusions

Inconel is a commonly employed materail in aerospace and marine industries but difficult to machine. This paper investigated the cutting force for milling process with two-phase experiments. Experiments in phase I are designed based on Taguchi method. Three parameters, the spindle speed, the feedrate per cutting edge and the depth of cut, are used for designing experiments. The results showed that the first two parameters are more important in the cutting force. A phase II exhaust experiments to find a modified cutting force model is designed with spindle speed set from 400 to 800 rpm while the feedrate per cutting edge is set from 0.04 to 0.08 mm/tooth. The results showed that

(1) There exists a strong size effect in milling Inconel 718. The cutting force changed with the rotating displacement of spindle. Specific cutting force is larger at small thickness of cut and become smaller when the thickness of cut increases. This size effect is critical for machining super alloys similar to Inconel.
(2) A modified cutting force model for milling Inconel 718 is derived as shown in equations 4 and 5 that take spindle speed and feedrate per cutting edge into account. Mean square error of the fitted model is about 4% and compensates the inaccuracy of conventional cutting force model.

Acknowledgements

The authors want to express their gratefulness to staffs at the Intelligent Machinery Center of the Industrial Technology Research Institute (ITRI), Taiwan for their assistance on the experimental equipment. Funding for this research has been supported by the Ministry of Science and Technology, Taiwan under contracts MOST104-3011-E-005-002, MOST105-2218-E-194-004 and MOST106-2218-E-194-002.

References

1. M. Rahman, W.K.H.Seah, T.T.Teo, J. of Matrs. Proc. Tech., **63**(1-3), 199-204 (1997)
2. I.A. Choudhury, M.A. El-Baradie, Proc. Inst. Mech. Engrs. Part B, **212**(3), 195-206 (1998)
3. M.E. Martellotti, Trans. of the ASME, **63**, 677-700 (1941)
4. A. Benson, *Basic Mechanical Engineering*, IV: Machine Tool, http://slideplayer.com/slide/6083347/
5. J. Tlusty, P. MacNeil, Annals of CIRP, **24**, 21-25 (1975)

6. M.C. Shaw, *Metal Cutting Principles,* 2nd ed., (Oxford Univ. Press, 2004)
7. H.Z. Li, J. of Matrs. Proc. Tech., **180**(1-3), 296-304 (2006)
8. M. Nalbant, A. Altin, H.Gökkaya, Materials and Design, **28**(5), 1719-1724 (2007)
9. Y.S. Liao, H.M. Lin, J.H. Wang, J. of Matrs. Proc. Tech., **201**(1-3), 460-465 (2008)
10. D.G. Thakur, B. Ramamoorthy, L. Vijayaraghavan, Intl. J. Adv. Mfg. Tech., **59**(5-8), 483-489 (2012)
11. D.M. D'Addona, S.J. Raykar, M.M. Narke, Procedia CIRP **62**, 269-274 (2017)
12. A. Sharman, R.C. Dewes, D.K. Aspinwall, J. of Matrs. Proc. Tech., **118**(1), 29-35 (2001)
13. A. Devillez, G.Le Coz, S.Dominiak, D.Dudzinski, J. of Matrs. Proc. Tech., **211**(10), 1590-1598 (2011)
14. G. Tauguchi, *Taguchi on Robust Technology Development: Bringing Quality Engineering Upstream* (ASME Press, 1993)

Experimental study and finite element analysis on shear lag effect of thin-walled curved box beam under vehicle loads

Hailin Lu, Heng Cai[a], Zheng Tang and Zijun Nan

School of Resource and Civil Engineering, Wuhan Institute of Technology, 430074, Wuhan, China

Abstract. Shear lag effects of curved box beam under vehicle loads are investigated by using three-dimensional finite element method, where 4 parameters of vehicle loads, load size, vehicle speed, vehicle load position, load types, are considered. The change rules of stress distribution and shear lag coefficients of upper flange at mid-span are obtained when the loads move to the mid-span. The results indicate that under vehicle loads, the peak shear lag coefficients is at the junction between the flange and web, shear lag effect is prominent, shear lag effect is greatly influenced by vehicle speed and vehicle load position, while load size and load types almost don't affect shear lag coefficients but do affect the stress. The model experiment of a cantilever curved box beam is carried out to compare with finite element analysis, and the error between them is small, which testify the validity and reliability of finite element model.

1 Introduction

Thin-walled box beam is a common structure type in bridge construction. Because of shear lag effect [1], stress of flange is non-uniformed distributed along width, lots of researches of shear lag effect under static loads have been studied home and abroad so far [2-6], where concentrated loads and distributed loads are utilized most of the time. Unlike straight box beam, curved box beam considered curvature radius will produce strong coupled effect of bending and torsion, making it difficult to compute its deformation and internal force, meanwhile, the impact of vehicle loads and vertical seismic excitation can intensify it if under dynamic loads, therefore it is very necessary to study shear effect of curved box beam under dynamic loads. In this paper, 3D finite element analysis is employed to study the influence of four parameters on shear lag effect under vehicle loads, experiment is performed to verify the validity of the curved box beam models.

2 Curved box beam model

In the paper, the material properties and structural dimensions of curved box beam are shown in the literature [7], the material parameters are as following: Young's modulus E=3000MPa, Poisson's ratio $\mu = 0.385$, mass density $\rho = 1180\text{kg}/\text{m}^3$. Structural parameters: thickness of bottom and upper flange

[a] Corresponding author : 2319944517@qq.com

t_1=t_2=8mm, thickness of web t_w=8mm, height h =100mm, curvature radius $r = 2\text{m}$, curvature angle $\theta = 30^\circ$.

In the finite element models, step loads are utilized to make moving loads move from one side of curved box beam to the other side at the speed of 1m/s .Constant force, F=20N, is chosen to simulate vehicle loads to make shear lag more prominent. The load action mode on the curved box beam is shown as figure1.

In present work, curved box beams are analyzed by the 3D finite element method, using 4-nodes shell elements. The finite element model of curved box beam is shown as figure 2, there are 2204 elements and 2242 nodes, as for boundary conditions, Ux, Uy, Uz and Uy, Uz are restricted at the junction between web and bottom flange in one side, while Ux, Uy and Uy are restricted in the other side.

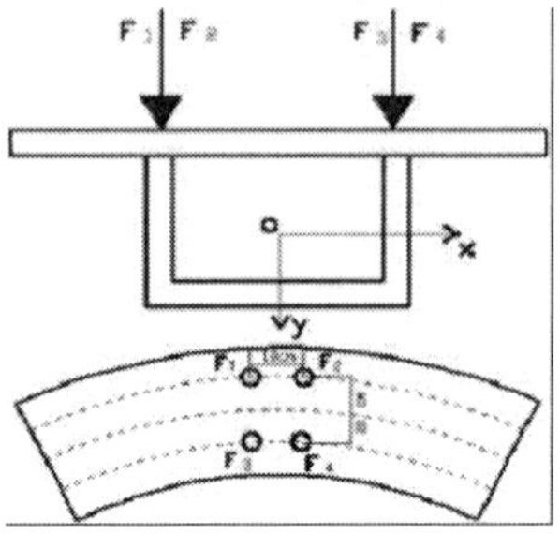

Figure 1. The load action mode on the box beam.

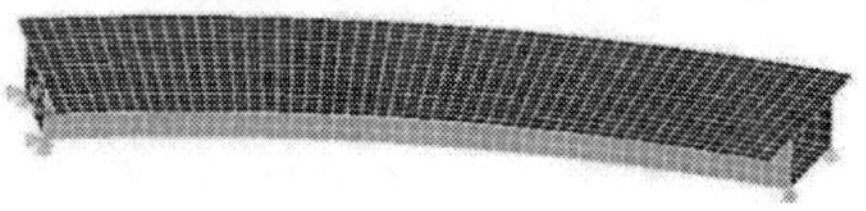

Figure 2. Curved box beam model.

3 Parametric analysis

3D finite element analysis is employed so as to reveal shear lag effect under vehicle loads, where 4 parameters, load size, vehicle speed, vehicle load position, load types, are considered . In view of this, the following values are considered: a. F=20N, 40N, 60N, 80N, 100N; b. V=0.1m/s, 0.3m/s, 0.5m/s, 0.7m/s, 1m/s; c. Move along the mid-line, move along outside, move along inside; d. Constant force, simple harmonic force.

The shear lag effect is most prominent when the loads are applied at the mid-span, the Shear lag coefficient is shown as Eq.(1)[1]

$$\lambda = \frac{\sigma}{\bar{\sigma}} \tag{1}$$

Where σ is the real stress value, $\bar{\sigma}$ is the stress computed by elementary beam theory.

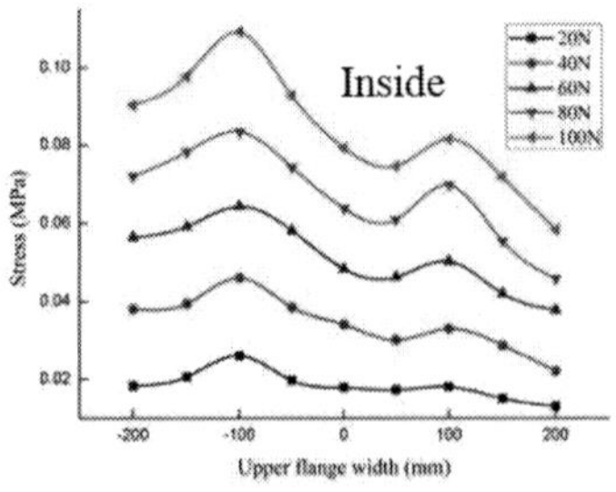

(a) Stress

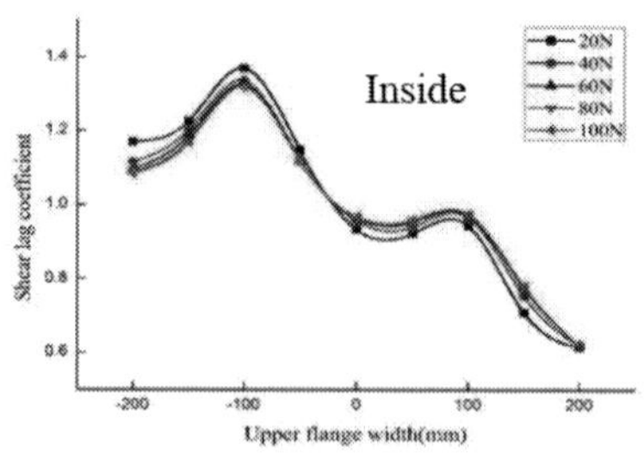

(b) Shear lag coefficient

Figure 3. Shear lag effect under different load size.

In the figure 3, the stress peak is at the junction between upper flange and web, and it gradually decreases along two side, the maximum shear lag coefficient reaches 1.37. However, the stress gets larger while shear lag coefficients hardly change with the load size doubles, which indicates that load size have limited influence on shear lag effect.

As is shown in figure 4, the stress become larger with the speed ranging from 0.1m/s to 0.5m/s, and then it decreases with the speed ranging from 0.5m/s to 1m/s. There is no certain rules for shear lag coefficient with the variation of speed, the shear lag effect is most non-prominent at the speed of 0.7m/s.

(a) Stress (b) Shear lag coefficient

Figure 4. Shear lag effect under different vehicle speed.

It can be easily found from figure 5 that vehicle load position has a great influence on shear lag effect. Inside shear lag effect is more prominent when the loads move along the inside of upper flange, on the contary, the outside shear lag effect is more prominent, the reason for this is that there is a bigger eccentricity if loads move along mid-line or ouside.

In Fig.6, compared with constant force, although simple harmonic force creates smaller stress on upper flange, but the shear lag coefficients between them are almost the same, which indicates that it is feasible to simplify harmonic force as constant force when shear lag effect of curved box beam under vehicle loads is studied .

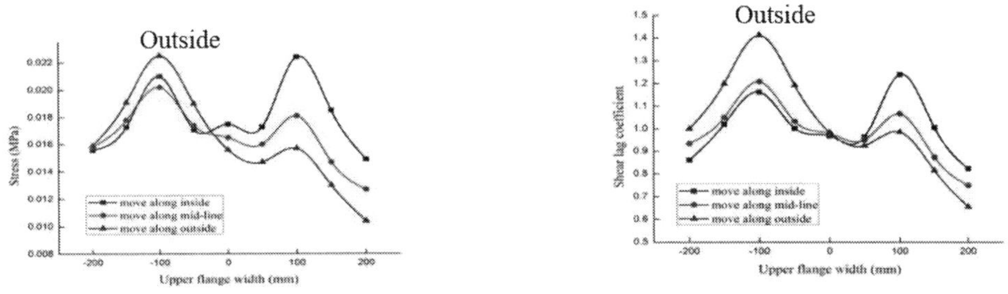

(a) Stress (b) Shear lag coefficient

Figure 5. Shear lag effect under different vehicle load position.

(a) Stress (b) Shear lag coefficient

Figure 6. Shear lag effect under different load types.

4 Experimental study

In present work, model test of cantilever curved box beam is carried out to verify the validity and reliability of finite element model, figure 7 shows the measuring points of upper flange, there are 1cm and *L*/2 from section I and section II to the fixed end respectively, 11 measuring points are laid out on the two sections separately with 5 between two webs and 3 on each cantilever slab.

The results are shown as figure 8, it can be found that the relative error between ANSYS and experiment is small at most of measuring points.

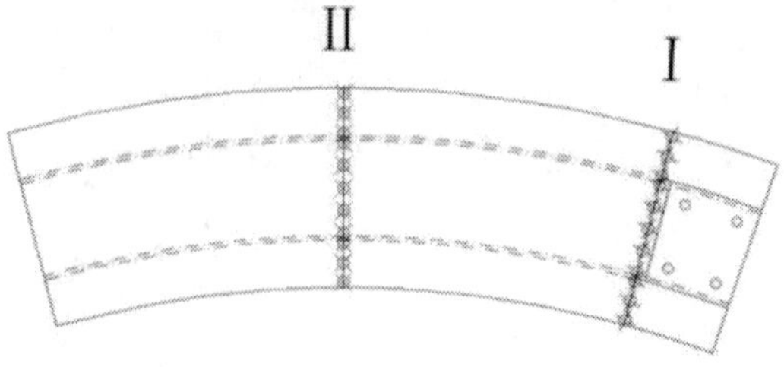

Figure 7. Measuring point of upper flange.

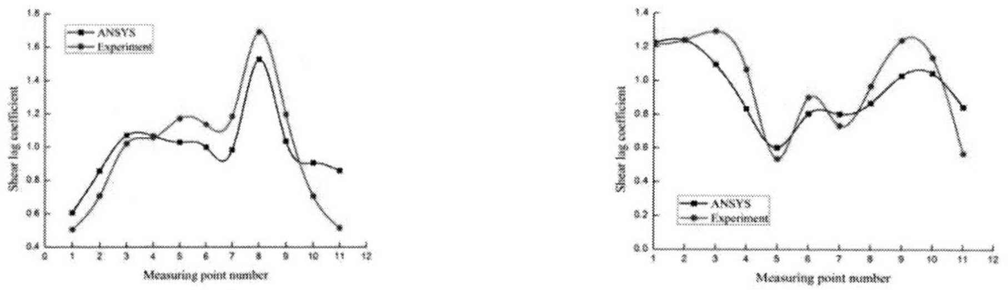

(a) Section I (b) Section II

Figure 8. Shear lag coefficients of section I and section II on upper flange.

5 Conclusions

The following conclusions are acquired:

1) The shear lag effect is prominent in the single-box single cell curved box girder at upper flange under vehicle loads.

2) vehicle speed and vehicle load position both have a great influence on shear lag effect, while load size and load types don't affect shear lag coefficient but affect stress.

3) In general, compared with inside shear lag effect, outside shear lag effect is more prominent especially when the loads move along outside, which indicates that the ouside of upper flange is damaged more easily under vehicle loads.

4) the finite element analysis approximately agrees well with experiment results, which verifies the validity and reliability of curved box beam models.

References

1. H.F. Xiang and L.C. Fan, *Advanced Theory of Bridge Structures* (China communication Press, Beijing, China 2008)
2. Y.H. Zhang, J Struct Eng-Asce, **138**(10), 1279-1284 (2012)
3. W.Q. Wu, S. Wan, J.S. Ye and T.Y. Fang, China Civil Engineering Journal, **37**(9), 31-36 (2004)
4. Y.H. Zhang, Y.R. Hu and L.X. Lin, China Civil Engineering Journal, **48**(6), 44-50 (2015)

5. Z.B. Lin and J. Zhao, Eng Struct, **41**, 90-97 (2012)
6. W.B. Zhou, L.Z. Jiang, Z.J. Liu and X.J. Liu, J Cent South Univ, **9**, 2650-2655 (2012)
7. H.L. Lu, *Theoretical Analysis And Experimental Study on Shear-Lag Effect of Prestress Concrete Curved Box Beam* (Tianjin University, Tianjin, China, 2007)

Study on the technology of grouting and reinforcing the foundation of a railway bridge above shallow old mine goaf

Hui-Quan Sui[1,a] and Chun-Yuan Bait[2]

[1]*Department of Architecture & Civil Engineering, Dalian Nationalities University, China*
[2]*Jin Yuan survey and Design Institute of Dalian Development Zone, China*

Abstract. The city fast track railway bridge studied in this paper is located above an old mined out area. There is a 1500m^2 underground goaf in the construction site with a depth of 30.8~53.2m, the seam thickness is 2.0~2.5m and the dip angle is 50~63 degrees. Although the mine has been abandoned for many years, there are still a lot of underground cavities. Collapse pits occurred on the surface occasionally till this day, especially during the rainy season. In order to make the design safe and reliable, drilling pressure grouting technique is proposed to fill and reinforce the foundation of the site. Proposed technique is tested by several experiments in simulated underground mined out area for the foundation treatment. The technical design, construction technology, grouting and reinforcement effect of the project are studied and summarized. The results of the experiment showed a stable and safe bridge foundation has been achieved. The proposed technique was then applied to on site experiment and pasted all tests.

1 Engineering background

A city fast track railway bridge to be designed is located on an old goaf area of an abandoned asbestos mine. This bridge is designed as a continuous span reinforced concrete slab bridge, with a length of 300 meters and a width of 8.8 meters. The main bridge adopts a single-box double-chamber reinforced concrete box girder structure, with a one shape abutment and a Y shape pier. The bridge foundation adopts excavated cast-in-place pile; the pier foundation pile diameter is 3.20m, with design load value of 15000kN. Abutment pile diameter is 2.0m, with a design load value of 6000kN. The bridge is designed to match A Class safe building classification.

The major issue facing in this project is the foundation part, the bottom of the site directly on an abandoned asbestos mine which was mined with well lane mining technique. The ground depth of the original mining area is relatively small (33.0m~200m), which belongs to the shallow mining class. The asbestos mine was operated with lane well technique in the upper area, and the dry filling method was used in the deep area. Although mining operation has been ceased for many years, due to the existence of the shallow mined out area, collapsing pits still occurs till this date, especially during rainy season. In order to make the design safe and reliable, drilling pressure grouting technique is proposed to fill and reinforce the foundation of the site.

[a] Corresponding author : 1184757853@qq.com

2 Site condition

The structure of the abandoned mine area is illustrated in Figure 1. As shown can be seen, the designed bridge is directly above the mine out area. There is a 1500m^2 underground goaf in the construction site with a depth of 30.8~53.2m, the seam thickness is 2.0~2.5m and the dip angle is 50~63 degrees. Lane column mining technique was used, mining ratio is around 7:3, and mining range is 230 meters.

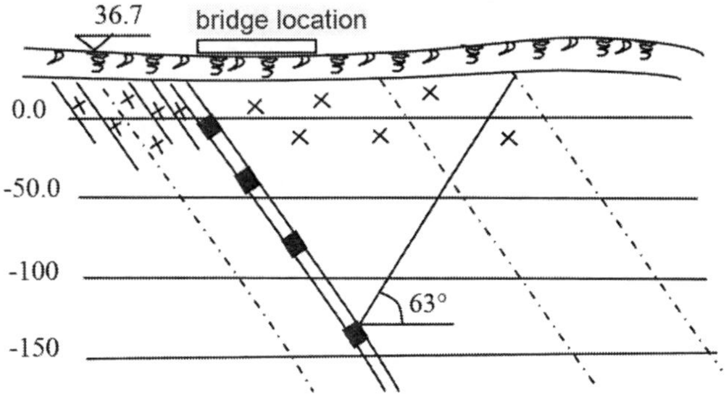

Figure 1. The relative position of the proposed bridge and the goaf.

The formation of the site is analyzed from top to bottom with on-site drilling technique, the results show as following:

- Filling soil （*Q4ml*）, loose structure, mainly consists of quartzite fragments, slag and silty clay composition, the thickness is around 2.8~3.8m.
- Gravel Soil（*Q3dpl*）, medium density structure, mainly consists of quartzite rock gravel cracks filled with silty clay component; the thickness of this layer is around 1.8 meters.
- All Weathered diabase（$\beta\mu$）, loose structure, this layer is very dry and could be affected by water, the thickness is around 2.3 meters to 4.4 meters.
- Strongly weathered diabase（$\beta\mu$）, fractured structure, joint fissure development, the thickness of this layer is around 2.5 meters to 6.8 meters.
- Medium weathered diabase（$\beta\mu$）, the structure is relatively complete, joint fissure development, the thickness of this layer is around 9.5 meters to 14.9 meters.
- Micro-weathered disabuse（$\beta\mu$）, the structure of this layer is hard with fine grain, medium joint fissure development, the thickness of this layer around 7.0 meters to 28.3 meters.
- Ore layer（*Py1*）, the layer consists of granular and block structure, the lower part has been mined and the structure is heavily broken. The thickness of this layer is around 1.5meters to 7.8 meters.
- Dolomite（*Zg*）, this layer is the bottom of the mine out area, the integrity of the structure is maintained. The main physical and mechanical properties of the overlying mine out area is demonstrated in Table 1 as following:

Table 1. The main physical and mechanical properties of the overlying mine out area

Lithology	Density (g/cm3)	Elastic modulus (103Mpa)	Compressive strength (Mpa)	Tensile strenth (Mpa)	Poisson's ration (μ)	Cohesion (Mpa)	Internal friction angle (φ)	Wave Speed vp(m/s)
Diabase	2.9	12.33	78.04	13.33	0.34	16.18	63	5637
Ore	2.6	6.62	65.98	4.51	0.34	8.63		4626
Medium-weathered Diabase	2.86	8.7	37.75	2・25	0.313	4.61	45	4008

3 Similar material simulation tests

In order to analyze the stability and physical property of the structure of the mine out area, similar material is used in the simulation experiment, including quartz sand, barite and mica as aggregate, gypsum as compound. Water and other appropriate retarder were also added. In addition to the physical and mechanical properties of the mining area, Newton similarity criterion was also considered to reconstruct an experimental model with a 1:30 ratio. The deformation pattern of surrounding rock and ground surface are simulated after mining of underground seam. The stability of overlying strata and the suitability of building are evaluated, providing technical support for grouting reinforcement construction.

3.1 Mined out area surface deformation analysis

- Surface slab damage: Surface slab damage mainly occurs within the mined out area, it is caused by the alteration between the diabase layer and dolomitic limestone layer. This alteration deteriorates the physical integrity of the structure, which makes the layer brittle. With the mining layer been mined out, displacement and deformation occurs in surrounding rock and the roof layer. New balance formed between remained layers with a pressure arch structure. Any damage or fractured of the roof or the pressure arch will produce a wide range of further fragmentary damage. [1-5]
- Surface slab fall off. Surface slab falls off occurs when the direct and indirect roof of the mined out area reaches a certain value and the roof bent the limit.
- Roof fissure damage: the fissure zone located above the fall off zone, and is mainly consist of cracking damage and fracture damage. The thickness of the fall off zone and fissure zone is calculated based on following formula. Consider the Height of the fall of zone and fissure zone is H, then

$$H = \frac{hM}{Rh + S} \tag{1}$$

Where: h---Vertical Height of the Mined out area,
M---Mining layer Thickness,
R&S----Coefficient related to rock formation.
R,S Value for each related formation is shown in following Table2.

Table 2. R and S values of caving zone and fracture zone

R		S	
Falling zone	Fissure zone	Falling zone	Fissure zone
0.103~0.082	0.041	3.325~2.660	1.330

3.2 Non-continuous deformation on the surface of mined out area

Non-continuous deformation on the surface occurs in the mining area where the mining depth is less than 45 meters. These deformations usually happened in two different types as following.

- Surface Cracks：This type of deformation often takes place above the tensile deformation zone above the mining area. Surface cracks can extend from a few meters to tens of meters with a width of 1-2 meters, depth of 20 meters.
- Bead-like collapse pit：Bead-like collapse pit usually occurs within the shallow mining area. When underground cracking reaches the surface, surface rock deforms under pressure. As a consequence, a series of funnel shape rock deformation take places. This type of deformation happens along the mining track with a diameter of 1 to 2 meters.

3.3 Formation principles of cracks and movement in the shallow surface

Two major aspects are considered as the main cause of the deformation: First, the mining depth is relatively shallow, exposed rocks are highly weathered. And the integrity of supporting layer is compromised. In the mined out area, the roof fall off occurs due to lack of support. This type of deformation extends to the surface in an arc form. When it reaches the surface, a wedge body takes places. During rainy season, the shallow rock mass is saturated with water, the wedge body gradually sinks with this change. The weathered rock layer can not support the sinking wedge body, thus collapsing pits will occurs on the surface. Second: when the mined out area becomes large enough, the lack of support will cause the roof rock to bend and deform. Collapse pits will also take place in these areas. These two principles are demonstrated in following Figure 2.

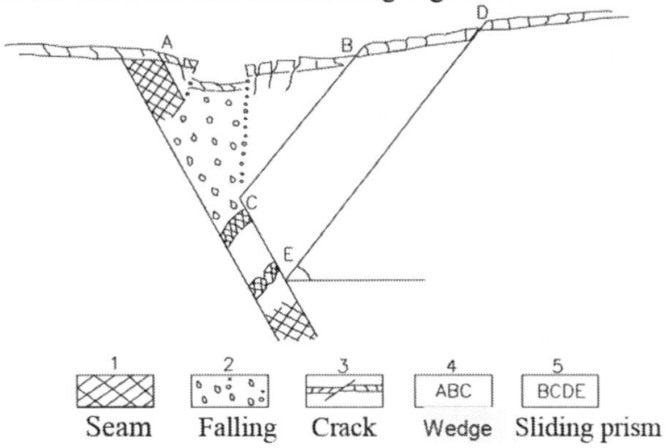

Figure 2. Formation principles of cracks and movement in the shallow surface

Lane column mining, pillar strength is the main factor affecting the ground collapse and destruction. The main forms of pillar failure include shear failure and tensile failure, i.e. tension fracture produced by pillar compression, which is similar to the split failure of uniaxial compression test. The strength of the pillars is calculated according to the following formula:

$$\sigma_p = \sigma_{p1}(C + D \cdot W / h) \quad (2)$$

Where: σ_{p1}--- when w/h =1, pillar strength
w/h---the pillar width / height

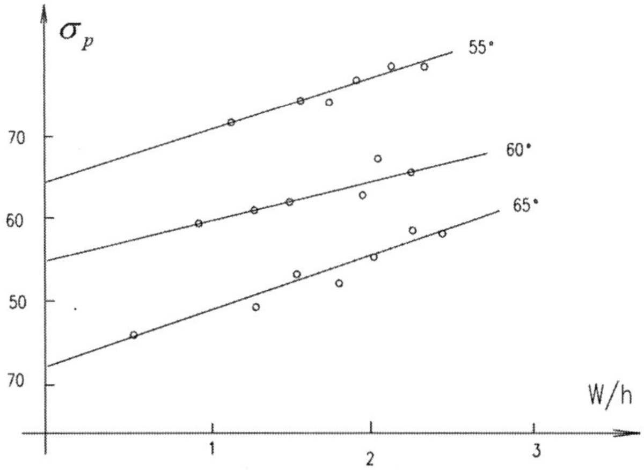

Figure 3. σ_p-W/h regression curve

According to the data of experimental and field observation, the linear regression analysis can be carried out to calculate the coefficient of the formula(2) for strength of the pillar under the corresponding dip angle, as shown in Figures 3 and Table 3.

Table 3. Results of regression analysis

group	n	$\sigma_{p1}(Mpa)$	C	D	standard deviation	correlation coefficient	$t_{0.06}$	correlation
65°	7	49.1	0.875	0.144	1.04	0.972	2.015	significantly
60°	6	60.3	0.933	0.082	1.304	0.880	2.132	significantly
55°	6	71.9	0.888	0.10	0.687	0.973	2.132	significantly

4 Field experiment

Based on the comprehensive analysis of the data of surface movement, combined with the observation and the results of the simulation experiment, it is concluded that the influence of surface subsidence and the deformation is basically stable for mining area with a depth greater than 45 meters. In the shallow area where the depth is less than 45 meters, the rock formation is less stable, and grouting reinforcement techniques need to be used to support the area.

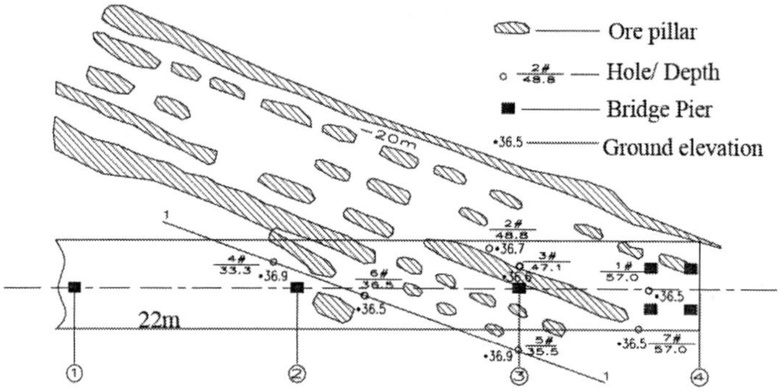

Figure 4. Floor plan of grouting hole layout

Seven grouting hole are designed as shown in Figure 4. The hole 4# 、5# 、6# are along a straight line above the mined out area, and is relatively higher than the other four grouting holes. As the goaf is full of water, in order to improve the effect of grouting, to achieve the function of drainage and diversion, the detailed grouting procedure is as following: when grouting in hole number 4# 、5# 、6#, drainage is carried out in the hole number 2#. When grouting in drill hole number 2# 、3# 、1#, drainage is carried out in the hole number 7#. And when above procedure finishes, grouting the last drill hole number 7#.The intermittent injection method is used; each grouting hole is injected in three different stages. In the first stage, slurry with a water-cement ratio of 2:1 is used. In the second stage, the ration changed to 1.5:1, and in the last stage, the ration is 1:1. By dividing the procedure to three stages can expand the slurry diffusion radius, improve the slurry transport distance and also improve the filling reinforcement effect. The total injection amount of seven grouting holes is 1528 m^3.

5 Grouting effect

The total grouting operation takes 18 days. 7 days after the finishing date, drill method is used to test the out come of the grouting operation. Leak test and cross-hole in-situ acoustic velocity measurement are both carried out to test the integrity of the grouting part. [6-8]

According to the results of the drilling in-situ wave velocity test, it is not difficult to see there are two abnormalities in hole1#~2#; 3#~4#; and 5#~6# hole has an anomaly. The curve reflects the obvious of the recess, the value of the low wave velocity and abnormal wave velocity can be seen in the following Table 4. The velocity of sound in goaf area is about 3000/s~3500/s which is close to the velocity of surrounding rock, as shown in Figure 5.

According to the drilling data, the upper anomaly area is the main leakage section of the micro-weathered diabase, and the lower part is mainly the mined-out area.

Table 4. Comparative analysis of wave velocity

Test group	Abnormal band (m)		Lithology		Wave velocity (m/s)	
	I	II	I	II	I	II
1#~2#	31~37	40~42	diabase	Mined out area	3250~3500	3260~3370
3#~4#	26~33	35~38	diabase	Mined out area	3250~3510	3510~3520
5#~6#	39~50		diabase		3250~3520	

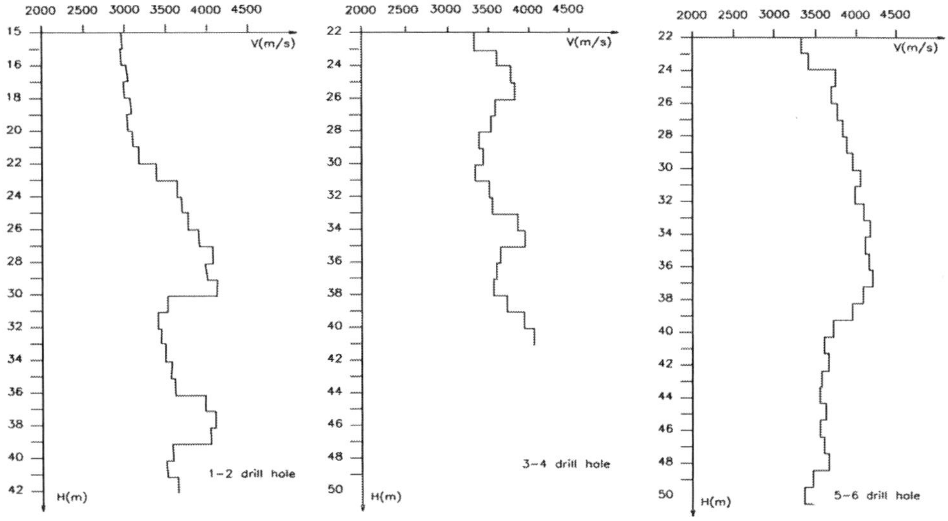

Figure 5. Cross-hole sound velocity V-H curve

The test result of DPT and SPT on site in the typical rock strata is shown in Table 5.

Table 5. The test result of DPT&SPT

Name of rock and soil	Standard values of bearing capacity of rock and soil layer（kPa）
Plain Filling Soils (Q_4^{ml})	80
granular soil (Q_3^{dpl})	280
Fully weathered diabase ($\beta\mu$)	240
Strongly weathered diabase ($\beta\mu$)	320
Moderately weathered diabase ($\beta\mu$)	2500
weak weathered diabase ($\beta\mu$)	4200

Through the observation of the core of the borehole, the collapse of the mined out area has been formed with surrounding pillar. No excess gap in the grouting area, core surface and gap already been filled with injection material. The result shows that the fissures are contained and the filling effect is good. The drilling sampling rate is between 50%~80%, including a large quantity of grouting materials, which reflects the successful of the grouting procedure. Therefore, the grouting reinforcement project achieves desired effect.

6 Discussion and Conclusions

The study shows that the underground layout and rock formation in a mined out area is not stable enough to support new construction project. Therefore, a thorough stability analysis is carried out and following reinforcement approach is proposed. The proposed grouting reinforcement technique is first tested in similar material simulation experiment for physical and mechanical parameters. With the test results and the consideration of surface crack and collapse pit in the construction site, 7 different grouting holes with precise positions are selected. And in order to provide support for the construction site, the intermittent injection method is used; each grouting hole is injected in three different stages with filling slurry in different water to cement ratio.

It is also concluded that in mined out area, due to the hollow space from the mining operating Non-continuous deformation of the surround rock such as surface cracking and roof rock fall off will take places, especially in rainy season. This poses a threat to buildings and personal safety. Therefore, foundation reinforcement is vital for construction project directly above mined out area.

Further filed experiment, filed operation and following up drilling test shows that the proposed grouting reinforcement technique achieves expected filling and reinforcement objective. The drilling sample test results also shows that the fissures are contained and the filling reinforcement can provide sufficient and safe support for the city rail bridge project.

References

1. H. Tan, Subgrade Engineering, **1**, 214 (2014)
2. H. Sui and Z. Wang, Chinese Journal of Geotechnical Engineering, **23**, 510-512 (2001)
3. H. Sui, S. Wang and H. Wang, Progress in Safety Science and Technology, **4**, 220-223 (Beijing, China, 2004)
4. G. Guo, X. Miao and Z. Zhang, Science Technology and Engineering, **5**, 44 (2002)
5. W. Li and Y. Dai, Shanxi Architecture, **39**(16), 76 (2013)
6. Y. Zhang, W. He and L. Zhang, Journal of Engineering Geology, **18**, 297 (2010)
7. S. Zu, China Resources Comprehensive Utilization, **31**, 58 (2013)
8. H. Liu, X. An, J. Shen and Y. Ni, Journal of Bioresources and Bioproducts, **1**(2), 55 (2016)

Assessment of traffic-induced vibrations of Ozanam Building of Adamson University

Earl Shadd D. Araña, Leira Liz A. Bisain, Rae Cecil N. Acedera, John Elmer C. Guillermo and Albert A. Griño, Jr.[a]

Civil Engineering Department, Adamson University, Philippines

Abstract. Vibration is one of the main factors for fatigue in structures which can cause damages such as cracks and in critical cases could result to collapse. Long term contact to vibrations can induce the cracks to be transmitted through the structural members of the building that can compromise the structural integrity of the structure. The Ozanam Building of Adamson University is situated near the Romualdez Street, which is considered as one of the main routes of cargo trucks and other heavy vehicles from the Manila North Harbor Port. It was observed that cracks were visible on the façade of the building. For that reason, the structural integrity of Ozanam Building was evaluated against the traffic-induced vibrations using vibration meter. It was found out that the maximum peak particle velocity present in Ozanam Building is 8.60 mm/s which means that the building is safe and below the threshold value for risk in structural damage. Thus, vibration due to traffic has no significant effect that can cause structural damage to the building. In addition, preventive measures were discussed in this paper to reduce the traffic-induced vibrations to an acceptable level.

1 Introduction

1.1 Background of the study

Vibration is one of the main factors for fatigue in structures. Produced by different sources, vibrations propagate from one medium to the other. Long-term contact to vibrations can cause damages in buildings, minor effects such as cracks, which in critical cases could result in collapse. Also, these vibrations can cause errors to laboratory experimentation which are operating in Micro and Nano Scales.

According to Hajek [1], it is usual for large transportation agencies to receive complaints from residents living near highways about annoying highway noise. Complaints about vibration caused by highway traffic are less frequent, but they do occur. Typically, vibration complaints are also accompanied by complaints about excessive highway noise. Many transportation agencies do not have applicable guidelines for the resolution of vibration complains.

The vibrations can be measured by using a vibration meter that usually measures the peak particle velocity. This instrument can also be used to know the amplitude and frequency range of vibrations.

[a] Corresponding author : aagrino@gmail.com

The peak particle velocity is the maximum speed of a particle as it oscillates about a point of equilibrium that was moved by a passing wave.

According to Hunaidi [2], like most vibration problems, traffic vibrations can be characterized by source-path-receiver scenario, which is shown in the Figure 1.Vehicle contact with irregularities in the road surface (i.e. potholes, cracks and uneven manhole covers) induces dynamic loads on the pavement. These loads generate stress waves, which propagates in the soil, eventually reaching the foundations of adjacent buildings and causing them to vibrate. Heavy vehicles such as buses and trucks mainly cause traffic vibrations. Passenger cars and light trucks rarely induce vibrations that are perceptible in buildings.

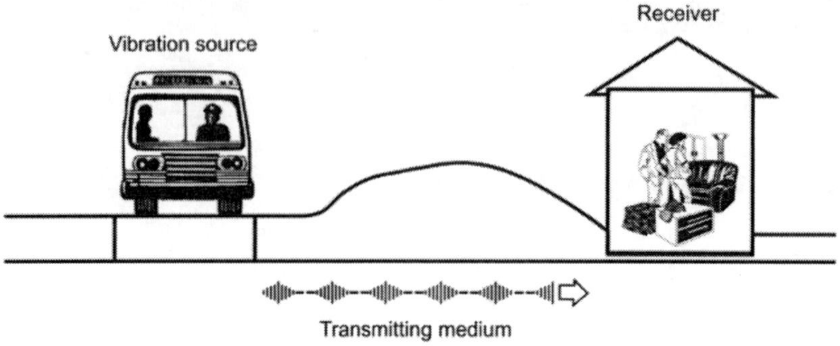

Figure 1. Traffic vibrations in source-path-receiver scenario (Hunaidi, 2007)

On a study presented by Hajek [1], the condition of the pavement surface is the decisive contributor to ground-borne vibration. It is also the most significant factor that can be controlled by highway agencies. Discrete pavement discontinuities, such as stepped transverse cracks exceeding about 4mm, appear to be significant enough to overshadow the effect of random surface roughness and result in specific sources of vibration. Heavy vehicles produce higher ground-borne vibration because of the larger mass acting on the pavement, and typically higher sound levels. Trucks equipped with steel suspension can produce higher dynamic loads and vibrations compared to trucks equipped with air suspension.

The Ozanam Building of Adamson University houses the College of Engineering and its departments. The said building is continuously experiencing vibrations due to the flow of cargo trucks and traffic. These trucks are going and coming from the Pier located at Manila, carrying imported and exported goods.

Thus, the researchers partook on a study of the evaluation of the traffic-induced vibrations in Ozanam Building in Adamson University.

1.2 Objectives of the study

1.2.1 General objective

The objective of this study is to evaluate the traffic-induced vibrations occurring on the Ozanam Building of Adamson University.

1.2.2 Specific objectives

1. To determine the peak hour in which vehicles that cause traffic-induced vibrations pass through the Romualdez Street.
2. To determine the peak particle velocity and the effects of the vibration to the Ozanam Building in reference to the Transport and Road Research Laboratory (U.K.) and the British Standards Institute;

3. To determine the location in the Ozanam Building that experiences the greatest intensity of vibrations using the vibration meter and total station; and
4. To determine if there are any visible damages on the building that is due to the vibrations induced by the traffic at Romualdez Street.

1.3 Significance of the study

This study is helpful in improving the structural facilities of the Ozanam Building of Adamson University. This also ensures the safety of the students occupying the building. Most occupants of the OZ Building are engineering students. They are taught that safety, welfare and public interest of the people is the paramount priority of the designers and thus, this study ensures their safety.

This study also helps the quality of the life of the socially disadvantaged because of the knowledge that this paper contributes to the civil engineering community. It might lead to advancements that ensures the protection of the occupants of buildings against traffic-induced vibrations.

Not only does the Adamson University community benefit from this study, but also all buildings within the vicinity of Adamson University and Romualdez Street. This study helps structural designers and contractors build structures, which can withstand vibrations and other seismic events. This research also contributes knowledge on future studies since that assessment of buildings integrity is considered uncommon like any other studies.

2 Review of related literature

2.1 Peak particle velocity

The peak particle has been established as the best measure of the damage of potential of ground vibrations. The US Bureau of Mines [3] originally suggested a limit of 50 mm/s as being safe for residences but in fact the problems occur below this level and Hajek now suggest 12.5 mm/s as a safe limit for older properties.

The Ozanam is considered as an old property since it has been constructed for a long time. Therefore, any peak particle velocity reading from the vibration meter that is greater than 12.5 mm/s should be understood that the Ozanam Building is in the risk of potential damages.

2.2 Assessment of ground-borne vibrations

According to Hunaidi and Tremblay [4], the process of assessing a potential ground-borne vibration problem consists of establishing vibration levels at the receiver and comparing them with acceptance criteria. The study found out that highway traffic induced ground-borne vibration can be perceptible in very extreme circumstances. However, typically, the feeling of vibration reported by residents living near highways is due to low frequency sound levels produced by trucks that can cause resonance of light exterior or interior building and room elements. Ground-borne vibration induced by highway traffic can be effectively controlled by the maintenance of smooth roadway surfaces [1]. Also, the study shows a graph about the relationship between the vibrations of the establishment compared to the distance from the center line of the lane.

As seen in Figure 2 that the Peak Vertical Particle Velocity decreases exponentially with distance from the centerline of the lane as well as the acceptable threshold value for perception, historical building and architectural damages.

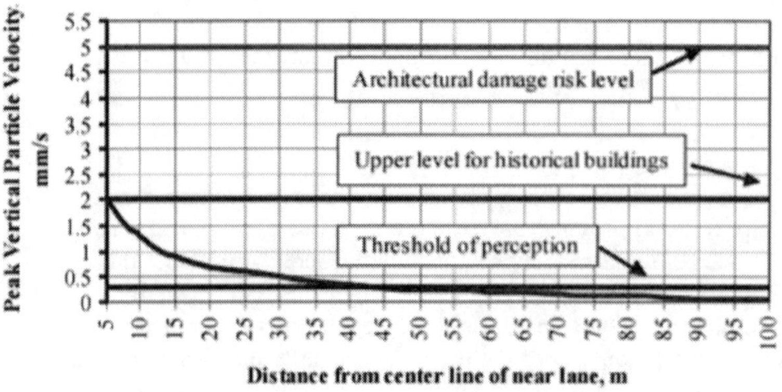

Figure 2. Maximum highway truck vibration level versus distance (Hajek, 2006)

2.3 Transport and road research laboratory in United Kingdom

The researchers from the Transport and Road Research Laboratory (TRRL) in United Kingdom carried out an extensive work about traffic induced vibrations and they developed a guidelines for assessing vibrations caused by traffic. Whiffin and Leonard [5] proposed the guidelines as shown in Table 1 to assess the effects of vibration on people and buildings.

Table 1. Effects of Vibration on People and Buildings (Whiffin and Leonard, 1971)

Peak Particle Velocity (mm/s)	Human Reaction	Effects on Buildings
0 to 0.15	Imperceptible	Unlikely to cause damage of any type
0.15 to 0.30	Threshold of Perception	Unlikely to cause damage of any type
2.0	Vibrations Perceptible	Effects ancient and historical monuments
2.5	Continuous exposure to vibration is annoying	Virtually no risk of architectural damage to normal buildings
5.0	Vibrations annoying to people in the buildings	Threshold for risk of architectural damage in house with plastered walls and ceilings
10 to 15	Continuous vibrations are unpleasant and unacceptable	Would cause architectural damage and possibly minor structural damage

2.4 Damages to structures due to traffic-induced vibrations

According to Hunaidi [6], house owners may complain about damage induced by traffic vibrations, such as cracks in walls and ceilings, separation of masonry blocks, and cracks in foundation. However, vibration levels are rarely high enough to be the direct cause of this damage, though they could contribute to the process of deterioration from other causes. Building components usually have residual strains as a result of uneven soil movement, moisture and temperature cycles, poor maintenance or past renovations and repairs. Therefore, small vibration levels induced by traffic could trigger damage by "topping up" residual strains. Consequently, it is difficult to establish a vibration level that may cause building damage and, therefore, controversy continues to surround the issue. In some cases, when a building is subject to vibration for many years, fatigue damage (i.e. that caused by repeated loading) may occur if the induced stresses in the building are high enough. In addition to damage caused directly by vibration, indirect damage may result from differential movements caused by soil settlement due to densification. Loose sandy soils are particularly susceptible to densification when subjected to vibration.

Several countries have adopted standards for evaluating the effect of vibration on buildings. No such national standards exist here in the Philippines. The most stringent vibration guide value specified in published standards for damage to house is more than 30 times the human perception level. Occupants would therefore find potentially damaging vibrations to be extremely annoying because of their very high level.

According to Hunaidi [7], solutions and preventive strategies that have been suggested to reduce vibration to an acceptable level include periodic maintenance of road surfaces, control of traffic flow and speed, improvement of the road structure, sufficient distance between roads and buildings, building isolation systems and screening of vibration using in-ground barriers. Some of these measures have proven to be effective. However, there is no easy way to completely remove the vibration due to traffic.

According to Mathur [8], there are no easy ways to mitigate traffic vibrations. There are, however, a limited number of options available to reduce vibration to an acceptable level include soil improvement, reduction of road traffic-induced vibrations by layered soil bags, geofoam trench method, tuned mass damper and fluid vicious dampers.

The researchers use these guidelines in formulating a recommendation on how to prevent damages to the Ozanam Building or to reduce the vibration to an acceptable level.

3 Methodology

3.1 Data collection

According to Jaswal [9], collection of data would require building wave barriers to isolate vibration sources by digging trenches around the building boundary. Wave barriers are expensive to build and takes a lot of space for implication thus it is not viable in urban areas. The researchers followed the said principle considering that most of the students occupy the Ozanam Building and would rarely be unoccupied.

The researchers determined the Peak Hour on which the vehicles caused traffic-induced vibrations. This determined the time on which the researchers conducted the vibration meter reading. Safety concerns were always present when personnel are working near or adjacent to a highway. Safety considerations included the workers as well as the motoring public. Construction safety has received extensive attention, but this research focused on personnel working near or adjacent to the highway, which were involved in conducting the activities such as: collecting data, setting up data collection equipment or administrating roadside interviews.

The observers positioned clearly in order to see the traffic they were counting. Trucks, buses, parked cars or signs, which typically blocked vantage points, were avoided. They were located away from the traveled way for safety of the counter as well as affecting the drivers. The observer wore safety vest every time it gets near traffic. Seated at a parked car, provided protection from the elements and provided safety for the observer; however, it was located in such a way that it did not interfere with the typical traffic movements. A sign indicated that traffic counting was in progress satisfied most driver's curiosity, which avoided unwanted interruptions.

As with any field collection of traffic data, care was considered in protecting the observer at all times. The researchers wore reflective vests, and acted in a manner that does not distract motorists or influence their driving behavior.

3.1.1 Determination of peak hour

The researchers determined the time on which the data needed in this study were being collected. The researchers allotted two (2) days in determining the peak hour factor. The Peak Hour Factor (PHF) measured the relationship between the peak 15-minute interval of traffic compared to the total vehicle volume over the entire Peak Hour. It represented how constant vehicle volumes were, during the peak hour. The PHF is a ratio of the total hourly traffic volume against the busiest 15-minute interval.

The researchers tallied-in the number of vehicles that were travelling at high-speeds and the vehicles that yielded vibrations. In guaranteeing, the researchers conducted the procedure in the Ozanam Carpark, which is a safe place and is more effective due to the proximity to the Romualdez Street.

The researchers used statistical methods such as mean and standard deviation on the determination of peak hour due to multiple numbers of trials and observations. The mean for every time interval was used in the computation of the PHF.

The following formulas were used in this analysis:

$$MEAN = \overline{x} = \left(\sum_{i=0}^{n} x_i\right)/n \tag{1}$$

$$STANDARD\ DEVIATION = \sigma = \sqrt{\sum_{i=0}^{n} \frac{(x_i - \overline{x})^2}{n-1}} \tag{2}$$

$$PHF = \frac{V}{4x\ Vp} \tag{3}$$

Where: x_i = vehicle count per time interval
n = number of trials
V_p = Peak 15-minute Volume
V = Peak Hourly Volume
PHF = Peak Hour Factor

3.1.2 Set-up of the vibration meter

For proper evaluation of the effect of building vibrations induced by road traffic, measured vibrations were undistorted together with data processing and analysis that followed in the established procedures. Instrumentation for the measurement of vibration signals as shown in Figure 3, which includes vibration sensors, signal conditioners, and recording equipment, had sufficient resolution and sensitivity. Measurements were gathered at locations where the vibration levels reflect the purpose of the evaluation. In evaluating the effect of vibrations with respect to human annoyance, measurements were taken at locations where the vibration level found out greatest, normally at the midpoints of floors. On wood floors, the measurement points were located near joists to avoid local resonance of individual floor panels.

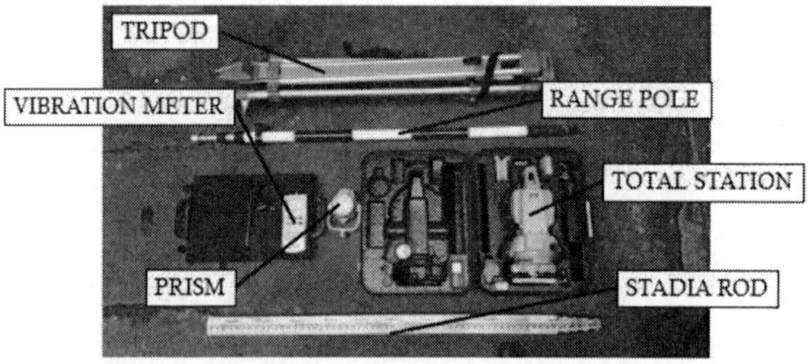

Figure 3. Equipments and Apparatus used for data collection.

The researchers collected data in the corners of the Ozanam Building. The Lutron Vibration Meter's vibration sensor was mounted on the walls of the points of data collection. The points of data collection where chosen because the researchers took in the priority to evaluate the structural integrity of the Ozanam building. The gathering of data using the vibration meter lasted for at least a quarter of an hour within the peak hour.

After computation of the latitudes and departures, the researchers plot the coordinates in a plane and connect the columns from one another in order to determine the dimensions of the Ozanam Building. This shall now be added to the dimension of the setback and the sidewalk.

3.2 Data analysis

In evaluating the effect of vibrations on a building, measurements were made on the foundation and on the ground close to the building on the side facing the road. Vibration sensors were mounted using methods that can faithfully transmit to the transducer—the actual motion of the ground or building components over the frequency range of interest. Since the mounting method was not suspected of distorting the motion, therefore alternative methods were not tested.

The degree of detail required in the analysis of the vibration signals depends on the nature and purpose of the investigation. For a preliminary evaluation, it might be sufficient to find the peak of the vibration signal and to determine the predominant frequency of vibration by counting the number of negative and positive peaks in a given time interval. For an in-depth evaluation, advanced analysis methods are necessary, such as one-third-octave frequency band analysis, frequency-weighting according to established human response curves, and spectral analysis.

This study was considered a preliminary investigation because it was the first study of its kind that was conducted on the Adamson University. As written in the related literature of this paper, the researchers focused on the determination of the Peak Particle Velocity because of the functionality of the vibration meter and in order to avoid further complications.

The data collected was plotted in a graph showing the relationship between the Peak Particle Velocity (PPV) and the distance from the Romualdez Street.

4 Results and discussion

4.1 Peak hour analysis

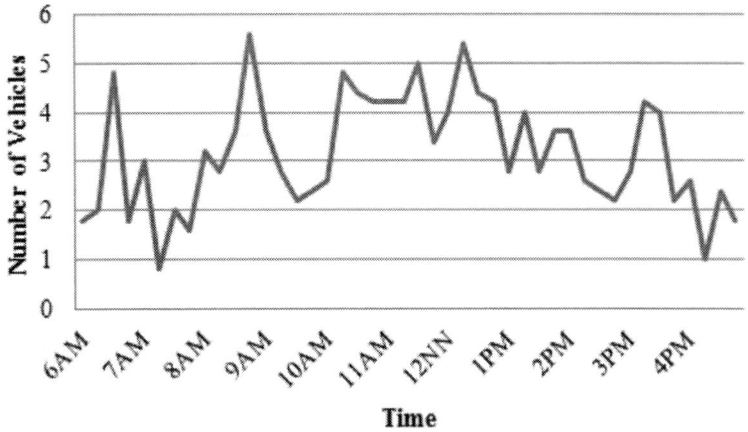

Figure 4. Variation of the number of vehicles traversing Romualdez St. from 6 AM to 5 PM

As shown in Figure 4, the variation of the number of heavy vehicles traversing Romualdez St with respect to time. The data obtained from this figure is important in order to compute and generate a graph for the peak hour factor as shown in Figure 5. The peak hour analysis showed that the ideal time interval to conduct the study is from 11:00 AM to 12:00 NN with a peak hour factor of 0.96. This value is close to one which means that the traffic volume at that particular time interval is almost constant.

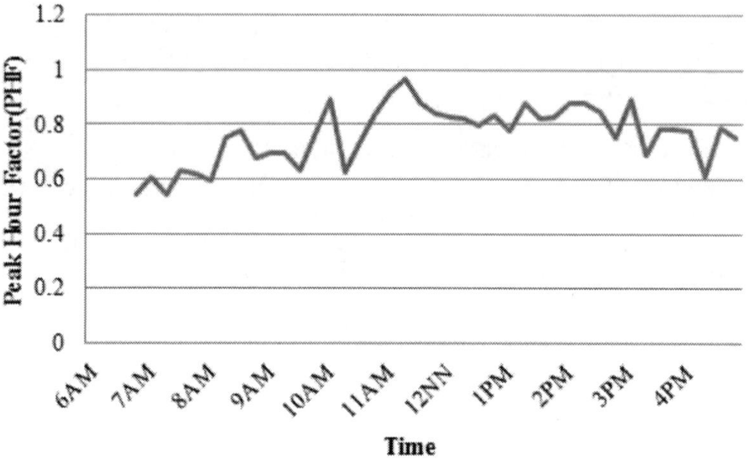

Figure 5. Variation of the Peak Hour Factor (PHF) from 6 AM to 5 PM

4.2 Distances of the columns of Ozanam building to Romualdez street

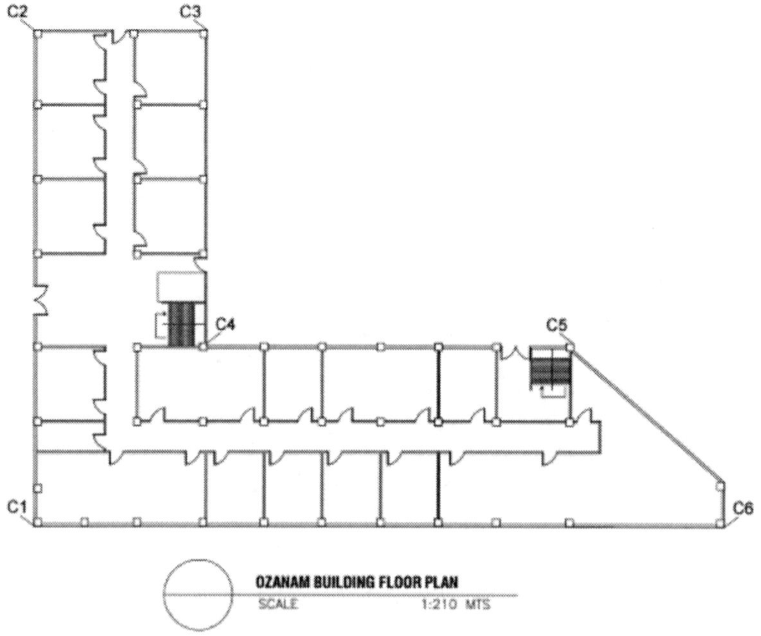

Figure 6. Columniation of the Ozanam Building

A total number of six (6) columns was considered to represent the ground floor area within the perimeter of Ozanam building as shown in Figure 6. The data obtained from the total station was able to yield the perpendicular distances of the columns with respect to the center line of the road in Romualdez Street. The setback between the property line of the Adamson University and the Ozanam

Building is found to be 3.80m. The road width of the Romualdez Street is found to be 10 meters as shown in Figure 7. In addition to this, the sidewalk width along the Romualdez Street near the Ozanam Building has a width of 2.70m.

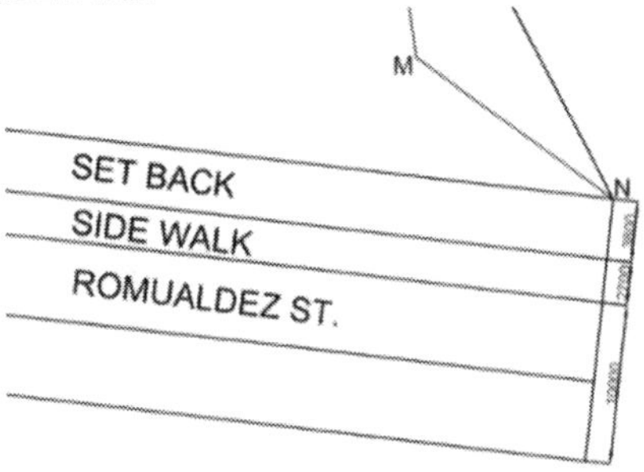

Figure 7. Widths of Setback, Sidewalk and the Romualdez Street

Table 2. Perpendicular distances of columns from the centerline of the road in Romualdez Street

Column Number	Distance (m)
1	11.50
2	60.80
3	56.10
4	30.05
5	31.07
6	11.50

The perpendicular distances of the column of Ozanam Building as shown in Table 2 was obtained by adding the computed departure using the total station, the setback, side walk, and half of the road width of the Romualdez Street.

4.3 Peak particle velocity (mm/s) of the columns of the Ozanam building

In order to predict the damage response of structures due to construction vibrations, it is desirable to perform a dynamic mathematical analysis of the structure. However, due to the numerous structural and environmental variables involved, even approximate analyses are difficult and expensive. Such analyses can seldom be accomplished within the scope of normal projects. Older structures are especially difficult to model dynamically because the building components are typically in varying states of deterioration, and previous settlements and movements in the structure often have redistributed the loads and stresses into unknown patterns.

When historic or landmark designated buildings are involved the cultural, social, economic, political and architectural importance of such buildings must also be considered in setting vibration limits. The possibility of damaging a unique historic building must be weighed against the increased cost of construction operations when low maximum permissible vibration limits are specified. As such, the selection of the proper vibration criterion for a historic structure becomes an economic decision as well as a technical one.

In this study, researchers considered only the first floor of the Ozanam building because vibrations originates directly on the ground floor and it propagates throughout the building.

Table 3. Summary of the Peak Particle Velocity(mm/s) of the Columns in Ozanam Building

Column Number	Peak Particle Velocity (mm/s)	Remarks
1	8.6	Threshold for risk of architectural damage
2	1.8	Unlikely to cause damage of any type
3	1.8	Unlikely to cause damage of any type

4	2.0	Effects ancient and historical monuments
5	3.1	Virtually no risk of architectural damage to normal buildings
6	7.8	Threshold for risk of architectural damage

The Lutron Vibration Meter with a model number VB-801HA was used to obtain the required peak particle velocities of each column in Ozanam Building. The following data as shown in Table 3 was obtained during experimentation:

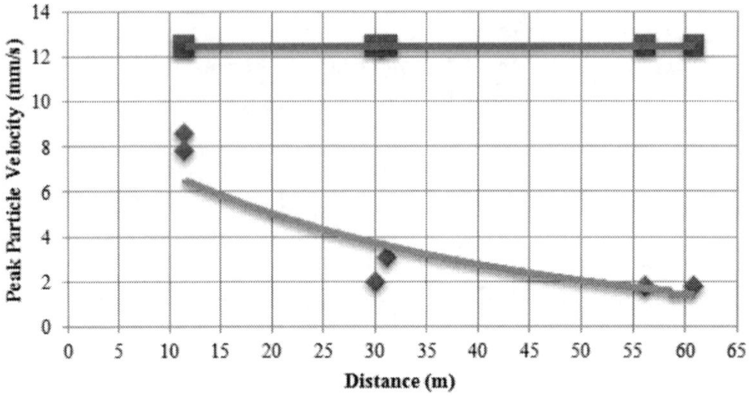

Figure 8. The relationship of peak particle velocity and distance of the columns from the centerline of the road in Romualdez Street.

As shown in Figure 8, the maximum peak particle velocity can be found on Column 1 with a value of 8.6 mm/s approximately 11.50m from the centerline of the road in Romualdez Street. This signifies that even though the Ozanam Building is exposed to such kinds of vibrations, it is still safe from any possible structural damages because according to the Traffic and Road Research Laboratory (U.K.), the minimum allowable peak particle velocity of older properties is at 12.5 millimetres per second.

4.4 Visible cracks on the building that is due to vibrations induced by traffic at Romualdez street

Upon close inspection of the Ozanam Building, there were no visible damages on the building that is due to vibrations induced by traffic at Romualdez Streeet in the interior. However, in the exterior or façade of the building, there are visible damages that might be caused by traffic-induced vibrations as shown in Figure 9 and 10.

Figure 9. Cracks near the Column 6 of Ozanam Building

In relation with the study Jakubczyk-Galczynska and Jankowski [10], when it comes to the effects of traffic-induced vibrations, they may be and small and imperceptible; however, they may as well become the reason of a major malfunction or even lead to collapse of a structure. The most visible effects of traffics are chaps and scratches on the plaster covering of the building, falling of paint, and plaster, crack on the elements of the structure or even building malfunction and collapses.

Figure 10. Cracks on the façade of the Ozanam Building

Another study by Tomazevic [11] states the traffic vibrations did not cause visible damage to the walls. They caused propagation of damage (cracks) in the case where the walls have been previously damaged. Traffic-induced vibrations caused invisible changes of the walls' structure, such as microscopic crushing of mortar and consequent softening and disintegration of masonry. These changes resulted into the measured changes in dynamic properties of the walls (natural vibration frequencies) as well as in measure changes in compressive and lateral strains under the same loading conditions.

5 Conclusion and recommendation

5.1 Conclusion

Based on careful observation and evaluation of the results obtained from the actual field-testing using vibration meter and total station, the researchers concluded the following:

- The peak hour in which vehicles that cause traffic-induced vibrations pass through the Romualdez Street was found to be at 11:00 AM to 12:00 NN with a Peak Hour Factor of 0.96.
- The researchers found out that the peak particle velocity due to traffic-induced vibrations in Ozanam Building of Adamson University was 8.6 millimetre per second in Column 1 approximately 11.50 m from the centerline of the road in Romualdez Street. This implies that the building is safe since the maximum threshold value is 12.5 mm/s. Thus, vibration due to traffic has no significant effect that can cause structural damage to the building.
- As for the damages due to traffic-induced vibrations, there has been a lot of cracks that can be seen in the facade of Ozanam Building along the Romualdez Street. This can be seen on the plaster walls of the Ozanam Building around the Second Floor.

5.2 Recommendation

The researchers recommend the future prospects who want to undergo related studies on traffic-induced vibrations to consider the following:

- The vibration meter used in this study has a frequency range of 10 Hz. This means that the meter is reading with a reads the peak particle velocity every one-tenth of a second. The researchers recommend using a more sophisticated vibration meter. To be specific, it is recommended using the Lutron Vibration Meter Model No. VB-8212, which is a portable type of a vibration meter like the one used in this study, but more sophisticated because of the larger sensor.
- For the Ozanam Building of Adamson University, the Adamson University - Physical Facilities Office (AdU-PFO) should consider retrofitting of the columns with special focus on those who are near the Romualdez Street. Good design is important, but even the best design is useless if not implemented properly. For this reason, it is advisable during the entire design and retrofitting process to monitor the myriad details that may affect vibration by careful review of relevant design, construction and shop drawings, and by repeated site inspection in the course of retrofitting.
- As for the visible damages as seen in the façade of the Romualdez Street, the AdU-PFO should take in regards using a Pressurize Concrete Epoxy Injector seal the cracks. This is for consideration because the researchers do not know if the cracks are traversing the cross section of the plaster walls. If so, the researchers recommend removing and replacing new plaster walls on the façade of the Ozanam Building facing the Romualdez Street.
- Ideally, vibration measurements should also be performed after the facility has been retrofitted so that conformance with vibration specification can be verified and any residual problems can be identified and resolved. With this in mind, the researchers recommend conducting another evaluation of the traffic-induced vibrations after retrofitting of columns and sealing/replacing of cracked walls to verify if the actions taken by the Physical Facilities Office of Adamson University minimized the vibrations to an acceptable level.

References

1. J.J. Hajek, C.T. Blaney and D.K. Hein, Mitigation of highway traffic-induced vibration, *In Annual Conference of the Transportation Association of Canada Charlottetown, Prince Edward Island, Session on Quiet Pavements: Reducing Noise and Vibration* (2006)
2. O. Hunaidi, *Traffic Vibrations in Buildings*, Construction Technology Update No. 39, Institute of Research in Construction (1985)
3. H.R. Nicholls, C.F. Johnson, W.L. Duvall, USBM, *Blasting Vibrations and Their Effects on Structures*
4. O. Hunaidi and M. Tremblay, Canadian Journal of Civil Engineering, **24**(5), 736-753 (1997)
5. A. C. Whiffin, D. R. Leonard, TRRL, *A survey of Traffic-Induced Vibrations* (1971)
6. M.O. Al-Hunaidi, Journal of Low Frequency Noise, Vibration and Active Control, **15**(1), 25-42 (1996)
7. O. Hunaidi, J.H. Rainer and G. Pernica, *Proceedings of 2nd International Symposium on Transport Noise and Vibration,* 103-108 (St. Petersburg, Russia, 1994)
8. S. Mathur, *A Study on Traffic Induced Vibrations* (2013)
9. H. Jaswal, *Analysis of Micro-Vibrations in Buildings* (2007)
10. A. Jakubczyk-Galczyńska and R. Jankowski, *The 9th International Conference Environmental Engineering* (Vilnius, Lithuania, 2014)
11. M. Tomazevic, *The Influence of Traffic-Induced Vibrations on Historic Stone Masonry Buildings*, (2006)

Necessity for diversity of commercial facilities for daily shopping in residential planning[a]

Pingli Chen[1,b] and Yungchi Yen[2]

[1]*Department of Urban Planning and Landscape, National Quemoy University, Taiwan*
[2]*Center for National Land Development Research, National Cheng Kung University, Taiwan*

Abstract. Since the 1990s, the government administrations have thought that supermarkets could replace traditional retail markets as the commercial facilities for daily shopping in Taiwan. However, traditional retail markets remained prosperous. This paper aims to investigate the followings: 1) whether traditional retail markets still play an important role in residents' daily shopping and 2) whether supermarkets and traditional retail markets are actually complementary rather than competitive as commercial facilities for daily shopping. The examination is based on the data regarding married women's daily shopping behaviours in a planned residential area in Taipei from a questionnaire carried out in 2007. Several things are concluded below. First, consumers do not definitely choose the new types of facilities for daily shopping. Second, supermarkets in residential areas may be complementary to traditional retail markets in residents' daily shopping. Third, traditional retail market plays a role beyond acting as the exchange space for goods. Fourth, even though cars are popular in Taiwan, married women still prefer to conduct their daily shopping inside or nearby their neighbourhood unit by walking. This paper suggests it is necessary to maintain diversity rather than to anticipate or limit the type of commercial facilities available for daily shopping in residential planning.

1 Introduction

Mainstream supermarket chains and hypermarkets such as Carrefour emerged in Taiwan beginning in the 1980s and, originally, government administrators or local planning boards thought supermarkets would be able to replace traditional retail markets. For example, Article 7 of the Taipei Retail Market Management Regulations was amended to develop planned market-lands into public supermarkets in 1990. However, in many cities and rural areas, traditional retail markets remained prosperous.

Many studies with different perspectives have been conducted such as studies focused on the use of hypermarkets [1] and on consumers' habits and interrelated activities in a large downtown shopping centre [2]. In this study we focus on the daily shopping activities of married women to observe their behaviours in traditional retail markets and in the new type of community commercial facilities such as supermarkets and hypermarkets, the former having existed before the introduction of modern urban planning theories and the latter having increased in prevalence since the 1980s. We aim 1) to discuss

[a] The draft version of this paper has been presented at the 3th International Conference on Built Environment in Developing Countries, Malaysia, 2011.

[b] Corresponding author : piyocu@gmail.com

how married women conduct their daily shopping and how characteristics such as their occupation and children influence them in this activity and 2) to clarify the roles of traditional retail markets, supermarkets, and hypermarkets in married women's daily shopping activities, and 3) to discuss the issues of commercial facilities for daily shopping in residential planning in the future.

The second section deals with the relationships between neighbourhood units, daily shopping and married women by considering different bodies of literature. In the third section, we briefly present an overview of our case study area, the Minsheng Community in Taipei city, anddescribe our research methodology. The fourth section analyses the features of married women's daily shopping activities. Finally, some conclusions are drawn in the last section. We do this not only by reporting the findings but also by discussing their implications for the future of planning in residential environments.

2 Neighbourhood units, daily shopping, and married women

The neighbourhood unit is considered to be a popular residential design paradigm, which was conceived by Perry[3] and has been widely used as a design concept in many countries [4]. In Perry's description of the physical attributes of the neighbourhood, local shops were important elements in a neighbourhood unit. Conceptually, these local shops were expected to be on the edge of neighbourhoods, preferably at traffic junctions and close to similar districts of nearby neighbourhoods.

This approach was used in new town developments in many countries. When it was applied in England and Japan, the arrangement of the neighbourhoods was changed from Perry's original plan. The Dudley Report, published by the Ministry of Health in Britain, recommended that the main neighbourhood shopping activities should be within the neighbourhood and should form part of the neighbourhood centre [5]. A similar idea was commonly applied in new town plans in Japan after World War II.

In Taiwan, after World War II ended, the first new town named Chung-Hsin New Village was developed in 1957. The plan of Chung-Hsin New Village referred to the cases in Britain. In Chung-Hsin New Village, traditional retail markets were established in the centre of neighbourhood units for the residents' daily shopping. After this experimental new town developed, several facilities were legislated to be arranged in neighbourhood units as public facilities in the Urban Planning Act of 1964 in Taiwan. The "market" was one of those facilities, and it aimed to satisfy the residents' daily shopping needs.

Before supermarkets appeared in Taiwan, the term "market" was equivalent to the traditional retail markets. As is literally stated by the word 'traditional', traditional retail markets are areas where booths and stores are located and people sell goods in an age-old way, face-to-face. Every booth and store is small in scale and only provides specific goods, e.g., the butcher, greengrocer, florists or grocery stores. These booths and stores have a complicated relationship in that they complement and compete with each other in the market where they are located.

As supermarkets became increasingly popular, the consequences started to change. There was an official research project regarding community facilities planning undertaken in 1989 by the Department of Urban and Housing Development in CEPD. The study claimed that traditional retail markets and supermarkets should be considered together when developing a residential community because people had changed their everyday shopping behaviours[6]. Although this claim was based on case studies of private sector development rather than on the shopping behaviours or needs of the residents, the idea "to replace the traditional retail markets with supermarkets" has gradually permeated the bureaucracy. The amendment of the Taipei Retail Market Management Regulation mentioned above reflects this phenomenon. Supermarkets and their suppliers, are emerging as key policy and political debates in many other developing countries [7].

According to Christaller's central place theory, accessibility can be used to explain phenomenon related to daily shopping because grocery and household items are defined as low order goods. Daily shopping is often regarded as a local activity. However, such a local activity is actually an important part of our everyday lives and sometimes not as simple as Christaller states. For example, Wang and Lo examined the preferences of Chinese immigrants for the fast-growing Chinese supermarkets versus the competing mainstream supermarket chains and found that ethnic affinity had a stronger effect on immigrants' choice of shopping venue than economic rationality such as considerations of accessibility[8]. The authors therefore claimed that grocery shopping has socio-cultural meaning for immigrants and that the social use of ethnic shopping spaces indicates that immigrants are not only consumers in ethnic shopping places but also participants in this unique ethnic retail environment.

Actually, activities such as domestic duties and childcare are part of the daily routine for married women and influence their movements in everyday lives. Therefore, married women may have different life patterns than married men. It was discovered that married women have to make more trips and spend more time buying necessities on weekdays than married men do[9]. Their study also showed that married women with babies or infants will go out more frequently but spend a shorter time out than other married women do.

Beyond these discussions, we speculate that such a "taken-for-granted" activity is more intricate and interesting than previously thought and should be discussed in more detail. Actually, many studies starts to rethink the role of traditional retail markets[10] [11]. We will examine whether it is appropriate to replace the traditional retail markets with supermarkets.

3 Study methodology

3.1 Summary of case study area

We chose the Minsheng Community, a planned residential area in Taipei city, as our case study area for the following reasons: 1) it was well-planned with the help of foreign planners in 1960s and offered new ideas about how a good residential area should be designed during that period; 2) the population in this community underwent a high-growth period during 1970s-1980s, and it is now in a stable state in terms of private development, population growth, and the life patterns of the residents; and 3) there are sufficient and various facilities for everyday life in this area and it is appropriate for observing the shopping behaviours and preferences of residents. Figure 1 shows the location of the Minsheng Community in Taipei City.

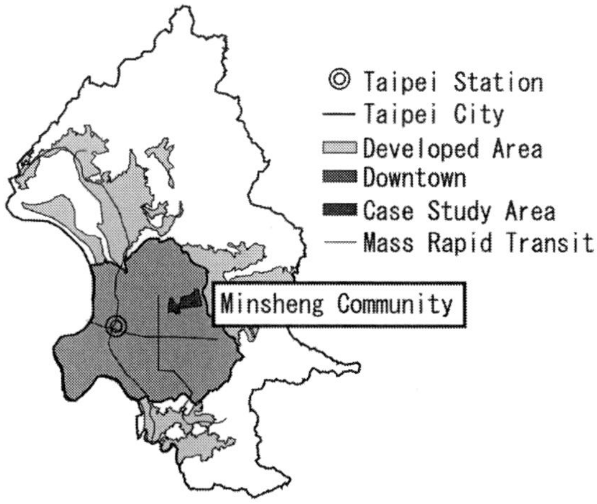

Figure 1. Location of the Minsheng Community in Taipei.

The Minsheng Community was developed in the 1960s as the model project for residential land adjustment, with help from the United States in terms of money and professional knowledge of planning. Because it was close to downtown Taipei City, the housing was mainly in medium-rise apartment. In the initial plan, the Minsheng Community was separated into seven neighbourhood units. The size of the neighbourhood unit was about 8 to 16 ha, which is small due to the high density in this area. There were two centre areas set in the Minsheng Community and therefore, two neighbourhood clusters that are formed by three to four neighbourhood units can be identified. Figure 2 shows the original detail plan of the Minsheng Community in 1968. The detail plan has changed for several times and version of 1988 was the latest one when our study started.

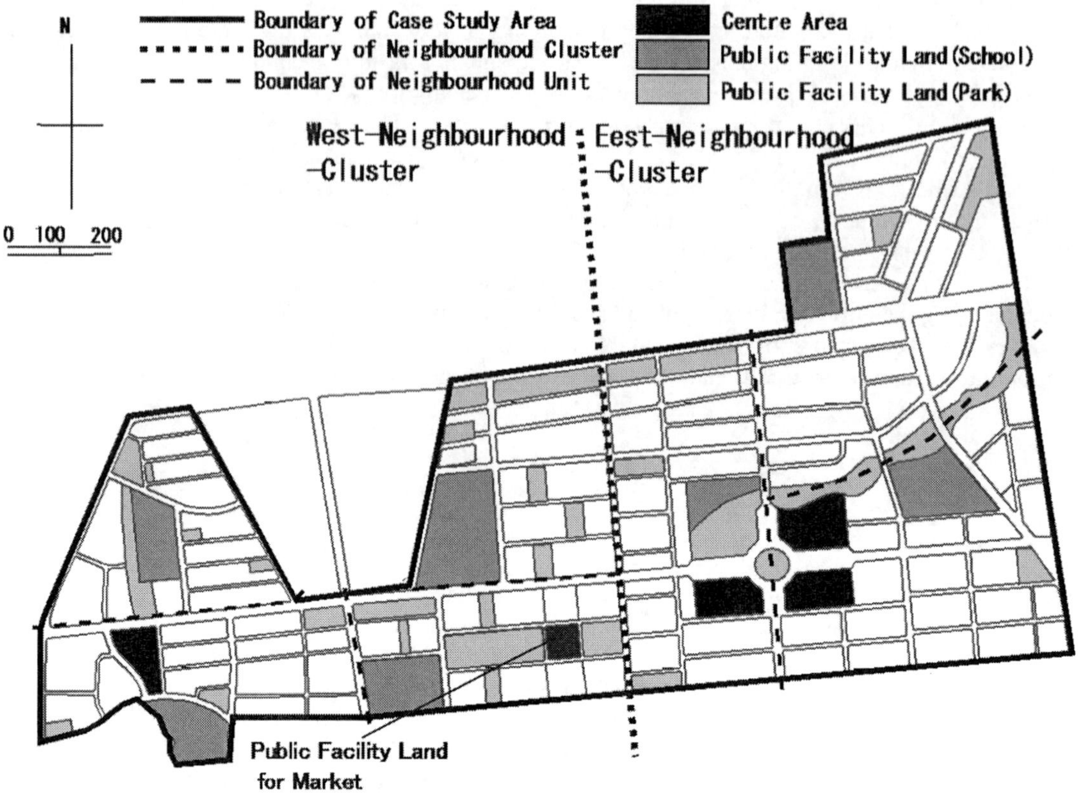

Figure 2. The original detail plan of the Minsheng Community in 1968.

Living facilities were arranged at two levels: the neighbourhood unit level and the neighbourhood cluster level. At the neighbourhood unit level, there was one primary school and several parks. The other facilities, including those for daily shopping, were arranged at the junction of the neighbourhood units to serve more people at the neighbourhood cluster level. Due to the small size of the neighbourhood units, both neighbourhood units and neighbourhood clusters were in walking distance of necessary facilities. The terrain was flat and it was possible to walk to the facilities at the neighbourhood cluster level or to go there by bicycle. Although supermarkets were still a new type of shopping facility in many countries and had not appeared in Taiwan yet in the 1960s, the commercial facilities for daily shopping in the Minsheng Community were assumed to be supermarkets due to the suggestion of the foreign experts. However, to pander to the shopping habits of residents at that time, a block of public facility land for traditional retail market in the middle of the Minsheng Community was planned.

After 40 years, the landscape has changed in the Minsheng Community. As a result of loose zoning regulations and pressure to development land and property in Taipei City, several condominiums have been renovated from medium-rise to high-rise, and there are now banks, clothing stores, restaurants, and sundry shops located not only at central areas but also along the main streets. The hypermarket even appears in residential areas. These changes are actually beyond the scope of the planners' expectations.

The residents here are middle-class. The population in 2006 was 55,608, and 20,080 households were located there. A population pyramid of the case study area is shown in Figure 3. Although the Minsheng Community has developed over 40 years, many young people live there and the population is not over-saturated with old people as we usually see in some suburban 'old' new town areas. The reason many move to this community could be that the second-hand housing market is active in

Taiwan, and the community is attractive because of its good location within TaipeiCity. The proportion of those who are over 65 years old here is 12.0%, which is close to the rate of 11.64% in Taipei City. This fact also reflects the number of new families moving into the Minsheng Community. However, because of the large number of elderly citizens in Taiwan, established communities such as the Minsheng Community face a demographic shift, and this indicates the importance of considering the needs of residents in different stages of the family life cycle.

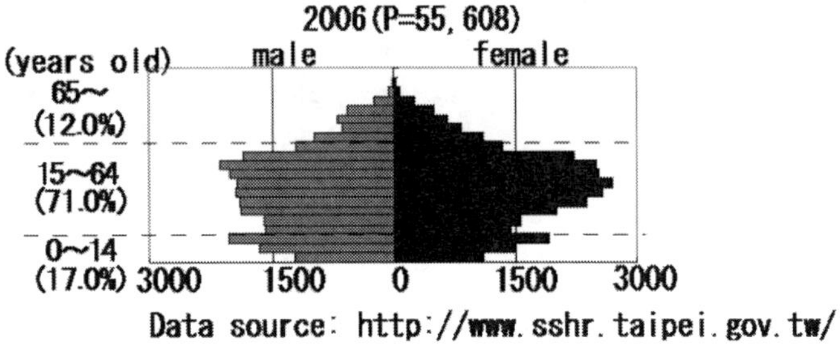

Figure 3. Population pyramid of case study area.

3.2 Commercial facilities for daily shopping in case study area

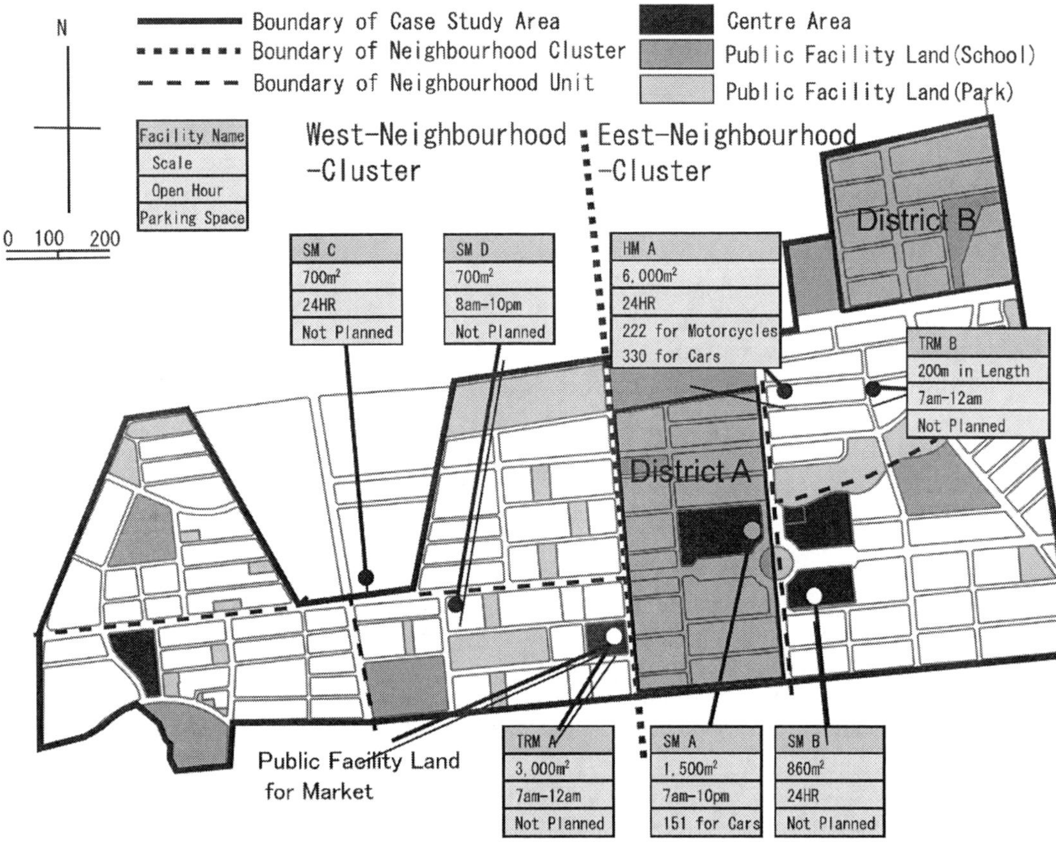

Figure 4. The information and location of commercial facilities.

In this study, we focus on three kinds of commercial facilities for daily shopping: traditional retail markets, supermarkets, and hypermarkets. From the pilot survey, which we will explain in the next section, we found that most married women do their daily shopping inside of the Minsheng

Community. Therefore, we will chiefly describe the commercial facilities for daily shopping inside of the community. During our survey time period in 2007, two traditional retail markets, five supermarkets, and one hypermarket were located in the Minsheng Community. The information and location of these facilities are shown in Figure 4.

Traditional retail market A (TRM A) was planned while traditional retail market B(TRM B) was spontaneous. The former is a building with several booths inside. The latter, however, is formed by shops on the first floor of the buildings and booths in front of these buildings along the residential streets. Most boothes and shops in these two markets are open in the morning and close before 14:00. Before the first supermarket appeared in 1989, these two traditional retail markets supplied the residents with what they needed in their everyday lives.

Due to the growth of new commercial facilities for daily shopping such as supermarkets and hypermarkets in the urban residential area since the 1990s, the environment for daily shopping in the Minsheng Community has greatly changed. Two supermarkets are located in the east centre area where commercial facility placement was anticipated in the plan.Supermarket A (SM A), the publicly operated chain supermarket that opened in 1991, and Supermarket B (SM B), the Japanese-based chain supermarket that opened in 1994. Besides these two supermarkets, Supermarket C (SM C, based in Hong Kong, opened in 1989)and Supermarket D (SM D, based in Taiwan, opened in 2003) are located along the Minsheng East Road, the main street across from our case study area. The Hypermarket A (HM A) is located in the east neighbourhood cluster.

3.3 Research methodology

In this study we focus on married women's daily shopping activities. Methodologically, we use a mixed approach to the case study. A survey questionnaire provides quantitative data on married women's daily shopping behaviours in the case study area, and an interview survey provides narratives of married women's daily shopping experiences, their attitudes, their preferences, and their explanations of their behaviours.

The reasons why we choose married women are detailed below. First, married women are usually considered to take care of their families. They perform activities such as housework, child-rearing, looking after aging parents, and purchasing necessities for their families as part of their gender role in many countries. Although the number of women working outside of the home has increased, some studies show that most married women are still responsible for the domestic duties and childcare in Japan, Korea, Taiwan, the United States, and even in Europe [8-13]. The second reason for choosing married women is related to the phenomenon above. It is said that married women's radii are the basis for arranging facilities for everyday lives in residential planning because they are responsible for daily shopping and usually move around[14]. So, we expected that married women would be appropriate in our study because our main purpose is to address the issues of commercial facilities for daily shopping in residential planning.

The questionnaire survey consists of questions concerning how married women use the commercial facilities for daily shopping, including frequency, transportation issues, and items purchased. The survey was conducted from January 15-21, 2007, with a pilot survey conducted earlier to identify the question items and formulate the questions for the formal survey questionnaire. A random sampling technique was applied as follows. First, we chose two neighbourhood units in the eastern neighbourhood-cluster in our case study area - one is close to the centre area (District A) and another is far from it (District B). The location of District A and District B is shown in Figure 4. Then, we listed all of the address numbers (including the floors) in these two neighbourhood units by site investigation. Third, we took each household at an interval of three numbers as our samples and chose 1,168 households. All of the married women in these chosen households were asked to answer the questionnaire. We ultimately gathered 162 samples for our analysis. Because the data necessary for our study are mostly categorical data, we have used the Chi-square test and Fisher's exact test to analyse the relationships between married women's attributes and their activities.

The interview survey was held during January 2008. We used snowball sampling to choose eight married women in District A and five married women in District B as our interviewees. The attributes of the interviewees are shown in Table 1.

Table 1. Attributes of interviewees.

Residential District		District A		District B	
Occupation		Career woman	Housewife	Career woman	Housewife
Family Life Cycle	Child-rearing Period	**A1**	**A2, A3**	**B1**	**B2**
	Post-child-rearing Period	**A4**	**A5, A6**	**B3**	**B4**
	Elderly Period	**A7**	**A8**		**B5**

4 Results

4.1 Individual and family attributes

In this study, we assume that married women's everyday lives are constrained by their occupation and family life cycle according to the theoretical arguments of family life cycle [8, 15-18].

We define the categories of occupation as 'career women' and 'housewives'. Career women are defined as the women who have paid jobs while housewives do not. We define the categories of family life cycle as 'child-rearing period', 'post-child-rearing period', and 'elderly period'. The 'child-rearing period' women mean the women who have at least one young child. Our study defined a young child as ranging in age from newborns to elementary school students (0-12 years old). The feature in this period is the double pressure from domestic duties and child-rearing responsibilities. Although married men have begun to become conscious of the importance in sharing housework with their wives, married women are still mainly responsible for these burdens in households in Taiwan. The 'post-child-rearing period' women mean those who are under 65 years old and their youngest child is older than 13 years old. The married women in this period still have to take care of their children, but the burden is lighter compared to the child-rearing period. The 'elderly period' for women refers to those who are over 65 years old. During this period, married women are relieved of the burden of child rearing (although some have started to take care of their grandchildren). However, as their health condition becomes feeble, they may have different needs in their daily lives. The distribution of the respondents is shown in Table 2.

Table 2. Cross table of occupation and family life cycle.

		Occupation		**Total**
		Housewives	Career Women	
Family Life Cycle	Child-rearing Period	16	45	61(37.7%)
	Post-child-rearing Period	44	36	80(49.4%)
	Elderly Period	18	3	21(12.9%)
Total		78(48.1%)	84(51.9%)	162(100%)

χ^2=25.11, df=2, P=0.000

Career women and housewives comprise approximately the same proportion of the study population. In the family life cycle, about half of the respondents were in the post-child-rearing period, and respondents in the child-rearing period are also well-represented. Of the respondents, 12.9% were in the elderly period, which is close to the prevalence of elderly individuals in our case study area. During the child-rearing period, career women are more prevalent than housewives. The percentage of career women becomes lower as the family life cycle period increases. The number of respondents in District A is slightly higher than in District B. The distribution of the family life cycle and occupation in these two districts is not significantly different based on the results of the Chi-square test (Table 3 and Table 4).

Table 3. Cross table of occupation and residential district.

		Occupation		Total
		Housewives	Career Women	
Residential District	District A	47	44	91(56.2%)
	District B	31	40	71(43.8%)
Total		78(48.1%)	84(51.9%)	162(100%)

χ^2=1.02, df=1, P=0.313

Table 4. Cross table of family life cycle and residential district.

		Occupation			Total
		Child-rearing Period	Post-child-rearing Period	Elderly Period	
Residential District	District A	34	43	14	91(56.2%)
	District B	27	37	7	71(43.8%)
Total		61(37.7%)	80(49.4%)	21(12.9%)	162(100%)

χ^2=1.14, df=2, P=0.567

4.2 The location and types of commercial facilities for daily shopping

Commercial facilities in residential areas are arranged to satisfy daily shopping needs. Here, we focus on food items and household items, which are purchased frequently in the facilities of interest to support family's daily life. We divide food items into perishables and non-perishables. The former are those with a short shelf life that are likely to decay or go bad quickly such as meat, vegetable, fish and fruit; the latter are those with a relatively long shelf life such as canned food, frozen food and beverage. Household items include tissues, detergents, cleaning goods, and personal hygiene products. Figure 5 shows the results of the survey of commercial facilities for daily shopping. Most of the respondents purchase food items and household items in the Minsheng Community rather than outside of the community, which is similar to the results of the pilot survey.

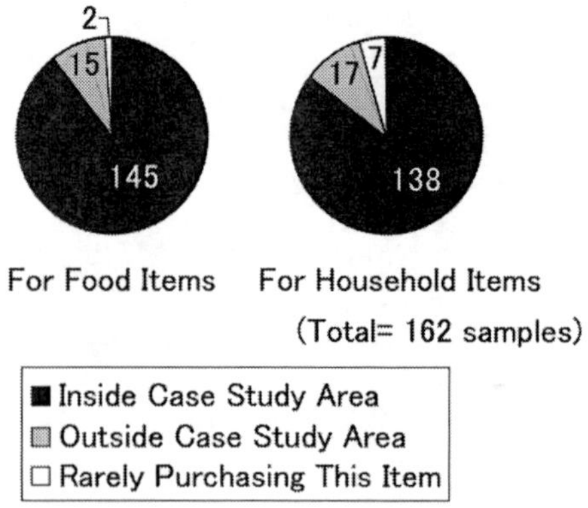

Figure 5. Location of daily shopping.

The married women in the Minsheng Community have a tendency to choose more than one commercial facilities for daily shopping. Table 5 shows how many respondents visit specific facilities inside the Minsheng Community. An interesting finding is that although TRM B and HM A are not planned commercial facilities, they are used by most respondents in this survey. The respondents use these two facilities in separate ways, purchasing perishables at TRM B and purchasing non-perishables and household items at HM A.

Although half users of HM A purchase perishables there, these people also purchase these items at TRM B. This means that hypermarkets and traditional retail markets may be complementary facilities because they have different advantages that are considered by married women conducting their daily shopping. However, HM A appears to be competitive with SM A and SM B because all of these stores provide non-perishables and household items. Due to less than 25% of the respondents use SM C and SM D, we will only focus on TRM A, TRM B, SM A, SM B, HM A below.

Table 5. The numbers and percentages of respondents who use specific facilities.

	Total		Perishables		Non-perishables		Household items	
	N	% (=N/162)	N	% (=N/162)	N	% (=N/162)	N	% (=N/162)
TRM A	64	40%	44	27%	13	8%	1	1%
TRM B	140	86%	127	78%	31	19%	6	4%
SM A	112	69%	42	26%	62	38%	30	19%
SM B	96	59%	45	28%	51	31%	21	13%
SM C	21	13%	11	7%	14	9%	10	6%
SM D	41	25%	4	2%	35	22%	36	22%
HM A	149	92%	79	49%	108	67%	114	70%

4.3 The frequency of commercial facilities for daily shopping

Figure 6 indicates how frequently respondents visit specific facilities. The frequencies of use of TRM B and HM A are also higher than those of other facilities. We can say that, for the married women in the Minsheng Community, these two facilities are the major ones for their daily shopping.

Although most of the respondents use TRM B regardless of their family life cycle or occupation, we noticed that married women with different characteristics have different tendencies related to their frequency of using TRM B. Married women in the elderly period use TRM B more frequently than those in the child-rearing period do. Additionally, housewives use TRM B more frequently than career women do. This tendency to use one market over the other implies that women's child rearing responsibility and their paid work constrain the frequency with which they purchased perishables in a traditional retail market.

From the interview survey, we found that although career women and women in child-rearing period appear to shop at the TRM B, the causes of this phenomenon are a bit different. In contract with the supermarkets and hypermarket in our case study area, TRM B has limited open hours in the morning. Some interviewees with careers indicated that they could only go to the traditional retail market on the weekend (interviewee A4 and B3) because it is tiring to go shopping in the early morning and then go to work. The interviewee with a baby (A2) states that she can only go on the weekends when her husband can take care of their baby because the traditional retail market is too crowded to go to with baby. Another interviewee with a baby, A3, even indicated that she does not go to traditional retail markets because it is very dangerous to buy products there when she has to push a baby pram. On the contrary, A3 likes to visit Matsusei, the supermarket. She likes to visit as it provides a clean, orderly, and safe shopping environment for her and her baby. According to the narratives, the constraint to career women is time limitations, but the constraint to women in the child-rearing period seems to be the unfriendly shopping environment of the traditional retail market.

From the interview survey, we found there is a different meaning of friendly vs. unfriendly shopping circumstances for the married women in a post-child-rearing period or elderly period. For these married women, traditional retail markets are friendlier than the new commercial facilities. The respondents are accustomed to going to traditional retail markets (especially the TRM B) and have gone for many decades. This habit has been founded on several things: the freshness of the perishables, the diversity of choices, and the familiarity and trust gleaned from the face-to-face selling style. Additionally, many interviewees indicated that going for daily shopping in traditional retail markets is not only a habitual activity of shopping itself, but also it is a way to keep in touch with people because they may meet neighbours there or chat with familiar booth or shops owners. The customers and the owners sometimes exchange their information and even get the benefit of interaction from information exchanges. By such a mechanism, they build a relationship. For example, A8 said she introduced a man to the daughter of one customer at a cooperative where A8 often visits, though the two young people had no resulting relationship. These married women in the post-child-rearing period or the elderly period also visit the new commercial facilities. They enjoy the cleanliness and the order there, but they rarely experience human touch and friendliness there as they have when shopping in traditional retail markets.

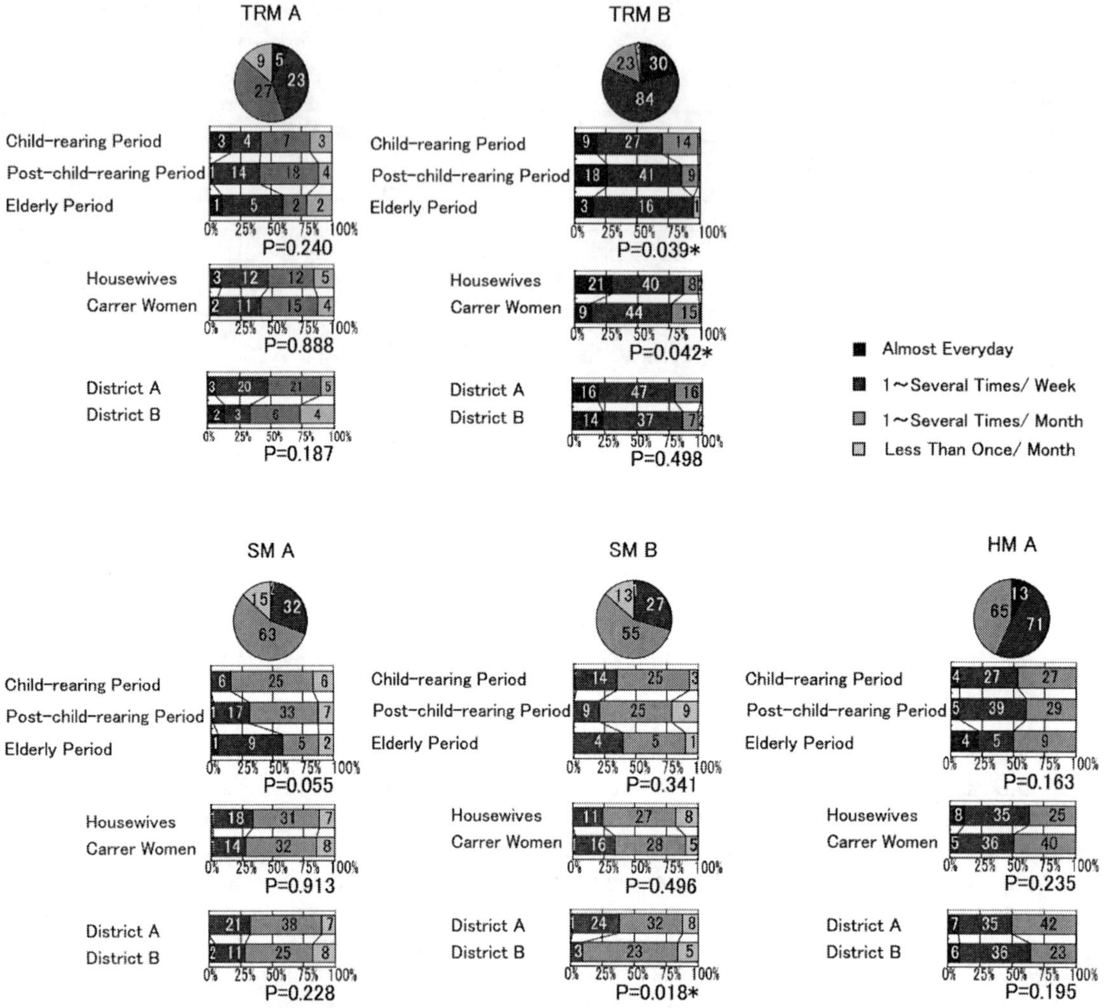

Figure 6. Frequency of visiting specific facilities (Only those who visit are calculated).

4.4 The transportation to commercial facilities for daily shopping

As indicated by Figure 7, we can see that most of the respondents walk to these facilities. Married women in the child-rearing and post-child-rearing periods, however, also show a tendency to go by way of other forms of transportation such as bicycle or motorcycle. Those who live in District A usually walk because they are close to the centre area, whereas those who live in District B may walk or go by bicycle or motorcycle due to the relatively long distance to the commercial facilities. However, no matter what occupation these married women pursue, which family life cycle they are in, or where they live, we found that a very small percentage of them go for their daily shopping by car.

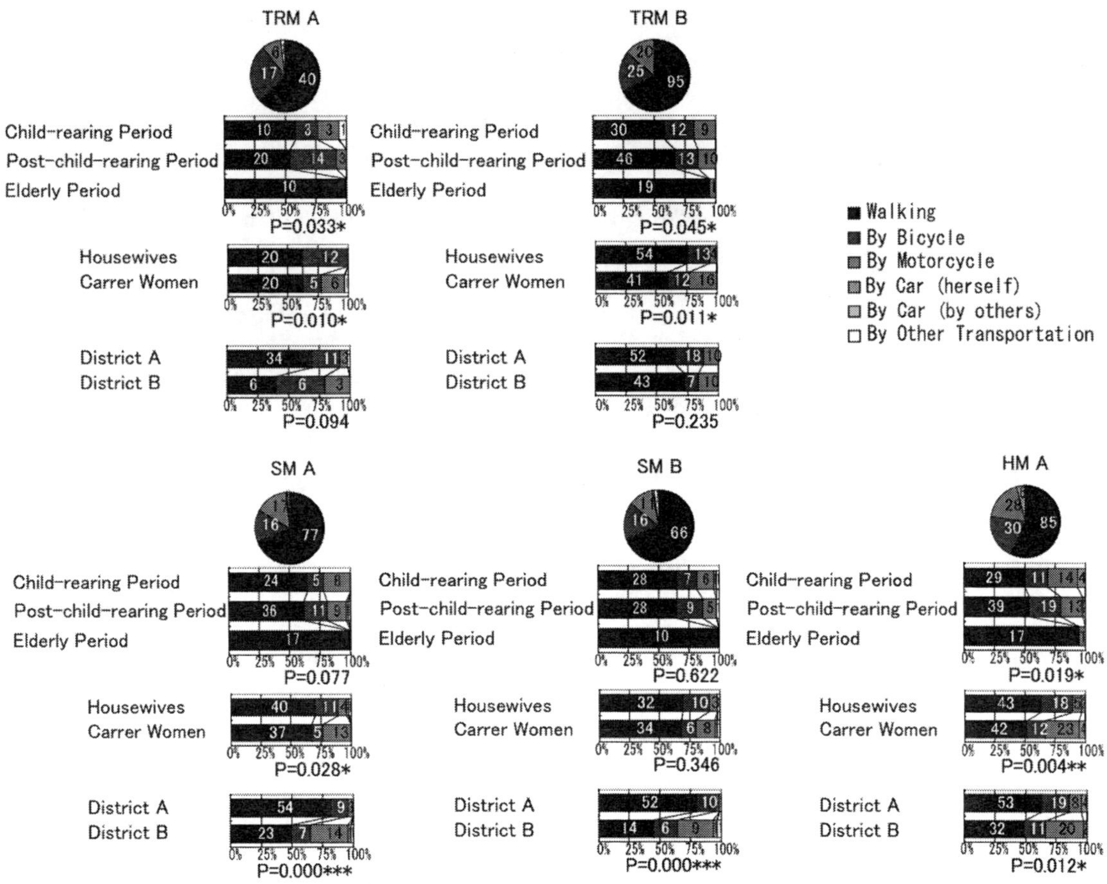

Figure 7. Transportation of visiting specific facilities (Only those who visit are calculated).

As determined from the interview survey, most of the interviewees indicate that they go shopping by walking because the places they go are close to their home. Considering the burden of moving cars, the difficulty of finding a parking space and the various mechanisms of time cost, they prefer to go by walking. Some interviewees mentioned the idea of environmental protection (A1, A2, B1, A5, A6, B4, B5) and the role of exercise for walking to conduct their daily shopping (A5, A6, B5).

5 Conclusion

From the analyses in section 4, we can conclude several interesting things below.

First, even if the type of shops where people can conduct their daily shopping changes, it does not mean that consumers will definitely choose the new types of facilities such as supermarkets or hypermarkets. The married women in the Minsheng Community have a tendency to choose more than one commercial facilities for daily shopping at the neighbourhood-cluster level. They still go to traditional retail markets for perishables and go to supermarkets or hypermarkets for non-perishables and household items at the same time. This means that traditional retail markets generally maintain strong advantages as the site for purchasing perishables, whereas supermarkets and hypermarkets play important roles in providing household items. We also found that married women with different characteristics have similar tendencies when choosing facilities but have different tendencies regarding their frequency of use. The different frequency of use comes from the limitations of married women's attributes. Although they have the same tendency in terms of frequency of use, career women have different reasons for reducing the frequency of their visits to the famous traditional retail markets from married women in the child-rearing period. Because different types of commercial facilities play different roles in daily shopping, and because different kinds of married women have different needs in choosing commercial facilities even for daily shopping, it is inappropriate to have only one of types of commercial facility when planning a residential area. Additionally, it is important to keep the diversity of commercial facilities for daily shopping to satisfy differing needs of residents rather than to anticipate or limit the type of community commercial facilities in residential planning.

Second, we find that hypermarkets in residential areas may not threaten traditional retail markets and may be complementary to them, though they clearly are competitive with supermarkets and may cause their decline. Indeed, hypermarkets play an important role in providing non-perishables and household items. From the viewpoint of increasing the diversity of commercial facilities for daily shopping, however, it may not be suitable to place hypermarkets in residential areas. At the very least, adequate discussions are necessary.

Third, from the narratives of the interviewees, we find that the traditional retail market plays a role beyond acting as the exchange space for goods. It is also a social space in married women's daily lives. This kind of social functioning is extremely valuable in an isolated society. It is important that planners, since Perry first introduced his residential plan, want to meet the needs of communities. In Japan, some supermarkets have noticed their shortcomings in terms of impersonality and have started to apply face-to-face methods into their selling strategies. It is encouraging to see diverse styles of interaction in commercial facilities. However, it does not mean that supermarkets can replace all the functions of traditional retail markets, and actually, such a replacement is not necessary.

Fourth, even though cars are popular in Taiwan, married women still prefer to conduct their daily shopping inside or nearby their neighbourhood unit by walking. This interesting finding implies a new meaning of neighbourhood unit in the context of environmental protection to reduce CO2. The third and the fourth findings imply the necessity to re-evaluate the neighbourhood unit and the planned/unplanned facilities therein.

Finally, it is an interesting finding that both of the favourite facilities in our case study are unplanned ones. There seems to be a gap between the residents' preferences and the idea of planners. Although it is not thoroughly discussed in this study, we suggest that further research is needed to fill the gap . Furthermore, we suggest that some research should be undertaken to identify the role of public facility lands for markets in the future.

References

1. K.K. Chu, L.M. He and H.H.Lin, Journal of the Financial Bank of Taiwan, **36**(8), 49-67 (2000)
2. H. Asada, K. Nagai and A. Konno, Journal of the City Planning Institute of Japan, **26**, 889-894 (1991)
3. C.A. Perry, *The Neighborhood Unit in Regional Survey of New York and Its Environs* (Tokyo, 1975)
4. T. Banerijee and W. Baer, *Beyond the NeighborhoodUnit: Residential Environments and Public Policy*(U.S.A., 1984)
5. A. Goss, Town Planning Review, **32**, 66-82 (1961)
6. Department of Urban and Housing Development in CEPD, *A Research on improving the quality of environment in urban residential community and planning of community facilities*(Taiwan, 1989)
7. T. Reardon and R. Hopkins, The European Journal of Development Research, **18**(4), 522-545 (2006)
8. L. Wang and L. Lo, Environment and Planning A, **39**, 684-699 (2007)
9. Y. Arai, T. Kawaguchi, K. Okamoto and H. Kamiya, Journal of the City Planning Institute of Japan, **24**, 373-378 (1989)
10. S. Gonzalez and P. Waley, Antipode, **45**(4), 965-983 (2013)
11. J. Smith, D. Maye and B. Ilbery, Area, **46**(2), 122-128 (2014)
12. K. Sechiyama, *Gender* (Japan, 1994)

13. Department of Budget, Accounting and Statistics, Taipei Government,*Taipei City Statistical Yearbook 2006* (Taiwan, 2007)
14. G.S. Becker, *A Treatise on the Family* (U.S.A., 1991)
15. S. Hanson and P. Hanson, Transportation, **10**, 165-183 (1981)
16. N. Gergson and M. Lowe, Sociological Review, **41**(3), 475-505(1993)
17. H. Dohi and S. Suzuki, *New Architecture Series 20 Residential Area Plan* (Japan, 1985)
18. E.M. Duvall, *Family Development* (U.S.A., 1971)
19. K. Morioka, *Changes of Modern Family* (Japan, 1993)
20. T. Sakamoto and K. Kinoshita, *Family Sociology- Fundamentals and Applications* (Japan, 2001)
21. M. Yamashita, M. Baba, C. Miyahara and Y. Sakurai, Journal of the City Planning Institute of Japan, **41**(3), 659-664 (2006)

Decoding network patterns for urban disaster prevention by comparing Neihu district of Taipei and Sumida district of Tokyo

Hung-Wen Chang[a] and Wei-I Lee

Architecture Department, National Taiwan University of Science and Technology, Taiwan

Abstract. In this study, we performed routes network transport and emergency shelters capacity rate analyses to determine the accessibility and efficacy of urban patterns, and established a quantitative method for supplying priorities for actions of "Sendai Framework for Disaster Risk Reduction 2015-2030". By comparing two case studies, we used Space Syntax to develop two important indicators, Rn and CR, to present geographic information and hazard risk in a physical environment. This research also found potential function of Rn and decoded some patterns for urban planners or decision makers as follows ：The most efficient configuration of the road network was not in the old areas of these two case studies because the several turns decreased the connectivity of the networks. And the CR indicator shown other findings about the quality of public facilities and services as follows：The service capacity of the emergency shelters was surveyed to indicate a higher correlation of residents population and preparedness security for disaster management. Therefore, with finding some risks that had not been encountered before, we addressed this proposed method is feasible and reliable to enhance the disaster preparedness for action regarding the 4th priority of "Sendai Framework for Disaster Risk Reduction 2015-2030".

1 Approach to priorities for actions of "sendai framework for disaster risk reduction 2015–2030"

The Sendai Framework for Disaster Risk Reduction 2015–2030 [1], which was formulated by the United Nations Office for Disaster Risk Reduction (UNISDR) after the Japan earthquake and tsunami in 2011, outlines seven clear targets and four priorities for action to prevent new and reduce existing disaster risks: (1) understanding disaster risk; (2) strengthening disaster risk governance to manage disaster risk; (3) investing in disaster reduction for resilience; and (4) enhancing disaster preparedness for effective response and to "Build Back Better" for recovery, rehabilitation, and reconstruction. The framework aims to substantially reduce disaster risk and potential loss of lives, livelihoods, and health, in addition to economic, physical, social, cultural, and environmental assets of persons, businesses, communities, and countries. The 4th priority for action is related to urban planning, prevention plans, and city governance (Table 1). An increased disaster risk indicates a need to strengthen disaster preparedness, take action in anticipation of events, and ensure that facilities are in place for effective

[a] Corresponding author : ai3098@gmail.com

response and recovery in all phases of a disaster, including through integrating disaster risk reduction with development measures. Thus, the vulnerabilities arising from a disaster can be managed by reducing the risk of disasters (Figure 1).

Table 1. The relationship between priorities for action of Sendai Framework for Disaster Risk Reduction 2015-2030 and disaster management phase.

Priorities	Action
1	Understanding disaster risk
2	Strengthening disaster risk governance to manage disaster risk
3	Investing in disaster risk reduction for resilience
4	Enhancing disaster preparedness for effective response and to "Build Back Better" in recovery, rehabilitation and reconstruction

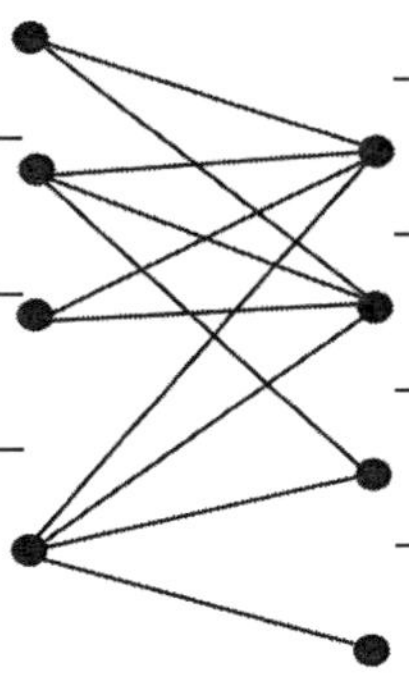

Disaster Management Phase
Mitigation
Preparedness
Response
Recovery

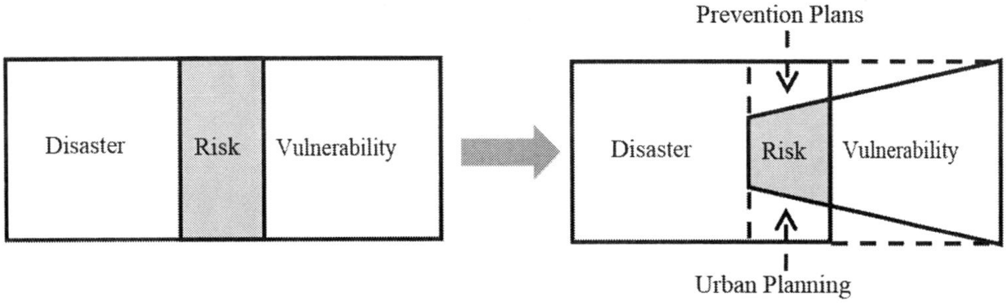

Figure 1. The role of prevention plans and urban planning in risk reduce.

2 Rapid diagnostic tests for urban resilience and literature review

Coordination of public infrastructure is an integral part of any town and country planning process, and aims to provide unlimited uptime of all systems, enabling the fulfilment of basic needs during disaster response, threat to public safety, damage or even destruction of the surrounding buildings. Jetel has described approval of land use plans, regulatory plans and decision making using the concept of disaster prevention as an effective strategy for city governance [2]. A major task in city governance is resource allocation. To understand the dangerous properties in high-risk areas and improve the evacuation equipment, enhancing the ability to respond to disasters is critical [3]. David Alexander [4] (Figure 2) and Wei-I Lee [5-7] reported the need to evaluate urban disaster prevention strategies under given circumstances. They also expressed the need to designate routes and evacuation shelters. The former should guide the evacuees away from the source of danger. This paper investigates the relationship between the emergency infrastructure for urban disaster prevention and priority 4 of the Sendai framework (Table 2) and reports that evacuation routes and shelters are the most important infrastructure in urban disaster prevention, and can help in saving a large number of lives.

Table 2. The relationship between the emergency infrastructure of urban disaster prevention and 4th priority for action of Sendai Framework for Disaster Risk Reduction 2015-2030.

National and local levels	Emergency Infrastructure					
Item	Evacuation Roads	Shelters	Fire Stations	Hospitals	Relief Supplies	Police Stations
Achievement -1	○	○	○	○	○	○
Achievement -2			○			
Achievement -3	○		○	○		○
Achievement -4		○				
Achievement -5	○	○			○	
Achievement -6	○	○				
Achievement -7	○	○				
Achievement -8	○	○			○	
Achievement -9						○
Achievement -10	○	○				
Achievement -11	○					
Achievement -12	○	○	○	○		○
Achievement -13	○	○				
Achievement -14		○		○		
Achievement -15		○		○	○	
Achievement -16			○			○
○:selected Sum	**10**	**11**	**5**	**5**	**4**	**5**

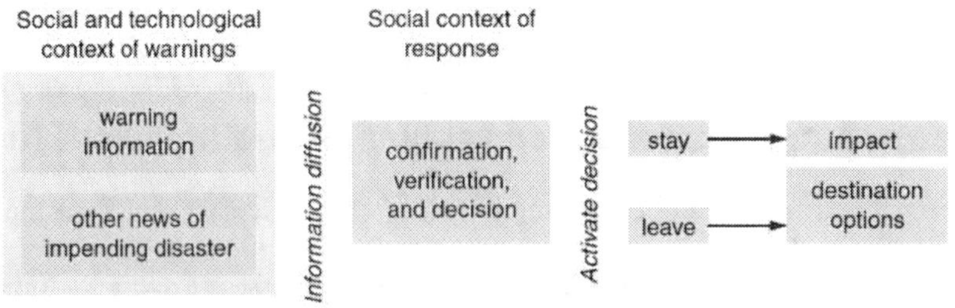

Figure 2. Evacuation plans have be made under given circumstances.(Reference from David Alexander's Principles in Emergency Planning and Management)

The details of achievements regarding the 4th priority of the Sendai Framework for Disaster Risk Reduction are as follows:

1. To update disaster preparedness and contingency policies, plans, and programs.
2. People-centric multihazard early warning systems, and disaster risk and emergency communication mechanisms.
3. Resilience of transportation and telecommunication infrastructure, educational facilities, and hospitals.
4. To establish community centers for the promotion of public awareness.
5. To establish or strengthen funding mechanisms and procedures for relief assistance and planning.
6. To train the existing workforce and voluntary workers to ensure better response in emergencies.

7. To ensure the continuity of operations and planning, including social and economic recovery, and basic services in the post-disaster phase.
8. To ensure rapid and effective response to disasters and resulting displacement, including access to safe shelters, essential food, and nonfood relief supplies.
9. To promote the cooperation of diverse institutions, multiple authorities, and related stakeholders at all levels.
10. To develop capacities for reducing disaster risk in the short, medium, and long term, such as by land use planning and improvement of structural standards.
11. To develop guidelines for the preparedness of disaster reconstruction, such as for land use planning and improvement of structural standards.
12. To consider the relocation of public facilities and infrastructure to appropriate areas outside the risk range.
13. To improve the ability of local authorities to evacuate persons living in disaster-prone areas.
14. To establish mechanisms of case registry and databases of mortality to improve the prevention of morbidity.
15. To enhance recovery schemes to provide psychosocial support and mental health services to all people in need.
16. To review and strengthen the national laws and procedures on international cooperation based on DRR guidelines.

A problem in the achievement of the "Sendai Framework for Disaster Risk Reduction 2015–2030" is the fact that no evaluation criteria are available. In this paper, an indicator to understand the risk levels in city environment is thus developed. The paper also contributes to the trend of urban resilience and risk management. Some major safety performance evaluation approaches are identified, which include expectation function, risk assessment, statistical quality control, price deflation, engineering economic factors, system analysis, artificial intelligence, and systems theory [8]. Based on these approaches, quantitative and qualitative models are proposed. However, the quantitative models focus on measuring indicators such as frequency, severity, percentages, relative weight, and economic gains and losses of the safety program. Moreover, an evaluation procedure for the evacuation routes and shelters is proposed for use by urban planners or the government.

Jianan Liu reported that an urban spatial structure is a methodology to understand land use and human behavior. Moreover, he proposed to use space syntax and geographic information systems [9]. In this study, a network of evacuation routes was evaluated by using space syntax.

Chang Hsueh Sheng and Liao Chin Hsien also used space syntax to study the relationship between the efficiency of a path and the location of shelters [10]. Moreover, they investigated the challenges in the planning of emergency shelter locations during disaster events. Previous research has been hindered by the inability to consider both the service capacity and mobility of the road network and by the lack of a method for assigning evacuees to emergency shelters after a disaster causes road closures. Therefore, they proposed considering the network and limited scope of serviceability of the emergency shelters. This paper analyzed the advantages of examining the service value of shelters in our case study.

Liao Chin Hsien and Chang Hsueh Sheng applied and aggregated landslide points, land use data, and socioeconomic indicators integrated with a mapping method to analyze the relationship between spatial structures and spatial patterns of disaggregate and aggregate levels in natural hazard areas [11]. Moreover, they sorted out regional maps to present the hazard locations. In this study, space syntax was used to evaluate the evacuation routes and capacity rate to examine the shelters' service value, and feedback to mapping the high-risk areas or low-risk areas. This paper thus proposes a tool for the 4th priority for action of the "Sendai Framework for Disaster Risk Reduction 2015–2030."

3 Methodology

The disaster prevention plans should be evaluated through spatial structure analysis and location configuration assessment to offer quantitative information for decision making. This paper presents a process to set up a frame(Figure 3) to rethink the arrangement of infrastructure and relief supplies. Moreover, it provides a tool as a spatial analytical method to understand the risk of a disaster for developing a new inspection system to manage disaster risk and urban resilience. The procedure and indicators are as follows:

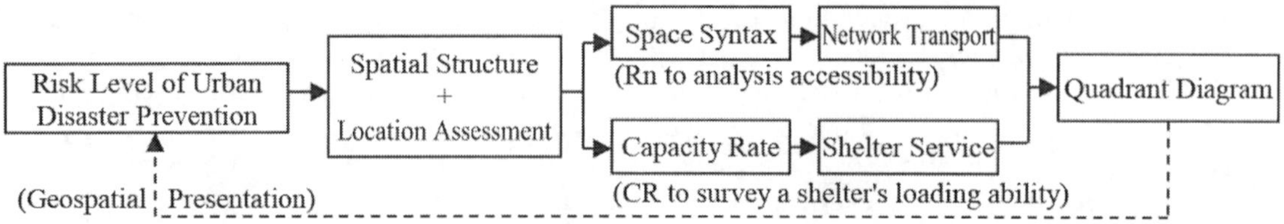

Figure 3. Framework of this research

3.1 Space Syntax for network transport analysis

On the basis of the ideas and procedures for understanding spatial topology, space syntax analysis has been used to examine the influence of spatial configuration on human behavior and analyze the factors important for ensuring the accessibility and efficacy of the route network in spatial configuration [12]. Ji-Soo Lee and Won-Hwa Hong reported some characteristics in temporary disaster relief facilities by using an integration (Rn) indicator [13]. Rn as a tool for measuring the level of accessibility to a movement potential or how easy it is to get to a segment from other segments [14]. The comparative value representing the accessibility of location was obtained by comparing the mean values of the shortest route (i.e., relative depth) from one unit to another in the road system. A high value of Rn for a unit indicates a highly accessible location, implying that the location of that unit is highly efficient and accessible in the road system [15].

The highly accessible route is the core section of topology in an urban route system that reaches the other roads with the fewest corners. In the "Depth Map" software used for space syntax analysis, higher and lower accessibility are indicated by red and blue, respectively (Figure 4). Details of the formula are as follows:

Formula: $Rn = 1/RRA$; $RRA = RA/DK$

Rn : Relative accessibility of whole district

RRA : Real comparative value of asymmetry

RA : Value of asymmetry $= 2(d-1)/k-2$

DK : Assumption of the relative depth of symmetry

d : Mean relative depth of the location

k : Number of units, n-1

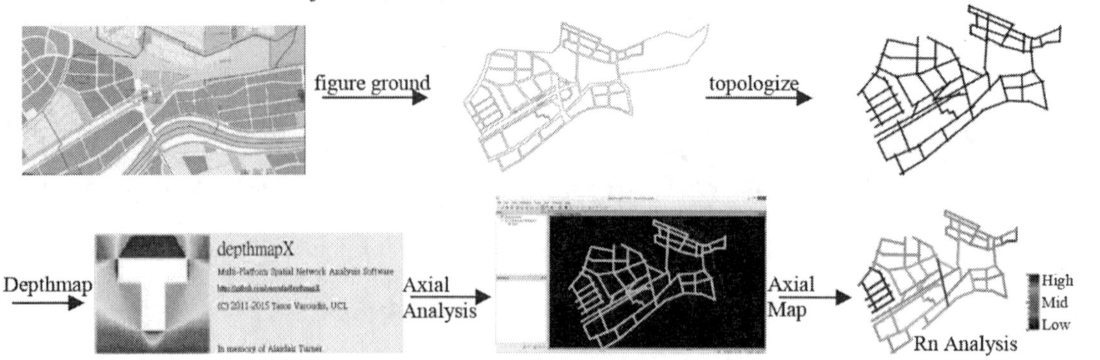

Figure 4. The analysis process in Depthmap X software.

3.2 Capacity rate for emergency shelter service survey

Evacuation routes and centers must be designated and preliminary shelters must be distributed during the short-term aftermath. There must be provisions for traffic control, maintenance of public order, and restriction of access to dangerous areas or places where emergency personnel are working [4]. Therefore, surveying the loading of capacity rate in every emergency shelter can determine whether the emergency shelters cover the service area [16]. In this study, we invested the percentage of capacity rate from a shelter and developed an indicator "CR" to determine the loading ability of shelters. In the formula, the denominator is the total number of people in a locality or village, and the numerator is the number of people a shelter can accommodate in an emergency. The formula is given as follows:

Formula: *CR (Capacity Rate) =APA/TPA*

APA：acceptable maximum people amount of a shelter's capacity

TPA：total people amount of a shelter's service cover area

3.3 Quadrant diagram of performance evaluation in hazard risk

In this study, we considered a quadrant diagram that divides the background into four parts. The X-axis (Rn indicator) and Y-axis (CR indicator) were used to draw data that contained two measurements. The measurement target was specified to facilitate environmental investigation using the quadrant diagram as a hazard measurement tool (Figure 5). In disaster management, the quadrant diagram is commonly used as a statistical method because the quadrant pattern represents the integration of two measurement definitions. Thus, if both the measured indicators are higher than the average then they can be defined in the first quadrant, which is the region with the highest spatial efficiency. Broadly, we can define the relatively low-risk region as the first quadrant, and the medium-risk region as the second and fourth quadrants. If one of Rn or CR is above the average then the risk can be defined in the second or fourth quadrant, which are areas with high spatial efficiency. However, if both of the measured indicators are below the average then the risk can be defined in the third quadrant, which is the area with the lowest spatial efficiency. Although disasters occur unpredictably, we must reconsider resource allocation and safety management in the areas of the third quadrant because they are at high risk relative to the overall urban system.

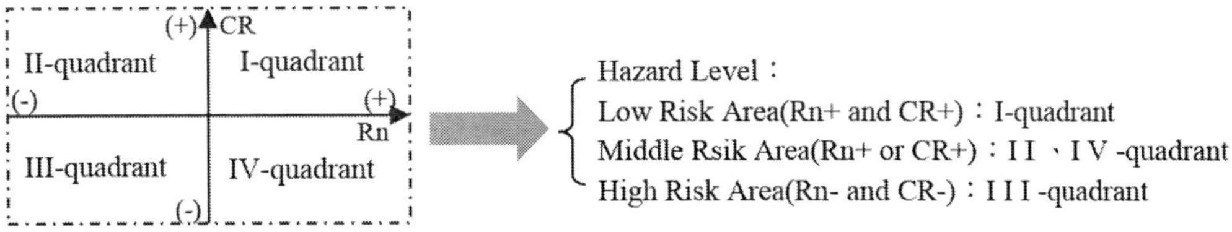

Figure 5. Quadrant diagram depicting hazard levels.

4 Empirical case study analysis

From a methodological point of view, this study translated a conceptual disaster prevention schema, an organizational urban infrastructure, and complex spatial relationships into quantitative spatial hazard metrics. Moreover, the study compared various real environmental configurations for different districts and cities. The method and conceptual framework described here can be applied to understanding the differences in disaster management decision making in urban networks. In Japan, the city governments have planned several disaster prevention mechanisms, as is the case in other East Asian countries such as Taiwan. In this study, empirical analysis of two similar urban regions was conducted and the quadrant diagram was developed. On the basis of space syntax and capacity rate, the Rn and CR indicators were used to evaluate the accessibility and efficacy of evacuation routes and service performance of shelters. In addition, Neihu and Sumida districts were chosen for the case study and compared based on the same environment characteristics, including old communities and new business districts, organic street shapes, and geometric boulevards. The differences between hazard risks and their influence on human behavior toward disaster management were the important concerns addressed in this research. Furthermore, using this information to support decision making for urban resilience and city governance was studied.

4.1 Rn indicator in Space Syntax case study of Neihu and Sumida districts

In this section, the routes in Neihu and Sumida districts were evaluated according to the Rn indicator. First, we found some obvious topological differences in the spatial configuration by using space syntax. In the case of Neihu District, the new regions had recently been developed and their topology was geometric. Space syntax proved that the distribution of the routes with high integration was concentrated. However, the old regions had been developed earlier and their topology was organic. Any road system with low integration should be allocated alternate disaster prevention plans to keep its roads useful in case of emergency.

By contrast, in Sumida District, the routes in old regions were highly integrated and concentrated in the topology of the spatial pattern. A ranking methodology using Rn as an indicator shows the maximum efficiency and integrated area in an urban system, and also presents the low-interactivity spots. Thus, the Rn indicator is a powerful indicator to describe the topology and spatial cognition of network configuration(Figure 6). Thus, this research studied the similar urban spaces of Neihu District in Taipei and Sumida District in Tokyo based on the properties of Rn and decoded some patterns for urban planners or decision makers as follows:

(1) The Rn indicator was used to determine the geometric characteristics in the new areas of these two districts; however, the results were not the same. The metrics in Neihu had some straight routes that provided an organized direction for evacuation. However, in Sumida, the central roads (red lines in the space syntax axial map) with high Rn values controlled the old area as a hierarchical network by which the residents or government might be aware of the location of important paths.

(2) The most efficient configuration of the road network was not in the old areas of these two districts because the several turns decreased the connectivity of the networks.

Therefore, in this study, we used the superiority of the Rn indicator as a metric to identify the location of high- or low-transport effectiveness areas. We calculated every path in a neighborhood or village as a unit for comparison with others. There were 39 units in Neihu (Table 3)and 26 units in Sumida(Table 4) to fit in the quadrant diagram. We calculated the average values as a mean index to identify the high- and low-efficacy areas. The low-efficacy area has high risk because it provides low transport effectiveness for people to evacuate, which is not favorable for urban disaster prevention plans [17]. The mean Rn indicator was Rn = 1.080 in Neihu and Rn = 1.249 in Sumida. This paper used Rn as the X-dimension to classify the higher or lower quantities in the case studies.

Table 3. Rn indicator of Neihu District

No.	Rn(average)	Segments	No.	Rn(average)	Segments	No.	Rn(average)	Segments
1	1.264	53	14	0.731	101	27	1.191	162
2	0.897	105	15	1.02	68	28	1.281	58
3	0.845	95	16	1.017	65	29	1.208	118
4	1.377	44	17	1.132	40	30	1.012	52
5	1.374	33	18	1.188	50	31	0.944	55
6	1.267	69	19	1.289	42	32	0.753	85
7	1.159	65	20	1.347	13	33	1.083	61
8	1.256	25	21	1.055	48	34	0.917	79
9	1.022	57	22	1.122	52	35	1.09	27
10	1.001	42	23	0.996	30	36	0.967	75
11	0.754	81	24	0.711	79	37	1.391	33
12	1.053	40	25	1.259	43	38	1.163	87
13	0.681	130	26	1.175	31	39	1.117	53

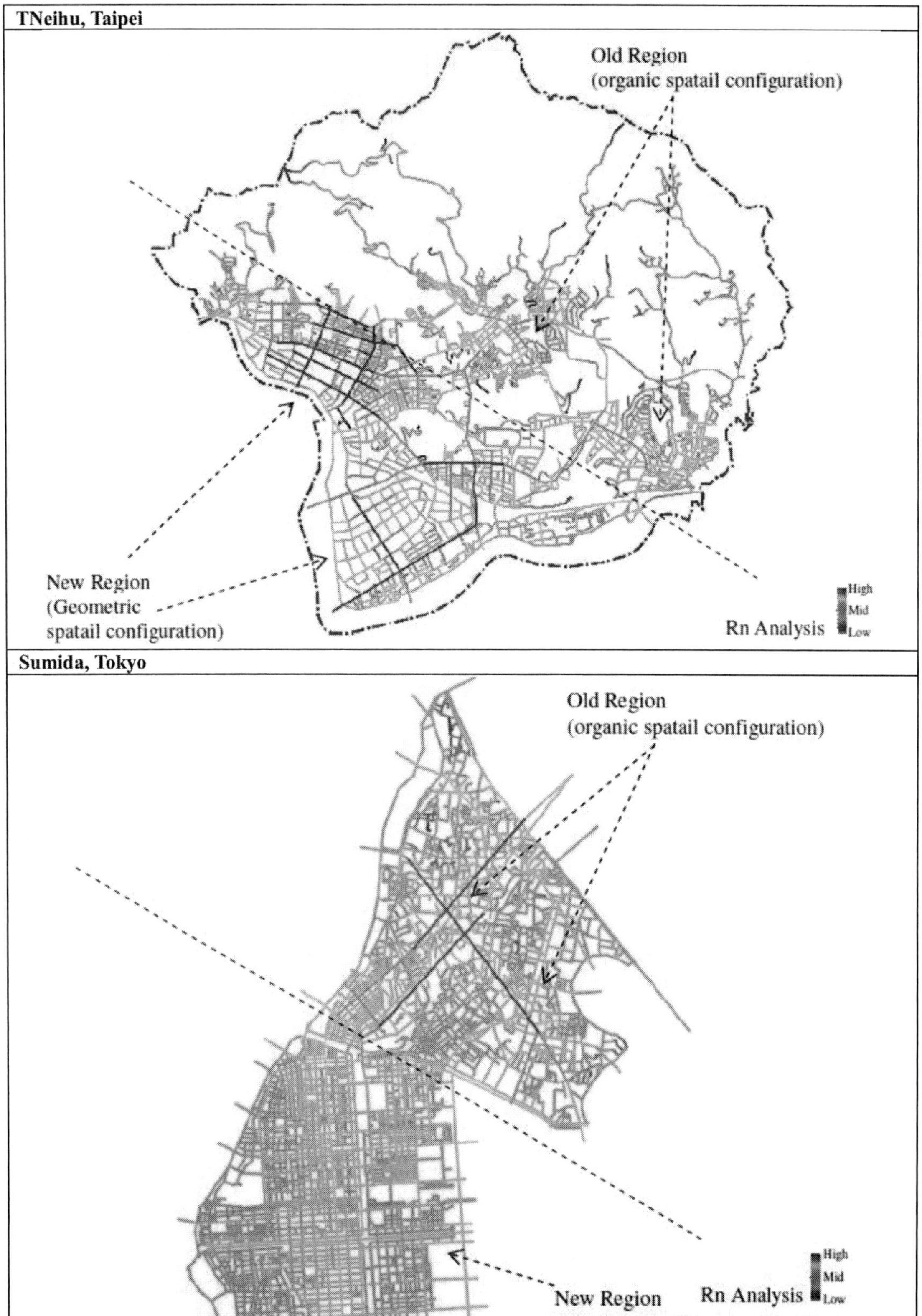

Figure 6. Axial analysis mapping of Neihu and Sumida districts by Space Syntax.

Table 4. Rn indicator of Sumida District

No.	Rn(average)	Segments	No.	Rn(average)	Segments	No.	Rn(average)	Segments
1	1.161	88	10	1.258	79	19	1.08	59
2	1.182	35	11	1.278	50	20	1.002	200
3	1.291	63	12	1.349	38	21	1.298	102
4	1.328	27	13	1.238	23	22	1.247	151
5	1.307	26	14	1.337	27	23	1.313	85
6	1.182	37	15	1.334	39	24	1.217	343
7	1.114	38	16	1.392	48	25	1.186	260
8	1.283	54	17	1.396	160	26	1.155	93
9	1.276	72	18	1.261	446			

4.2 CR indicator in capacity rate case study of Neihu and Sumida districts

In addition to transportation, the real environmental factors, such as the shelters and their capacities, must be considered. After classifying the degree of Rn indicator in Neihu and Sumida, we studied the capacity requirement to ensure the ability of shelters. For example, if people evacuate as fast as possible but there are few shelters with sufficient space, the disaster prevention plans cannot be efficient. Therefore, we surveyed the shelters' capacity and transferred a measurement metric called "CR" to determine the capacity of the shelters. In this section, we use an objective measure of capacity to investigate the shelters and understand the urban resilience of infrastructure during disaster response operations.

This paper calculated the capacity of each shelter (CR indicator) in a neighborhood or village as a unit for comparison with other shelters. We obtained 39 CR values in Neihu (Table 5)and 26 values in Sumida(Table 6) to fit in the quadrant diagram. This paper used the average value as a mean index to identify the high and low efficacy areas. The mean CR indicator was CR = 1.36% in Neihu and CR = 21.93% in Sumida. Moreover, we found some interesting differences between the management of cities and government, including the quality of public facilities and services. These findings were as follows:

(1) The service capacity of the emergency shelters in Sumida was 16 times higher than that in Neihu. This finding indicates the service level of public infrastructure between different cities or countries. Although this may not be a problem in urban planning and disaster prevention plans, it is a safety consideration to provide sufficient shelters or more capacity for disaster prevention and risk control [18].

(2) With respect to the correlation between residents and the shelters' capacity, the survey indicated a higher correlation in Sumida District ($R^2 = 0.0491$) than in Neihu ($R^2 = 0.0101$) considering similar population (the population is 287,760 in Neihu and 266,356 in Sumida). The survey also presents another way to complement urban resilience or disaster response resolution in case of severe disasters such as earthquakes.

Table 5. CR indicator of Neihu District

No.	TPA	APA	CR(%)	No.	TPA	APA	CR(%)	No.	TPA	APA	CR(%)
1	7541	81	1.08	14	7295	52	0.71	27	4014	31	0.76
2	9651	100	1.04	15	6221	85	1.37	28	9002	149	1.65
3	6869	71	1.04	16	8813	130	1.47	29	5706	94	1.65
4	6441	128	1.99	17	7900	181	2.29	30	8387	111	1.32
5	5995	197	3.29	18	8464	139	1.65	31	6541	77	1.18
6	10063	105	1.04	19	8089	129	1.60	32	5832	52	0.90
7	9744	157	1.62	20	4801	71	1.48	33	10485	244	2.33
8	6104	50	0.83	21	11990	180	1.50	34	9651	134	1.39
9	8605	71	0.83	22	8153	73	0.90	35	5997	84	1.39
10	8360	114	1.37	23	9258	83	0.90	36	1406	16	1.17
11	5757	68	1.18	24	5044	52	1.04	37	6410	151	2.36

12	5572	46	0.83	25	8263	79	0.96	38	8995	145	1.61
13	7091	52	0.74	26	7025	67	0.96	39	6225	112	1.81

Table 6. CR indicator of Sumida District

No.	TPA	APA	CR(%)	No.	TPA	APA	CR(%)	No.	TPA	APA	CR(%)
1	7268	909	12.51	10	8749	1425	16.29	19	6707	4866	72.56
2	4526	566	12.51	11	6625	1450	21.89	20	18213	5696	31.27
3	12948	1246	9.62	12	5129	1123	21.89	21	10405	1841	17.69
4	5676	674	11.87	13	5156	552	10.71	22	13141	3794	28.87
5	7537	895	11.87	14	11022	1267	11.50	23	10737	2671	24.88
6	10122	777	7.68	15	8136	999	12.28	24	22006	5304	24.10
7	1544	190	12.28	16	9840	1322	13.43	25	20693	8249	39.86
8	8364	3123	37.34	17	15315	3758	24.54	26	2295	492	21.42
9	10264	3832	37.34	18	23938	5908	24.68				

4.3 Hidden differences between hazard risks using quadrant diagram

We combined the two aforementioned indicators(Figure 7 and 8) and created a hazard map with geographic and disaster risk information that can be used as a rating for the 4th priority of the Sendai Framework for Disaster Risk Reduction. This hazard map can be used as a basis for an increase in disaster management capacity or resource allocation in future. Ultimately, we hope to enrich the 4th priority of the Sendai Framework for Disaster Risk Reduction by using a quantified indicator for disaster management and resilience. Using this assessment method, the following conclusions can be drawn regarding urban governance:

(1) For the low risk area, the spatial networks in Neihu were linear; however, Sumida was separated into two cores, both located in the old and new regions. Thus, in Neihu(Figure 9), only the new region had efficient evacuation ability, whereas in Sumida(Figure 10), the ability was balanced in both the old and new regions.

(2) For the high risk area, two districts with the same result were located at the bordering area. Although it is based on different geographical environments, it can be explained by a simple factor. Any land close to urban zoning and human activities with high land use requires more infrastructure such as effective routes or public open spaces. The high risk area is a residential and business hot spot, with low disaster prevention preparedness.

(3) The efficacy patterns in the old and new regions or organic and geometric spatial configurations were studied. There was no fixed or correct solution for the urban disaster prevention plans. This also provided another viewpoint on the theory of modern urban planning.

(4) The reasons or causes that make these area present high or low risk in Figure 9 and 10 is the important issue for urban planning and disaster prevention policy. And this research spots the high risk area of villages in Neihu and Sumida with the same spatial character through hazard quadrant diagram and risk mapping. In Neihu of Taipei, the high risk area are located nearby the hills and mountains that make the low efficient of accessibility with curved roads (the Rn indicator is 0.681~0.753 and the average is 1.080). Of course, the service radius of shelters nearby hills and mountains are much longer than others and lead to low capacity rate (the CR indicator is 0.71~0.83 and the average is 1.36). Identically, in Sumida of Tokyo, the high risk area are located the edge of a district and closed to rivers to cause low efficient of accessibility with blind alley (the Rn indicator is 1.002~1.161 and the average is 1.249). And these area are also earlier-development communities with crowded people but less land for public facilities. This reason caused the low capacity rate, too (the CR indicator is 12.51~10.71 and the average is 21.96). This research addressed a particular and useful methodology to recognize the service level of urban disaster prevention planning for evacuation roads and shelters.

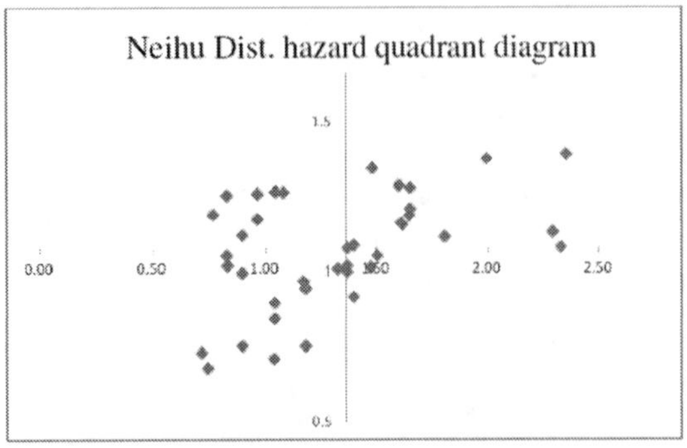

Figure 7. Quadrant Diagram of Neihu District

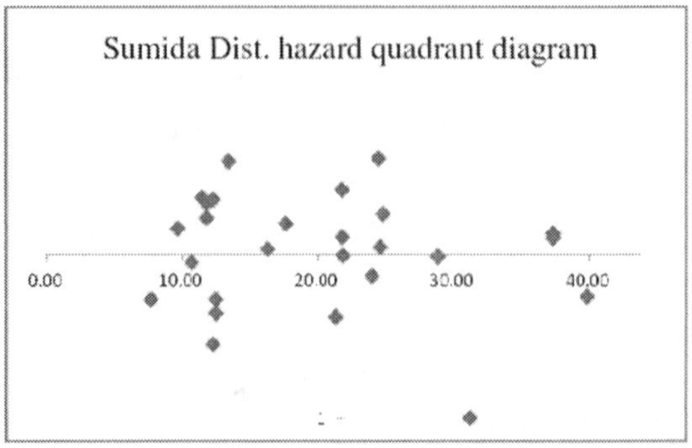

Figure 8. Quadrant Diagram of Sumida District

I-quadrant：35.9%
I I 、I V-quadrant：33.3%
I I I-quadrant：30.8%

I-quadrant：19.2%
I I 、I V-quadrant：57.7%
I I I-quadrant：23.1%

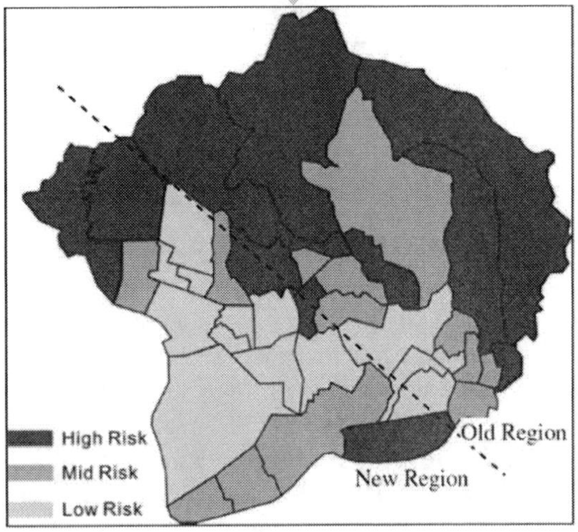

Figure 9. Hazard risk mapping of Neihu District

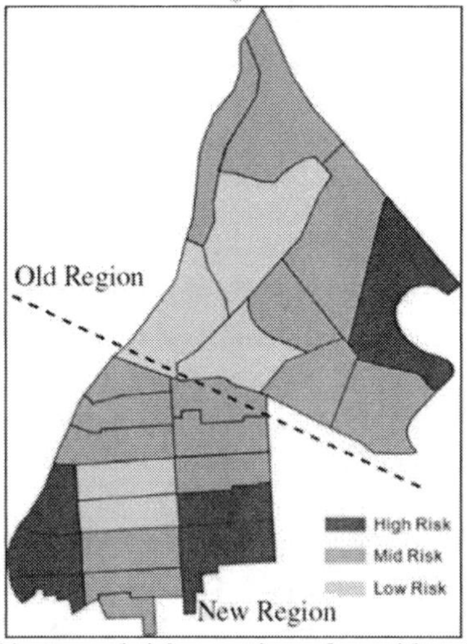

Figure 10. Hazard risk mapping of Sumida District

4.4 Discussion

The quadrant diagram is a strategic step and layout for implementing the Sendai Framework for Disaster Risk Reduction. This paper aims to propose a specific assessment method that is feasible and integrated in the 4th priority for action of the Sendai Framework for Disaster Risk Reduction. The proposed method is feasible and reliable and is not necessarily the only reference for the 4th priority. However, we hope to obtain some innovative research results to transform and enhance the disaster preparedness and social security of cities. Through a comprehensive risk survey of the urban road network and emergency shelters, we found some risks that had not been encountered before. This is also a good basis for action regarding the 4th priority, which is expected to lead other researchers to focus on the 4th priority, and enrich and complete the entire disaster reduction program. Thus, this paper presents an empirical study to report a new and innovative assessment methodology for the 4th priority. Moreover, this paper provides a comprehensive indicator for rapid risk assessment and diagnosing risk factors for the 4th priority under limited government resources and public infrastructure. Through this assessment method, we hope to establish a geographic hazard information tool to guide the government and disaster management units to adjust or review the urban planning mechanism, and thus provide a more comprehensive urban safety network.

Through this research, we suggest government, urban planners and scholars to rethink the original value of disaster prevention planning. Not only making disaster prevention policies but also modifying regular with quantitative analysis methodology. In these high risk area of this research, we also want to present government to add or build more emergency shelters in parks or open space. On the other hand, the new building permits should be requested to plan evacuation plans for public safety. Like large development plans and reconstruction projects, the evacuation shelters have to be added in first floor of sites to create public contribution to society. Of course, these evacuation shelters have to connect to important roads and lane in urban disaster prevention plans according to Space Syntax analysis.

5 Conclusion

Disasters have become an important urban problem because of their unpredictability and heavy casualties in the event of a disaster. For any country and city, people's security and social stability is a basic task and requirement. Even with large amounts of human resources or social costs, it is difficult to completely prevent a disaster. Although there is no solution to this problem, disaster prevention can be improved by using the 4th priority of the Sendai Framework for Disaster Risk Reduction. This paper reveals the hidden risks in cities and decodes the correlation factors, in addition to providing a scientifically quantifiable solution to the qualitative problem of disaster management.

This paper presents a hazard management method that recognizes the urban network patterns and is useful for promoting or upgrading the disaster prevention plans. In future studies, we intend to expand the study to other public facility problems. In addition to urban road networks and emergency shelters, we incorporate a comprehensive survey of other disaster prevention space systems to facilitate the spatial construction for a more safe urban resilience. This paper contributes to the implementation of the 4th priority and other human behaviors or evacuation studies. Moreover, we intend to study human behaviors and asylum to facilitate the implementation of the 4th and other priorities of the Sendai Framework for Disaster Risk Reduction and contribute to the qualitative and quantitative study of disaster management.

References

1. UNISDR, *Sendai Framework for Disaster Risk Reduction 2015-2030,* Third UN World Conference on Disaster Risk Reduction (2015)
2. V. Jetel, 8th International Scientific Conference Building Defects, MATEC Web of Conferences, **93**, 03009 (2017)
3. B. Zhou, 13th Global Congress on Manufacturing and Management, MATEC Web Conference **100**, 02062 (2017)
4. D. Alexander, Dunedin Academic Press, **150**, 180-183 (2002)
5. W.Y. Lee, S.T. Chien and S.H. Lee, *The Plans on Disaster Prevention System of Urban Planning in Taipei City*, The Department of Urban Development of Taipei City Government (1997)
6. W.Y. Lee and M.J. Ho, *The Handbook of Disaster Prevention in Urban Planning*, Architecture and Building Research Institute of Ministry of the Interior, Taiwan (2000)
7. W.Y. Lee, *To Study the Correlation and Spatial Planning of Rescue Locations and Prevention Routes in Central Area of Taipei City*, The Department of Urban Development of Taipei City Government (2001)
8. K.A. Adebiyi, O.E Charles- Owaba and M.A. Waheed, Disaster Prevention and Management: An International Journal, **16**(2), 178-187 (2007)
9. J. Liu, D. Wu, F. Hidetosi and W. Gao, Open Journal of Civil Engineering, **5**, 97-108 (2015)
10. H.S. Chang and C.H. Liao, Nat Hazards, **76**, 1551–1571 (2015)

11. C.H. Liao and H.S. Chang, *3rd International Conference on Environmental Science and Information Application Technology, Procedia Environmental Sciences*, **10**, 1092-1097 (2011)
12. J.B. Sheu and C. Pen, Transportmetrica A: Transport Science, **11**(3), 210-242 (2015)
13. J.S. Lee and W.H. Hong, Human and Ecological Risk Assessment: An International Journal, **21**(6), 1968-200 (2015)
14. B. Hillier and T. Stonor, City Planning Institute of Japan, **Special Issue：Future of Urban Space and Humanity**, 284-288 (2010)
15. B. Jiang and I. Omer, Transactions in GIS, **11**(6), 943-960 (2007)
16. F.T. Lin, *23rd International Fire Chiefs' Association of Asia General Conference*, Taipei, Taiwan (2004)
17. Transportation Research Board Washington, D.C., *The Role of Transit in Emergency Evacuation*, Committee on the Role of Public Transportation in Emergency Evacuation, Special Report 294, 43-73 (2008)
18. F. Kilci, B.Y. Kara and B. Bozkaya, European Journal of Operational Research, **243**(1), 323-332 (2015)

Low socioeconomic status and visual health behaviors as factors in childhood myopia development

Cheng-Jong Lee[1], Yu-Ting Hsiao[1], Philip M. M. Sced[2,a] and Shi-Huang Lee[3]

[1]*Department of Sport and Health Management, Da-Yeh University, Taiwan*
[2]*Department of English, Da-Yeh University, Taiwan*
[3]*Department of Ophthalmology, Taichung Tzu Chi Hospital, Taiwan*

Abstract. Although many studies have discussed visual acuity (VA), the childhood myopia epidemic has not been alleviated and requires further investigation. This study, therefore, investigates the differences in children's VA among various family structures and the effects of visual health attitudes (VHA) and visual health behaviors (VHB) on VA. The children in low socioeconomic status (SES) families in Changhua County, Taiwan are the main research subjects. Using purposive sampling, the study conducts a questionnaire survey and a VA examination on low SES children, undergoing counseling at two social welfare organizations in Taiwan. The VA examination was conducted at the two social welfare organizations by an ophthalmologist from Tzu Chi Hospital, Taichung, between October 2016 and December 2016. 265 questionnaires were returned. Significantly, over 60% of the samples of low SES children were not living with their parents. Of these children, the highest proportion (35.8%) lived with their mothers; the second-highest (31.3%), lived with their grandparents; and the third highest (29.4%), lived with their fathers. 86.4% of the low SES children had poor (below 1.0) VA. The study, therefore, concludes that schools should focus on low SES students, and should focus on improving the VHA of the youngest students.

1 Introduction

The World Health Organization (WHO) estimates that there are 246 million visually impaired people worldwide and that 90% of these people are of low socioeconomic status (SES) [1]. According to the WHO, 43% of moderate to severe visual impairment is caused by uncorrected refractive errors, such as myopia, hyperopia, astigmatism and presbyopia. Although, the WHO reports that worldwide socioeconomic development and concerted public health action have brought about a general decrease in visual impairment worldwide, there is a scholarly consensus that myopia is increasing worldwide and is approaching epidemic proportions in the industrialized nations of East Asia [2-7]. Lee et al. [8] report a prevalence of myopia as high as 86.1% among young Taiwanese men aged 18 to 24 years.

Although myopia and other refractive error can be corrected with the aid of glasses, contact lenses and refractive surgeries, myopes will likely suffer to some degree from lost educational and employment opportunities, lost economic gain, and impaired quality of life [9, 10]. Moreover, myopes are at a higher risk of blindness in old age. The increased prevalence of myopia will lead to increases

[a] Corresponding author : philsced@mail.dyu.edu.tw

in the numbers of blind people, whose health care needs pose a serious socioeconomic burden to families and societies.

1.1 Environmental factors

Myopia has traditionally been considered a hereditary condition, and many previous studies indicate that children with myopia are more likely to have a family history of myopia [11-15]. However, there is a growing scholarly consensus that both environmental and hereditary factors play a role in the development of myopia. To cite a few of the studies that argue for the importance of environmental factors in myopia development, Xiang et al. found that a high percentage of children in Guangzhou become myopic by the age of 15, irrespective of the myopic status of their parents [16]. He et al. reported that, in China, school children in urban settings exhibit a higher prevalence of myopia than children in rural settings [17], suggesting that environmental factors are involved in the children's myopia development, because the genetic background of the children living in the urban and rural areas is similar. And Rose et al. found that a lower prevalence of myopia among Chinese children raised in Sydney than among Chinese children living in Singapore was associated with increased hours of outdoor activities [18].

1.2 Reading habits and visual health behaviors

One study [18] found that a lower prevalence of myopia among Chinese children raised in Sydney than among Chinese children living in Singapore was associated with increased hours of outdoor activities. Animal studies have shown that light levels as high as 15,000 to 30,000 lux either in the form of artificial laboratory light or natural daylight retard experimental myopia in chicks, guinea pigs and monkeys [10]. Based on these results from animal models, it is hypothesized that high light levels outdoors trigger the release of dopamine, an ocular growth inhibitor that inhibits myopic development.

Lyu et al. [4] investigated the reading habits, visual behaviors and living environments of elementary school children in Chaoyang District, Beijing and found that myopia was associated with reading posture, reading distance, and font size. You et al. [19] note that dim illumination during reading is an additional factor in the development of myopia. Does this association of reading habits and visual behaviors with myopia suggest that parents and teachers who promote healthy visual health behavior (VBH) in children may be able to help reduce the children's risk of myopia?

More research is needed to investigate whether unhealthy VHB such as not doing sufficient daytime outdoor activity, obsessively watching electronic devices (e.g. smartphones, computer monitors or TV screens) and continuously studying in dimly lit environments are factors in the development of myopia. Further study is also needed to investigate whether there is a relationship between unhealthy VHB and SES.

Previous studies have shown that high SES children exhibit higher incidences of myopia [19]. This study will focus on low SES children, investigating the healthiness of their VHB and probing the hypothesis that unhealthy VHB are causatives factor in myopia development. This study may, therefore, shed light on the question of whether parents who create healthy study environments and discipline their children to practice healthy VHB can reduce their children's risk of developing myopia.

2 Materials and methods

This study passed a review (CRREC-105-074) by the Institutional Review Board (IRB) of China Medical University Hospital, Taichung, Taiwan. The visual acuity (VA) examinations were conducted by an ophthalmologist from Tzu Chi Hospital, Taichung, Taiwan, between October 2016 and December 2016. Before any children were surveyed, both the children and their parents had signed a consent form, indicating that they understood the purpose and contents of the survey and that they were willing to answer the questions. The study was, therefore, carried out according to the regulations of the IRB. Children in low SES families counseled by three social welfare organizations in Taiwan were the research subjects. The research subjects included children who were in single-parent families, foster families, or families in which parental responsibility fell on grandparents, relatives, foreign spouses, and parents with physical or mental disorders.

2.1 Research tools

To anticipate the reliability of the questionnaire, a preliminary test using the questionnaire was conducted on students counselled by the social welfare organizations. A purposive sampling study and an examination of VA were also conducted as part of the preliminary study in October 2016. 50 questionnaires were returned in total. The test results indicated that both research variables were over .70, showing a reliable result and a high interior consistency. 330 formal questionnaires were distributed in December 2016 and 265 questionnaires were returned. The scale reliability was above .70, showing high consistency.

3 Hypotheses

Based on the widely utilized Theory of Planned Behavior, which suggests that people's behaviors can be predicted based on their attitudes [20], the purpose of this study was to investigate the relationship between VHB and visual health attitudes (VHA). The study's hypotheses follow.
1. The low SES children's VHA will exhibit a difference across demographic variables.
2. The low SES children's VHB will exhibit a difference across demographic variables.
3. VHA will have a positive impact on VHB.
4. VHB will have a positive impact on VA.

4 Results

4.1 Samples structural analysis

Structural analysis was carried out by SPSS 18.0. The demographic statistics in the basic data of the sample are shown by gender. There are 130 (49.1%) males and 135 (50.9%) females. 105 (39.6%) students live with their father and mother and 160 (60.4%) students do not, among whom, the highest proportion live with the mother (29.1%). Significantly, of the samples of low SES children, over 60%, therefore, do not live with their parents. Of the low SES children, the highest proportion (35.8%) are raised by their mothers, the second highest proportion (31.3%), by their grandparents, and the third highest (29.4%), by their fathers. The proportion of children raised by their grandparents is, therefore, higher than the proportion raised by their fathers.

4.2 VA testing

Table 1. Eyes classified by unaided Visual Acuity (VA)

VA	Case No.	Percentage	Accumulated Percentage
Over 1.0 (normal)	92	34.7	34.7
0.8-0.9 (mild vision loss)	26	9.8	44.5
0.4-0.7 (ametropia)	79	29.8	74.3
Under 0.3 (Severe ametropia)	68	25.7	100.0
Total	265	100.0	

This study passed a review by the Institutional Review Board. The study conducted a questionnaire and performed a VA examination on low SES children. The VA examination used a standard visual scale to measure uncorrected defective VA. Any uncorrected VA less than 0.9 without other eye diseases was regarded as poor VA. The worst eye would be measured if both eyes exhibited poor VA, and the right eye would be measured if the eyes were of the same VA. The result of the examination is shown in table 1. Approximately 65.3% of the student's VA is under 1.0.

4.3 Test of hypotheses

The result of the ANOVA analysis, using school grades as the independent variable and VHB and VHA as dependent variables, showed that there was a significant difference ($p<.05$) in VHA. As shown in Table 2, first graders scored lower than students in other grades in terms of "correct" VHA. The most probable reason for this low score is not the negligence of the teacher, but rather the naturally lower cognitive ability of the first graders, which impedes them from understanding the importance and benefits of visual health, despite the teacher's best efforts in the classroom. The visual health knowledge the first graders are taught does not turn into "correct" VHA. Therefore, families, schools and social welfare institutions should carefully consider how better to inculcate correct VHA in first graders, who are most in danger of losing at the start line.

Table 2. Single factor variance analysis of grade of students

Variable	Grade	Mean	Standard Deviation	F-value	PostHoc (Scheffe)
Visual Health Attitude (VHA)	(1) First Grade	2.190	0.722	5.169*	(2) > (1) (4) > (1) (5) > (1) (6) > (1)
	(2) Second Grade	2.855	0.618		
	(3) Third Grade	2.712	0.644		
	(4) Fourth Grade	2.881	0.629		
	(5) Fifth Grade	2.793	0.590		
	(6) Sixth Grade	2.923	0.481		

* p<.05

A significant difference between the various family structures emerged regarding the questionnaire item "I have a balanced diet, and am not a picky eater". As shown in Table 3, the students who live with their father have a better-balanced diet than the students who live with their mother. A plausible explanation for this may be that fathers tend to discipline their children more strictly.

Table 3. Analysis of the differences between various family structures

Question	Family Structure	Mean	Standard Deviation	F-value	PostHoc (Scheffe)
I have a balanced diet and am not a picky eater.	(1) Live with parents	2.619	0.944	4.694*	(2) > (3)
	(2) Live with father	3.088	0.949		
	(3) Live with mother	2.402	1.016		
	(4) Others	2.605	1.027		

* p<.05

As shown in Table 4, VHB and VHA exhibit no significant difference between male and female.

Table 4. Analysis of the differences between genders

Variable	Gender	Mean	Standard deviation	t-value	p-value
Visual Health Attitude (VHA)	male	2.796	0.542	.229	.819
	female	2.778	0.694		
Visual Health Behavior (VHB)	male	2.522	0.636	.682	.496
	female	2.468	0.646		

The children whose VA was over 1.0 and between 0.4 to 0.7 had significant differences in 3 questions related to VHB, as shown in table 5. The children with good VA had established the habit of resting the eyes by taking a 10-minute break after reading for 30 minutes, but the children whose VA was between 0.4 to 0.7 did not pay attention to resting the eyes, because they could see clearly, felt no myopia, and wanted to finish homework as soon as possible. The children with good VA had "correct" VHA and would do outdoor sports on weekdays and holidays. The low SES children did not have bright fluorescent lamps and table lamps, lived in poor environments and did not, in some cases, even have a desk. The majority of low SES children did not have a dedicated desk lamp and only turned on a lamp to write or read under weak light in order to save money, forcing them to maintain an unhealthy reading distance. The children whose VA was between 0.4 to 0.7 would squint their eyes to read, making it harder for the ciliary muscles of they eye to adjust. The children whose VA was between 0.4 to 0.7 could read clearly if they squinted their eyes and they did not think they had problems, because they experienced little adverse effect in their daily lives. They also tended to ignore VHB, feeling that wearing glasses made no difference to them, and that they did not have myopia.

Table 5. Analysis of the differences in visual health behavior (VHB) for various visual acuities (VA)

VHB	VA	Vision Level	Mean	Standard deviation	PostHoc (Scheffe)
A. I look into the distance and rest for 10 minutes after reading for 30 minutes.	(1) over 1.0	2.625	1.099	2.994*	(1) > (3)
	(2)0.8-1.0	2.313	1.026		
	(3)0.4-0.7	2.012	1.006		
	(4) under 0.3	2.236	0.980		
B. I spend at least 120 minutes outdoors every day.	(1) over 1.0	2.968	1.031	4.814*	(1) > (3)
	(2)0.8-1.0	2.525	1.003		
	(3)0.4-0.7	2.189	1.013		
	(4) under 0.3	2.454	0.9588		
C. I maintain a reading distance of over 35 cm while reading or doing homework.	(1) over 1.0	2.968	0.999	3.695*	(1) > (3)
	(2)0.8-1.0	2.606	0.966		
	(3)0.4-0.7	2.303	1.078		
	(4) under 0.3	2.436	0.976		

* $p < .05$

5 Discussion

The assumptions in this study were analyzed using SPSS and AMOS statistical software, yielding the following results:

1. H1: The VHA of low SES children will exhibit a difference across the demographic variables.
   As shown in Table 3, there was a significant difference between the school grades in regard to VHA; thus, H1 was feasible.
2. H2: The VHB of low SES children will exhibit a difference across the demographic variables.
   As shown in table 4, there was a significant difference between various family structures in regard to VHB (e.g. not a picky eater); thus, H2 was feasible.
3. H3: Student's VHA will have a positive impact on VHB.
   As shown in Table 5, the regression coefficient of VHA to VA was .406, achieving a significance level of $p<.05$; thus, H3 was feasible. B5, "I take a rest instead of seeing an ophthalmologist when I can't see clearly," was the most effective question regarding VHA (regression coefficient .746).
4. H4: VHB will have a positive impact on VA.

As shown in Table 5, there was a significant difference ($p<.05$) in regression coefficient (.224) of VHB to VA; thus, H4 was feasible, and VHB had a positive impact on VA. C2 "I look into the distance and rest for 10 minutes after reading, watching TV, playing games, playing smartphone or computer for 30 minutes" was the most effective question regarding VHB (regression coefficient .657).

65.3% of the low SES children's vision was below 0.9, which is an increase of 19.18% on the defective vision rate in Taiwan (46.12%). A failure to maintain a healthy book-to-eye distance when reading seemed to be the main cause of myopia onset. The children with poor VA had worse VHA, leading to unhealthy VHB and ultimately to myopia. A further cause of myopia onset might be the ubiquity of addictive electronic devices such as smartphones and tablets. Los SES parents or guardians be too busy working to take care of their children, who consequently do not develop adequate self-control to use such electronic devices responsibly. In some cases, the children may even be encouraged to spend excessive amounts of time passively watching these devices, a phenomenon known informally in Taiwan as "3C babysitting". This phenomenon may also occur in families where children are raised by grandparents, who are less likely to be strict about visual health. Families with foreign spouses also tend to be low SES and, therefore, may be too busy working to care for their children or may resort to "grandparenting" or even "3C babysitting".

6 Findings

According to the results of the study, "correct" VHA led to "correct" VHB, and correct VHB ensured good VA for students. Although myopia and other ametropias can be corrected by glasses and refractive surgery, myopia still causes a certain loss of education and employment opportunities as well as economic loss and general loss of quality of life [9, 10]. Low SES children need more concern from parents, guardians and schools, in addition to better visual conditions to facilitate the reversal of their low SES. It is also expected that ophthalmologists, optometrists, and social workers will be more active in assisting with the visual health care of low SES children.

7 Implications

The results showed that junior students had bad VHA, implying there is still a deficiency in both the quality and quantity of visual health education in schools. To guide the next generation of students and parents towards deeply-rooted and correct VHA, schools should be active in providing effective, understandable, multi-language visual health guidelines for junior students and their families.

8 Limitations

1. Because the questionnaire and VA examination needed to have a parent or guardian's signature and the students' consent before filling in and testing, it was not possible to obtain test results from all the low SES elementary school students in Changhua County.
2. A purposive sampling study was conducted on the results of the VA examination of the low SES children who were counseled by social welfare organizations in Changhua County. We were not able to collect the data for those who were not counseled by social welfare organizations.
3. Because a guidance counselor read the questions to and helped each of the children complete the questionnaire, it is possible that error crept into the results due to the children's being unduly influenced by the counselor or to the children's failing to comprehend the questions.

9 Suggestions for future research

The VA examination and questionnaire were conducted using quantitative research, which required the respondents to answer on a scale of 1 to 4. The respondents might have had reservations about or compromised or even randomly filled in their answers. There are few studies on the visual health care of low SES children and, as a result, there was little literature and few study results that could be verified or analyzed. There is a need for more research on myopia among low SES children generally, but there is also a lack of research using qualitative methods to study children with poor VA, obtaining data through in-depth interviews.

References

1. World Health Organization, http://www.who.int/mediacentre/factsheets/fs282/en/ (Accessed 21 Jan 2017)
2. L.L.K. Lin, Y.F. Shih, C.K. Hsiao and C.J. Chen, Annals Academy of Medicine Singapore, **33**(1), 27-33 (2004)
3. S. Vitale, R.D. Sperduto and F.L. Ferris, Arch Ophthalmol, **127**(12), 1632-1639 (2009)
4. Y. Lyu, H. Zhang, Y. Gong, D. Wang, T. Chen, X. Guo, S. Yang, D. Liu and M. Kang, China. Jpn J Ophthalmol, **59**, 421-429 (2015)
5. D.C. Tsai, L.J. Lin, N. Huang, C.C. Hsu, S.Y. Chen, A.W.H. Chiu and C.J.L. Liu, Clin Exp Ophthalmol, **43**, 612-620 (2015)
6. I. G. Morgan, K. Ohno-Matsui and S.M. Saw, Myopia. Lancet., **379**, 1739-1748 (2012)
7. P.C. Wu, C.L. Tsai, H.L. Wu, Y.H. Yang and H.K. Kuo, Ophthalmology, **120**, 1080-1085 (2013)
8. Y.Y. Lee, C.T. Lo, S.J. Sheu and L.T. Yin, Ophthalmic Epidemiol, **22**, 66-73 (2015)
9. B. Holden, P. Sankaridurg, E. Smith, T. Aller, M. Jong and M. He, Eye, **28**, 142-146 (2014)
10. D. Ramamurthy, S.Y. Lin Chua and S.M. Saw, Clin Exp Optom, **98**, 497-506 (2015)
11. L.T. Lim, Y. Gong, E. Y. Ah-kee, G. Xiao, X. Zhang and S. Yu, Ophthalmol Eye Dis, **6**, 31-35 (2014)
12. I. Morgan and K. Rose, Prog Retin Eye Res., **24**, 1-38 (2005)
13. D.O. Mutti, G.L. Mitchell, M.L. Moeschberger, L.A. Jones and K. Zadnik, Invest Ophthalmol Vis Sci., **43**, 3633-3640 (2002)
14. S.M. Saw, J. Katz, O.D. Schein, S.J. Chew and T.K. Chan, Epidemiol Rev., **18**, 175-187 (1996)
15. D.O. Mutti, K. Zadnik and A.J. Adams, Invest Ophthalmol Vis Sci., **37**, 952-957 (1996)
16. F. Xiang, M. He and I.G. Morgan, Optom Vis Sci., **89**, 1487-1496 (2012)
17. M. He, Y. Zheng and F. Xiang, Optom Vis Sci., **86**, 40-44 (2009)
18. K. A. Rose, I.G. Morgan, W. Smith, G. Burlutsky, P. Mitchell and S.M. Saw, Arch Ophthalmol, **126**, 527-530 (2008)
19. Q.S. You, L.J. Wu, J.L. Duan, Y.X. Luo, L.J. Liu, X. Li, Q. Gao, W. Wang, L. Xu, J.B. Jonas and X.H. Guo, PLoS One, doi: 10.1371/journal.pone.0052668 (2012)
20. M. Fishbein and I. Ajzen, *Predicting and Changing Behavior: The Reasoned Action Approach* (New York: Taylor & Francis, 2010)

MATEC Web of Conferences 169, 01046 (2018) https://doi.org/10.1051/matecconf/201816901046
IMETI 2017

Modeling and simulation of brain herniation caused by subdural hematoma using finite-element method

Ke-Chun Huang[1,7], Furen Xiao[1,2], I-Jen Chiang[1,3], Yi-Long Chen[4,6], Yi-Hsin Tsai[1,2], Jau-Min Wong[1] and Chun-Chih Liao[1,5,a]

[1]Institute of Biomedical Engineering, National Taiwan University, Taiwan
[2]Institute of Clinical Neuroscience, National Taiwan University Hospital, Taiwan
[3]Graduate Institute of Medical Informatics, Taipei Medical University, Taiwan
[4]Institute of Biomedical Engineering, National Yang-Ming University, Taiwan
[5]Department of Neurosurgery, Taipei Hospital, Taiwan
[6]Department of Neurosurgery, Taipei City Hospital, Taiwan
[7]Biomedical Engineering Office, Taipei City Hospital, Taiwan

Abstract. Brain shift and herniation are important signs of increased intracranial pressure (ICP) caused by hematomas or other types of intracranial mass. We propose a novel finite-element model that can be deformed in response to increased ICP. The half sphere model of the brain is partially divided into two compartments by the intact mid-sagittal plane, allowing subfalcine herniation. A 40 mm circle in the center of its equatorial plane allows transtentorial herniation. We perform a single load step, structural static analysis, simulating a left-sided subdural hematoma (SDH) compressing the cerebral hemispheres from the outer surface of the left hemisphere. Subfalcine and transtentorial brain herniations are reproduced and visualized. The Poisson's ratio represents the tightness of the brain and the pressure load represents the ICP. There is a linear relationship between maximal deformation and the pressure load. The maximal deformation at the basal circumference and that at the basal midline closely resembles the maximal thickness of the SDH and the midline shift. We have developed a simple finite-element model that can simulate brain shift and herniation caused by pressure loads exerted on its surface by a mass. The experimental results correlate well with clinical observation on patients with acute and chronic SDH.

1 Introduction

Traumatic brain injury (TBI) is an important cause of mortality and severe disability worldwide. Pathophysiologically, TBI can be classified under the categories of primary and secondary injury, which are considered as stages in the evolution of head trauma [1]. By definition, "primary" head injury occurs at impact and may involve neural or vascular elements of the brain. Although surgeons can remove some intracranial hematoma, there is no treatment that can reverse primary injury to brain parenchyma and prevention is the only thing that helps. On the other hand, "secondary" injury is the result of physiological processes that are initiated following impact or soon thereafter. Caused by

[a] Corresponding author : d95548001@ntu.edu.tw

systemic or intracranial conditions, secondary injury may aggravate primary brain injury, but it can be ameliorated by taking specific measures.

After trauma, the edematous and deformed brain can produce increased intracranial pressure (ICP), resulting in brain shift, which is an important manifestation of secondary injury. The injured cerebral hemisphere may herniate through the subfalcine space (SFS) to the other side, resulting in subfalcine herniation (SFH), appearing as midline shift (MLS) on brain images such as computed tomography (CT) studies [2]. With further elevation of the ICP, the medial part of the cerebrum may herniate through the tentorial incisura (TI), resulting in transtentorial herniation (TTH) and brain stem compression. On images, one can find changes in the shape of the brain stem or effacement of the cisternal spaces around it [2].

Biomechanics is devoted to the analysis, measurement and modeling of the effects which are measured quantitatively under various mechanical loading situations in biological systems. The special discipline of biomechanics which is concerned with injury caused by mechanical interaction is denoted as trauma biomechanics [3]. We have proposed methods that can automatically measure the magnitude of MLS with symmetry-based and landmark-based methods [4-6]. To our knowledge, there has been no biomechanical model dedicated to simulating brain herniation in the literature.

Being the most important tool in trauma biomechanics, the finite-element (FE) method has been widely used in determining head injury criteria for motor vehicle accident and sports-related injuries. In addition to cadàver experiments based on global head kinematics, complex FE models of the head composed of different tissue types or complex geometry have been proposed to better describe regional response of the brain to primary injury [7-9].

The goal of this study is to develop the simplest 3-dimensional FE model that can be applied to quantitatively simulate brain deformation, specifically brain herniation, caused by an intracranial mass with increased ICP. The resulting brain deformation predicted by our model is then evaluated and compared to our previous results derived from real patients.

2 Methods

2.1 Finite-element model specification

We have constructed a simple model of the supratentorial brain [10]. The model assumes the shape of the upper half of a sphere with a diameter of 160 mm and a volume of 1072 ml. The outer surface of this model simulates the convexity surfaces of the cerebral hemispheres, while the irregular basal surface of the cerebrum is simplified and represented by a flat equatorial plane in this model. A 40 mm circle is removed from the center of the basal surface to simulate the midbrain, the upper part of the brain stem occupying the TI.

To model the partially connected cerebral hemispheres, our model is divided into two compartments by the intact mid-sagittal plane (iMSP). Each cerebral "hemisphere", which actually assumes the shape of quarter sphere, has three surfaces: the convexity, the basal and the interhemispheric surfaces. A semicircle 80 mm in diameter is removed from the iMSP to simulate the SFS. The center of this semicircle is located 20 mm anterior to the center of the basal plane. The posterior ends of SFS and the TI meet at the falcotentorial junction, which is adjacent to the most posterior aspect of the midbrain.

We generate the 3-D FE model of the brain as described above on ANSYS 10.0 ED environment using solid modeling. The X, Y, and Z axes of the Cartesian coordinate system refer to the right-left, anterior-posterior, and cranial-caudal (superior-inferior) anatomical axes, respectively, and the iMSP is represented by the X-Z plane. The completed model of the cerebral hemispheres has only 12 keypoints, 21 lines, 12 areas and 2 volumes. Each cerebral hemisphere has 5 areas in its outer surface, including a convexity surface that is divided into two areas, and a basal surface that is divided into the skull base/tentorium portion and the TI portion, which is subdivided into two areas. The iMSP between the hemispheres is divided into two areas: the falx and the SFS. We have validated this model on CT craniometric data obtained from 50 patients with brain concussion [10].

Because the model is only used to simulate deformation of the brain within the cranial cavity caused by intracranial lesions, it consists of only one type of linear elastic material: the brain itself. We omit all extracranial materials, such as the dura, the skull and the scalp, which must be considered in modeling primary brain injuries. Under increased ICP, the cerebrospinal fluid (CSF) spaces such as the ventricles and the cisterns are compressed, so we also neglect all CSF spaces in the model. The specific material properties of the brain are still controversial. From the literature, the Young's modulus (YM) of the brain is taken as 10000 Pa [11] and the Poisson's ratio (PR) is taken as 0.45 [12].

We choose the SOLID187 element, a higher order 3-D, 10-node tetrahedral element, to construct our model. This element has a quadratic displacement behavior and is well suited to modeling irregular meshes [13]. The FE model is meshed automatically at "SmartSize" level 5 using the built-in tool of ANSYS, generating 1391 nodes and 776 elements.

2.2 Experimental setup

We perform a single load step, static structural analysis of our brain model simulating a left-sided subdural hematoma (SDH) compressing the cerebral hemispheres. We apply uniform compressive pressure loads on the convexity surfaces of the left cerebral hemisphere, simulating increased ICP caused by the SDH. Since the falx has a much higher YM than the brain [14], we consider it immobile in all degree of freedoms (DOFs). Displacements of the convexity surface of right cerebral hemispheres are also constrained to 0 in all DOFs, simulating the rigid skull covering the brain.

On the basal surface of the brain model, only the TI portion is allowed to move in all directions. The remaining basal surface of the brain lies on the bony skull base anteriorly and the tentorium cerebelli posteriorly. The former structure is fixed to the bone and cannot move. Similar to the falx, the tentorium is also a tough structure and is also considered immobile in our model. As a result, parts of the basal surface of the brain other than the TI are only allowed to move in X- and Y-directions and their displacements are constrained to 0 in Z-direction. The number of surface load conditions is 66 and the number of constraints is 974.

We evaluate the deformation of our model in two sets of experiments. In the first one, we apply different pressure loads ranged from 5 mmHg (667Pa) to 80 mmHg (10664Pa) using material properties derived from the literature. In the second one, we apply a 20-mmHg pressure load, slightly higher than the upper normal limit of ICP [15], to our model having different PR values ranged from 0.25 to 0.49. Different values of YM are also tested. We choose the large displacement static analysis mode to solve our problem. Using the postprocessor of the ANSYS system, the nodal solutions for the displacements are listed and analyzed. Then, the deformed shapes are plotted for evaluation. The displacement over nodes located at the basal circumference (BCC, 40 nodes), the basal midline (BML, 25 nodes), and the horizontal line across the tentorial incisura (HTI, 9 nodes) are recorded. These three groups of nodes are employed as "local approximates" of the maximal displacements of the entire model. Their displacement vector sums as well as the X-, Y- and Z-components are compared to the corresponding global maxima to see if the latter can be estimated.

3 Results

3.1 Brain deformation under different pressure loads using material properties derived from the literature

We apply different pressure ranged from 5 mmHg (667Pa) to 80 mmHg (10664Pa) to the model having a YM of 10000 and a PR of 0.45. Brain deformation after applying a 20-mmHg pressure load is shown in the upper middle part of Fig. 1, using a cutaway view from the anterior-inferior aspect using coronal (Y-Z) cutting planes 20 mm anterior to the center of the basal planes through the center of the subfalcine space. A large proportion of the compressed hemisphere, extending from the outer surface all the way to the SFS, has large displacement, with the subfalcine region where clinicians measure the MLS having the largest.

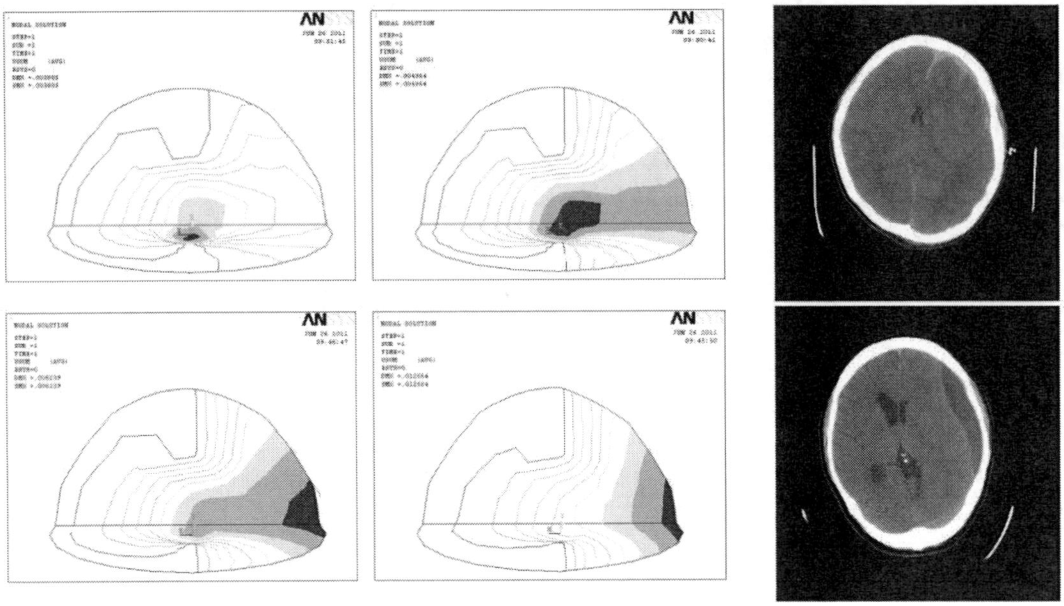

Figure 1. Contour plots showing deformation of our brain model with a Young's modulus of 10000 and Poisson's ratios of 0.49 (upper left), 0.45 (upper middle), 0.42 (lower left) and 0.30 (lower middle) after applying a 20-mmHg pressure load. Cutaway views through the center of the subfalcine space are used to display the most deformed regions (shaded). The upper two plots resemble brain compression by acute subdural hematoma, as demonstrated by the axial (X-Y plane) computed tomographic image (upper right) near the basal surface; while the lower two plots resemble brain compression by chronic subdural hematoma (lower right).

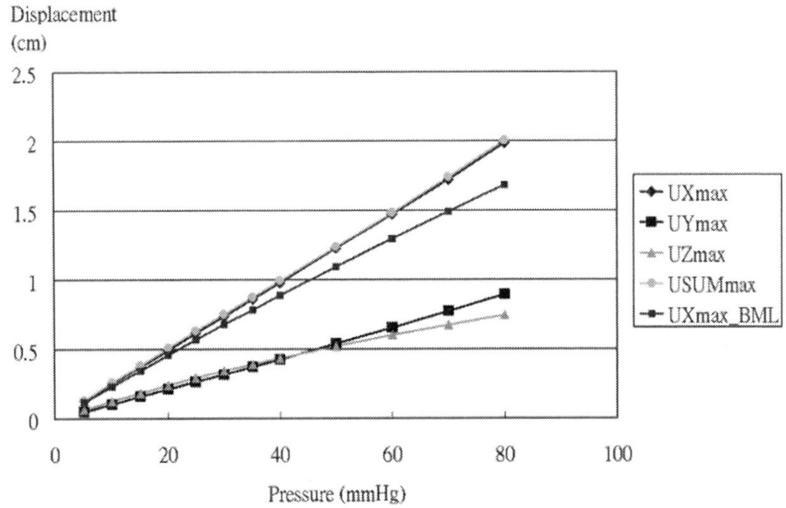

Figure 2. The largest X, Y, Z and total displacements (UXmax, UYmax, UZmax and USUMmax) and the largest X displacement along the basal midline (UXmax_BML) versus pressure load in our brain model with a Young's modulus of 10000 and a Poisson's ratio of 0.45.

The magnitude of brain displacement versus pressure load is summarized in Fig. 2. The largest X displacements (UXmax) in our brain model range from 1.27 mm with a pressure load of 5 mmHg to 19.86 mm with a pressure load of 80 mmHg. The largest Y displacements (UYmax) and the largest Z displacements (UZmax) range from 0.53 mm to 8.99 mm and from 0.65 to 7.49 mm, respectively. The largest total displacements (USUMmax), the vector sums of UXmax, UYmax and UZmax, range from 1.32 mm to 20.05 mm. UXmax along the basal midline of the model (UXmax_BML), which is similar to the MLS, range from 1.19 mm to 16.86 mm. The relationships between these five parameters and the pressure load are almost linear. Nevertheless, UZmax seems to rise slightly slower as the pressure load increases. The two largest parameters, UXmax and USUMmax, increase by 2.5 mm with every 10 mmHg increase in pressure. A 5-mm largest displacement in our brain model (4.97 mm for UXmax and 5.13 mm for USUMmax) occurs with a pressure load of 20 mmHg, which is the treatment threshold recommended in the clinical practice guideline [16]. The difference between UXmax and UXmax_BML ranges from 0.08 mm to 3.00 mm, being less than 1 mm with pressure loads under 40 mmHg.

The accuracy of the local approximates, which are easily visualized on patient images, versus pressure load is summarized in Fig. 3. UYmax along the basal circumference (UYmax_BCC) accurately describes the global UYmax, having ratios larger than 0.99 under all conditions. UZmax along the horizontal line across the tentorial incisura (UZmax_HTI) also accurately describes the global UZmax under all conditions. All UZmax_HTI/UZmax ratios are larger than 0.97. The ratio between MLS (UXmax_BML) and UXmax is 0.93 with a pressure load of 5 mmHg, and then decreases to 0.91 with a pressure load of 40 mmHg. This ratio drops further with even larger pressure loads. The largest X displacement along the BCC (UXmax_BCC) is about 0.89 times UXmax with pressure loads larger than 20 mmHg.

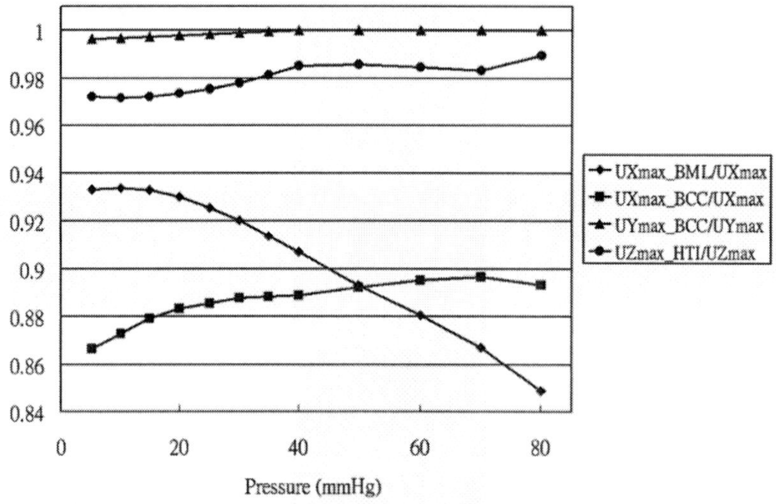

Figure 3. The accuracy of the local approximates versus pressure load in our brain model with a Young's modulus of 10000 and a Poisson's ratio of 0.45. Abbreviations: BML, basal midline; BCC, basal circumference; HTI, horizontal line across the tentorial incisura.

3.2 Brain deformation with different material properties

Displacement of our brain model with different PRs after applying a 20-mmHg pressure load is shown in Fig. 1, using cutaway views through the center of the subfalcine space. Values of 0.30, 0.42, 0.45 and 0.49 are used as examples of very low, low, "normal", and high PRs. With a very low PR, the region with the largest displacement in the brain model only occurs at the anterolateral aspect of the BCC. Away from this region, the displacement decreases as the distance increases. With "nearly normal" PRs, a large proportion of the compressed hemisphere has the largest displacement. The region extended from the outer surface to the SFS, showing significant SFH compared to brain shift at the BCC. However, the region with the largest displacement moves abruptly from brain surface to the subfalcine region with PRs larger than 0.434. With a very large PR, the area with the largest displacement is localized inside the TI, showing TTH as the predominant mode of brain shift.

On the other hand, displacement of our brain model is inversely related to the YM given the same PR and the same pressure load. The relative brain deformation is not affected by the YM, so the contour plots with different YMs are not shown.

The magnitude of brain displacement versus different PRs is summarized in Fig. 4. UXmax in our brain model ranges from 14.88 mm with a PR of 0.25 to 3.21 mm with a PR of 0.49. Likewise, UYmax ranges from 8.16 mm to 1.02 mm. On the other hand, UZmax decreases from 7.82 mm to the minimal value of 2.40 mm with a PR of 0.44, then increases gradually to 2.87 mm with larger PRs. USUMmax decreases from 14.91 mm with a PR of 0.49 to 5.64 mm with a PR of 0.43, then slowly decreases to 3.81 mm with a PR of 0.49. The slower decrease in USUMmax is attributed to gradual increase in UZmax.

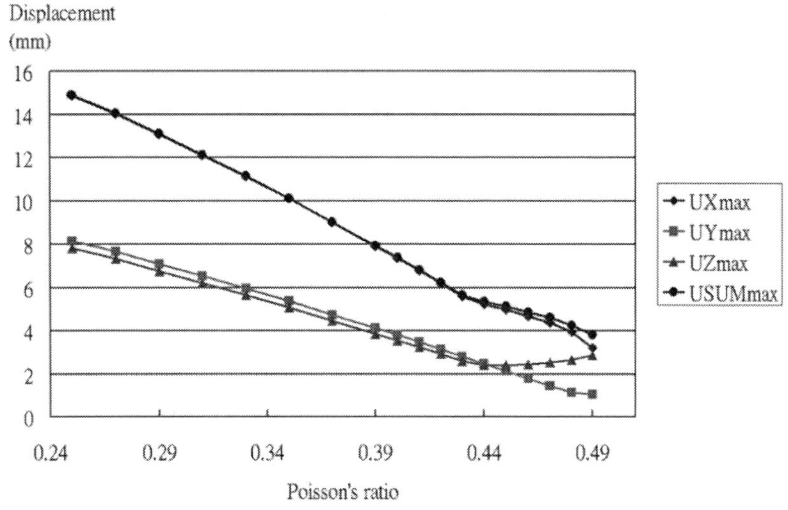

Figure 4. The largest X, Y, Z and total displacements (UXmax, UYmax, UZmax and USUMmax) versus Poisson's ratio (PR) in our brain model with a Young's modulus of 10000 and a pressure load of 20 mmHg.

The values and accuracy of the local approximates versus PR is summarized in Fig. 5. In the left part, the absolute values of UXmax, UXmax_BCC and UXmax_BML are plotted against PR. The relationship between UXmax_BCC and PR is linear, while the ratio between UXmax_BML and UXmax_BCC appears to be exponential. UXmax is always close to the larger values among UXmax_BML and UXmax_BCC, with the largest absolute difference of only 0.35 mm. The percentage error is about 5% with larger PRs, when UXmax is approximated by UXmax_BML.

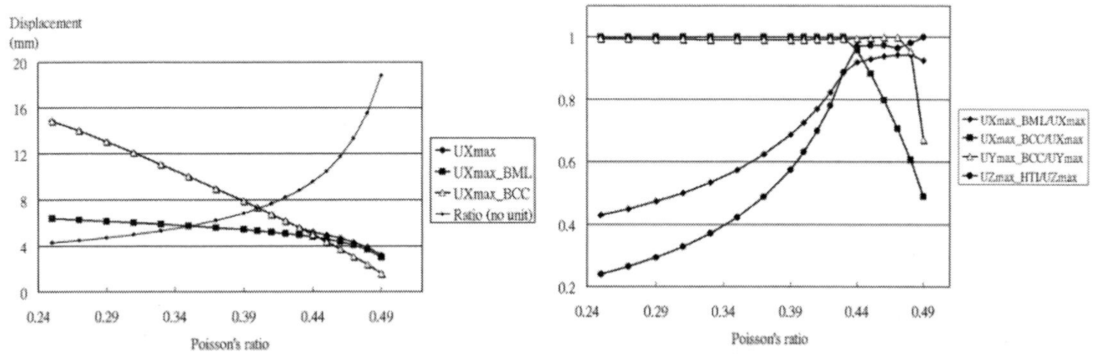

Figure 5. The values (left) and accuracy (right) of local approximates for describing the largest displacements versus Poisson's ratio (PR) in our brain model with Young's modulus of 10000 and pressure load of 20 mmHg.

In the right part of Fig. 5, the accuracy of the local approximates versus PR is plotted. UYmax_BCC accurately approximate the global UYmax with PRs smaller than 0.48. The ratios between UYmax_BCC and UYmax are 0.95 and 0.67 with PRs of 0.48 and 0.49, respectively. The ratios between UZmax_HTI and UZmax are larger than 0.97 with PRs larger than 0.434, revealing the central position of the UZmax nodes. UZmax_HTI becomes a poor approximate when the node having UZmax is at the surface of the compressed cerebral hemisphere. Likewise, the MLS (UXmax_BML) can approximate UXmax better with larger PRs, when the node having UXmax is near the TI. On the other hand, UXmax_BCC can approximate the UXmax with PRs smaller than 0.434. The ratio between UXmax_BCC and UXmax decreased to 0.96 with a PR of 0.44, then decreased further and is surpassed by UXmax_BML/UXmax with larger PRs.

4 Discussion

We have proposed a simplified FE model that can be applied to simulating brain shift caused by an SDH. Our model is constructed with a rather small number of nodes and elements, but is able to simulate brain displacement and herniation in real patients. The pressure loads we tested covered almost all ICP conditions encountered clinically and the resulting patterns of brain shift appear similar to those appeared on images in patients with acute or chronic SDH, as shown in Fig. 1.

4.1 Meaning of each parameter

In the first part of our experiment, our model with "standard" biomechanical parameters is subject to different pressure loads. Pressure gradient across different compartments of the intracranial space does exist under certain conditions such as posterior fossa tumor [17], but the ICP in patients with intracranial hematoma usually equalize within hours. Clinically, the pressure is almost always considered uniform within the dura, and it is acceptable to place the monitor anywhere. To date, there is no specific recommendation regarding the best location of the ICP monitor in patients with severe traumatic brain injury [18]. We also follow this approach by assuming that there is only one equalized pressure value within the intracranial space.

In the second part of our experiment, models with different PRs are subject to a pressure load slightly higher than normal ICP. We have tested these models with different pressure loads, and the relationship between the maximal displacements and the pressure appear to be linear in the range between 5 and 80 mmHg, obviating the need for testing with multiple load conditions.

With all PRs, the UXmax node at the outer surface of the compressed cerebral hemisphere always locates at the BCC. Therefore, UXmax_BCC can be regarded as the point with maximal SDH thickness. Brain models with low PR have MLS is much smaller than the UXmax_BCC. They behave like atrophic brains compressed by chronic SDH, a condition commonly seen in aged patients. On the other hand, models with very high PR resembles the behavior of tight or "angry" brains, where small volumes of SDH, represented by small UXmax_BCCs, causes larger brain deformation at its center and squeezes it through the TI, as noted in the autopsy and image findings with TTH [15]. From the regression formula derived from Fig. 5, the thickness of the SDH ranges between 0.15 mm with a PR of 0.50 and 29 mm with a PR of 0, under an ICP of 20 mmHg, concordant with clinical observations.

Although many complex constitutive models had been proposed previously, Wittek et al. had pointed out that the choice of the brain tissue constitutive model, when used with an appropriate finite deformation solution, does not affect the accuracy of computed displacements, and therefore a simple linear elastic model for the brain tissue is sufficient [19]. Our results support their opinion. In our experiments, we are able to use a single global variable, the PR of the simple linear elastic model, to model the "tightness" of the whole supratentorial brain to produce realistic deformations. There is no other biomechanical model in the literature that can simulate both acute and chronic SDH without any change in its geometry to accommodate the loss of brain volume.

The "slower" UZmax change to pressure load is related to the relatively coarse meshing of our model. In this paper, the number of nodes/elements is 1391/776. Using a finer mesh consisted of 12450 modes and 8263 elements generated with a full version of ANSYS, UZmax changes almost linearly with pressure load up to 70 mmHg.

4.2 Clinical implications: subfalcine and transtentorial herniations

We have carried out several studies on automatic recognition of the MLS, the evidence of SFH on brain CT images [4-6]. Using visual inspection, we use a quadratic curve to model the deformed midline at the foramen of Monro, near the base of the cerebrum. In the simple FE model proposed in this paper, the shapes of all deformed basal midlines under different test conditions assume the shape of quadratic curves as we observed on patient images. This model may provide a biomechanical basis for modeling the deformed midline and quantifying the MLS.

We set the value of PR as 0.45 for the simple linear elastic model in this study [12]. To verify this value, we measured the thickness of the SDH and the corresponding MLS on initial CT images in 12 surgically-treated acute SDH patients younger than 50 years of age, collected from our previous studies [5, 20]. All patients showed signs of increased ICP. Regression analysis showed a linear relationship with a Pearson's correlation coefficient (R2) of 0.956. The MLS is about 0.96 times the thickness of SDH (95% confidence interval: 0.83 to 1.10). This ratio between the MLS and SDH thickness is very close to our FE model with a YM of 10000 and a PR of 0.45, which have MLS/SDH ratio between 0.95 and 1.03 with pressure loads larger than 30 mmHg, when surgical decompression is necessary. To fully validate our model, large-scale studies incorporating CT images and corresponding initial ICP data measured intra-operatively are required.

In a model with PR close to 0.45, the region with the largest brain shift extends from the anterolateral region of the affected cerebral hemisphere toward the subfalcine region. This region is decompressed by current practice of unilateral front-parieto-temporal craniectomy for acute SDH with or without brain contusion [16, 21]. From our FE experiments, we also show that MLS is a good estimator of USUMmax in models with medium to large PRs (Fig. 5, right). In addition, the MLS with a 20 mmHg pressure load never exceeds 7 mm even with very low PRs (Fig. 5, left), showing the robustness of MLS as a sign of increased ICP, especially compared to the shift at the brain surface, or the size of the mass.

In addition to SFH, our model also demonstrates TTH, shift of the brain region adjacent to the TI. Compared to the SFS, the TI is relatively small in area. As a result, qualitative and quantitative description of TTH is much more difficult than describing the MLS [2]. Change in brain stem width had been proposed as a possible sign of compression [22]. In our experiments, the UXmax_HTI, behaves similar to the MLS, i.e. UXmax_BML with different ICPs and PRs.

On the other hand, UZmax does increase with PRs larger than 0.44 (Fig. 4). Our model is more consistent with the classic observation, which describes TTH as downward movement of the brain stem. However, the magnitude of UZmax_HTI is rather small. Even with very large ICP and very large PR, the maximal UZmax_HTI never exceeds 7 mm, which is far smaller than the largest possible MLS. The result is concordant to the case report by Ropper in 1993 [23]. Using MRI, the author is able to demonstrate the 2-mm downward displacement and the 13-mm horizontal displacement of the upper brain stem.

4.3 Limitations of this study

Our study has several limitations. The simple FE model is an "average" across subjects and cannot be used to provide the detailed deformation of each part of the brain on specific patient without modification. To simplify our model, the internal structures of the brain are omitted. As a result, detailed point-to-point correspondence, even to the nonlinearly transformed anatomical atlas, is not possible. We assume the PR of the model to be constant, while the brain may change its PR over time, either locally or globally, over time when it is compressed by the hematoma.

Acknowledgment

This work was supported by the Department of Health, Taipei City Government (10601-62-072) and National Taiwan University Hospital (UN105-002).

References

1. L.M. Liau, M. Bergsneider and D.P. Becker, *Neurological Surgery. 4th Ed.*, 1549-1594 (WB Saunders, Philadelphia, 1996)
2. M.R. Bullock, R. Chesnut, J. Ghajar, D. Gordon, R. Hartl, D.W. Newell, F. Servadei, B.C. Walters and J.E. Wilberger, Neurosurgery, **58**(Suppl 3), S2-62 (2006)
3. K.U. Schmitt, P.F. Niederer, M.H. Muser and F. Walz, *Trauma Biomechanics, 2nd Ed.* (Springer Berlin, 2007)
4. C.C. Liao, F. Xiao, J.M. Wong and I.J. Chiang, Biomed Eng Appl Basis Comm, **18**, 305-311 (2006)
5. C.C. Liao, F. Xiao, J.M. Wong and I.J. Chiang, Comput Biol Med, **40**, 331-339 (2010)
6. F. Xiao, I.J. Chiang, J.M. Wong, Y.H. Tsai, K.C. Huang and C.C. Liao, Comput Biol Med, **41**, 756-762 (2011)
7. T.J. Horgan and M.D. Gilchrist, Int J Crash, **8**, 353-366 (2003)
8. R.J. Cloots, H.M. Gervaise, J.A. van Dommelen and M.G. Geers, Ann Biomed Eng, **36**, 1203-1215 (2008)
9. N. Yoganandan, J. Li, J. Zhang and F.A. Pintar, Role of Falx on Brain Stress-Strain Responses. Jiangyue Zhang and Frank A. Pintar. In: Andre Kamkin, Irina Kiseleva (Ed.) *Mechanosensitivity of the Nervous System* (Springer Science+Business Media B.V., Dordrecht, Netherlands, 2009)
10. K.C. Huang, F. Xiao, I.J. Chiang, Y.L. Chen, Y.H. Tsai, J.M. Wong and C.C. Liao, AASRI Procedia, **6**, 95-102 (2014)
11. A. Pena, M.D. Bolton, H. Whitehouse and J.D. Pickard, Neurosurgery, **45**, 107-118 (1999)
12. H. Takizawa, K. Sugiura, M. Baba and J.D. Miller, Neurol Med Chir (Tokyo), **34**, 65-69 (1994)
13. N. Nakasone, T.A. Stolarski and S. Yoshimoto, *Engineering analysis with ANSYS software* (Elsevier Butterworth-Heinemann, Oxford, United Kingdom, 2007)
14. M.I. Miga, D.W. Roberts, F.E. Kennedy, L.A. Platenik, A. Hartov, K.E. Lunn and K.D. Paulsen, Neurosurgery, **49**, 75-85 (2001)
15. M.S. Greenberg, *Handbook of Neurosurgery, 4th ed.* (Greenberg Graphics, Lakeland, FL, 1997)
16. M.R. Bullock, R. Chesnut, J. Ghajar, D. Gordon, R. Hartl, D.W. Newell, F. Servadei, B.C. Walters and J.E. Wilberger, Neurosurgery, **58**(Suppl 3), S16-24 (2006)
17. G.E. Cold and N. Juul (Eds.), *Monitoring of cerebral and spinal haemodynamics during neurosurgery* (Springer-Verlag, Berlin, 2008)
18. S.L. Bratton, R.M. Chestnut, J. Ghajar, F.F. McConnell Hammond, O.A. Harris, R. Hartl, G.T. Manley, A. Nemecek, D.W. Newell, G. Rosenthal, J. Schouten, L. Shutter, S.D. Timmons, J.S. Ullman, W. Videtta, J.E. Wilberger and D.W. Wright, J Neurotrauma, **24**(Suppl 1), S45-54 (2007)
19. A. Wittek, T. Hawkins and K. Miller, Biomech Model Mechanobiol, **8**, 77-84 (2009)
20. C.C. Liao, F. Xiao, J.M. Wong and I.J. Chiang, Comput Med Imaging Graph, **34**, 563-571 (2010)
21. A.P. Huang, Y.K. Tu, Y.H. Tsai, Y.S. Chen, W.C. Hong, C.C. Yang, L.T. Kuo, I.C. Su, S.H. Huang and S.J. Huang, J Neurotrauma, **25**, 1347-1354 (2008)
22. H.M. Liu, Y.K. Tu and C.T. Su, J Trauma, **38**, 330-333 (1995)
23. A.H. Ropper, J Neurol Neurosurg Psychiatry, **56**, 932-935 (1993)

AUTHOR INDEX

AUTHOR INDEX